Yanpeng Zhang, Feng Wen, and Min Xiao

Quantum Control of Multi-Wave Mixing

Related Titles

Prather, D.W., Shi, S., Sharkawy, A., Murakowski, J., Schneider, G.

Photonic Crystals

Theory, Applications and Fabrication

2009

ISBN: 978-0-470-27803-1

Ishikawa, H.

Ultrafast All-Optical Signal Processing Devices

2008

ISBN: 978-0-470-75869-4

Saleh, B.E., Teich, M.C.

Fundamentals of Photonics, Second Edition

2nd Edition

2007

ISBN: 978-0-471-35832-9

Stegeman, G.I., Stegeman, R.F.

Nonlinear Optics

Phenomena, Materials, and Devices

2012

ISBN: 978-1-118-07272-1

Mandel, P.

Nonlinear Optics

An Analytical Approach

2010

ISBN: 978-3-527-40923-5

Chen, C., Sasaki, T., Li, R., Wu, Y., Lin, Z., Mori, Y., Hu, Z., Wang, J., Yoshimura, M., Kaneda, Y.

Nonlinear Optical Borate Crystals

Principles and Applications

2012

Print ISBN: 978-3-527-41009-5

Also available in digital formats.

Yanpeng Zhang, Feng Wen, and Min Xiao

Quantum Control of Multi-Wave Mixing

WILEY-VCH
Verlag GmbH & Co. KGaA

高等教育出版社
HIGHER EDUCATION PRESS

The Authors

Prof. Yanpeng Zhang
Xi'an Jiaotong University
Department of Electronics
No. 28, Xianning West Road
Department of Electronics
710049 Xi'an, Shaanxi
China
ypzhang@mail.xjtu.edu.cn

Dr. Feng Wen
Xi'an Jiaotong University
Department of Electronics
No.28, Xianning West Road
Department of Electronics
710049 Xi'an, Shaanxi
China

Prof. Min Xiao
University of Arkansas
Department of Physics
PHYS 204
Department of Physics
United States

Cover
Geometry of the preparation of the entangled photon source experiment. The pump in purple is focused on an atomic group and couples with probe and conjugate fields located symmetrically at a small angle in respect to the pump axis, then the generated FWM and SWM photons are entangled in the frequency and spatial space.

Additional book-related material for lecturers can be found online at www.wiley-vch.de and www.wiley.com.

Library of Congress Card No.: applied for

British Library Cataloguing-in-Publication Data
A catalogue record for this book is available from the British Library.

Bibliographic information published by the Deutsche Nationalbibliothek
The Deutsche Nationalbibliothek lists this publication in the Deutsche Nationalbibliografie; detailed bibliographic data are available on the Internet at <http://dnb.d-nb.de>.

Published by Wiley-VCH GmbH & Co.KGaA; Boschstr. 12, 69469 Weinheim, Germany, under exclusive license granted by Higher Education Press Limited Company for all media and languages excluding Chinese and throughout the world excluding Mainland China and with non-exclusive license versions in Mainland China.

Print ISBN: 978-3-527-41189-4
ePDF ISBN: 978-3-527-67239-4
ePub ISBN: 978-3-527-67238-7
mobi ISBN: 978-3-527-67237-0
oBook ISBN: 978-3-527-67236-3

Cover Design Adam-Design, Weinheim
Typesetting Laserwords Private Limited, Chennai, India
Printing and Binding Markono Print Media Pte Ltd, Singapore

Printed on acid-free paper

Contents

Preface

It is widely agreed that the optical properties of materials change drastically in systems where the superpositions of quantum states are coherently excited. Many interesting scientific discoveries and technical applications have been made with nonlinear optical effects in several kinds of nonlinear materials. Examples of such effects include the modification of absorptive properties resulting in electromagnetically induced transparency (EIT) and lasing without population inversion, as well as the modification of dispersive properties to give a resonantly enhanced index of refraction accompanied by vanishing absorption. There are already several excellent general textbooks covering various aspects of nonlinear optics, including "*Nonlinear Optics*" by R. W. Boyd, "*Nonlinear Optics*" by Y. R. Shen, "*Quantum Electronics*" by A. Yariv, "*Nonlinear Fiber Optics*" by G. P. Agrawal, "*Photons and Nonlinear Optics*" by D. N. Klyshko, and so on. Although these textbooks have provided solid foundations for readers to understand various nonlinear optical processes, some comprehensive and deep knowledge on a certain nonlinear optical effect is essential for studying and researching. With the intention of giving the special knowledge and recent progress about multi-wave mixing (MWM) effect, the authors have published two earlier monographs, "Multi-Wave Mixing Processes" and "Coherent Control of Four-Wave Mixing" in 2009 and 2011, respectively, which covers the experimental and theoretical studies of several topics related to MWM processes previously done in authors' groups. The topics covered in the two monographs include difference-frequency femtosecond and sum-frequency attosecond beats of four-wave mixing (FWM) processes; heterodyne detections of FWM, six-wave mixing (SWM), and eight-wave mixing (EWM) processes; the Raman, Raman–Rayleigh, Rayleigh–Brillouin, and coexisting Raman–Rayleigh–Brillouin-enhanced polarization beats; high-order correlation functions of different noisy fields on the femto- and attosecond polarization beats, and heterodyne/homodyne detections of the ultrafast third-order polarization beats.

This new monograph is built on the previous works and extends them significantly. The intention is to present all the additional and new works done in recent years in authors' groups. Also, many added latest results, extended detailed calculations, and more deep discussions are cited from the already published papers, which can help readers better appreciate the interesting nonlinear optical phenomena.

Besides showing more results on controls and interactions between MWM processes in hot atomic media, this monograph also presents and discusses several novel types of spatial solitons and two-photon fluorescence in the FWM process, which are completely new phenomena in multi-level atomic systems.

Chapter 1 gives the outline of this monograph and introduce some basic concepts frequently used in the later chapters, such as the suppression and enhancement conditions of MWM, two- and three-photon fluorescence, MWM in ring optical cavity, photonic band gap, and MWM with Rydberg blockade, and so on. Chapter 2 gives the extended results on the EIT-assisted MWM which includes not only the coexistence of FWM, SWM, and EWM signals in multi-EIT windows, but also the interplay and interference among the multiple MWM processes. Chapter 3 gives the results on the ac-Stark effect and Autler–Townes (AT) splitting in the MWM process, which can be accurately explained by multi-dressed state theory. Chapter 4 presents the switching methods between enhancement and suppression of MWM in both frequency and spatial domains, and the evolution of the dressed effects from pure enhancement into pure suppression when the probe detuning as well as the powers of the dressing and probe fields are changed. Chapter 5 shows the polarization-controlled MWM processes in multiple Zeeman sublevels system, in which the generated MWM signals can be modified and controlled, and the dark-state effects can be controlled to evolve from pure enhancement into pure suppression in the MWM processes via the polarization states of the laser beams. Chapter 6 presents the demonstration of the modification and control of the two-photon fluorescence process and three-photon correlation in MWM by manipulating the dark-state or EIT windows. On the other hand, the vacuum-induced Rabi splitting and optical bistability in a coupled atom-cavity system is also included. Chapter 7 shows the forming of electromagnetically induced grating (EIG) and electromagnetically induced lattice (EIL) in FWM, which can lead to spatial shift and splitting of laser beams as well as several novel types of spatial solitons of FWM signals, such as gap, dipole, vortex, and surface solitons with the generated FWM beams in different regions in experimental parametric space. Chapter 8 presents the observations of the prototype investigation of all-optical switches, routers, and space demultiplexer by the FWM process.

The authors believe that the current monograph treats some special topics of quantum controls of MWM and can be useful to researchers who are interested in the related fields. Several features presented here are distinctly different from and more advanced than the previously reported works. For example, the authors have shown the two-photon ac-Stark effect and AT splitting of MWM can be used to determine the energy-level shift of the atom. Also, theoretical calculations are in good agreement with the experimentally measured results in demonstrating the two phenomena in MWM processes. Efficient spatial–temporal interference between FWM and SWM signals generated in a four-level atomic system has been carefully investigated, which exhibits controllable interactions between two different (third- and fifth-) order nonlinear optical processes. Evolutions of the enhancement and suppression of MWM signals under various dressing schemes are experimentally investigated, and can be explained in detail with dressing state theory. Such controllable high-order nonlinear optical processes can be used for designing new devices for all-optical communication and quantum information processing.

The authors also experimentally compare the intensity spectra of probe transmission, FWM, and fluorescence signals under dressing effects, in which the two-photon fluorescence signal with ultranarrow line width much less than the Doppler-free EIT window is obtained. Moreover, the three photons of the two coexisting FWM signals and the probe signal in a double-lambda-type system are experimentally found to be strongly correlated, or anticorrelated with each other. Such ultranarrow linewidth fluorescence, strongly correlated or anticorrelated photons and controllable cavity mode splitting as well as the optical bistability process could have potential applications in optical communication and quantum information processing.

The authors also experimentally demonstrate that by arranging the strong pump and coupling laser beams in specially designed spatial configurations (to satisfy phase-matching conditions for efficient MWM processes), the generated MWM signals can be spatially shifted and split controllably by the cross-phase modulation (XPM) in the Kerr nonlinear medium. Therefore, the periodic splitting of the metastable energy level and periodic refractive index of the medium is experimentally obtained by the spatially periodic interfered pattern of dressing fields. Moreover, in the propagation of FWM or probe beams, when the spatial diffraction is balanced by XPM, the spatial beam profiles of the beams can become stable to form spatial optical solitons. For different geometrical configuration and experimental parameters (such as laser powers, frequency detunings, and temperature), novel gap, vortex, dipole, and surface soliton have been shown to experimentally appear in the multi-level atomic system in vapor cells. These studies have opened the door for achieving all-optical controlled spatial switch, routing, and soliton communications with an ultrashort response time.

This monograph serves as a reference book intended for scientists, researchers, graduate students, and advanced undergraduates in nonlinear optics and related fields.

We take this opportunity to thank all the researchers and collaborators who have worked on the research projects as described in this book.

Yanpeng Zhang, Feng Wen, and *Min Xiao*

1 Introduction

The subjects of this book center mainly around three topics. The first topic (Chapters 2–5) covers the quantum interference of coexistent four-wave mixing (FWM) processes because the generated FWM signal can be selectively enhanced or suppressed via electromagnetically induced transparency (EIT) windows and induced atomic coherence. Both enhancement and suppression of dressed FWM can be observed directly by scanning the dressing field instead of the probe field. With specially designed spatial patterns and phase-matching conditions for laser beams, coexisting FWM and six-wave mixing (SWM) processes can also be generated very efficiently. Also, three dual-dressed schemes (nested, sequential, and parallel) of coexisting FWM have been studied. Frequency, spatial, and temporal interferences that occur between two different wave mixing processes for the relative phase between different multi-wave mixing (MWM) processes is modulated. In such cases, the FWM and SWM signals are modulated with phase difference. By regulating the laser beam, the constructive and destructive interference can be selected, and then ac-Stark effect and Autler–Townes (AT) splitting are observed. Furthermore, by manipulating the polarization states of the laser beams, the MWM processes can also be modified and controlled. The second topic (Chapter 6) relates to nonclassical properties of the MWM process; the correlation or anticorrelation between two coexisting FWM signals; and comparison among the depths of the probe transmission, FWM, and the two-photon fluorescence signals in the same nonlinear process. Especially, one can switch from bright to dark states in the FWM and fluorescence channels with the relative phase modulated from 0 to $-\pi$. The relationship of vacuum Rabi splitting (VRS) and optical bistability (OB) of cavity MWM signals and methods to control VRS and OB are also included. The third topic (Chapters 7 and 8) relates to the interplays in frequency and spatial domains of MWM processes induced by atomic coherence in multi-level atomic systems. The generated two-dimensional surface solitons, multi-component dipole, and vortex vector solitons of the MWM signal accompanying the formation of electromagnetically induced lattice (EIL) are presented in Chapter 7. Finally, an application of spatial displacements and splitting of the probe and generated FWM beams, that is, all-optical spatial routing, switching, and demultiplexer are shown and investigated in Chapter 8. Experimental results are presented and compared with the theoretical calculations

Quantum Control of Multi-Wave Mixing, First Edition. Yanpeng Zhang, Feng Wen, and Min Xiao.

throughout the book. In this book, the emphasis is on the work done by the authors' groups in the past few years. Some of the work presented in this book are built on our previous books "Multi-wave Mixing Processes" and "Coherent Control of Four-Wave Mixing," published by Higher Education Press and Springer 2009 and 2011, respectively, where we have mainly discussed the coexistence and interactions between efficient MWM processes enhanced by atomic coherence in multi-level atomic systems in the former, and in the latter the control in frequency and spatial domains of FWM processes induced by atomic coherence in multi-level atomic systems. Before starting the main topics of this book, some basic physical concepts and mathematical techniques, which are useful and needed in the later chapters, are briefly introduced and discussed in this introduction chapter.

1.1 Suppression and Enhancement Conditions of the FWM Process

In this section, the dressed state theory with suppression and enhancement of FWM, dark-state theory in high-order nonlinear processes is introduced. Specifically, in Section 1.1.1, we discuss different dressing states including singly dressing and doubly dressing states. And the AT splitting for FWM by scanning the detuning of the probe field induced by the dressing effect will also be introduced. In Section 1.1.2, the same phenomenon is investigated using the quantum interference method. Also, we discuss that the high-order nonlinear processes can be effectively controlled by the dark state. In Section 1.1.3, we focus on suppression and enhancement of the FWM process by scanning the detuning of the dressing field. In addition, the interaction between dressing fields is introduced in different dressing schemes.

1.1.1 Dressed State Theory

The most successful applications of resonant systems is in the two-photon rather than single-photon transition process because EIT [1, 2] forming in a two-photon interference process can reduce the linear absorption of a probe beam with a strong coupling beam resonant with the up-level transition. For example, enhanced MWM processes due to two-photon Raman resonances [3, 4] have been experimentally demonstrated in several multi-level atomic systems [5–7]. The keys in such enhanced nonlinear optical processes include the enhanced nonlinear susceptibilities due to induced atomic coherence and slowed laser beam propagation in the atomic medium [8–10], as well as greatly reduced linear absorption of the generated optical fields due to EIT [11]. By changing the strength [12], detuning [13], polarization [7], and phase [14] of the dressing fields, the EIT window and nonlinear susceptibilities can be effectively modulated. Therefore, the generated MWM signal can be enhanced or suppressed selectively. On the other hand, AT splitting of the MWM [15–17] signal can be controlled effectively at the same time.

In order to describe more precisely what is the physics meaning of the AT splitting and suppression or enhancement of the wave mixing process mentioned, we introduce the dressing state theory first and use it to briefly describe the relations between the AT splitting and the suppression or enhancement of MWM.

In Figure 1.1(a), the external dressing field $\boldsymbol{E}_3$ and self-dressing field $\boldsymbol{E}_2$ ($\boldsymbol{E}_2'$) dress the energy level $|1\rangle$ simultaneously. First, $\boldsymbol{E}_2$ ($\boldsymbol{E}_2'$) splits the state $|1\rangle$ to create the primary dressed states $|\pm\rangle$ written as $|\pm\rangle = \sin\theta_1|1\rangle + \cos\theta_1|2\rangle$. We can obtain the eigenvalues of $|+\rangle$ and $|-\rangle$ as $\lambda_+ = (\Delta_2 + \sqrt{\Delta_2^2 + 4|G_2|^2})/2$ and $\lambda_- = (\Delta_2 - \sqrt{\Delta_2^2 + 4|G_2|^2})/2$ (measured from level $|1\rangle$), respectively, as shown in Figure 1.1(b). Next, $\boldsymbol{E}_3$ further splits $|+\rangle$ or $|-\rangle$ to create the secondary dressed states $|+\pm\rangle$ or $|-\pm\rangle$ determined by the detuning of $\boldsymbol{E}_3$, as shown in Figure 1.1(c) and (d). For instance, if $\boldsymbol{E}_3$ couples with the upper dressed state $|+\rangle$, then secondary dressed states are given as $|+\pm\rangle = \sin\theta_2|+\rangle + \cos\theta_2|3\rangle$ (Figure 1.1(c)), where $\sin\theta_1 = -a_1/a_2$, $\cos\theta_1 = G_2^b/a_2$, $\sin\theta_2 = -a_3/a_4$, $\cos\theta_2 = G_3/a_4$, $a_1 = \Delta_2 - \lambda_\pm$, $a_2 = \sqrt{a_1^2 + |G_2^b|^2}$, $a_3 = \Delta_3 - \lambda_+ - \lambda_{+\pm}$, $a_4 = \sqrt{a_3^2 + |G_3|^2}$, and $G_2^b = G_2 + G_2'$. The eigenvalues of $|++\rangle$ and $|+-\rangle$ are $\lambda_{++} = (\Delta_3' + \sqrt{\Delta_3'^2 + 4|G_3|^2})/2$ and $\lambda_{+-} = (\Delta_3' - \sqrt{\Delta_3'^2 + 4|G_3|^2})/2$ (measured from level $|+\rangle$), respectively, where $\Delta_3' = \Delta_3 - \lambda_+$.

By scanning the detuning of the probe field (Δ_1), the primary AT splitting for the FWM signal (Figure 1.2(a)), or primary and secondary AT splitting (Figure 1.2(b)) can be obtained. In Figure 1.2(a), the AT splitting that results from the field $\boldsymbol{E}_2$ ($\boldsymbol{E}_2'$) has two bright states (two peaks) and one dark state (the dip at $\Delta_1 = 0$), and the splitting separation $\Delta_{\lambda 1}$ ($\Delta_{\lambda 1} = \lambda_+ - \lambda_-$) gets larger with the power of $\boldsymbol{E}_2$ ($\boldsymbol{E}_2'$) increasing. In addition, if $\boldsymbol{E}_3$ also couples $|-\rangle$, Figure 1.2(b) presents the primary and secondary AT splitting for different coupling field power P_3. It is obvious that the secondary AT splitting $\Delta_{\lambda 2}$ ($\Delta_{\lambda 2} = \lambda_{++} - \lambda_{+-}$) (Figure 1.2(b)) gets larger with increasing power of $\boldsymbol{E}_3$.

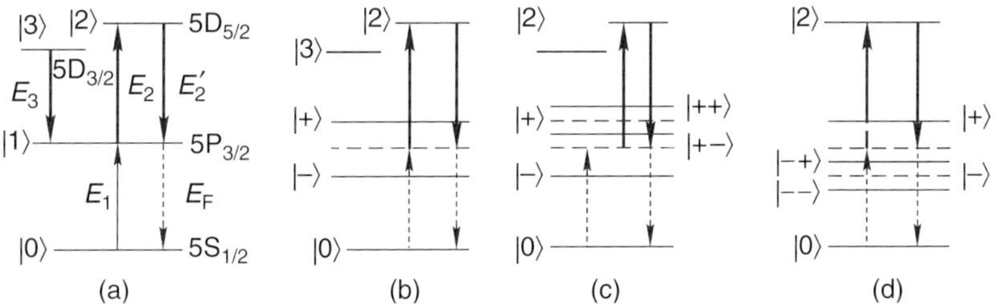

Figure 1.1 The diagrams of (a) a Y-type four-level, (b) the singly-dressed state, and (c) and (d) the doubly-dressed state.

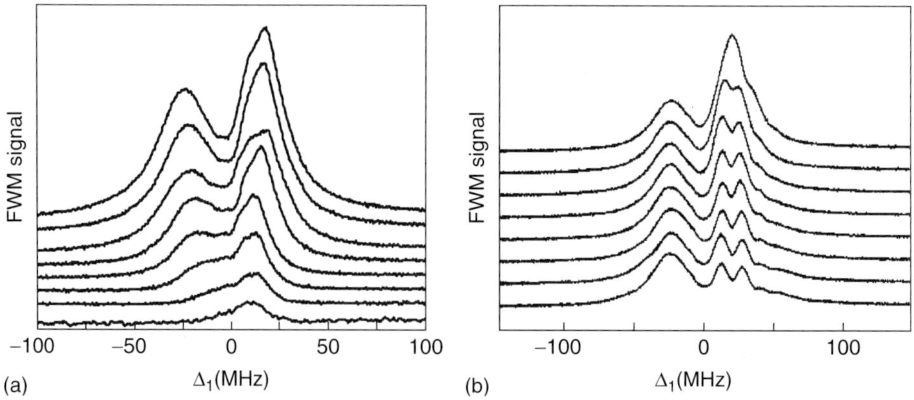

Figure 1.2 (a) Measured primary AT splitting and (b) primary and secondary AT splitting of FWM signals.

1.1.2
Dark-State Theory in MWM Processes

In the following, we discuss in detail the dark-state theory in three-level, Y-type four-level, and K-type five-level. The evolution of the FWM signal is fitted well with the theoretical calculation [5, 14].

Well known in quantum optics, the phenomenon of coherent dark states [18] is based on a superposition of long-lived system eigenstates that decouple from the light field. The dark state can be used to make a resonant, opaque medium transparent by means of quantum interference. Associated with the induced transparency is a dramatic modification of the refractive properties; therefore, it has found numerous applications. Prominent examples are EIT and lasing without inversion [19, 20], subrecoil laser cooling [21], and ultrasensitive magnetometers [22]. The possibility of coherently controlling the propagation of quantum light pulses via dark-state polaritons opens up interesting applications involving the generation of nonclassical states in atomic ensembles (squeezed or entangled states), reversible quantum memories for light waves [23–25], and high-resolution spectroscopy [26–28]. Furthermore, the combination of the present technique with studies on few-photon nonlinear optics [29–31] can be used, in principle, for processing of quantum information stored in collective excitations of matter. In the following, we discuss how high-order nonlinear processes can be effectively controlled by the dark state.

First, let us consider how a simple FWM process in a Ξ-type three-level could be manipulated and analyzed with the dark-state theory developed in this section. As shown in Figure 1.3, in the interaction picture, the Hamilton of a coupled atom-field system can be written as $H = \hbar(G_1|1\rangle\langle 0|\exp(-i\Delta_1 t/\hbar) + G_2|1\rangle\langle 2|\exp(-i\Delta_2 t/\hbar) + \mathrm{HC})$, and the wave function of an atom in bare state is $|\psi\rangle = c_0(t)|0\rangle + c_1(t)|1\rangle + c_2(t)|2\rangle$. If the coupled system of the laser field and atom is

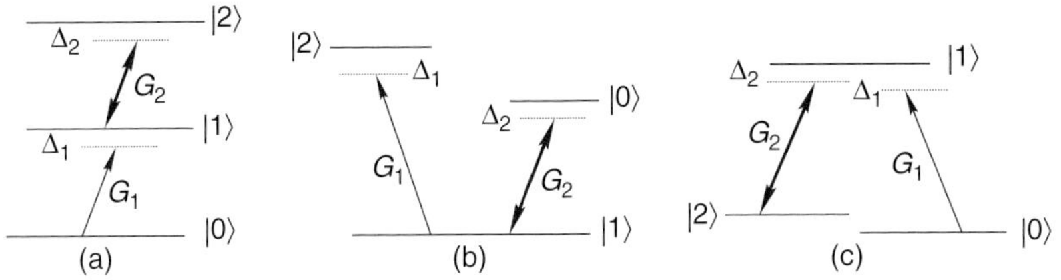

Figure 1.3 The diagrams of (a) a Ξ-type three-level, (b) a V-type three-level, and (c) a Λ-type three-level atomic systems, respectively.

interacting with a bright state, the bright state collapses into $|1\rangle$ and its corresponding eigenvalue is $\sqrt{G_1^2+G_2^2}$. Thus, we can obtain the coefficients of the bright state as $C_0 = G_1 \exp(-i\Delta_1 t/\hbar)/\sqrt{G_1^2+G_2^2}$, $C_2 = G_2 \exp(-i\Delta_2 t/\hbar)/\sqrt{G_1^2+G_2^2}$, and $C_1 = 0$. Furthermore, as $|1\rangle$ and $|B\rangle$ are orthogonal with each other, the state of the system can be expanded into the linear combination of $|1\rangle$ and $|B\rangle$.

Next, when the coupled system is interacting with a dark state, the bright state turns into $|1\rangle$ and its corresponding eigenvalue is null. Therefore, with the equation of $H|\psi\rangle = 0$, we can easily get the coefficients in the expression of dark state as $C_0 = G_1 \exp(-i\Delta_1 t/\hbar)/\sqrt{G_1^2+G_2^2}$, $C_2 = -G_2 \exp(-i\Delta_2 t/\hbar)/\sqrt{G_1^2+G_2^2}$, and $C_1 = 0$.

From the expression of dark state, we can see that C_2 is leading to or lagging from C_0 by a π phase factor, so $|0\rangle$ and $|2\rangle$ can produce destructive interference and dark state resulting in EIT. However, in bright state, C_2 is in-phase with C_0; therefore, state $|0\rangle$ has constructive interference with state $|2\rangle$, which leads to electromagnetically induced absorption (EIA). Furthermore, by changing the dressing field's detuning and intensities, and the interaction times (phase), the bright and dark states can be controlled to evolve or, in other words, transform into each other.

In the above-mentioned switch between bright and dark states in ladder-type three-level systems, we have found that the FWM process also is significantly influenced. We have in experiment demonstrated the evolution of FWM, which is fitted well with the theoretical calculation based on the dark-state theory. Furthermore, more than one external dressing field is usually used, as shown in Figure 1.4, so more than one dark state emerges in such multi-level systems. In the following, we give a brief description on how multi-dark states can be obtained with state superposition theory.

When the dressing fields $\boldsymbol{E}_2$ and $\boldsymbol{E}_4$ are all strong sufficiently in Y-type four-level systems, their dressing effects on the FWM process will be affected by the quantum interference. We can obtain the Hamilton in the interaction picture,

$$\begin{aligned} \mathrm{H} = & -\hbar(\Delta_4|4\rangle\langle 4| + \Delta_2|2\rangle\langle 2| - \Delta_1|0\rangle\langle 0|) \\ & -\hbar(G_1|0\rangle\langle 1| + G_2|2\rangle\langle 1| + G_4|4\rangle\langle 1| + \mathrm{HC}) \end{aligned} \tag{1.1}$$

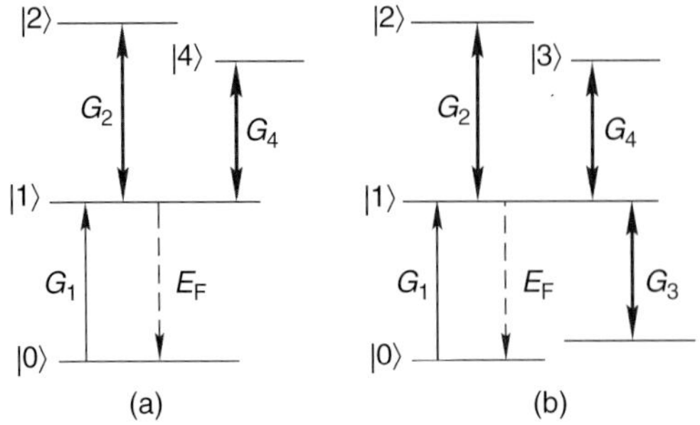

Figure 1.4 The diagrams of (a) a Y-type four-level and (b) a K-type five-level atomic systems.

Under the resonant conditions of $G_2, G_4 \gg G_1$, $\Delta_1 + \Delta_2 = 0$, and $\Delta_2 = 0$, we can get three independent dark states:

$$|D1\rangle = \frac{G_2|0\rangle - G_1|2\rangle}{\sqrt{|G_1|^2 + |G_2|^2}} \simeq |0\rangle - \frac{G_1}{G_2}|2\rangle \tag{1.2}$$

$$|D2\rangle = \frac{G_4|0\rangle - G_1|4\rangle}{\sqrt{|G_4|^2 + |G_1|^2}} \simeq |0\rangle - \frac{G_1}{G_4}|3\rangle \tag{1.3}$$

$$|D4\rangle = \frac{G_4|2\rangle - G_2|4\rangle}{\sqrt{|G_2|^2 + |G_4|^2}} = \cos\theta|2\rangle - \sin\theta|3\rangle \tag{1.4}$$

where $\cos\theta = G_4/\sqrt{|G_2|^2 + |G_4|^2}$ and $\sin\theta = G_2/\sqrt{|G_2|^2 + |G_4|^2}$. Therefore, the total dark-state amplitude is then given by

$$|D\rangle = |D1\rangle + |D2\rangle + |D4\rangle = 2|0\rangle + \left(\cos\theta - \frac{G_1}{G_2}\right)|2\rangle - \left(\frac{G_1}{G_4} + \sin\theta\right)|4\rangle \tag{1.5}$$

Utilizing the same calculation process, we can easily get the dark state in the K-type five-level atomic system:

$$|D'\rangle = 3|0\rangle - \frac{G_1}{G_2}|2\rangle - \left(\frac{G_1}{G_4} - \cos^*\theta\right)|4\rangle - \left(\frac{G_1}{G_3} + \sin^*\theta\right)|3\rangle \tag{1.6}$$

where $\cos^*\theta = G_3/\sqrt{|G_3|^2 + |G_4|^2}$ and $\sin^*\theta = G_4/\sqrt{|G_3|^2 + |G_4|^2}$.

On the other hand, in order to understand how the three independent dark states interfere with each other in Figure 1.4(a), we calculate the probability of an atom locating at dark state as

$$\begin{aligned}|\langle D|\varphi\rangle|^2 = {}& 4\rho_{00} + \left(\cos\theta - \frac{G_1}{G_2}\right)^2 \rho_{22} + \left(\frac{G_1}{G_4} + \sin\theta\right)^2 \rho_{44} \\ &+4\mathrm{Re}\left(\left(\cos\theta - \frac{G_1}{G_2}\right)\rho_{20} - \left(\frac{G_1}{G_4} + \sin\theta\right)\rho_{40}\right) \\ &-4\mathrm{Re}\left(\left(\frac{G_1}{G_4} + \sin\theta\right)\left(\cos\theta - \frac{G_1}{G_2}\right)^* \rho_{42}\right)\end{aligned} \tag{1.7}$$

where $|\varphi\rangle = c_0|0\rangle + c_1|1\rangle + c_2|2\rangle + c_3|4\rangle$ is the wave function of the atom in its bare-state basis; $\rho_{ii}(i = 1, 2, 3, 4)$ is the probability of the one atom populating in state $|i\rangle$; $\rho_{ij}(i, j = 1, 2, 3, 4)$ is proportional to the dipole moment between $|i\rangle$ and $|j\rangle$. Among them, the populations are assumed to be $\rho_{00} \approx 1$, $\rho_{22} \approx 0$, and $\rho_{44} \approx 0$ in Eq. (1.5). By exploring the quantum interference among the three states $|0\rangle$, $|2\rangle$, and $|4\rangle$, the intensity of the FWM signal is obtained as $I = |N'\mu\rho_{10}^{(3)}|^2$, where $N' = N(1 - |\langle D|\psi\rangle|^2)$ is the weight factor describing the number density of atoms not in dark states, with N being the total particle number density. Therefore, it is obvious that the quantum interference effect can exert an influence on the FWM signals via changing the effective number density particle participating in the FWM process.

As only the dark states $|D2\rangle$ and $|D4\rangle$ are related to the field $\boldsymbol{E}_4$, with G_4 increasing, both the probability amplitudes of the states $|D2\rangle$ and $|D4\rangle$ will grow and the theoretical result of $|\langle D2|\psi\rangle|^2 > |\langle D4|\psi\rangle|^2$ means that the probability amplitude of state $|D2\rangle$ is stronger than that of $|D4\rangle$. So the controls of the two dark states will bring different results in the FWM processes.

1.1.3 Suppression and Enhancement Conditions

In the previous section, we have discussed singly- and doubly-dressed MWM process in a multi-level atomic system by scanning the detuning of the dressing field [32, 33]. However, the interactions between the dressed states are not involved. In this section, with such an interaction introduced, not only three kinds of doubly-dressed schemes (nested-, parallel-, and sequential-cascade modes) of MWMs are analyzed and discussed in depth but the transition between the bright and dark states is also given.

As shown in Figure 1.5, the fields $\boldsymbol{E}_3$ (Figure 1.5(a)) and $\boldsymbol{E}_2$ (Figure 1.5(b)) dress the energy level $|1\rangle$ to form a Λ-type or Ξ-type singly-dressed system via the perturbation subchains $\rho_{10} \xrightarrow{-\omega_3} \rho_{30} \xrightarrow{\omega_3} \rho_{10}$ and $\rho_{10} \xrightarrow{\omega_2} \rho_{20} \xrightarrow{-\omega_2} \rho_{10}$, respectively. Therefore, we get the evolution of a singly-dressed FWM and its corresponding probe transmission signal shown in Figure 1.6. Comparing the two dressed systems, we can see that the transition sequence is almost opposite, which can be explained using the dark-state theory. With $\Delta_1 < 0$, the signals are resonant with the left dark state and right bright state in Λ-type singly-dressed case (Figure 1.6(a); however, the distribution order of dark and bright state is reverse in the Ξ-type level structure as shown in Figure 1.6(b). One symmetric center at $\Delta_1 = 0$ is observed in such cases.

Next, we introduce the doubly-dressed effect on the generation process of FWM and probe transmission signals. It is generally divided into parallel-, nested-, and sequential-cascade modes. First, in the parallel-type doubly-dressed case, as shown in Figure 1.7(a), the fields $\boldsymbol{E}_3$ and $\boldsymbol{E}_4$ dress different level $|1\rangle$ and $|0\rangle$, via the subchains $\rho_{10} \xrightarrow{-\omega_3} \rho_{30} \xrightarrow{\omega_3} \rho_{10}$ $(\rho_{(G_3\pm)0})$ and $\rho_{20} \xrightarrow{-\omega_4} \rho_{24} \xrightarrow{\omega_4} \rho_{20}$ $(\rho_{2(G_4\pm)})$, respectively. This means that the dressing effects of $\boldsymbol{E}_3$ and $\boldsymbol{E}_4$ lie parallel. In the parallel-cascade case, the profiles (dashed lines in Figure 1.8(a)) are induced by one

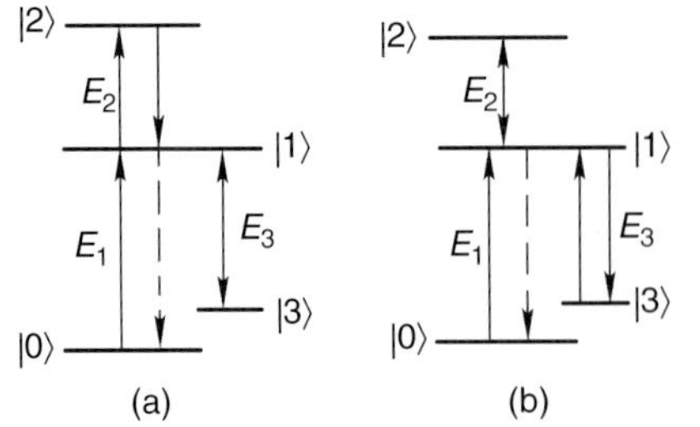

Figure 1.5 (a) Λ- and (b) Ξ-type singly-dressed FWM.

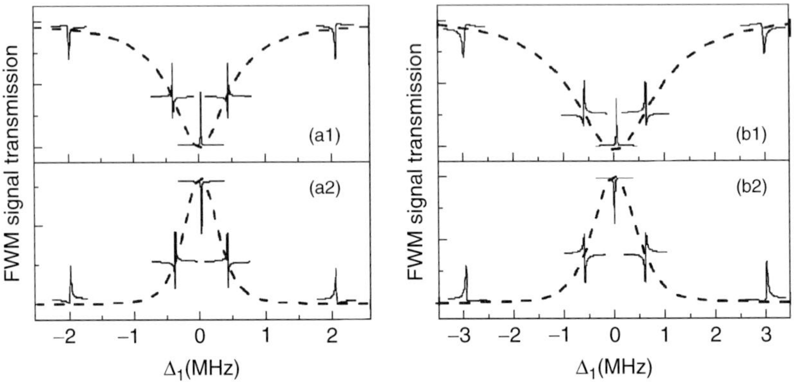

Figure 1.6 (a1) and (b1) Singly-dressed Λ-type and Ξ-type probe transmission signals. (a2) and (b2) the corresponding FWM signals. The dashed curves in (a) and (b) are the probe transmission and the corresponding pure FWM signals without $\boldsymbol{E}_3$ or $\boldsymbol{E}_2$, respectively.

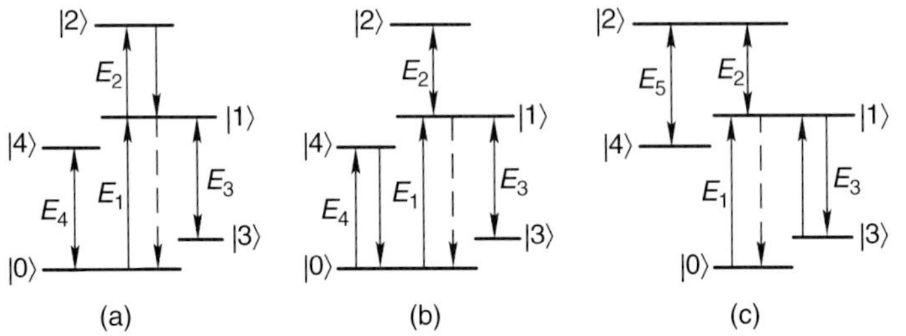

Figure 1.7 (a) Parallel-cascade, (b) sequential-cascade, and (c) nested-cascade mode doubly-dressed FWMs, respectively.

dressing field, and the transition between bright and dark states (in Figure 1.8(a)) is caused by the other dressing field. So, the two dressing fields in the parallel-dressed cascade mode have no interaction. Only one symmetric center at $\Delta_1 = 0$ is observed in such cases too.

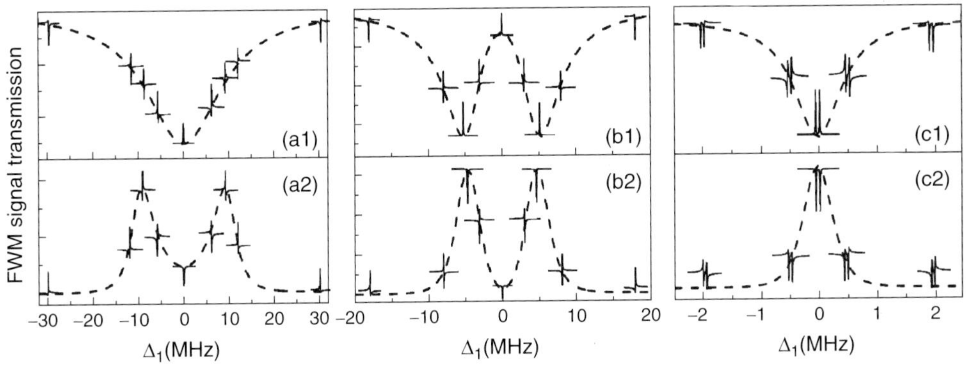

Figure 1.8 (a1), (b1), and (c1) Parallel-cascade, sequential-cascade, and nested-cascade doubly-dressed probe transmission signal, respectively. (a2), (b2), and (c2) the corresponding FWM signal. The dashed curves in (a), (b), and (c) are the probe transmission and the corresponding pure FWM signals without $\boldsymbol{E}_3$ or $\boldsymbol{E}_5$, respectively.

In Figure 1.7(b), the FWM signal is in sequential-cascade doubly-dressed mode (dressed by $\boldsymbol{E}_2$ and $\boldsymbol{E}_3$), in which the fields $\boldsymbol{E}_2$ and $\boldsymbol{E}_3$ dress the same level $|1\rangle$ via the subchain $\rho_{10} \xrightarrow{\omega_2} \rho_{20} \xrightarrow{-\omega_2} \rho_{10} \xrightarrow{-\omega_3} \rho_{30} \xrightarrow{\omega_3} \rho_{10}$. This means that the dressing effects of $\boldsymbol{E}_2$ ($\rho_{10} \xrightarrow{\omega_2} \rho_{20} \xrightarrow{-\omega_2} \rho_{10}$) and $\boldsymbol{E}_3$ ($\rho_{10} \xrightarrow{-\omega_3} \rho_{30} \xrightarrow{\omega_3} \rho_{10}$) join together sequentially; therefore, we call this type of dressing mode as sequential-cascade mode. The profiles of the FWM (i.e., AT splitting) and probe transmission (i.e., EIT) (the dashed lines in Figure 1.8(b)) show the double-peak and double-dip structure versus Δ_1 with $\boldsymbol{E}_2$ blocked. The transition between bright and dark states (in Figure 1.8(b)) is caused by $\boldsymbol{E}_2$. Both the signals in Figure 1.8(b) show three symmetric centers that reflect the interaction between two dressing fields. In the comparison between the sequential-cascade mode and parallel-cascade mode, the transition between the bright and dark states reveals interaction between the two dressing fields in the former mode, while no interaction is revealed in the latter mode.

In Figure 1.7(c), the FWM signal is in the nested doubly-dressed mode (inner-dressing field $\boldsymbol{E}_2$ and outer-dressing field $\boldsymbol{E}_5$), in which $\boldsymbol{E}_5$ dresses the same level $|1\rangle$ after the inner-dressing field $\boldsymbol{E}_2$ having dressed it. This means the dressing effect of $\boldsymbol{E}_5$ ($\rho_{20} \xrightarrow{-\omega_5} \rho_{40} \xrightarrow{\omega_5} \rho_{20}$) is nested in that of $\boldsymbol{E}_2$ ($\rho_{10} \xrightarrow{\omega_2} \rho_{20}$ and $\rho_{20} \xrightarrow{-\omega_2} \rho_{10}$). By scanning the detuning of the inner-dressing field $\boldsymbol{E}_2$, the dashed lines in Figure 1.8(c) show one AT splitting structure in the pure FWM signal and EIT window in probe transmission signal. In such cases, there is only one symmetric center when the inner-dressing field is scanned, but three symmetric centers when the outer-dressing field is scanned, as reflects the strong interaction between two dressing fields. In comparison between the nested-cascade and sequential-cascade dressing modes, the interaction between the two dressing fields in the former mode is stronger, and reflected as dual bright state and dual dark state in the evolution of the FWM signal and probe transmission.

1.2 Fluorescence in MWM

Traditional fluorescence is defined as emitted light by an object, which usually has a longer wavelength than the absorbed radiation. However, in case those two photons are absorbed by one atom, the emission with shorter wavelength than the photon absorbed can be obtained. Furthermore, the emitted field may also be of the same wavelength as the absorbed radiation, termed *resonance fluorescence*.

This section provides a brief description of the two-photon fluorescence process that accompanies the MWM process. In our theoretical and experimental scheme, both MWM signals and fluorescence signals are transmitted in an EIT window. Therefore, compared with resonance fluorescence, the fluorescence process with MWM has several distinct differences and advantages. First, extremely narrow fluorescence signals (about 50 MHz) can be obtained in an open-cycle atomic system, as the generated fluorescence signals fall into EIT windows. Such fluorescence signals with extremely narrow line widths have not been reported before, either experimentally or theoretically. This will allow us to investigate the quantum correlation and narrow line width laser as the generated fluorescence signal is very high in coherence and monochromaticity. Second, by individually controlling the EIT windows, fluorescence signals can be clearly separated or superimposed selectively. Third, the amplitude of the fluorescence signal can be controlled by changing the intensity and frequency of pumping laser via dark states.

In order to give a clear physical description of the fluorescence process, we consider six types of level structures, two-level, V-, Λ-, and Ξ-type three-level, Y- and reversed Y-type four-level atomic systems, as shown in Figure 1.9. First, for the two-level system (Figure 1.9(a)), the single- and two-photon fluorescence signals R_1 and R_2 will be generated as a result of the spontaneous emission of photons from $|1\rangle$ to $|0\rangle$, which can be described by the Liouville pathways $(R_1)\ \rho_{00}^{(0)} \xrightarrow{\omega_1} \rho_{10}^{(1)} \xrightarrow{-\omega_1} \rho_{11}^{(2)}$ and $(R_2)\ \rho_{00}^{(0)} \xrightarrow{\omega_1} \rho_{10}^{(1)} \xrightarrow{-\omega_2} \rho_{00}^{(2)} \xrightarrow{\omega_2} \rho_{10}^{(3)} \xrightarrow{-\omega_1} \rho_{11}^{(4)}$, respectively. Next, for the V-type three-level system (Figure 1.9(b)), there exist two single-photon fluorescence signals R_1 (from $|1\rangle$ to $|0\rangle$) and R_2 (from $|2\rangle$ to $|0\rangle$), and two two-photon fluorescence signals R_3 (from $|1\rangle$ to $|0\rangle$) and R_4 (from $|2\rangle$ to $|0\rangle$), which can be described by the Liouville pathways $(R_1)\ \rho_{00}^{(0)} \xrightarrow{\omega_1} \rho_{10}^{(1)} \xrightarrow{-\omega_1} \rho_{11}^{(2)}$, $(R_2)\ \rho_{00}^{(0)} \xrightarrow{\omega_2} \rho_{20}^{(1)} \xrightarrow{-\omega_2} \rho_{22}^{(2)}$, $(R_3)\ \rho_{00}^{(0)} \xrightarrow{\omega_1} \rho_{10}^{(1)} \xrightarrow{-\omega_2} \rho_{12}^{(2)} \xrightarrow{\omega_2} \rho_{10}^{(3)} \xrightarrow{-\omega_1} \rho_{11}^{(4)}$, and (R_4) $\rho_{00}^{(0)} \xrightarrow{\omega_2} \rho_{20}^{(1)} \xrightarrow{-\omega_1} \rho_{21}^{(2)} \xrightarrow{\omega_1} \rho_{20}^{(3)} \xrightarrow{-\omega_2} \rho_{22}^{(4)}$. Then, for the Λ-type system three-level (Figure 1.9(c)), the decay of photons from $|1\rangle$ to $|0\rangle$ will generate single- and two-photon fluorescence signals R_1 and R_2, which can be described via the Liouville pathways $(R_1)\ \rho_{00}^{(0)} \xrightarrow{\omega_1} \rho_{10}^{(1)} \xrightarrow{-\omega_1} \rho_{11}^{(2)}$ and $(R_2)\ \rho_{00}^{(0)} \xrightarrow{\omega_1} \rho_{10}^{(1)} \xrightarrow{-\omega_2} \rho_{20}^{(2)} \xrightarrow{\omega_2} \rho_{10}^{(3)} \xrightarrow{-\omega_1} \rho_{11}^{(4)}$, respectively. While for the Ξ-type system (Figure 1.9(d)), there is only one single-photon fluorescence signal R_1 and one two-photon signal R_2, which can be represented by the Liouville pathways $(R_1)\ \rho_{00}^{(0)} \xrightarrow{\omega_1} \rho_{10}^{(1)} \xrightarrow{-\omega_1} \rho_{11}^{(2)}$ (spontaneous emission from $|1\rangle$ to $|0\rangle$) and $(R_2)\ \rho_{00}^{(0)} \xrightarrow{\omega_1} \rho_{10}^{(1)} \xrightarrow{\omega_2} \rho_{20}^{(2)} \xrightarrow{-\omega_1} \rho_{21}^{(3)} \xrightarrow{-\omega_2} \rho_{22}^{(4)}$ (spontaneous emission from $|2\rangle$ to $|1\rangle$), respectively. Further, for the Y-type four-level system (Figure 1.9 (e)), there is an additional two-photon fluorescence signal R_3

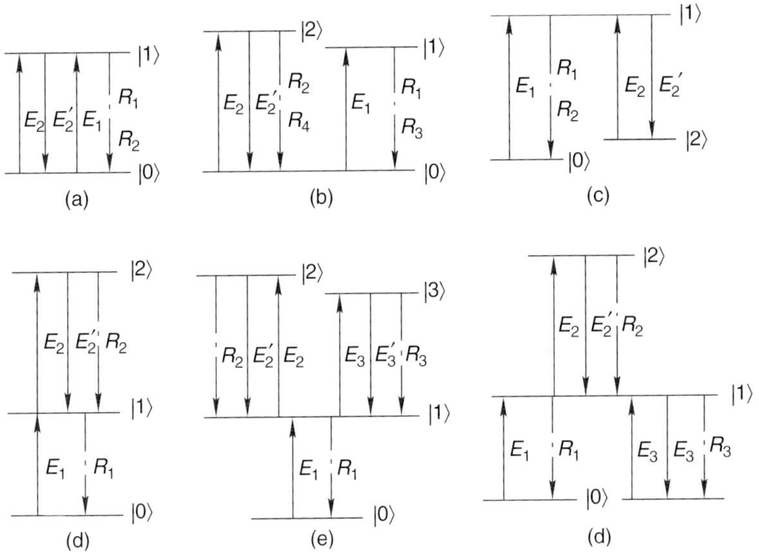

Figure 1.9 The diagrams of (a) two-level, (b) a three-level V-type, (c) a three-level Λ-type, (d) a three-level Ξ-type, and (e) four-level Y-type atomic systems.

compared to the three-level Ξ-type system (Figure 1.9 (d)), which is the spontaneous emission from $|3\rangle$ to $|1\rangle$, and can be expressed by the Liouville pathway (R_3) $\rho_{00}^{(0)} \xrightarrow{\omega_1} \rho_{10}^{(1)} \xrightarrow{\omega_3} \rho_{30}^{(2)} \xrightarrow{-\omega_1} \rho_{31}^{(3)} \xrightarrow{-\omega_3} \rho_{33}^{(4)}$. Here, we only give the corresponding doubly-dressed density matrix elements for the Y-type four-level system as follows:

$$\rho_{11}^{(2)} = \frac{-|G_1|^2}{\Gamma_{11}(d_1 + |G_2|^2/d_2 + |G_3|^2/d_3)} \tag{1.8}$$

$$\rho_{22}^{(4)} = \frac{|G_1|^2|G_2|^2}{\Gamma_{22}d_1d_4(d_2 + |G_2|^2/(d_1 + |G_3|^2/d_3))} \tag{1.9}$$

$$\rho_{33}^{(4)} = \frac{|G_1|^2|G_3|^2}{\Gamma_{33}d_1d_5(d_3 + |G_3|^2/(d_1 + |G_2|^2/d_2))} \tag{1.10}$$

where $d_1 = \Gamma_{10} + i\Delta_1$, $d_2 = \Gamma_{20} + i(\Delta_1 + \Delta_2)$, $d_3 = \Gamma_{30} + i(\Delta_1 + \Delta_3)$, $d_4 = \Gamma_{21} + i\Delta_2$, and $d_5 = \Gamma_{31} + i\Delta_3$ with frequency detuning $\Delta_i = \Omega_i - \omega_i$ (Ω_i is the resonance frequency of the transition driven by E_i) and transverse relaxation rate Γ_{ij} between $|i\rangle$ and $|j\rangle$. Finally, the reversed Y-type four-level system (Figure 1.9(f)), will also generate an additional two-photon fluorescence signal R_3, compared with the Ξ-type system, which can be presented as (R_3) $\rho_{00}^{(0)} \xrightarrow{\omega_1} \rho_{10}^{(1)} \xrightarrow{-\omega_3} \rho_{30}^{(2)} \xrightarrow{\omega_3} \rho_{10}^{(3)} \xrightarrow{-\omega_1} \rho_{11}^{(4)}$ (spontaneous emission from $|1\rangle$ to $|0\rangle$).

By scanning Δ_1, the probe transmission, FWM and fluorescence signals (Figure 1.10(a)) in the Y-type system are obtained. In the fluorescence signals, the global background represents the single-photon fluorescence R_1 ($\rho_{11}^{(2)}$), and other two small sharp peaks on it are derived from the two-photon fluorescence

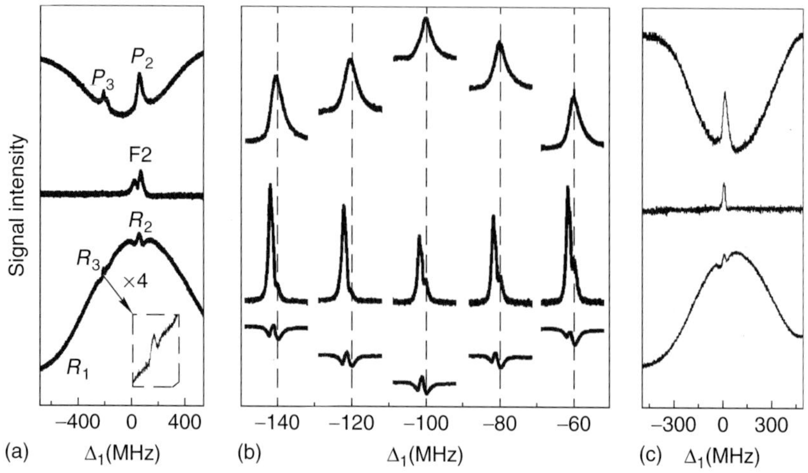

Figure 1.10 Measured probe transmission (upper curves), MWM (middle curves), and fluorescence (bottom curves) ((a) and (c)) versus probe detuning Δ_1, (b) versus Δ_2 at discrete Δ_1 in Y-type ((a) and (b)) and inverted Y-type (c) four-level atomic system.

signals R_2 ($\rho_{22}^{(4)}$) and R_3 ($\rho_{33}^{(4)}$). Figure 1.10(b) represents the measured signals by scanning Δ_2 at discrete Δ_1. Similarly, for the fluorescence signal, the global profile expresses the dressed R_1 ($\rho_{11}^{(2)}$) by E_3; the dip lower than the corresponding baseline represents further dressed R_1 ($\rho_{11}^{(2)}$) by E_2, and the peak within the dip is the two-photon fluorescence signal R_2 ($\rho_{22}^{(4)}$). Furthermore, Figure 1.10(c) represents the probe transmission, SWM, and fluorescence versus Δ_1 in the reversed Y-type system. Similarly, the global background represents the fluorescence R_1 and R_3 ($\rho_{11}^{(2)}$ and $\rho_{11}^{(4)}$), and the small sharp peak on it is the two-photon fluorescence R_2 ($\rho_{22}^{(4)}$).

Although generated simultaneously, the two-photon fluorescence signal and the FWM signals have different behaviors and the difference can be seen from the Liouville pathway. First, the FWM signal is caused by the atomic coherence effect, while the fluorescence signal is induced by spontaneous decay of photons pumped to the upper levels. Second, the direction of the FWM signal is determined by the limitations of the phase-matching conditions, but the fluorescence signal is not directional. Third, the FWM process follows the closed-loop path while the fluorescence process does not.

1.3
MWM Process in Ring Optical Cavity

In this section, we consider the MWM process in the ring cavity. First, a brief discussion on the relation between atom-cavity coupling strength and cavity mode splitting, and how the intensities of input and output are affected by atom-cavity coupling strength are given. Next, noise squeezing with nonlinear media in ring cavity is studied. As an important entanglement light source, squeezed light field with FWM process is briefly discussed.

1.3.1
High-Order Cavity Mode Splitting with MWM Process

VRS has been reported when single two-level atom or N two-level atoms are strongly coupled with cavity mode. The frequency separation of the VRS are $2g$ and $2G$ in the two atom-cavity systems, respectively, with g and $G = g\sqrt{N}$ being single- and multi-atom coupling strength, respectively. The coherently prepared atom-cavity system will also result in the intracavity EIT results, OB, and multi-stability behavior, in which close contact between VRS and OB behavior can be deduced.

In this section, we investigate the relationship between the VRS and OB of the generated MWM cavity mode, achieve the control of VRS and OB simultaneously by the coherent control of dark and bright states, and get the inclined VRS.

As shown in Figure 1.11, the coupled atom-cavity system consisting of Rubidium atoms confined in a four-mirror-formed ring cavity, in which only the generated FWM (SWM) signal can form the cavity mode. Figure 1.12 gives the transmission spectra of the FWM (E_1, E_2, E'_2) cavity mode, containing the splitting positions and heights of the multi-modes. For an empty cavity, the cavity transmission spectra has Lorentzian shape and equal mode spacing (free spectral range Δ_{FSR}), shown as the dashed curves in Figure 1.12. For the coupled atom-cavity system, not only is the zero-order longitudinal mode split with symmetrical center $\Delta_1 = 0$ (Figure 1.12(a1)) but high-order modes are also split by the cavity field when the coupling strength $g\sqrt{N}$ is increased to near or larger than Δ_{FSR}, shown as the solid curves in Figure 1.12(a2) to (a4).

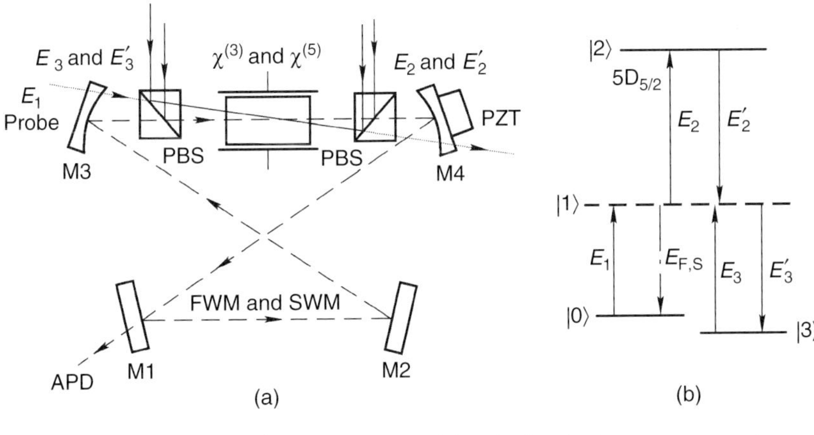

Figure 1.11 (a) A scheme of a ring cavity containing the four-level atoms to generate FWM and SWM processes. (b) Scheme of the four-level atomic system.

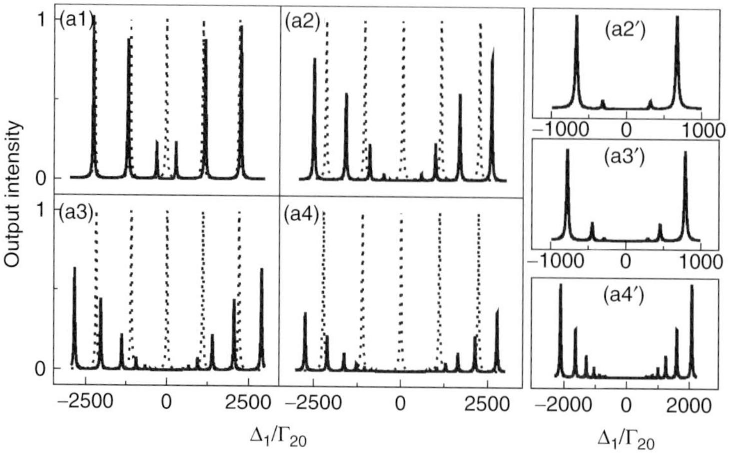

Figure 1.12 (a1)–(a4) Solid curves are transmission spectra of the generated FWM cavity mode with *N* increasing. Dashed curves are the transmission spectra of empty cavity. The illustrations (a2′)–(a4′) FWM cavity mode transmissions in smaller regions corresponding to (a2)–(a4).

Next, the relationship between the input–output intensities under steady-state condition is shown in Figure 1.13, which displays OB behavior of the transmitted FWM cavity mode influenced by dark state. In detail, Figure 1.13(a) illustrates the output intensity of the FWM cavity mode versus probe input intensity (I_i) and detuning (Δ_1), which represents the modulation of VRS in frequency domain and input–output relationship (OB behavior) simultaneously. With I_i increasing, the spectrum of the transmitted FWM cavity mode expands rapidly when Δ_1 is scanned, as shown in Figure 1.13(b). Figure 1.13(c) obviously shows OB hysteresis cycle and the threshold of the optical parametric amplification (OPA) FWM process, and reveals that the increasing $|\Delta_1|$ can result in a significant change in the OB behavior with the increase in the right OB threshold value. Finally, Figure 1.13(d) displays the OB threshold values at different Δ_1, in which the left OB threshold value shifts slowly while the right one shifts sharply. Moreover, Figure 1.13(d) directly predicts there is no OB at or close to the position of dark state ($\Delta_1/\Gamma_{20} = 0$), which results from disappearing linear dispersion at $\Delta_1/\Gamma_{20} = 0$ by the interference between two possible absorption channels $|0\rangle \rightarrow |\pm\rangle$ induced by g^2N.

1.3.2 Squeezed Noise Power with MWM

The quantum correlations have been presented in signal and idler lights from type I and type II [34, 35] OPA inside an optical cavity, which have recently generated increased interests. Moreover, the FWM or SWM processes, which are assisted by the EIT window in multi-level atomic systems, are efficient sources for squeezed radiation and correlated photons. Recent experiments have demonstrated the slowing down [36–38], storage, and retrieval [39, 40] of squeezed states of light

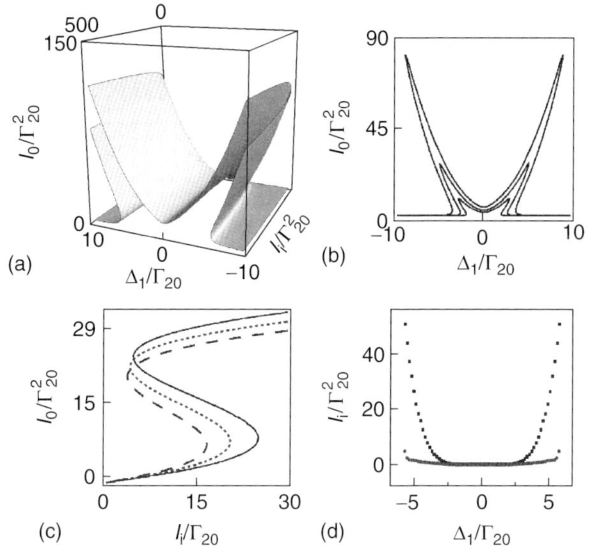

Figure 1.13 Observed input–output intensity relationship of the system. (a) FWM transmission output changes with both probe detuning and probe input intensity; (b) FWM transmission output with probe detuning at different probe input intensity; (c) OB at different probe detuning; and (d) probe input intensity versus probe detuning with the dots of up branch standing for the right OB threshold of the OPA FWM process and dots of down branch standing for the left one.

through EIT in multi-level atomic systems [2, 41, 42], which are important to the implementing of quantum network protocol [43, 44].

In this section, we theoretically investigate the quantum correlations (noise power) between the generated FWM and SWM signals inside an optical cavity, in which the generated FWM signal can be considered as the input field to participate in SWM nonlinear process, and vice versa. This interaction Hamiltonian in the processes can be expressed as $H_{\text{int}} = i\hbar\kappa\widehat{a}_{\text{F}}^{+}\widehat{a}_{\text{S}}^{+} + \text{HC}$, where $\kappa = \chi^{(3)}E_2^2 + \chi^{(5)}E_2^2E_3^2$ is the nonlinear coupling coefficient proportional to the susceptibilities ($\chi^{(3)}$ and $\chi^{(5)}$) and to the amplitudes (E_2 and E_3) of the other two strong coherent fields, and $\widehat{a}_{\text{F}}$ and $\widehat{a}_{\text{S}}$ are the annihilation operators of the generated FWM and SWM fields, respectively. For an arbitrary annihilation operator, it can be described as $\widehat{a} = (\widehat{x} + i\widehat{y})/2$ with two quadrature components, that is, the amplitude $\widehat{x}$ and phase $\widehat{y}$, which satisfy the canonical commutation relation $[\widehat{x}, \widehat{y}] = 2i$.

The equations of motion for the two cavity modes ($\widehat{a}_{\text{F}}$ and $\widehat{a}_{\text{S}}$) can be obtained by solving the Langevin equation:

$$\frac{d\widehat{a}_{\text{F}}}{dt} = -i\left(\Delta - \frac{ck_{\text{F}}}{n_{0\text{F}}}n_{1\text{F}}\right)\widehat{a}_{\text{F}} - (\gamma + \gamma_c)\widehat{a}_{\text{F}} + \kappa\widehat{a}_{\text{S}}^{+} + \sqrt{2\gamma}\widehat{a}_{\text{F}}^{\text{in}} + \sqrt{2\gamma_c}\widehat{c}_{\text{F}} \tag{1.11}$$

$$\frac{d\widehat{a}_S}{dt} = -i\left(\Delta - \frac{ck_S}{n_{0S}} n_{1S}\right)\widehat{a}_S - (\gamma + \gamma_c)\widehat{a}_S + \kappa\widehat{a}_F^{+} + \sqrt{2\gamma}\widehat{a}_S^{\rm in} + \sqrt{2\gamma_c}\widehat{c}_S \tag{1.12}$$

where $\widehat{a}_F^{\rm in}$ ($\widehat{a}_S^{\rm in}$) and $\widehat{a}_F$ ($\widehat{a}_S$) represent input and intracavity FWM (SWM) modes; $\widehat{c}_F$ ($\widehat{c}_S$) is the vacuum mode coupled to the FWM (SWM) mode; Δ is cavity detuning; γ represents the same decay rate of the transition generating FWM (SWM) signal, and γ_c represents other losses in cavity. The items n_{0F} (n_{0S}) and n_{1F} (n_{1S}) are the linear refractive indices at ω_F (ω_S) of the FWM (SWM) signal in vacuum and medium, respectively. In our energy level system, $k_F = k_S = \omega_1 n_0/c$ with $\omega_F = \omega_S = \omega_1$ and $n_{0F} = n_{0S} = n_0$, and so $n_{1F} = n_{1S} = \sqrt{1 + \mathrm{Re}\chi^{(1)}_{F(S)}}$, where the linear susceptibility is given by $\chi^{(1)}_{F(S)} = D\rho^{(1)}_{10F(S)}$ with $D = N\mu^2_{F(S)}/(\hbar\varepsilon_0 G_{F(S)})$, with $\mu_{F(S)}$ being the dipole momentum of the transition generating $\boldsymbol{E}_F$ ($\boldsymbol{E}_S$). Here, we can control the quantum correlations by controlling dark and bright states introduced by $n_{1F}(n_{1S})$.

As shown by the dashed line in Figure 1.14, when other coherent fields are blocked, there is no FWM or SWM process generated, so the quantum correlations between the FWM and SWM output modes correspond to the shot-noise limit (SNL). However, when the coherent fields are injected into the cavity, the noise power spectra of the quantum correlation of the quadrature of the FWM and SWM output fields are observed as the solid lines in Figure 1.14, which show that the amplitude correlation and phase anticorrelation become nosier (Figure 1.14(a)), while the degrees of the amplitude anticorrelation and phase correlation increases. In all these phenomena, the Heisenberg uncertainty relationship is satisfied.

1.3.3
Three-Mode Continuous-Variable Entanglement with MWM

Quantum entanglement has attracted great interest in recent years as it is the central resource for the applications of quantum communication and computation. Bipartite continuous-variable (CV) entanglement was experimentally obtained by Ou *et al.* [45–47] and multi-partite entangled state was also produced by using one

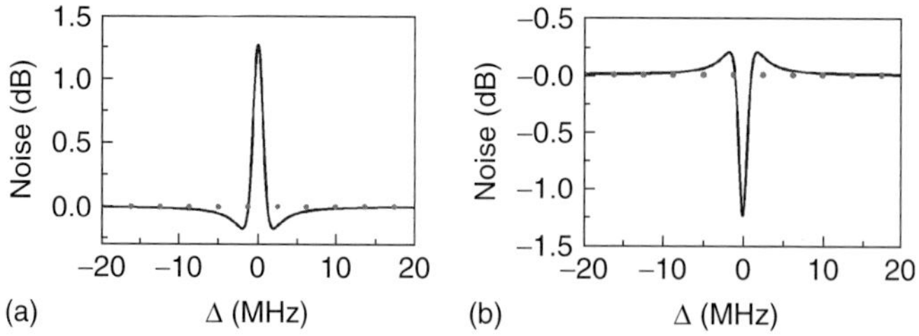

Figure 1.14 (a) and (b) The noise power spectra of the quantum correlations of the quadratures versus Δ. (a) For amplitude correlation and phase anticorrelation. (b) For amplitude anticorrelation and phase correlation.

single-mode squeezed state and linear optics [48–50]. In this section, we consider the quantum nature of the pump in above threshold OPA.

Here, we consider two beams, a pump field ($\widehat{a}_{\mathrm{p}}$) with frequency ω_{p} and a probe field ($\widehat{a}_1$) (considered as a weak injected field) with frequency ω_1, entering into the optical cavity. The Stokes ($\widehat{a}_{\mathrm{S}}$) and anti-Stokes ($\widehat{a}_{\mathrm{AS}}$) fields with frequencies ω_{S} and ω_{AS} can be generated simultaneously by the third-order nonlinear process. To produce a Stokes photon and an anti-Stokes photon, there must annihilate a pump photon and a probe photon. For this scheme, the Hamiltonian of the free modes in the optical cavity can be written as $H_1 = \hbar\omega_{\mathrm{p}}\widehat{a}_{\mathrm{p}}^{+}\widehat{a}_{\mathrm{p}} + \hbar\omega_1\widehat{a}_1^{+}\widehat{a}_1 + \hbar\omega_{\mathrm{S}}\widehat{a}_{\mathrm{S}}^{+}\widehat{a}_{\mathrm{S}} + \hbar\omega_{\mathrm{AS}}\widehat{a}_{\mathrm{AS}}^{+}\widehat{a}_{\mathrm{AS}}$, and the interaction Hamiltonian in this third-order nonlinear process can be written as $H_2 = i\hbar\kappa(\widehat{a}_{\mathrm{p}}\widehat{a}_1\widehat{a}_{\mathrm{S}}^{+}\widehat{a}_{\mathrm{AS}}^{+} - \widehat{a}_{\mathrm{p}}^{+}\widehat{a}_1^{+}\widehat{a}_{\mathrm{S}}a_{\mathrm{AS}})$, where κ is the effective coupling constant proportional to the third-order nonlinear susceptibility ($\chi^{(3)}$). Moreover, the Hamiltonian for the external inputs of the pump and probe field is given by $H_3 = i\hbar(\varepsilon_{\mathrm{p}}\widehat{a}_{\mathrm{p}}^{+} - \varepsilon_{\mathrm{p}}^{*}\widehat{a}_{\mathrm{p}}) + i\hbar(\varepsilon_1\widehat{a}_1^{+} - \varepsilon_1^{*}\widehat{a}_1)$, where ε_{p} and ε_1 are the classical pump and probe field amplitudes. The losses of the modes in cavity can be represented by $L_i\widehat{\rho} = \gamma_i(2\widehat{a}_i\widehat{\rho}\widehat{a}_i^{+} - \widehat{a}_i^{+}\widehat{a}_i\widehat{\rho} - \widehat{\rho}\widehat{a}_i^{+}\widehat{a}_i)$, $(i = \mathrm{p}, 1, \mathrm{S}, \mathrm{AS})$ where γ_i are the damping constants for these cavity modes, respectively. On the basis of the above-mentioned equations and the master equation, one can investigate the entanglement characteristics in our system. Meanwhile, we can also introduce the effect of dark state to control the entanglement between these cavity modes.

Moreover, we have considered a scheme to directly produce bright three-mode CV entanglement by the optical parametric oscillator (OPO) FWM process above threshold. In our scheme, not only three-mode entangled beams can be produced by third-order nonlinear process but the nonlinear interaction and conversion efficiency can be enhanced with dark state as well.

1.4 Photonic Band Gap

In our system generating MWM, many spatial effects can be induced, which will bring about a significant influence on the MWM signals. In this section, the formation of electromagnetically induced grating (EIG) and the EIL in the FWM and their various effects will be introduced. In Section 1.4.1, we discuss the spatially periodic dressing fields that lead to the periodic splitting of the metastable energy level and periodic refractive index of the medium, which varies over a length scale comparable to optical wavelengths of dressing fields. In Section 1.4.2, the two methods for calculating the photonic band gap (PBG) [51, 52], namely, the transfer matrix and plane wave expansion methods, are introduced. In Section 1.4.3, we focus on the nonlinear Talbot effect when the FWM is generated under the dressing effects of the EIG or EIL. And in Section 1.4.4, the soliton formation and dynamics in the propagation of the FWM signal are introduced, under the competition between third- and fifth-order nonlinearities.

1.4.1
Periodic Energy Level

In Figure 1.15, where only the strong fields E_2 and E_2' are turned on, a singly-dressed mode is formed. In the spatial interaction region, E_2 and E_2' will interfere with each other and create a periodic intensity distribution, which leads to periodic Rabi frequency amplitude $|G_{2t}(x)|^2 = G_{20}^2 + G_{20}'^2 + 2G_{20}G_{20}' \cos[2(k_2 \sin\theta)x]$. Therefore, the dressing effect changes periodically. Influenced by $|G_{2t}(x)|^2$, the naked state $|1\rangle$ will be split into two dressing states denoted as $|G_{2t}(x)+\rangle$ and $|G_{2t}(x)-\rangle$, as shown in Figure 1.15(c), locating in the two sides of $|1\rangle$, with their eigenfrequencies having offset from that of $|1\rangle$ determined by the pump detuning Δ_2 and spatial varying Rabi frequency amplitude $G_{2t}(x)$ as $\lambda_{G_t(x)\pm} = \Delta_2/2 \pm \sqrt{\Delta_2^2/4 + |G_{2t}(x)|^2}$. So the split energy levels $|G_{2t}(x)+\rangle$ and $|G_{2t}(x)-\rangle$ is spatial periodical along x-direction.

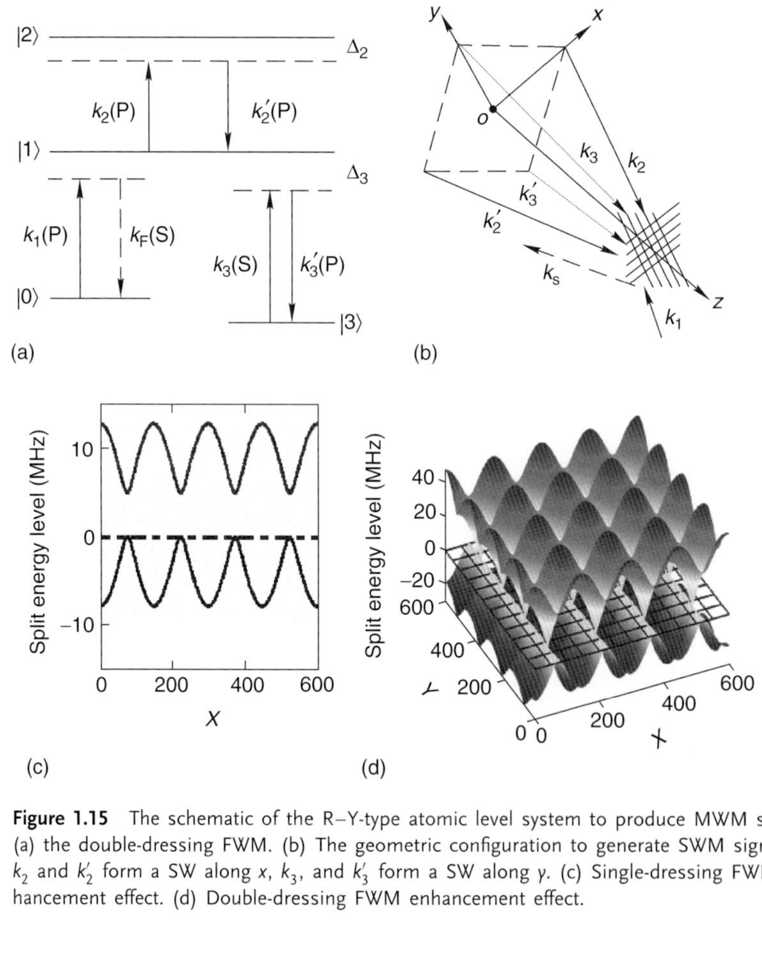

Figure 1.15 The schematic of the R–Y-type atomic level system to produce MWM signals: (a) the double-dressing FWM. (b) The geometric configuration to generate SWM signal; k_2 and k_2' form a SW along x, k_3, and k_3' form a SW along y. (c) Single-dressing FWM enhancement effect. (d) Double-dressing FWM enhancement effect.

As the refractive index of the probe E_1 is strongly dependent on the position of the dressed levels, the refractive index is spatial periodic, and EIG is obtained.

As shown in Figure 1.15(a) and (b), when E_3, E_3', E_2, and E_2' are all turned on, a doubly-dressed mode is formed, as shown in Figure 1.15. Such a configuration will lead to the spatial two-dimensional dressed energy states. First, the spatial interference between the inner-dressing fields E_3 and E_3' leads to the periodic dressing effect, and therefore splits the naked state $|1\rangle$ into two dressing states denoted as $|G_{3t}(y)+\rangle$ and $|G_{3t}(y)-\rangle$, with eigenfrequencies $\lambda_{G_{3t}(y)\pm} = -\Delta_3/2 \pm \sqrt{\Delta_3^2/4 + |G_{3t}(y)|^2}$ periodically varying along the y-direction. Next, the secondary dressing fields E_2 and E_2' will play a role and split the first-order dressing state $|G_{3t}(y)+\rangle$ ($|G_{3t}(y)-\rangle$) into two second-order dressing states $|G_{3t}(y)+G_{2t}(x)+\rangle$ and $|G_{3t}(y)+G_{2t}(x)-\rangle$ ($|G_{3t}(y)-G_{2t}(x)+\rangle$ and $|G_{3t}(y)-G_{2t}(x)-\rangle$), given that E_2 and E_2' are near the resonance with $|G_{3t}(y)+\rangle$ ($|G_{3t}(y)-\rangle$). Obviously, the eigenfrequencies of the two second-order dressing states $|G_{3t}(y)+G_{2t}(x)\pm\rangle$ and $|G_{3t}(y)-G_{2t}(x)\pm\rangle$ are periodical, as shown in Figure 1.15(d). Such two-dimensional dressed energy states can lead to a two-dimensional periodic refractive index, that is, the formation of EIL.

1.4.2 Method of Transfer Matrix

It is well known that a structure with a spatial periodical refractive index will give rise to a PBG [53], in which the incident filed will be strongly reflected. Two theoretical methods usually employed can be used to calculate the PBG of the EIG and EIL. One is the transfer matrix method [54], and in this method one periodic of EIG can be divided into many layers, and the refractive index in each layer is assumed to be constant. The relationship between the incident and outgoing light field of each layer is connected by a transfer matrix determined by the refractive index in the layer itself. Thus, the total transfer matrix T_Λ of one periodic in the EIG in the medium can be obtained by the product of each layer's, that is, $T_\Lambda(\Delta) = \sum_{j=1}^{N} T_j(\Delta)$, in which $T_j(\Delta)$ is the transfer matrix of the jth layer when we divide the whole period into N layers, and its dependence on the probe field detuning Δ is obvious. Furthermore, with the Bloch's theorem introduced, the PBG of EIG with periodicity Λ can be obtained numerically by solving the equation $m\pi/a + i\kappa' = \cos^{-1}\{\mathrm{Tr}[T_\Lambda(\Delta)]\}/\Lambda$, the left-hand sides of which mean that the real part of the Bloch wave vector is around the edge of the Brillouin region; the imaginary part is nonzero.

The other method is plane wave expansion. The spatially modulated total linear and nonlinear refractive index can be expressed as $n(\zeta) = n_0 + \delta n_1 \cos(2k_2\zeta) + \delta n_2 \cos(4k_2\zeta)$, where n_0 is the spatially uniform refractive index; δn_1 and δn_2 are the coefficients for spatially varying terms in the modulated index. By expanding the electrical field of the incident field into the combination of Bloch waves, and the spatially periodical index into Fourier series, then substituting these two expressions into the Maxwell equation, and solving the homogeneous equations, we can obtain $\kappa_\pm =$

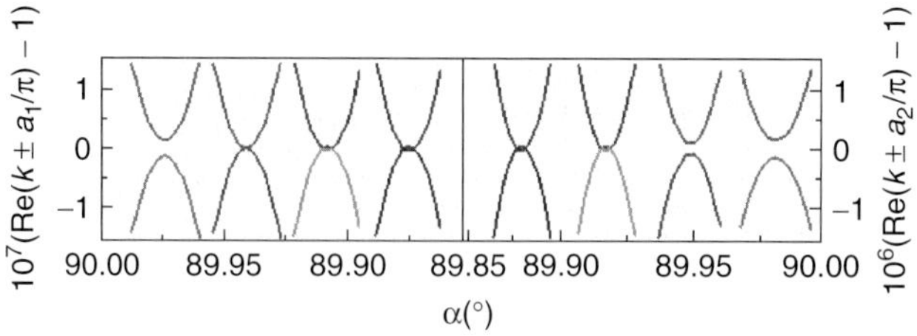

Figure 1.16 The photonic band gap changed with the incident angle.

$k_i \pm \sqrt{\{k_p^2[1+\chi_1+(\chi_2^2-\chi_3^2)\chi_2'] - k_i^2\}^2 - k_p^4[\chi_2\chi_3\chi_2' + (\chi_2^2+2\chi_2\chi_3)\chi_3']^2}/2k_i$, which is called *dispersion relationship* and reveals the relation between the Bloch wave vectors $\kappa_\pm$ and the probe field Δ_1. Here, χ_m and χ_m' $(m=1,2,3)$ are the Fourier coefficients of the susceptibility; k_i is the wave vector component projected on the EIG periodic orientation of the probe field. A region of Δ_1 can be found satisfying $\mathrm{Re}(k_\pm a_i/\pi - 1) = 0$, which is the PBG. In the expression, the PBG also depends on the incident angle and can be controlled as shown in Figure 1.16.

1.4.3
Nonlinear Talbot Effect

The Talbot effect [55–59], first observed by Talbot in 1836, is a near-field diffraction phenomenon in which the light field spatially imprinted with periodic structure can have self-imaging at certain periodic imaging planes (the so-called Talbot plane). Such a self-imaging effect holds a range of applications from image preprocessing and synthesis to optical computing [60]. In this section, we give an introduction on how to generate controllable Talbot self-imaging with FWM and SWM signals.

Since dressing effect can modify the FWM and SWM signals, we can construct the spatial periodic dressing effect to modify the usual spatial uniform MWM signals into spatial periodically dressed signals, with spatial periodic intensity distribution in the transverse dimension. The periodic pattern of the MWM (MWM) signals can be flexibly controlled by adjusting the atomic coherence. More specifically, the one- and two-dimensional periodic dressing effects are determined by the spatial periodic Rabi frequency $|G_{2t}(x)|^2$ and $|G_{2t}(x)|^2$ and $|G_{3t}(y)|^2$, respectively, similar to the case of the formation of EIG and EIL mentioned in Section 1.4.2. Taking the two-dimensional case, for instance, we can obtain the dressing states $|G_{3t}(y) + G_{2t}(x) \pm\rangle$ and $|G_{3t}(y) - G_{2t}(x) \pm\rangle$ with two-dimensional periodic varying frequencies $\lambda_{G_{3t}(y)+G_{2t}(x)\pm}$ and $\lambda_{G_{3t}(y)-G_{2t}(x)\pm}$. Then, the periodic enhancement condition and suppression condition can be obtained, and the FWM is modulated to get spatial periodic intensity distribution.

Next, we consider the propagation of the periodic FWM (SWM) signals. For simplicity, we define the signal output plane as $z=0$, and adopt the paraxial approximation. Therefore, the propagation of the MWM signals can be regarded as

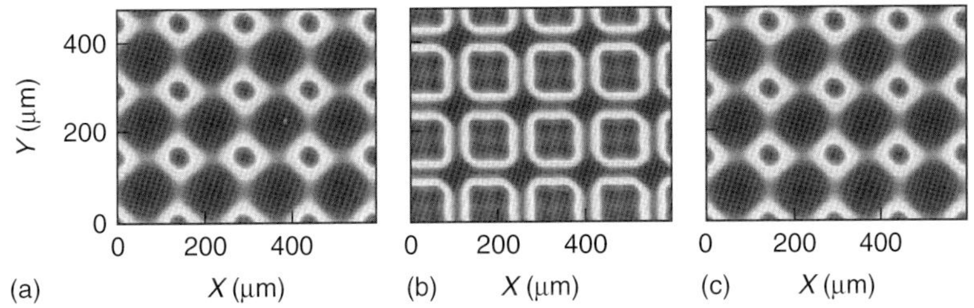

Figure 1.17 The three panels are the cut plots of the Talbot effect at (a) the initial place, (b) the quarter Talbot length, (c) and the Talbot plane.

a Fresnel diffraction process. In the perspective of Fourier optics, the transfer function of a Fresnel diffraction system with z as the propagation axis can be expressed as $H_F(\xi) = \exp(ik_{z0}z)\exp(-i\pi\lambda z\xi^2)$, where ξ is the spatial frequency and k_{z0} is the input signal wave vector projected on the z-axis. It is obvious that a signal with multi-ξ components can be distorted in the propagation because a ξ-quadratic phase is introduced to it in the diffraction. However, for the periodic MWM signal described in the preceding text, the self-imaging without distortion at certain propagation lengths can be demonstrated. In detail, the periodical electrical field distribution of the MWM signal $g_0(x,y)$ can be expanded into two-dimensional Fourier series as $g_0(x,y) = \sum_{m,n=-\infty}^{\infty} c_{m,n}\exp[i2\pi(nx/d_x + my/d_y)]$, and in spatial frequency domain can be further written as $G_0(\xi,\eta) = \sum_{m,n=-\infty}^{\infty} c_{m,n}\delta(\xi - n/d_x)\cdot\delta(\eta - m/d_y)$, where $c_{m,n}$ is the Fourier coefficient. So, after the Fresnel diffraction, multiplying the MWM signal and the transfer function in spatial frequency domain, we can obtain the MWM signal at a z distance as $G(\xi,\eta) = \sum_{m,n=-\infty}^{\infty} c_{m,n}\delta(\xi - n/d_x)\delta(\eta - m/d_y)\cdot \exp(ik_{z0}z)\exp(-i\phi_p)$, with $\phi_p = \pi\lambda_1 z[(n/d_x)^2 + (m/d_y)^2]$ being the phase introduced in the propagation, and k_{z0} being the projection of the MWM signal wave vector on z-axis. After inverse Fourier transformation, we can get the electrical field distribution in spatial domain as $g(x) = g_0(x)\exp(ik_{z0}z)$. In light of the fact that $|g(x)|^2 = |g_0(x)|^2$, we can say that the self-imaging of the MWM signals occurs at $z_T = 2qd_x^2/\lambda_1$, as shown in Figure 1.17(a) and (c), and $z_T|_{q=1}$ is defined as the first Talbot length.

1.4.4
Third- and Fifth-Order Nonlinearity

The EIG, EIL, and the Talbot effect mentioned earlier reveals the spatial modulation of the FWM signal. An interesting effect in the spatial modulation of light is the formation of spatial soliton, in which the nonlinearity is very important to balance with the diffraction. In this section, we introduce the competition between the third- and fifth-order nonlinearities, which is important for the soliton formation in the Kerr medium. In nonlinear optics [61, 62], the susceptibility χ connects the refractive index of the medium via the relation $n = \sqrt{1 + \mathrm{Re}(\chi)}$. In general, the susceptibility can be expressed as $\chi = \chi^{(1)} + \sum_{j=1}^{\infty}\chi^{(2j+1)}|E|^{2j}$, where E is the electrical field intensity incident into the medium and $\chi^{(m)}$ (m is odd number) is the *m*th order susceptibility. Taking into account the higher order terms, one can represent the refractive index as $n = n_1 + n_2 I + n_4 I^2 + \cdots$, that is, a power-law Kerr-type expression, with n_m being the nonlinear coefficients and $I = \varepsilon_0 c|E|^2/2$ being the optical intensity. Obviously, if nonlinear coefficients n_2 and n_4 have the same signs, the incident field under such nonlinearity will exhibit simply increasing strength of nonlinear self-action, namely, self-focusing for $n_2 > 0$ and $n_4 > 0$ self-defocusing for $n_2 < 0$ and $n_4 < 0$. A more interesting situation occurs if the two nonlinear contributions have opposite signs, that is, $n_2 n_4 < 0$, and this case is usually called the *competing* cubic-quintic (CQ) nonlinearity, which is widely used in solitonic science to seek stable higher order soliton solutions. For example, let $n_4 > 0$, the fifth order will introduce self-focusing. Then, for any sign of Kerr coefficient n_2, there will be a threshold power of light beam E, above which the higher order self-focusing will be dominant. In this case, the beam will collapse, similar to the case with pure third-order nonlinearity of $n_2 > 0$ and $n_4 = 0$. However, if $n_4 < 0$, then the collapse can be suppressed, for the incident light beam with sufficiently high intensity will experience an effective self-defocusing environment, that is, $\mathrm{d}n/\mathrm{d}I < 0$. Here, the threshold intensity is given by $I_{\mathrm{th}} = -n_2/2n_4$. That is the reason why CQ nonlinearity has been a charming research focus in the past decades.

To give one example, even though higher order solitons cannot stably exist in a Kerr medium, they have been numerically confirmed in a system with CQ nonlinearity. Experimentally, a CQ nonlinear dielectric response with positive cubic and negative quintic contributions has been observed in chalcogenide glasses, and in organic materials. However, in all these cases the quintic nonlinearity is accompanied by significant higher order multi-photon processes, such as two-photon absorption. The suppression of absorption and enhancement of a high-order nonlinear refractive index, in a system with atomic coherence can be employed to overcome this problem, and our dressed FWM system is an ideal system to implement this scheme.

In Chapter 7, weintroduce the competition between the positive third- and negative fifth-order nonlinearity in an atomic system with EIT, in which the fundamental soliton, dipole soliton, and vortex soliton can form and transform with each other. Also, the EIG, PBG, and discrete soliton forming with such CQ nonlinearity is introduced.

1.5 MWM with Rydberg Blockade

With the development of modern laser cooling and trapping techniques, the so-called ultracold Rydberg gases and plasmas have been experimentally created, in which the atomic or plasmic density is very high and the Rydberg interaction becomes very strong [63–66], which attracts more and more attention due to their wide range of applications, such as quantum computing and scalable quantum information processing [67, 68]. Among such applications, it has been proposed to realize the quantum logic gates by employing the sensitivity of the highly excited state energy to the interaction between neighboring Rydberg atoms [69, 70]. Also, the phenomena related to atomic coherence, such as coherent population trapping (CPT) [71], stimulated Raman adiabatic passage (STIRAP) [72] have been demonstrated in the Rydberg atomic assemble. Such progresses give the possibility of controlling the dressed MWM process by the Rydberg blockade.

It is well known that the probe transmission, and enhancement and suppression of FWM signal can be controlled by the dressing effect of the light field, which modifies the unperturbed levels significantly [32, 73, 74] and depends on the detuning of dressing field strongly. As the Rydberg blockade could shift energy levels, it can modulate the probe transmission and FWM signal via the modified dressing field detuning. Our work is focused on the MWM process under the modulation of both the dressing effect and Rydberg blockade.

In this section, we introduce the principle of how to control the probe transmission and MWM processes by the interaction between the dressing effect and primary blockade, or by that between the dressing effect and secondary blockade.

For two atoms with Rydberg levels, there exists considerable interaction between them, and therefore they constitute a diatomic system, in which the atomic coherence and FWM can occur. For two single atoms with the three-level subsystem ($|0\rangle \rightarrow |1\rangle \rightarrow |2\rangle$), $3 \times 3 = 9$ energy levels will be generated in the corresponding diatomic system, as shown in Figure 1.18(a). Owing to the long-range interaction potential between Rydberg atoms with high principal quantum number n, there will be an energy level shift from the unperturbed energy level. Such interaction-induced energy shift could be called *blockade effect*. In Figure 1.18(b), these energy level shift curves around $n = 70$ in rubidium are plotted. The internuclear distance R can be experimentally controlled by changing the atomic intensity, that is, the cooling conditions in magneto-optical trap (MOT) in ultracold gas or the

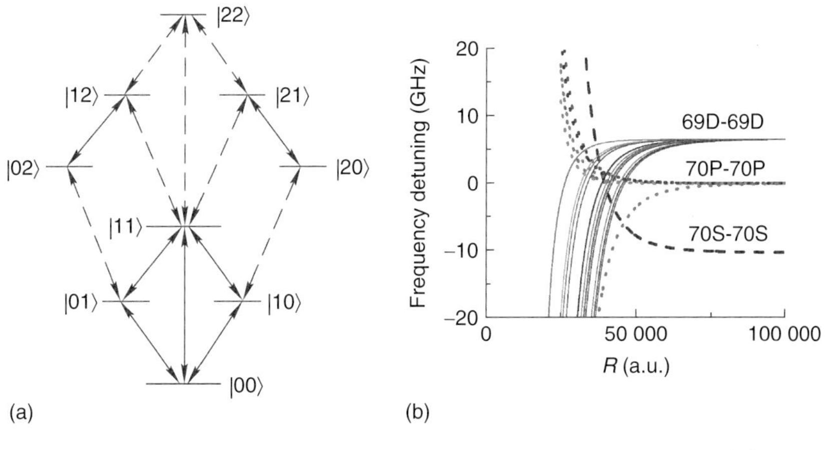

Figure 1.18 (a) The diatomic systems consisting of 3×3. (b) The energy level shift curves for different symmetries of 70S–70S (dashed lines), 70P–70P (dotted lines), and 69D–69D (solid lines) of rubidium.

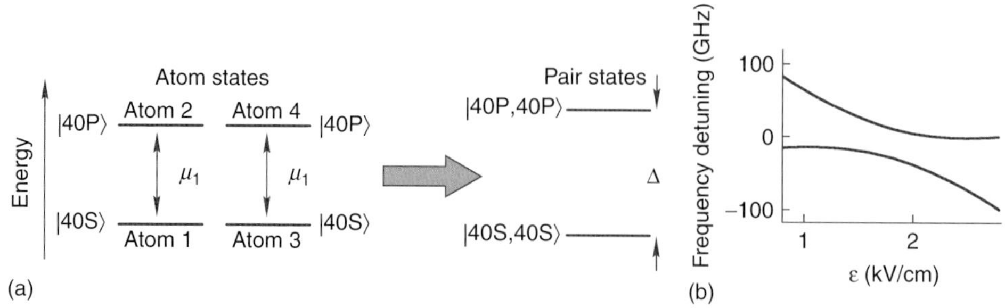

Figure 1.19 (a) The energy levels of a quadratomic system in atom states; μ_1 is the dipole matrix element between $|40S\rangle$ and $|40P\rangle$. The right-hand figure illustrates the corresponding energy levels in pair sates, in which Δ is the energy gap between $|40S, 40S\rangle$ and $|40P, 40P\rangle$. (b) The eigenfrequencies of the two levels $E_+(R)$ (the upper curve) and $E_-(R)$ (the lower curve) with both the primary and secondary Rydberg blockades considered, versus ε with R fixed.

temperature in hot atom vapor, and such level shift can effectively influence the dressing effect and therefore the MWM process.

Furthermore, if the Rydberg energy level of a diatomic system has a strong probability of the dipole transiting to that of another diatomic system, as shown in Figure 1.19(a), the perturbation will make the eigenfrequency shift again, as shown in Figure 1.19(a), in which Δ is the energy gap between the levels without transition perturbation. We can call this effect as the secondary blockade to distinguish from the first blockade in Figure 1.19(a). The eigenfrequencies of the perturbed system can be written as $E_\pm = \Delta_{|ss\rangle\pm}(\varepsilon) = \Delta(\varepsilon)/2 \pm \sqrt{(\Delta(\varepsilon)/2)^2 + (\mu_1^2/R^3)^2}$, which is obviously related to Δ and equivalent to the energy level shift after the twice blockade. Here, μ_1 is the transition dipole moment between two diatomic levels. Moreover, Δ can be changed by the external electric filed intensity ε. Therefore, the energy level shift based on primary and secondary Rydberg interactions can also be Δ-dependent, as shown in Figure 1.19(b), in which the upper curve is for $\Delta_{|ss\rangle+}(\varepsilon)$, and the lower for $\Delta_{|ss\rangle-}(\varepsilon)$. Therefore, the dressed FWM processes can be also controlled by tuning ε directly.

1.6 Summary

In this chapter, several methods and results in realizing quantum control of MWM have been introduced, including AT splitting of FWM with the role of dressing field; suppression, or enhancement of the MWM process via dark state; control of the VRS and OB behavior with dark-state polariton in MWM process; the formation and control of EIG, PBG, and Talbot effect in the FWM system; and the interaction between dressing effect and Rydberg blockade. Such quantum control of the MWM process can be achieved and have potential applications in optical devices and quantum information. In a word, the core control object in this book is the MWM process, and the most important tool in the control is dark state and EIT.

Problems

1.1 According to the content in this chapter, list what parameters can be used to modulate the EIT window and MWM process in the experiment.

1.2 Compare the resonance fluorescence and fluorescence accompanying with MWM process, and show that the differences between them and the advantages of the latter.

1.3 Summarize the formation condition of the photonic band gap in atomic medium.

1.4 Summarize how to control the probe transmission and MWM processes via the interaction between the dressing effect and primary Rydberg blockade.

References

1. Arimondo, E. (1996) Coherent population trapping in laser spectroscopy. *Prog. Opt.*, **35**, 257–354.
2. Harris, S.E. (1997) Electromagnetically induced transparency. *Phys. Today*, **50**, 36.
3. Lee, D. and Albrecht, A. (1985) A unified view of Raman, resonance Raman, and fluorescence spectroscopy (and their analogues in two-photon absorption). *Adv. Infrared Raman Spectrosc.*, **12**, 179–213.
4. Lotem, H., Lynch, R. Jr.,, and Bloembergen, N. (1976) Interference between Raman resonances in four-wave difference mixing. *Phys. Rev. A*, **14**, 1748.
5. Zhang, Y., Brown, A.W., and Xiao, M. (2007) Opening four-wave mixing and six-wave mixing channels via dual electromagnetically induced transparency windows. *Phys. Rev. Lett.*, **99**, 123603.
6. Zhang, Y., Li, P., Zheng, H., Wang, Z., Chen, H., Li, C., Zhang, R., and Xiao, M. (2011) Observation of Autler-Townes splitting in six-wave mixing. *Opt. Express*, **19**, 7769–7777.
7. Wang, Z., Zheng, H., Chen, H., Li, P., Sang, S., Lan, H., Li, C., and Zhang, Y. (2012) Polarized suppression and enhancement of six-wave mixing in electromagnetically induced transparency window. *IEEE J. Quantum Electron.*, **48**, 669–677.
8. Hau, L.V., Harris, S.E., Dutton, Z., and Behroozi, C.H. (1999) Light speed reduction to 17 metres per second in an ultracold atomic gas. *Nature*, **397**, 594–598.
9. Kash, M.M., Sautenkov, V.A., Zibrov, A.S., Hollberg, L., Welch, G.R., Lukin, M.D., Rostovtsev, Y., Fry, E.S., and Scully, M.O. (1999) Ultraslow group velocity and enhanced nonlinear optical effects in a coherently driven hot atomic gas. *Phys. Rev. Lett.*, **82**, 5229–5232.
10. Budker, D., Kimball, D.F., Rochester, S.M., and Yashchuk, V.V. (1999) Nonlinear magneto-optics and reduced group velocity of light in atomic vapor with slow ground state relaxation. *Phys. Rev. Lett.*, **83**, 1767–1770.
11. Li, Y., Jin, S., and Xiao, M. (1995) Observation of an electromagnetically induced change of absorption in multilevel rubidium atoms. *Phys. Rev. A*, **51**, 1754–1757.
12. Gea-Banacloche, J., Li, Y.-Q., Jin, S.-Z., and Xiao, M. (1995) Electromagnetically induced transparency in ladder-type inhomogeneously broadened media: theory and experiment. *Phys. Rev. A*, **51**, 576–584.
13. Li, Y.-Q. and Xiao, M. (1996) Enhancement of nondegenerate four-wave mixing based on electromagnetically induced transparency in rubidium atoms. *Opt. Lett.*, **21**, 1064–1066.
14. Zhang, Y., Khadka, U., Anderson, B., and Xiao, M. (2009) Temporal and spatial interference between four-wave mixing and six-wave mixing channels. *Phys. Rev. Lett.*, **102**, 13601.

15. Zhang, Y., Nie, Z., Wang, Z., Li, C., Wen, F., and Xiao, M. (2010) Evidence of Autler–Townes splitting in high-order nonlinear processes. *Opt. Lett.*, **35**, 3420–3422.
16. Li, J.H., Peng, J.C., and Chen, A.X. (2004) Optical multi-wave mixing process based on electromagnetically induced transparency. *Commun. Theor. Phys.*, **41**, 106–110.
17. Wang, Z., Zhang, Y., Chen, H., Wu, Z., Fu, Y., and Zheng, H. (2011) Enhancement and suppression of two coexisting six-wave-mixing processes. *Phys. Rev. A*, **84**, 013804.
18. Fleischhauer, M. and Lukin, M. (2000) Dark-state polaritons in electromagnetically induced transparency. *Phys. Rev. Lett.*, **84**, 5094–5097.
19. Scully, M., Zhu, S., and Fearn, H. (1992) Lasing without inversion. *Z. Phys. D: At. Mol. Clusters*, **22**, 471–481.
20. Kocharovskaya, O. (1992) Amplification and lasing without inversion. *Phys. Rep.*, **219**, 175–190.
21. Bardou, F., Bouchaud, J., Emile, O., Aspect, A., and Cohen-Tannoudji, C. (1994) Subrecoil laser cooling and Lévy flights. *Phys. Rev. Lett.*, **72**, 203–206.
22. Kulikov, I., Dowling, J.P., and Lee, H. (2003) Matter-Wave BEC Magnetometer: A Breakthrough in Ultra-Sensitive Magnetic Measurements.
23. Chaneliere, T., Matsukevich, D., Jenkins, S., Lan, S.Y., Kennedy, T., and Kuzmich, A. (2005) Storage and retrieval of single photons transmitted between remote quantum memories. *Nature*, **438**, 833–836.
24. Reim, K., Nunn, J., Lorenz, V., Sussman, B., Lee, K., Langford, N., Jaksch, D., and Walmsley, I. (2010) Towards high-speed optical quantum memories. *Nat. Photonics*, **4**, 218–221.
25. Bao, X.H., Qian, Y., Yang, J., Zhang, H., Chen, Z.B., Yang, T., and Pan, J.W. (2008) Generation of narrow-band polarization-entangled photon pairs for atomic quantum memories. *Phys. Rev. Lett.*, **101**, 190501.
26. Wineland, D.J., Bollinger, J.J., Itano, W.M., Moore, F.L., and Heinzen, D.J. (1992) Spin squeezing and reduced quantum noise in spectroscopy. *Phys. Rev. A*, **46**, R6797–R6800.
27. Wineland, D.J., Bollinger, J.J., Itano, W.M., and Heinzen, D.J. (1994) Squeezed atomic states and projection noise in spectroscopy. *Phys. Rev. A*, **50**, 67–88.
28. Huelga, S.F., Macchiavello, C., Pellizzari, T., Ekert, A.K., Plenio, M.B., and Cirac, J.I. (1997) Improvement of frequency standards with quantum entanglement. *Phys. Rev. Lett.*, **79**, 3865–3868.
29. Harris, S.E. and Yamamoto, Y. (1998) Photon switching by quantum interference. *Phys. Rev. Lett.*, **81**, 3611–3614.
30. Harris, S.E. and Hau, L.V. (1999) Nonlinear optics at low light levels. *Phys. Rev. Lett.*, **82**, 4611–4614.
31. Lukin, M.D. and Imamoğlu, A. (2000) Nonlinear optics and quantum entanglement of ultraslow single photons. *Phys. Rev. Lett.*, **84**, 1419–1422.
32. Li, C., Zhang, Y., Nie, Z., Du, Y., Wang, R., Song, J., and Xiao, M. (2010) Controlling enhancement and suppression of four-wave mixing via polarized light. *Phys. Rev. A*, **81**, 033801.
33. Li, C., Zheng, H., Zhang, Y., Nie, Z., Song, J., and Xiao, M. (2009) Observation of enhancement and suppression in four-wave mixing processes. *Appl. Phys. Lett.*, **95**, 041103.
34. Shih, Y., Sergienko, A., Rubin, M.H., Kiess, T., and Alley, C. (1994) Two-photon entanglement in type-II parametric down-conversion. *Phys. Rev. A*, **50**, 23.
35. Laurat, J., Coudreau, T., Keller, G., Treps, N., and Fabre, C. (2005) Effects of mode coupling on the generation of quadrature Einstein-Podolsky-Rosen entanglement in a type-II optical parametric oscillator below threshold. *Phys. Rev. A*, **71**, 022313.
36. Akamatsu, D., Yokoi, Y., Arikawa, M., Nagatsuka, S., Tanimura, T., Furusawa, A., and Kozuma, M. (2007) Ultraslow propagation of squeezed vacuum pulses with electromagnetically induced transparency. *Phys. Rev. Lett.*, **99**, 153602.
37. Arikawa, M., Honda, K., Akamatsu, D., Yokoil, Y., Akiba, K., Nagatsuka, S., Furusawa, A., and Kozuma, M. (2007)

Observation of electromagnetically induced transparency for a squeezed vacuum with the time domain method. *Opt. Express*, **15**, 11849–11854.

38. Hètet, G., Buchler, B.C., Glöeckl, O., Hsu, M.T.L., Akulshin, A.M., Bachor, H.A., and Lam, P.K. (2008) Delay of squeezing and entanglement using electromagnetically induced transparency in a vapour cell. *Opt. Express*, **16**, 7369–7381 Arxiv preprint arXiv:0803.2097.
39. Honda, K., Akamatsu, D., Arikawa, M., Yokoi, Y., Akiba, K., Nagatsuka, S., Tanimura, T., Furusawa, A., and Kozuma, M. (2008) Storage and retrieval of a squeezed vacuum. *Phys. Rev. Lett.*, **100**, 093601.
40. Appel, J., Figueroa, E., Korystov, D., Lobino, M., and Lvovsky, A.I. (2008) Quantum memory for squeezed light. *Phys. Rev. Lett.*, **100**, 093602.
41. Marangos, J. (1998) Electromagnetically induced transparency. *J. Mod. Opt.*, **45**, 471–503.
42. Fleischhauer, M., Imamoglu, A., and Marangos, J.P. (2005) Electromagnetically induced transparency: optics in coherent media. *Rev. Mod. Phys.*, **77**, 633–673.
43. Duan, L.M., Lukin, M., Cirac, I., and Zoller, P. (2001) Long-distance quantum communication with atomic ensembles and linear optics. *Nature*, **414**, 413–418 Arxiv preprint quant-ph/0105105.
44. Simon, C., de Riedmatten, H., Afzelius, M., Sangouard, N., Zbinden, H., and Gisin, N. (2007) Quantum repeaters with photon pair sources and multimode memories. *Phys. Rev. Lett.*, **98**, 190503.
45. Ou, Z., Pereira, S., and Kimble, H. (1992) Realization of the Einstein-Podolsky-Rosen paradox for continuous variables in nondegenerate parametric amplification. *Appl. Phys. B*, **55**, 265–278.
46. Zhang, Y., Wang, H., Li, X., Jing, J., Xie, C., and Peng, K. (2000) Experimental generation of bright two-mode quadrature squeezed light from a narrow-band nondegenerate optical parametric amplifier. *Phys. Rev. A*, **62**, 023813.
47. Tan, H.T., Zhu, S.Y., and Zubairy, M.S. (2005) Continuous-variable entanglement in a correlated spontaneous emission laser. *Phys. Rev. A*, **72**, 022305.
48. Tan, H., Li, G., and Zhu, S. (2007) Macroscopic three-mode squeezed and fully inseparable entangled beams from triply coupled intracavity Kerr nonlinearities. *Phys. Rev. A*, **75**, 063815.
49. Su, X., Tan, A., Jia, X., Zhang, J., Xie, C., and Peng, K. (2007) Experimental preparation of quadripartite cluster and Greenberger-Horne-Zeilinger entangled states for continuous variables. *Phys. Rev. Lett.*, **98**, 70502.
50. Zhang, J., Xie, C., and Peng, K. (2002) Controlled dense coding for continuous variables using three-particle entangled states. *Phys. Rev. A*, **66**, 32318–32318.
51. Rosberg, C.R., Neshev, D.N., Krolikowski, W., Mitchell, A., Vicencio, R.A., Molina, M.I., and Kivshar, Y.S. (2006) Observation of surface gap solitons in semi-infinite waveguide arrays. *Phys. Rev. Lett.*, **97**, 83901.
52. Yeh, P., Yariv, A., and Hong, C.S. (1977) Electromagnetic propagation in periodic stratified media. I. General theory. *J. Opt. Soc. Am.*, **67**, 423–438.
53. Yablonovitch, E. (1993) Photonic band-gap structures. *J. Opt. Soc. Am. B*, **10**, 283–283.
54. Born, M. and Wolf, E. (1980) *Principle of Optics* Chapter 10, Pergamon Press, New York.
55. Talbot, H.F. (1836) LXXVI. Facts relating to optical science. No. IV. *London Edinburgh Philos. Mag. J. Sci.*, **9**, 401–407.
56. Rayleigh, L. (1881) On copying diffraction-gratings, and on some phenomena connected therewith. *Philos. Mag.*, **11**, 196–205.
57. Zhang, Y., Wen, J., Zhu, S., and Xiao, M. (2010) Nonlinear Talbot effect. *Phys. Rev. Lett.*, **104**, 183901.
58. Wen, J., Zhang, Y., Zhu, S.N., and Xiao, M. (2011) Theory of nonlinear Talbot effect. *J. Opt. Soc. Am. B*, **28**, 275–280.
59. Wen, J., Du, S., Chen, H., and Xiao, M. (2011) Electromagnetically induced Talbot effect. *Appl. Phys. Lett.*, **98**, 081108.

60. Patorski, K. (1989) The self-imaging phenomenon and its applications. *Prog. Opt*, **27**, 1–108.
61. Shen, Y.R. (1984) *The Principles of Nonlinear Optics*, Vol. 1, Wiley-Interscience, New York, p. 575 p.
62. Boyd, R.W. (2003) *Nonlinear Optics*, Academic Press, San Diego, CA.
63. Singer, K., Stanojevic, J., Weidemüller, M., and Côté, R. (2005) Long-range interactions between alkali Rydberg atom pairs correlated to the ns–ns, np–np and nd–nd asymptotes. *J. Phys. B: At. Mol. Opt. Phys.*, **38**, S295.
64. Farooqi, S.M., Tong, D., Krishnan, S., Stanojevic, J., Zhang, Y., Ensher, J., Estrin, A., Boisseau, C., Côté, R., and Eyler, E. (2003) Long-range molecular resonances in a cold Rydberg gas. *Phys. Rev. Lett.*, **91**, 183002.
65. Stanojevic, J., Côté, R., Tong, D., Farooqi, S., Eyler, E., and Gould, P. (2006) Long-range Rydberg-Rydberg interactions and molecular resonances. *Eur. Phys. J. D*, **40**, 3–12.
66. Reinhard, A., Liebisch, T.C., Knuffman, B., and Raithel, G. (2007) Level shifts of rubidium Rydberg states due to binary interactions. *Phys. Rev. A*, **75**, 032712.
67. Weimer, H., Müller, M., Lesanovsky, I., Zoller, P., and Büchler, H.P. (2010) A Rydberg quantum simulator. *Nat. Phys.*, **6**, 382–388.
68. Saffman, M., Walker, T., and Mølmer, K. (2010) Quantum information with Rydberg atoms. *Rev. Mod. Phys.*, **82**, 2313.
69. Jaksch, D., Cirac, J., Zoller, P., Rolston, S., Côté, R., and Lukin, M. (2000) Fast quantum gates for neutral atoms. *Phys. Rev. Lett.*, **85**, 2208–2211Arxiv preprint quant-ph/0004038.
70. Protsenko, I.E., Reymond, G., Schlosser, N., and Grangier, P. (2002) Conditional quantum logic using two atomic qubits. *Phys. Rev. A*, **66**, 062306.
71. Schempp, H., Günter, G., Hofmann, C., Giese, C., Saliba, S., DePaola, B., Amthor, T., Weidemüller, M., Sevinçli, S., and Pohl, T. (2010) Coherent population trapping with controlled interparticle interactions. *Phys. Rev. Lett.*, **104**, 173602.
72. Møller, D., Madsen, L.B., and Mølmer, K. (2008) Quantum gates and multiparticle entanglement by Rydberg excitation blockade and adiabatic passage. *Phys. Rev. Lett.*, **100**, 170504.
73. Li, N., Zhao, Z., Chen, H., Li, P., Li, Y., Zhao, Y., Zhou, G., Jia, S., and Zhang, Y. (2012) Observation of dressed odd-order multi-wave mixing in five-level atomic medium. *Opt. Express*, **20**, 1912–1929.
74. Nie, Z., Zhang, Y., Zhao, Y., Yuan, C., Li, C., Tao, R., Si, J., and Gan, C. (2011) Enhancing and suppressing four – wave mixing in electromagnetically induced transparency window. *J. Raman Spectrosc.*, **42**, 1–4.

2
MWM Quantum Control via EIT

Highlights

Electromagnetically induced transparency window makes different higher-order nonlinear processes coexist, and even provide an effective means to study the interaction among them.

The generation of coexisting MWM processes in multi-level atomic systems is interesting. The generated MWM signals can be selectively enhanced and suppressed via an EIT window. Furthermore, the interplays between the coexisting generated MWM signals are quite fascinating. In this chapter, the EIT-assisted MWM and their various effects are introduced. First, we discuss the mechanism for multi-EIT windows to assist the coexistence of MWM signals. Then, the interaction and energy interchange between coexisting MWMs when the MWM is generated under various multi-dressed schemes are introduced. And, last, under the competition between third- and fifth-order nonlinearities, the coexisting seven distinguishable dressed odd-order MWM signals as well as temporal and spatial interference between two FWM signals are introduced. Such investigations of these multi-dressing schemes and interactions are useful to understand and control the generated high-order nonlinear optical signals and can be useful in pushing them into various applications, such as coherent quantum control, nonlinear optical spectroscopy, precise measurements, and quantum information processing.

2.1
Interference of Three MWM via EIT

When laser fields create a coherent superposition state between two ground-state sublevels of an atom, these laser beams experience reduced absorption or increased transmission, which is well known as coherent population trapping (CPT) [1–4]. Under this condition, the system falls into a dark state and the population is trapped within the two ground states. Similar to CPT, effects related to EIT [5–7] have been useful, as the medium becomes transparent and the probe beam is going through it without attenuation. This quantum interference effect has been widely investigated

Quantum Control of Multi-Wave Mixing, First Edition. Yanpeng Zhang, Feng Wen, and Min Xiao.

and applied to many fields such as ultraslow light [8], light storage [9–11], and quantum memory [12, 13]. EIT is also used for investigating MWM processes as the generated signals can transmit through the resonant atomic medium with little absorption under the EIT conditions. Not only can the FWM process be resonantly enhanced but the generated FWM signals can also be allowed to transmit through the atomic medium with little absorption [4, 14–18]. Moreover, the SWM process has been observed in a closed-cycle, four-level cold atomic system and doubly excited autoionizing Rydberg states [19, 20], besides the eight-wave mixing (EWM) process has been individually studied in multi-level atomic systems [21]. By choosing appropriate atomic-level schemes and driving fields, one can generate controllable nonlinearities with very interesting applications in designing novel nonlinear optical devices. On the other hand, in previously studied close-cycled (ladder-type, N-type, double-Λ-type, and folded) systems, FWM, SWM, and EWM processes cannot coexist in a given configuration; and different order nonlinearities can only be observed individually. In our recent studies, the coexistence of FWM, SWM, and EWM processes in an open five-level atomic system has been theoretically predicted [22], and have demonstrated that the generated FWM and SWM signals can coexist in an open (such as V-type, Y-type, and inverted Y-type) atomic system, in which the coexisting SWM signal can become comparable or even greater than the companion FWM by manipulating the atomic coherence [21, 23–25]. Moreover, the competition and interference between the two coexisting FWM processes have been studied via atomic coherence in a four-level atomic system [25].

In this section, we discuss the interplay and interference among two SWM and two FWM processes generated simultaneously in a five-level atomic system.

2.1.1 Experiment Setup

There are five energy levels (Figure 2.1(a)) from the ^{85}Rb atom involved in the experimental schemes. Coupling beams E_2 (frequency ω_2, wave vector $\mathbf{k}_2$, and Rabi frequency G_2) and E_2' (ω_2, $\mathbf{k'}_2$, and G_2') with the same frequency detuning Δ_2 connect the transition between $|1\rangle$ ($5P_{3/2}$) and $|2\rangle$ ($5D_{5/2}$), where $\Delta_i = \Omega_i - \omega_i$ and Ω_i is the atomic resonance frequency for the corresponding transition. Pumping beams E_4' (ω_3, $\mathbf{k}_3$, and G_3) and E_3' (ω_3, $\mathbf{k}_3'$, and G_3') with the same frequency detuning Δ_3 connect the transition between $|0\rangle$ ($5S_{1/2}(F=3)$) and $|3\rangle$ ($5P_{1/2}$). The additional coupling beam E_4 (ω_4, $\mathbf{k}_4$, Δ_4, and G_4) connects the transition of $|1\rangle$ ($5P_{3/2}$) to $|4\rangle$ ($5D_{3/2}$). A weak laser beam E_1 (ω_1, $\mathbf{k}_1$, Δ_1, and G_1) probes the transition between $|0\rangle$ ($5S_{1/2}(F=3)$) and $|1\rangle$ ($5P_{3/2}$) (Figure 2.1(a)). The laser beams are aligned spatially as shown in Figure 2.1(b), with five laser beams E_2, E_2', E_3, E_3', and E_4 propagating through the atomic medium in the same direction with small angles $\sim 0.3^\circ$ between them. The probe beam E_1 propagates in the opposite direction with a small angle.

The experiment is carried out on hot ^{85}Rb atoms by four external cavity diode lasers (ECDLs) with line widths of less than or equal to 1 MHz. The probe laser beam E_1, with a wavelength of 780.245 nm and a horizontal polarization from an ECDL has a power of 0.7 mW. The laser beams E_2 and E_2' with wavelength of 775.978 nm and

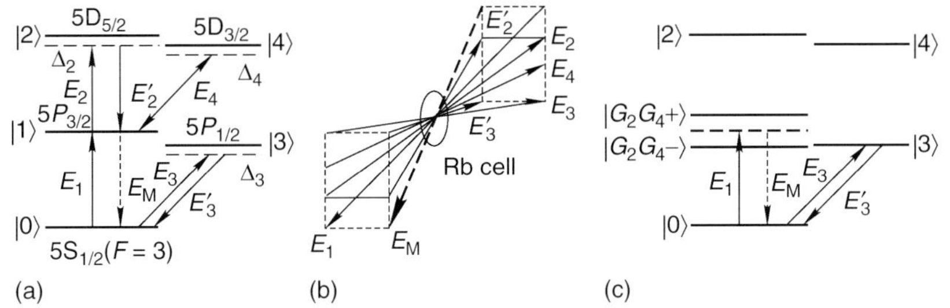

Figure 2.1 (a) The experimental atomic system. (b) Square-box-pattern beam geometry used in the experiment. (c) Dressed-state picture for the experimental atomic system.

a vertical polarization, are from another ECDL split, with powers of 35 and 5 mW, respectively. The beam E_2' is delayed by an amount τ using a computer-controlled stage. The laser beams E_3 and E_3', with a wavelength of 780.245 nm and a vertical polarization, are from the third ECDL split with an equal power of 15 mW. The laser beam E_4, with a wavelength of 776.157 nm and a vertical polarization from the fourth ECDL, has a power of 35 mW. The cell with a length of 5 cm is heated up to 60 °C and the density is about 2.5×10^{11} cm^{-3}. The optical depth of the atomic sample is about 43. Great care was taken in aligning the six laser beams with spatial overlaps and wave vector phase-matching conditions as shown in Figure 2.1(b). Under certain conditions, two FWM signals (E_{F1} with the phase-matching condition $\mathbf{k}_{F1} = \mathbf{k}_3 - \mathbf{k}_3' + \mathbf{k}_1$ via the path of $|0\rangle \xrightarrow{E_3} |3\rangle \xrightarrow{(E_3')*} |0\rangle \xrightarrow{E_1} |1\rangle$ and E_{F2} with $\mathbf{k}_{F2} = \mathbf{k}_1 + \mathbf{k}_2 - \mathbf{k}_2'$ by $|0\rangle \xrightarrow{E_1} |1\rangle \xrightarrow{E_2} |2\rangle \xrightarrow{(E_2')^*} |1\rangle$) and two SWM signals ($E_{S1}$ with $\mathbf{k}_{S1} = \mathbf{k}_3 - \mathbf{k}_3' + \mathbf{k}_1 + \mathbf{k}_2 - \mathbf{k}_2$ by $|0\rangle \xrightarrow{E_3} |3\rangle \xrightarrow{(E_3')^*} |0\rangle \xrightarrow{E_1} |1\rangle \xrightarrow{E_2} |2\rangle \xrightarrow{E_2^*} |1\rangle$ and E_{S2} with $\mathbf{k}_{S2} = \mathbf{k}_3 - \mathbf{k}_3' + \mathbf{k}_1 + \mathbf{k}_4 - \mathbf{k}_4$ by $|0\rangle \xrightarrow{E_3} |3\rangle \xrightarrow{E_3'*} |0\rangle \xrightarrow{E_1} |1\rangle \xrightarrow{E_4} |4\rangle \xrightarrow{E_4^*} |1\rangle$) are all in the direction of E_M, with the same frequency ω_1 and horizontal polarization at the right lower corner of Figure 2.1(b), and are detected by an avalanche photodiode detector. The transmitted probe beam is detected by a photodiode detector.

2.1.2 Basic Theory

The dressed-state picture can be used to describe this composite system. We consider the investigated atomic system as a combination of one V-type three-level subsystem ($|1\rangle - |0\rangle - |3\rangle$), one ladder-type three-level subsystem ($|0\rangle - |1\rangle - |2\rangle$), and two V-type four-level subsystems ($|2\rangle - |1\rangle - |0\rangle - |3\rangle$ and $|4\rangle - |1\rangle - |0\rangle - |3\rangle$) corresponding to each MWM process described above in the preceding text. Table 2.1 gives all possible phase-matched Liouville pathways for coexisting FWM and SWM processes.

First, without the beams E_2, E_2', and E_4, a pure FWM process E_{F1} will be generated via the Liouville pathway (F1) $\rho_{00}^{(0)} \xrightarrow{E_3} \rho_{30}^{(1)} \xrightarrow{(E_3')^*} \rho_{00}^{(2)} \xrightarrow{E_1} \rho_{10}^{(3)}$ in the V-type three-level subsystem ($|1\rangle - |0\rangle - |3\rangle$), where the density matrix

Table 2.1 Phase-matching conditions and Liouville pathway of coexisting FWM and SWM processes.

FWM processes	$\mathbf{k}_{F1} = \mathbf{k}_{DF1} = \mathbf{k}_3 - \mathbf{k}'_3 + \mathbf{k}_1$	
	$\rho_{00}^{(0)} \xrightarrow{E_3} \rho_{30}^{(1)} \xrightarrow{(E'_3)^*} \rho_{00}^{(2)} \xrightarrow{E_1} \rho_{10}^{(3)}$	(F1)
	$\rho_{00}^{(0)} \xrightarrow{E_3} \rho_{30}^{(1)} \xrightarrow{(E'_3)^*} \rho_{00}^{(2)} \xrightarrow{E_1} \rho_{G_2G_4\pm0}^{(3)}$	(DF1)
	$\mathbf{k}_{F2} = \mathbf{k}_{DF2} = \mathbf{k}_1 + \mathbf{k}_2 - \mathbf{k}'_2$	
	$\rho_{00}^{(0)} \xrightarrow{E_1} \rho_{10}^{(1)} \xrightarrow{E_2} \rho_{20}^{(2)} \xrightarrow{(E'_2)^*} \rho_{10}^{(3)}$	(F2)
	$\rho_{00}^{(0)} \xrightarrow{E_1} \rho_{G_2G_4\pm0}^{(1)} \xrightarrow{E_2} \rho_{20}^{(2)} \xrightarrow{(E'_2)^*} \rho_{G_2G_4\pm0}^{(3)}$	(DF2)
SWM processes	$\mathbf{k}_{S1} = \mathbf{k}_{DS1} = \mathbf{k}_3 - \mathbf{k}'_3 + \mathbf{k}_1 + \mathbf{k}_2 - \mathbf{k}_2$	
	$\rho_{00}^{(0)} \xrightarrow{E_3} \rho_{30}^{(1)} \xrightarrow{(E'_3)^*} \rho_{00}^{(2)} \xrightarrow{E_1} \rho_{10}^{(3)} \xrightarrow{E_2} \rho_{20}^{(4)} \xrightarrow{(E_2)^*} \rho_{10}^{(5)}$	(S1)
	$\rho_{00}^{(0)} \xrightarrow{E_3} \rho_{30}^{(1)} \xrightarrow{(E'_3)^*} \rho_{00}^{(2)} \xrightarrow{E_1} \rho_{G_2G_4\pm0}^{(3)} \xrightarrow{E_2} \rho_{20}^{(4)} \xrightarrow{(E_2)^*} \rho_{G_2G_4\pm0}^{(5)}$	(DS1)
	$\mathbf{k}_{S2} = \mathbf{k}_{DS2} = \mathbf{k}_3 - \mathbf{k}'_3 + \mathbf{k}_1 + \mathbf{k}_4 - \mathbf{k}_4$	
	$\rho_{00}^{(0)} \xrightarrow{E_3} \rho_{30}^{(1)} \xrightarrow{(E'_3)^*} \rho_{00}^{(2)} \xrightarrow{E_1} \rho_{10}^{(3)} \xrightarrow{E_4} \rho_{40}^{(4)} \xrightarrow{(E_4)^*} \rho_{10}^{(5)}$	(S2)
	$\rho_{00}^{(0)} \xrightarrow{E_3} \rho_{30}^{(1)} \xrightarrow{(E'_3)^*} \rho_{00}^{(2)} \xrightarrow{E_1} \rho_{G_2G_4\pm0}^{(3)} \xrightarrow{E_4} \rho_{40}^{(4)} \xrightarrow{(E_4)^*} \rho_{G_2G_4\pm0}^{(5)}$	(DS2)

element $\rho_{ik}^{(n)}$ represents the particle population at the state $|i\rangle$ when $i=k$ and represents the atomic coherence between the states $|i\rangle$ and $|k\rangle$ when $i \neq k$. (n) is the order number representing the iteration process in solving the density matrix equations. The corresponding third-order nonlinear density matrix element, which gives the leading contributions to the FWM process, is $\rho_{F1}^{(3)} = G_{F1}\exp[i(\mathbf{k}_{F1}\cdot\mathbf{r} - \omega_1 t)]/A_{F1} = |\rho_{F1}^{(3)}|\exp[i(\mathbf{k}_{F1}\cdot\mathbf{r} - \omega_1 t + \varphi_{31})]$, where $G_{F1} = -iG_1G_3(G'_3)^*$, $A_{F1} = \Gamma_{00}d_1d_3$, $d_1 = \Gamma_{10} + i\Delta_1$, and $d_3 = \Gamma_{30} + i\Delta_3$; and Γ_{ij} is the transverse relaxation rate between states $|i\rangle$ and $|j\rangle$. Then, when the beams E_2 and E_4 are turned on simultaneously, they will affect the energy level $|1\rangle$ to create two dressed states $|G_2G_4\pm\rangle$. So, the Liouville pathway of the pure FWM process E_{F1} can be modified as (DF1) $\rho_{00}^{(0)} \xrightarrow{E_3} \rho_{30}^{(1)} \xrightarrow{(E'_3)^*} \rho_{00}^{(2)} \xrightarrow{E_1} \rho_{G_2G_4\pm0}^{(3)}$. Furthermore, the density matrix element $\rho_{F1}^{(3)}$ related to the FWM process E_{F1} can be modified as $\rho_{DF1}^{(3)} = G_{F1}\exp[i(\mathbf{k}_{F1}\cdot\mathbf{r} - \omega_1 t)]/A_{DF1} = |\rho_{DF1}^{(3)}|\exp[i(\mathbf{k}_{F1}\cdot\mathbf{r} - \omega_1 t + \varphi'_{31})]$ via the pathway DF1, where $A_{DF1} = \Gamma_{00}d_3(d_1 + |G_2|^2/d_2 + |G_4|^2/d_4)$, $d_2 = \Gamma_{20} + i(\Delta_1 + \Delta_2)$, and $d_4 = \Gamma_{40} + i(\Delta_1 + \Delta_4)$.

Next, without the beams E_3, E'_3, and E_4, the second pure FWM process E_{F2} from the ladder-type three-level subsystem ($|0\rangle - |1\rangle - |2\rangle$) will be generated via the Liouville pathway (F2)$\rho_{00}^{(0)} \xrightarrow{E_1} \rho_{10}^{(1)} \xrightarrow{E_2} \rho_{20}^{(2)} \xrightarrow{(E'_2)^*} \rho_{10}^{(3)}$, which via the pathway F2 gives $\rho_{F2}^{(3)} = G_{F2}\exp\{i[\mathbf{k}_{F2}\times\mathbf{r} - (\omega_1 t + \omega_2\tau)]\}/A_{F2} = |\rho_{F2}^{(3)}|\exp\{i[\mathbf{k}_{F2}\times\mathbf{r} - (\omega_1 t + \omega_2\tau) + \varphi_{32}]\}$, where $G_{F2} = -iG_1G_2(G'_2)^*$ and $A_{F2} = d_1^2 d_2$. When the beam E_4 is turned on and the dressing effects of the beams E_2 and E_4 are considered simultaneously, the resulting dressed FWM process E_{DF2} can be described by the Liouville pathway (DF2) $\rho_{00}^{(0)} \xrightarrow{E_1} \rho_{G_2G_4\pm0}^{(1)} \xrightarrow{E_2} \rho_{20}^{(2)} \xrightarrow{(E'_2)^*} \rho_{G_2G_4\pm0}^{(3)}$ and the density matrix element

$$\rho^{(3)}_{DF2} = \frac{G_{F2}\exp\{i[\mathbf{k}_{F2}\cdot\mathbf{r} - (\omega_1 t + \omega_2\tau)]\}}{A_{DF2}} = |\rho^{(3)}_{DF2}|\exp\{i[\mathbf{k}_{F2}\cdot\mathbf{r} - (\omega_1 t + \omega_2\tau) + \varphi'_{32}]\}$$

where $A_{DF2} = d_2(d_1 + |G_2|^2/d_2 + |G_4|^2/d_4)^2$.

Now, we focus on the two SWM processes generated from the two V-type four-level subsystems ($|2\rangle - |1\rangle - |0\rangle - |3\rangle$ and $|4\rangle - |1\rangle - |0\rangle - |3\rangle$). On the one hand, when blocking the beams E'_2, E_4 and ignoring the dressing effect of the beam E_2, we can obtain the pure SWM process E_{S1} from the V-type four-level subsystem ($|2\rangle - |1\rangle - |0\rangle - |3\rangle$) via the Liouville pathway (S1)$\rho^{(0)}_{00} \xrightarrow{E_3} \rho^{(1)}_{30} \xrightarrow{(E'_3)^*} \rho^{(2)}_{00} \xrightarrow{E_1} \rho^{(3)}_{10} \xrightarrow{E_2} \rho^{(4)}_{20} \xrightarrow{(E_2)^*} \rho^{(5)}_{10}$. This pure SWM via the pathway S1 has the corresponding fifth-order nonlinear density matrix element $\rho^{(5)}_{S1} = G_{S1}\exp[i(\mathbf{k}_{S1}\cdot\mathbf{r} - \omega_1 t)]/A_{S1} = |\rho^{(5)}_{S1}|\exp[i(\mathbf{k}_{S1}\cdot\mathbf{r} - \omega_1 t + \varphi_{51})]$, where $G_{S1} = -iG_1G_3(G'_3)^*G_2(G_2)^*$ and $A_{S1} = \Gamma_{00}d_1^2d_2d_3$. On the other hand, when blocking the beams E'_2, E_2 and ignoring the dressing effect of E_4, we can obtain the other pure SWM process E_{S2} from the second V-type four-level subsystem ($|4\rangle - |1\rangle - |0\rangle - |3\rangle$). By the Liouville pathway (S2) $\rho^{(0)}_{00} \xrightarrow{E_3} \rho^{(1)}_{30} \xrightarrow{(E'_3)^*} \rho^{(2)}_{00} \xrightarrow{E_1} \rho^{(3)}_{10} \xrightarrow{E_4} \rho^{(4)}_{40} \xrightarrow{(E_4)^*} \rho^{(5)}_{10}$, we can obtain $\rho^{(5)}_{S2} = G_{S2}\exp[i(\mathbf{k}_{S2}\cdot\mathbf{r} - \omega_1 t)]/A_{S2} = |\rho^{(5)}_{S2}|\exp[i(\mathbf{k}_{S2}\cdot\mathbf{r} - \omega_1 t + \varphi_{52})]$ for E_{S2}, where $G_{S2} = -iG_1G_3(G'_3)^*G_4(G_4)^*$ and $A_{S2} = \Gamma_{00}d_1^2d_3d_4$. If only the beam E'_2 clocked and the dressing effects of both beams E_2 and E_4 were considered, the above-mentioned two pure SWM processes will coexist and dress each other. These processes can be described by the modified Liouville pathways $\rho^{(0)}_{00} \xrightarrow{E_3} \rho^{(1)}_{30} \xrightarrow{(E'_3)^*} \rho^{(2)}_{00} \xrightarrow{E_1} \rho^{(3)}_{G_2G_4\pm0} \xrightarrow{E_2} \rho^{(4)}_{20} \xrightarrow{(E_2)^*} \rho^{(5)}_{G_2G_4\pm0}$ (DS1) and $\rho^{(0)}_{00} \xrightarrow{E_3} \rho^{(1)}_{30} \xrightarrow{(E'_3)^*} \rho^{(2)}_{00} \xrightarrow{E_1} \rho^{(3)}_{G_2G_4\pm0} \xrightarrow{E_4} \rho^{(4)}_{40} \xrightarrow{(E_4)^*} \rho^{(5)}_{G_2G_4\pm0}$ (DS2), respectively. By the pathways DS1 and DS2, we can obtain $\rho^{(5)}_{DS1} = G_{S1}\exp[i(\mathbf{k}_{S1}\cdot\mathbf{r} - \omega_1 t)]/A_{DS1} = |\rho^{(5)}_{DS1}|\exp[i(\mathbf{k}_{S1}\cdot\mathbf{r} - \omega_1 t + \varphi'_{51})]$ for E_{DS1}, where $A_{DS1} = \Gamma_{00}d_2d_3(d_1 + |G_2|^2/d_2 + |G_4|^2/d_4)^2$ and $\rho^{(5)}_{DS2} = G_{S2}\exp[i(\mathbf{k}_{S2}\cdot\mathbf{r} - \omega_1 t)]/A_{DS2} = |\rho^{(5)}_{DS2}|\exp[i(\mathbf{k}_{S2}\cdot\mathbf{r} - \omega_1 t + \varphi'_{52})]$ for E_{DS2}, where $A_{DS2} = \Gamma_{00}d_3d_4(d'_1 + |G_2|^2/d_2 + |G_4|^2/d_4)^2$, $d'_1 = \Gamma_{10} + i\Delta'_1$, $\Delta'_1 = \Omega_1 - \omega_1$, $d_2 = \Gamma_{20} + i(\Delta'_1 + \Delta_2)$, and $d_4 = \Gamma_{40} + i(\Delta'_1 + \Delta_4)$.

2.1.3
Results and Discussions

When we investigate the interplay between generated MWM signals, the beam E'_2 is always turned off. Thus, there are two SWM (E_{S1} and E_{S2}) and one FWM (E_{F1}) signals propagating in the same direction simultaneously in the experiment. There exist several interesting physical processes in this composite system, which describe the interplay between these MWM processes. First, when one SWM process E_{S1} (or E_{S2}) and one FWM process E_{F1} are overlapped in frequency, the interplay between them is dominated by the contributions of the optical pumping effect of the beams E_3 and E'_3 with large frequency detuning. The second one is about the mutual dressing effects between two SWM processes E_{S1} and E_{S2} in the two V-type four-level subsystems, which perturb two SWM processes and modify the total amplitude of the SWM processes, especially when the two SWM signals are tuned together in frequency by adjusting frequency detuning. On the other hand, when the beam E_2' is turned on and the generated MWM signals (two SWM E_{S1}, E_{S2}, and one FWM E_{F2}) overlap in frequency, we investigate temporal and spatial interferences among these three nonlinear optical processes.

First, when the beam E_2' is blocked, the interplay between the coexistent MWM signals has been considered (Figure 2.2 and Figure 2.3). By two-photon Doppler-free configurations, there exist two ladder-type EIT windows in the five-level atomic system (Figure 2.1(a)). With the beam E_2 propagating in the opposite direction of the weak probe field E_1 in the $|0\rangle - |1\rangle - |2\rangle$ ladder-type subsystem, two-photon Doppler-free condition is satisfied. Thus, one EIT window (named EIT1, satisfying $\Delta_1 + \Delta_2 = 0$, the right peak in the lower curve of Figure 2.2(a2)) forms, which is induced by the beam E_2. Similarly, the other EIT window (named EIT2, satisfying $\Delta_1 + \Delta_4 = 0$, the right peak in the lower curve of Figure 2.2(a3)) appears in the $|0\rangle - |1\rangle - |4\rangle$ ladder-type subsystem induced by the beam E_4. Besides, in the $|1\rangle - |0\rangle - |3\rangle$ V-type three-level subsystem, because the beam E_1 propagates in the opposite direction to E_3 and E_3', the two-photon Doppler-free condition is not satisfied. But a similar saturated absorption will arise, which leads to the reduced absorption sharp peak in the probe transmission signal (the lower curve at F = 3 in Figure 2.2(b1)). Such a reduced absorption sharp peak can enhance FWM signals so that we refer to it as an EIT-like phenomenon.

On the other hand, there will be coexisting MWM processes in the above-mentioned EIT windows that generate signal beams at frequency ω_1. First, when beams E_1, E_3, and E_3' are turned on, an FWM signal E_{F1} (Figure 2.2(c1)) is generated with the phase-matching condition of $\mathbf{k}_{F1} = \mathbf{k}_3 - \mathbf{k}_3' + \mathbf{k}_1$ (Figure 2.2(d1)) as shown by the upper curve in Figure 2.2(a1), which falls into the V-type EIT-like window (the lower curve in Figure 2.2(a1)). Physically, the interference between the beams E_3' and E_3 produces a static population grating in the medium. Such a static grating diffracts E_1 and gives rise to the FWM signal E_{F1}. As the beams E_1 and $E_3 + E_3'$ are not satisfied by two-photon Doppler-free configurations, the FWM signal E_{F1} is not Doppler-free. Next, when the beam E_2 is used at one upper transition of $|1\rangle$ to $|2\rangle$, an SWM signal E_{S1} (Figure 2.2(c2)) satisfying the phase-matching condition of $\mathbf{k}_{S1} = \mathbf{k}_3 - \mathbf{k}_3' + \mathbf{k}_1 + \mathbf{k}_2 - \mathbf{k}_2$ (Figure 2.2(d2)) forms, as shown by the right peak of the upper curve in Figure 2.2(a2), which falls into the $|0\rangle - |1\rangle - |2\rangle$ ladder-type EIT window (the right peak of the lower curve in Figure 2.2(a2)). The SWM signal E_{S1} is Doppler-free as it is in the EIT window. Finally, when the beam E_4 is turned on and the beam E_2 is blocked, another SWM signal E_{S2} (Figure 2.2(c3)) satisfying the phase-matching condition of $\mathbf{k}_{S2} = \mathbf{k}_3 - \mathbf{k}_3' + \mathbf{k}_1 + \mathbf{k}_4 - \mathbf{k}_4$ (Figure 2.2(d3)) is generated, as shown by the right peak of the upper curve in Figure 2.2(a3), which falls into the $|0\rangle - |1\rangle - |4\rangle$ ladder-type EIT windows (the right peak of the lower curve in Figure 2.2(a3)). Also, the SWM signal E_{S2} is Doppler free as it is in the EIT window. Figure 2.2(a4) shows the generated MWM signals and the corresponding EIT windows when the above-mentioned five laser beams are turned on.

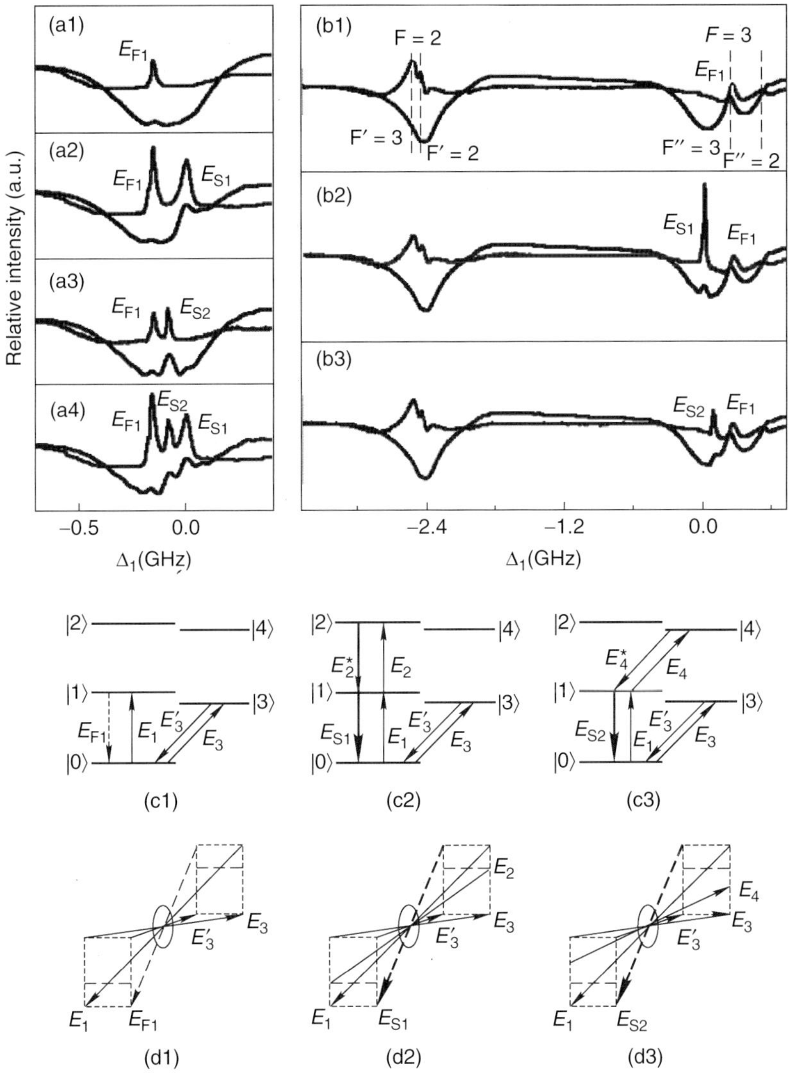

Figure 2.2 The different MWM signals (upper curves) and the probe transmission signals (lower curves) versus Δ_1 by blocking different laser beams. (a1) With E_2, E_2', and E_4 blocked, (a2) with E_2' and E_4 blocked, (a3) with E_2 and E_2' blocked, and (a4) with E_2' blocked. (b1)–(b3) The different MWM signals and the probe transmission signals corresponding to different hyperfine-level transitions (the left signals from ^{85}Rb$|5S_{1/2}, F=2\rangle$ and the right signals from ^{85}Rb$|5S_{1/2}, F=3\rangle$) versus Δ_1. The conditions of blocking laser beams are the same as in (a1)–(a3). (c1)–(c3) The simplified level diagrams of the MWM signals E_{F1}, E_{S1}, and E_{S2}. (d1)–(d3) The phase-matching diagrams of E_{F1}, E_{S1}, and E_{S2}. The thin and dotted line representing the MWM process (E_{F1}) is not Doppler free and the thick and long dashed line representing the MWM process (E_{S1} or E_{S2}) is Doppler free. Source: Adapted from Ref. [26].

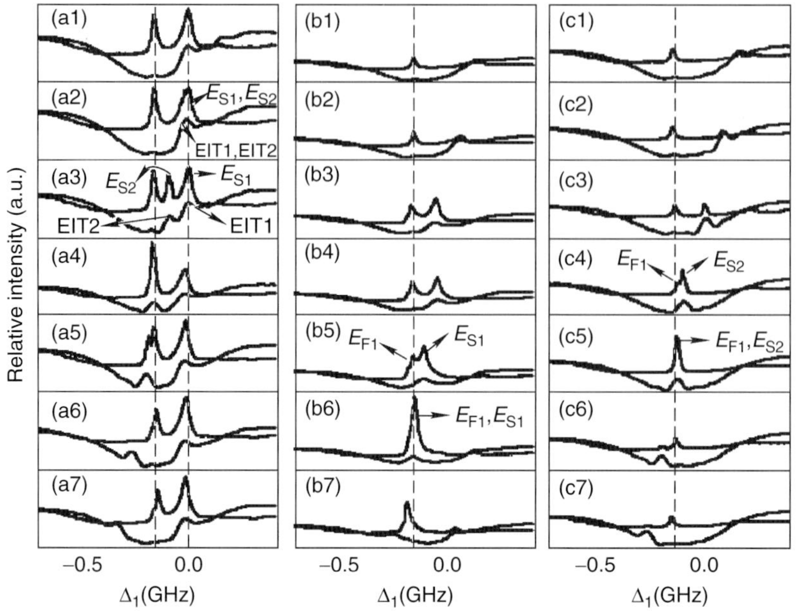

Figure 2.3 Interplay between two MWM signals. (a1)–(a7) Probe transmission signals (lower curves) and measured MWM signals (upper curves including the fixed E_{F1} peaks along the left dotted line, the fixed E_{S1} peaks along the right dotted line and the shifting E_{S2} peaks) versus Δ_1 for different Δ_4 values. (b1)–(b7) Probe transmission signals (lower curves) and measured MWM signals (upper curves including the fixed E_{F1} peaks along the dotted line and the shifting E_{S1} peaks) versus Δ_1 for different Δ_2 values. (c1)–(c7) Probe transmission signals (lower curves) and measured MWM signals (upper curves including the fixed E_{F1} peaks along the dotted line and the shifting E_{S2} peaks) versus Δ_1 for different Δ_4 values.

Figure 2.2(b1) shows two groups of FWM signals (upper curve) with the same phase-matching condition $\mathbf{k}_{F1} = \mathbf{k}_3 - \mathbf{k}'_3 + \mathbf{k}_1$ corresponding to the scanning of the probe beam E_1 (lower curve) around the transitions from $|5S_{1/2}, F=2\rangle$ to $|5P_{3/2}, F=1,2,3\rangle$ and from $|5S_{1/2}, F=3\rangle$ to $|5P_{3/2}, F=2,3,4\rangle$, respectively. The probe transmission spectrum shows the increased absorption in the left $|5S_{1/2}, F=2\rangle$ dip (the lower curve in Figure 2.2(b1)) due to the optical pumping effect of the beams E_3 and E'_3 from the ground-state $|5S_{1/2}, F=3\rangle$. The increased absorption of the probe transmission spectrum is regarded as EIA-like, which is beneficial to the generation of the FWM signal. Another effect of optical pumping is to generate multi-peak FWM signals corresponding to the hyperfine-level transition from $|5S_{1/2}, F=2\rangle$ to $|5P_{3/2}, F=1,2,3\rangle$ (the upper curve in Figure 2.2(b1)). Those peaks of the FWM signal from left to right correspond to $|5P_{3/2}, F'=3\rangle$ and $|5P_{3/2}, F'=2\rangle$ with a separation of about 63 MHz. More attractively, although without satisfying the Doppler-free condition for the FWM process E_{F1} due to the opposite propagation of beams E_1 and $E_3 + E'_3$ in the V-type subsystem, two sharp FWM signal peaks form at $|5S_{1/2}, F = 3\rangle$ because of enhanced atomic coherence in the EIT-like windows. Those two peaks correspond to $|5P_{1/2}, F'' = 3\rangle$ and $|5P_{1/2}, F'' = 2\rangle$ from left to right (the upper curve in Figure 2.2(b1)), whose separation is about 362 MHz, approximately equal to the energy spacing between $|5P_{1/2}, F'' = 3\rangle$ and $|5P_{1/2}, F'' = 2\rangle$. Figure 2.2(b2) or (b3)) shows the SWM signal E_{S1} (or E_{S2}) coexisting with the FWM signal E_{F1} at $|5S_{1/2}, F = 3\rangle$ when the E_2 (or E_4) beam is turned on.

In the following, we show the interplay and competition between EIT windows as well as generated MWM signals by changing different frequency detuning. In the five-level atomic system shown in Figure 2.1(a), the two EIT windows (EIT1 and EIT2) in the two ladder-type subsystems and the EIT-like window in the V-type subsystem will simultaneously form. As the EIT-like window is much weaker than the two EIT windows, as shown by the lower curves in Figure 2.3, we only investigate the interplay between the two EIT windows. As shown in Figure 2.3, by changing the frequency detuning Δ_4 with fixed frequency detuning Δ_2 and scanning the probe frequency detuning Δ_1, the two EIT windows can be controlled to overlap or separate. The lower curves in Figure 2.3(a1) to (a7) show the two modified EIT windows at different position of the probe transmission profile determined by the EIT condition $\Delta_1' = -\Delta_2$ and $\Delta_1 = -\Delta_4$, respectively. When the two EIT windows overlap (i.e., $\Delta_1 = \Delta_1'$) and dress each other, as shown in Figure 2.3(a2), the area of the EIT windows is slightly larger than that when the two EIT windows are far away (Figure 2.3(a3)).

The upper curves in Figure 2.3 show the interplay and competition between the generated MWM signals. We can identify and control them by selectively detuning different laser frequencies and blocking different laser beams. First, when we turn on five laser beams except for E_2' and only change the frequency detuning Δ_4, the SWM signals E_{S2} will shift from right to left through the SWM signals E_{S1} and the FWM signals E_{F1} as shown in Figure 2.3(a1) to (a7). Here, we only discuss the interplay between two SWM signals. In the expression $\rho_{DS1}^{(5)}$ of the dressed SWM process E_{DS1}, the beam E_2 is both the dressing field of the E_{DS1} signal and the field to generate it, so we refer to the dressing effect of the beam E_2 as self-dressing. However, the beam E_4 is only the dressing field, so we refer to the dressing effect of E_4 as external dressing. Similarly, for the second dressed SWM process E_{DS2}, E_4, and E_2 are the self-dressing field and the external-dressing field, respectively. The enhancement or suppression of the wave-mixing signal is induced by the external-dressing field. For the external dressing by the beam E_4 in E_{DS1}, according to $\rho_{DS1}^{(5)}$, we can obtain the suppression condition as $\Delta_1 = -\Delta_4$. Similarly, for the external dressing by the beam E_2 inE_{DS2}, the suppression condition is $\Delta_1' = -\Delta_2$ according to $\rho_{DS2}^{(5)}$. Especially, when comparing Figure 2.3(a2) with (a3), when two SWM signals E_{S1} and E_{S2} overlap (the upper curve in Figure 2.3(a2)), that is, satisfying $\Delta_1 = \Delta_1' = -\Delta_4 = -\Delta_2$, there exists a maximum suppression for both the SWM signals. Physically, when $\Delta_1 = \Delta_1' = -\Delta_4 = -\Delta_2$, as the beam E_1 touches the virtual energy level created by the dressing fields E_2 and E_4, as shown in Figure 2.1(c2), the mutual suppression of the two SWM signals is strongest. Here, two EIT windows corresponding to the two ladder subsystems $|0\rangle - |1\rangle - |2\rangle$

and $|0\rangle - |1\rangle - |4\rangle$ also overlap as two EIT conditions are simultaneously satisfied by $\Delta_1 = -\Delta_4 = \Delta_1' = -\Delta_2$ (the lower curve in Figure 2.3(a2)).

Next, when the beam E_4 is blocked (or the beam E_2 is blocked), we observe the interplay between the FWM signal E_{F1} and the SWM signal E_{S1} (or E_{S2}), as shown in the upper curves of Figure 2.3(b1) to (b7) (or Figure 2.3(c1) to (c7)). In the upper curves of Figure 2.3(b1) to (b7) (or Figure 2.3(c1) to (c7)), the fixed peaks correspond to the FWM signals E_{F1} and the shifting peaks correspond to the SWM signal E_{S1} for different frequency detuning Δ_2 with the beam E_4 blocked (or the SWM signal E_{S2} for different frequency detuning Δ_4 with the beam E_2 blocked). By comparing Figure 2.3(b5) with (b6) (or by comparing Figure 2.3(c4) with (c5)), one can see that the total intensity of the FWM signal E_{F1} and the SWM signal E_{S1} (or the SWM signal E_{S2}) is enhanced when they overlap each other. This is because the optical pumping effect of the beams E_3 and E_3' with large frequency detuning destroys the above-mentioned suppression condition. In detail, first, owing to the optical pumping effect of the beams E_3 and E_3' with large frequency detuning, the FWM signal E_{F1} is largely enhanced. Second, when the SWM signal E_{S1} (or the SWM signal E_{S2}) is tuned to overlap with E_{F1}, E_{S1} (or E_{S2}) is also maximally enhanced. Thus, the overlapped E_{F1} and E_{S1} (or E_{S2}) show the enhancement, as the increased intensity of E_{F1} and E_{S1} (or E_{S2}) caused by the optical pumping effect is larger than their reduced intensity caused by the mutual suppression.

Finally, when all laser beams including the beam E_2' are turned on and two EIT windows in the five-level atomic system overlap in frequency, temporal and spatial interferences between nonlinear optical processes are also investigated. On the one hand, the FWM signal E_{F2} is also Doppler-free and highly efficient as it is in the $|0\rangle - |1\rangle - |2\rangle$ EIT window. On the other hand, owing to $G_2 \gg G_2'$, the SWM signal E_{S1} can coexist with the FWM signal E_{F2} in the $|0\rangle - |1\rangle - |2\rangle$ EIT window. By changing the frequency detuning Δ_4, the $|0\rangle - |1\rangle - |4\rangle$ EIT window can be shifted toward the $|0\rangle - |1\rangle - |2\rangle$ EIT window. When the two EIT windows overlap, temporal and spatial interferences of two SWM (E_{S1}, E_{S2}) and one FWM (E_{F2}) can be observed. If we neglect the MWM processes that are either very weak or propagating in other directions, the coexisting one FWM and two SWM signals give the total detected intensity as $I(\tau, r) \propto |\rho^{(3)}_{DF2} + \rho^{(5)}_{DS1} + \rho^{(5)}_{DS2}|^2 = A_1 + A_2\cos(\varphi_{32}' - \varphi_{51}' + \varphi) + A_3\cos(\varphi_{32}' - \varphi_{52}' + \varphi)$, where $A_1 = |\rho^{(3)}_{F2}|^2 + |\rho^{(5)}_{S1}|^2 + |\rho^{(5)}_{S2}|^2 + 2|\rho^{(5)}_{S1}||\rho^{(5)}_{S2}|\cos(\varphi_{51}' - \varphi_{52}')$, $A_2 = 2|\rho^{(3)}_{F2}||\rho^{(5)}_{S2}|$, $A_3 = 2|\rho^{(3)}_{F2}||\rho^{(5)}_{S2}|$, $\varphi = \Delta\mathbf{k}\times\mathbf{r} - \omega_2\tau$, and $\Delta\mathbf{k} = (\mathbf{k}_2 - \mathbf{k}_2') - (\mathbf{k}_3 - \mathbf{k}_3')$. As the generated SWM signals E_{S1} and E_{S2} have a strong correlation characteristic, φ_{51}' in $\rho^{(5)}_{DS1}$ is approximately equal to φ_{52}' in $\rho^{(5)}_{DS2}$. Thus, the equation about $I(\tau,r)$ can be rewritten as $I(\tau,r) \propto A_1 + (A_2 + A_3)\cos(\varphi_{32}' - \varphi_5 + \varphi)$, where $\varphi_5 = \varphi_{51}' \approx \varphi_{52}'$.

From the equation $I(\tau,r)$, one can see that the total signal has an ultrafast time oscillation with a period of $2\pi/\omega_2$ and a spatial interference with a period of $2\pi/\Delta k$, which form a spatiotemporal interferogram. Figure 2.4 shows a three-dimensional interferogram pattern (Figure 2.4(a)) and its projections on time (Figure 2.4(b)). Figure 2.4(b) depicts a typical temporal interferogram with the temporal oscillation period of $2\pi/\omega_2 = 2.587$ fs corresponding to the $5P_{3/2}$ to $5D_{5/2}$ transition frequency of $\Omega_2 = 2.427\ \text{fs}^{-1}$ in ^{85}Rb. This gives a technique for precision measurement of

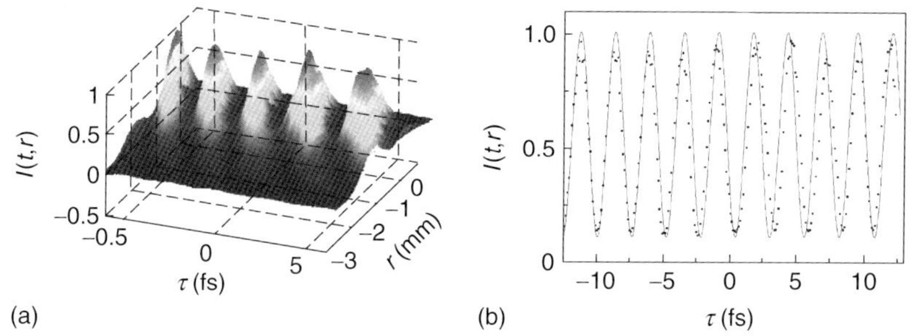

Figure 2.4 (a) A three-dimensional spatiotemporal interferogram of the total FWM and SWM signal intensity $I(\tau,r)$ versus time delay τ and transverse position τ. (b) Cross section of spatiotemporal interferogram on time plane ($r=0$) (square points are experimental data and the solid curve is the theoretically simulated result).

atomic transition frequency in optical wavelength range. The spatial interference is determined by Δk. Thus, one point two interference fringes form as shown by the structure along the r-direction in Figure 2.4(a) as we have $2\pi/\Delta \mathrm{k}=3$ mm in our experiment. The spatial interference pattern can change from constructive to destructive at the center of the beam profile ($r=0$) when the phase delay on the E_2' beam is varied. The solid curve in Figure 2.4(b) is the theoretical calculation from the full density matrix equations. It can be seen that the theoretical results are in good agreement with the experimental results.

2.1.4 Conclusion

In summary, several generated MWM processes can coexist in a five-level atomic system with carefully arranged laser beams. Two SWM signals (E_{S1} and E_{S2}) falling into each EIT window and one FWM signal (E_{F1}) in an EIT-like window can be tuned to overlap or separate by varying the corresponding frequency detuning. The experimental observation has clearly shown the interplay between these MWM signals, including the enhancement between the FWM process E_{F1} and any of the SWM processes because of the optical pumping effect and the mutual suppression of two SWM signals caused by the mutual dressing. On the other hand, when two EIT windows merge in frequency, the temporal and spatial interferences among the coexistent two SWM signals and the FWM signal E_{F2} are also discussed. The generated spatiotemporal interferogram with a femtosecond time scale can be used to determine the optical transition frequency. In this paper, different higher-order nonlinear optical processes coexist and compose correlative and coherent multi-channel signals together with the probe beam. As such, multi-channel signals can be used in the optical switch, multi-channel optical router, and optical logical calculation, it is important to understand and optimize higher-order nonlinear optical processes by the research on their interplay and interference. In addition, the work in this paper also opens the doors to other fields such as optical imaging storage and quantum information processing.

2.2
Observation of EWM via EIT

In this section, we show that, in an open-cycled multi-level atomic system, the FWM, SWM, and EWM signals can be simultaneously generated via transitions in different branches. These signals can be identified by blocking different incident laser beams and adjusting the frequency detuning of the incident laser beams. The atomic coherence effect is the key to enhancing the magnitude of the higher-order susceptibility to make it comparable with the lower order susceptibility, which is suppressed at the same time. Moreover, we focus on generating a highly efficient EWM process that can be enhanced by one EIT window and double optical pumping channels simultaneously in the open-cycle five-level atomic system.

2.2.1
Basic Theory

Coexisting MWM processes, as well as controllable MWM processes, were carried out in an atomic vapor of ^{87}Rb. Five energy levels ($5S_{1/2}(F=1)$, $5P_{3/2}$, $5D_{3/2}$, $5S_{1/2}(F=2)$, and $5P_{1/2}$) form an open *N*-type five-level atomic system, as shown in Figure 2.5(b). The six laser beams were carefully aligned as indicated in Figure 2.5(a). The atomic vapor cell was heated to 60 °C. In Figure 2.5(b), the probe laser beam $\boldsymbol{E}_1$ (with a wavelength near 780 nm from an ECDL, connecting the transition between $5S_{1/2}(F=1)-5P_{3/2}$) is horizontally polarized and has an intensity of $I_1 \approx 0.1 \times 10^4$ Wm^{-2}. Some significant parameters are wave vector $\mathbf{k}_i$, frequency ω_i, Rabi frequency G_i, and frequency detuning $\Delta_i = \Omega_i - \omega_i$ (Ω_i is the resonant frequency). The coupling laser beams $\boldsymbol{E}_2$ and $\boldsymbol{E}_2'$ (with intensity I_2, wavelength 776.16 nm connecting transition between $5P_{3/2}-5D_{3/2}$) are split from another ECDL, each with a vertical polarization. The pump laser beams $\boldsymbol{E}_3$ and $\boldsymbol{E}_3'$ (wavelength 780.24 nm connecting the transition between $5P_{3/2}-5S_{1/2}(F=2)$) are split from a cw Ti:sapphire laser with equal intensity ($I_3 \approx I_3'$), each with

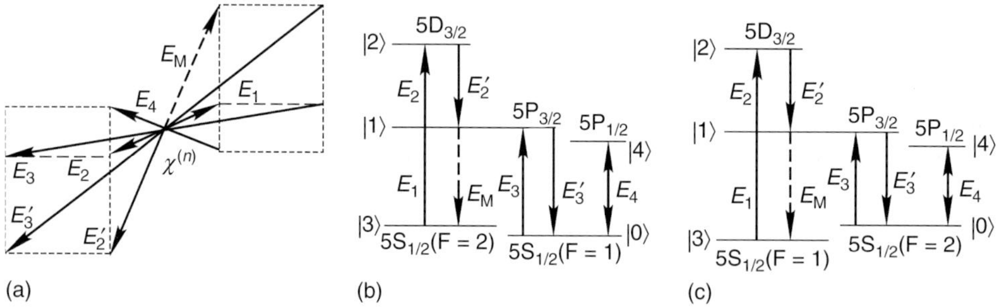

Figure 2.5 (a) Square-box-pattern beam geometry used in the experiment based on the phase-matching condition. The dash-dotted arrow is the generated MWM signals. (b) Five-level N-type atomic system of ^{87}Rb for generating MWM processes. (c) Another type of generating MWM process, which the incident laser beams connect different transitions of ^{87}Rb.

a vertical polarization. The pump laser beam E_4 (with intensity I_4, wavelength 794.97 nm connecting transition $5S_{1/2}(F=2) - 5P_{1/2}$) is from yet another ECDL, with a vertical polarization. Great care was taken in aligning the six laser beams with spatial overlaps inside the cell and phase-matching conditions with small angles ($\sim 0.3^\circ$) between them. The diameters at the vapor cell center for the pump and coupling beams are all about 0.5 mm, and the diameter of the probe beam (E_1) is about 0.3 mm.

The generated MWM signals with horizontal polarization are in the direction of E_M (at the lower right-hand corner of Figure 2.5(a)) and detected by an avalanche photo diode (APD). The transmitted probe beam is detected by a silicon photodiode as a reference. To experimentally demonstrate different wave-mixing processes, different coupling laser beams will be blocked during the experiments. In addition, FWM, SWM, and EWM processes can be made to be in similar magnitudes with appropriate conditions.

2.2.2 Experimental Results

With the subsystem (in Figure 2.5(b)) of four levels ($|0\rangle - |1\rangle - |2\rangle - |3\rangle$) and five laser fields ($\boldsymbol{E}_1$, $\boldsymbol{E}_2$, $\boldsymbol{E}_2'$, $\boldsymbol{E}_3$, and $\boldsymbol{E}_3'$), FWM and SWM signals at frequency ω_1 will be generated (Figure 2.6(a) to (g)). First, without $\boldsymbol{E}_3$ and $\boldsymbol{E}_3'$ beams, the strong coupling beam $\boldsymbol{E}_2$ (a power of 30 mW) and the weak coupling beam $\boldsymbol{E}_2'$ (a power of 3 mW) (i.e., $\boldsymbol{E}_2 \gg \boldsymbol{E}_2'$) together with the weak probe beam $\boldsymbol{E}_1$ generate a pure FWM signal $\boldsymbol{E}_{F1}$ (via the Liouville pathway $\rho_{00}^{(0)} \xrightarrow{\omega_1} \rho_{10}^{(1)} \xrightarrow{\omega_2} \rho_{20}^{(2)} \xrightarrow{-\omega_2} \rho_{10}^{(3)}$ according to the phase-matching condition $\mathbf{k}_{F1} = \mathbf{k}_1 + \mathbf{k}_2 - \mathbf{k}_2'$) at frequency ω_1, as shown in Figure 2.6(a) (top curve), which falls in the EIT window P (lower curve) and is very efficient. Next, $\boldsymbol{E}_3'$ is turned on ($\boldsymbol{E}_3$ still blocked), which will perturb the FWM process and create dressed states for level $|2\rangle$. The resulting dressed FWM (via the corrected Liouville pathway $\rho_{00}^{(0)} \xrightarrow{\omega_1} \rho_{G_3'\pm 0}^{(1)} \xrightarrow{\omega_2} \rho_{30}^{(2)} \xrightarrow{-\omega_2} \rho_{G_3'\pm 0}^{(3)}$ with $\mathbf{k}_{D1} = \mathbf{k}_1 + \mathbf{k}_2 - \mathbf{k}_2'$) signal is generated (Figure 2.6(b) labeled as D1) in the same direction as $\boldsymbol{E}_{D1}$ and in the same EIT window. Similarly, when $\boldsymbol{E}_3$ is turned on ($\boldsymbol{E}_3'$ blocked), a similar dressed FWM signal (via another corrected Liouville pathway $\rho_{00}^{(0)} \xrightarrow{\omega_1} \rho_{G_3\pm 0}^{(1)} \xrightarrow{\omega_2} \rho_{30}^{(2)} \xrightarrow{-\omega_2} \rho_{G_3\pm 0}^{(3)}$ with $\mathbf{k}_{D2} = \mathbf{k}_1 + \mathbf{k}_2 - \mathbf{k}_2'$) is generated (Figure 2.6(c) labeled as D2). From the phase-matching conditions, we get $\boldsymbol{k}_{F1} = \boldsymbol{k}_{D1} = \boldsymbol{k}_{D2}$.

Next, we consider the case with both $\boldsymbol{E}_3$ and $\boldsymbol{E}_3'$ on. Without both $\boldsymbol{E}_2$ and $\boldsymbol{E}_2'$ beams, a broad FWM signal $\boldsymbol{E}_{F2}$ (via $\rho_{00}^{(0)} \xrightarrow{\omega_1} \rho_{10}^{(1)} \xrightarrow{\omega_3} \rho_{30}^{(2)} \xrightarrow{-\omega_3} \rho_{10}^{(3)}$ with $\mathbf{k}_{F2} = \mathbf{k}_1 + \mathbf{k}_3 - \mathbf{k}_3'$) outside the EIT window is generated, which is due to the optical pumping by beams $\boldsymbol{E}_3$ and $\boldsymbol{E}_3'$. When either $\boldsymbol{E}_2$ or $\boldsymbol{E}_2'$ is blocked, both FWM and SWM processes can coexist and propagate in the same direction. When $\boldsymbol{E}_2'$ is turned on ($\boldsymbol{E}_2$ blocked), this system generates an FWM signal $\boldsymbol{E}_{F2}$ and efficient SWM signals ($\mathbf{k}_{S1} = \mathbf{k}_1 + \mathbf{k}_3 - \mathbf{k}_3' + \mathbf{k}_2' - \mathbf{k}_2'$), as shown in Figure 2.6(e) (labeled as S1). However, when $\boldsymbol{E}_2'$ is blocked instead of $\boldsymbol{E}_2$, there coexist an FWM signal $\boldsymbol{E}_{F2}$ and SWM signals ($\mathbf{k}_{S2} = \mathbf{k}_1 + \mathbf{k}_3 - \mathbf{k}_3' + \mathbf{k}_2 - \mathbf{k}_2$), which get very large as $\boldsymbol{E}_2$ is much stronger than $\boldsymbol{E}_2'$ (Figure 2.6(e) labeled as S2). These SWM

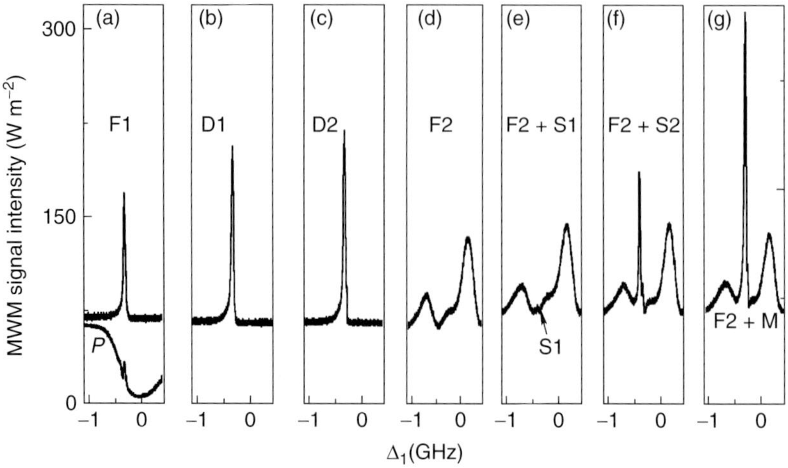

Figure 2.6 (a) Measured pure FWM1 signal (the upper curve F1) and probe transmission (the bottom curve P with one ladder-type EIT window) versus the probe detuning Δ_1. (b) Measured dressed FWM1 signals versus Δ_1 with $\boldsymbol{E}_3$ blocked. (c) Measured dressed FWM1 signals versus Δ_1 with $\boldsymbol{E}'_3$ blocked. (d) Measured pure FWM2 signal. (e) Measured dressed FWM2 signals versus Δ_1 with $\boldsymbol{E}_2$ blocked. (f) Measured dressed FWM2 signals versus Δ_1 with $\boldsymbol{E}'_2$ blocked. (g) Measured total MWM signals. The experimental parameters are $I_1 = 0.1 \times 10^4$ Wm^{-2}, $I_2 = 5.8 \times 10^4$ Wm^{-2}, $I'_2 = 2.0 \times 10^4$ Wm^{-2}, $I_3 = 0.7 \times 10^4$ Wm^{-2}, $I'_3 = 0.6 \times 10^4$ Wm^{-2}, and $I_4 = 0$. Source: Adapted from Ref. [27].

processes are generated via either $\rho_{00}^{(0)} \xrightarrow{\omega_1} \rho_{10}^{(1)} \xrightarrow{-\omega_3} \rho_{30}^{(2)} \xrightarrow{\omega_3} \rho_{10}^{(3)} \xrightarrow{-\omega_2} \rho_{20}^{(4)} \xrightarrow{\omega_2} \rho_{10}^{(5)}$ or $\rho_{00}^{(0)} \xrightarrow{\omega_1} \rho_{10}^{(1)} \xrightarrow{-\omega_3} \rho_{30}^{(2)} \xrightarrow{\omega_3} \rho_{10}^{(3)} \xrightarrow{\omega_2} \rho_{20}^{(4)} \xrightarrow{-\omega_2} \rho_{10}^{(5)}$.

Finally, with all five laser beams ($\boldsymbol{E}_1$, $\boldsymbol{E}_2$, $\boldsymbol{E}'_2$, $\boldsymbol{E}_3$, and $\boldsymbol{E}'_3$) on, especially with $\boldsymbol{E}_2 \gg \boldsymbol{E}'_2$, both the above-mentioned FWM and SWM processes can exist and propagate in the same direction, which can be shown by the fact that the total generated MWM signal (as shown in Figure 2.6(f), labeled as M) is larger than the pure FWM signal F1 (in Figure 2.6(a)). This system with all five laser beams on as shown in Figure 2.5(a) can be considered as dressed FWM for the three-level ladder system ($|0\rangle - |1\rangle - |2\rangle$) by two coupling beams $\boldsymbol{E}_3$ and $\boldsymbol{E}'_3$, which should only reduce the FWM signal at the on-resonance condition. So, the additional signal strength in M (compared to the FWM signal F1) must be due to the additional SWM signals (proportional to S1 and S2 in Figure 2.6(d) and (e)). More strikingly, one can easily see that the ratio of SWM/FWM strengths can be adjusted by controlling the strength of the $\boldsymbol{E}'_2$ beam. When $\boldsymbol{E}'_2$ has an intensity similar to $\boldsymbol{E}_2$, only a very small percentage of SWM signal exists, as the efficient FWM process dominates. As the intensity of $\boldsymbol{E}'_2$ decreases, the large FWM signal gets suppressed, and the percentage of SWM increases, until the FWM process is completely turned off when $\boldsymbol{E}'_2 \rightarrow 0$.

Now, let us consider two different optical pumping effects (Figure 2.7) before discussing the EWM process. When fields $\boldsymbol{E}_2$, $\boldsymbol{E}'_2$, $\boldsymbol{E}_3$, and $\boldsymbol{E}'_3$ are blocked but $\boldsymbol{E}_1$ and $\boldsymbol{E}_4$ are turned on. Figure 2.7(a1) and (b1) shows two groups of pump transmission signals corresponding to the transitions from $^{87}\mathrm{Ru}|5\mathrm{S}_{1/2}, \mathrm{F}=1\rangle$ to

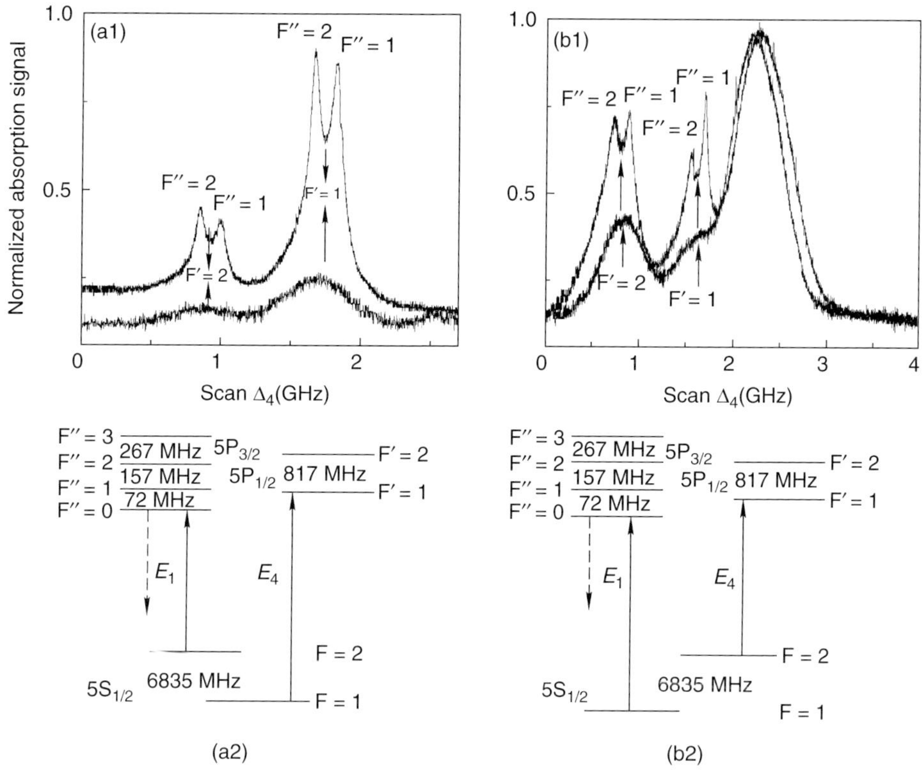

Figure 2.7 (a1) and (b1) The Doppler absorption signal corresponding to different hyperfine-level transitions (the (a1) signals from $^{87}\mathrm{Rb}|5\mathrm{S}_{1/2},\mathrm{F}=1\rangle$ and the (b1) signals from $^{87}\mathrm{Rb}|5\mathrm{S}_{1/2},\mathrm{F}=2\rangle$) versus Δ_4. (a2) and (b2) Relevant $5\mathrm{P}_{3/2}-5\mathrm{S}_{1/2}-5\mathrm{P}_{1/2}$ V-type subsystem of ^{87}Rb with $\boldsymbol{E}_1$, $\boldsymbol{E}_4$ driving different transitions. The other parameters are $I_1=0.1\times10^4\ \mathrm{Wm}^{-2}$ and $I_4=0.2\times10^4\ \mathrm{Wm}^{-2}$. Source: Adapted from Ref. [27].

$|5\mathrm{P}_{1/2},\mathrm{F}'=1,2\rangle$ and $^{87}\mathrm{Rb}|5\mathrm{S}_{1/2},\mathrm{F}=2\rangle$ to $|5\mathrm{P}_{1/2},\mathrm{F}'=1,2\rangle$, respectively. In the $5\mathrm{P}_{3/2}-5\mathrm{S}_{1/2}(\mathrm{F}=1$ and $\mathrm{F}=2)-5\mathrm{P}_{1/2}$ quasi-V-type four-level subsystem, beam $\boldsymbol{E}_1$, which transitions from $^{87}\mathrm{Rb}|5\mathrm{S}_{1/2},\mathrm{F}=2\rangle$ to $|5\mathrm{P}_{3/2},\mathrm{F}''=0,1,2,3\rangle$, is a strong laser beam. $\boldsymbol{E}_4$, which drives the transition $^{87}\mathrm{Ru}|5\mathrm{S}_{1/2},\mathrm{F}=1\rangle$ to $|5\mathrm{P}_{1/2},\mathrm{F}'=1,2\rangle$, propagates in the direction opposite to that of beam $\boldsymbol{E}_1$. The different Doppler absorption signals (lower curves and upper curves in Figure 2.7(a1)) versus Δ_4 by blocking or turning on $\boldsymbol{E}_1$. When $\boldsymbol{E}_1$ is blocked, scanning Δ_4, the absorption curve is shown as the lower curve in Figure 2.7(a1) with the distance between the two peaks of 817 MHz, which is produced by transitions from the state $5\mathrm{S}_{1/2}(\mathrm{F}=1)$ to the hyperfine of $5\mathrm{P}_{1/2}$, according to $\mathrm{F}'=1$ and $\mathrm{F}'=2$. When turning on $\boldsymbol{E}_1$, a similar saturated absorption will form, which can be shown by the fact that the generated Doppler absorption signal (upper curves as shown in Figure 2.7(a1)) is larger than the other Doppler absorption signal (lower curves as shown in Figure 2.7(a1)).

As we all know, optical pumping is a process in which absorption of light produces a population of the energy levels different from the Boltzmann distribution. In Figure 2.7(a2), the electrons excited by $\boldsymbol{E}_1$ from the lower state $|5S_{1/2}, F=2\rangle$ can transition to the higher state $|5P_{3/2}, F''=1,2,3\rangle$. Later, there is a "relaxation process," in which excited electrons are driven back to the ground state $|5S_{1/2}, F=1\rangle$. In other words, we easily realize that the pump transmission signals represent the absorption increased in the ^{87}Rb $|5S_{1/2}, F=1\rangle$ dip (the upper curve), which have the contribution from $\boldsymbol{E}_1$ optical pumping of the population from the ground state $|5S_{1/2}, F=2\rangle$, as shown in Figure 2.7(a1).

On the other hand, the transitions from $|5S_{1/2}, F=2\rangle$ to $|5P_{3/2}, F''=1,2,3\rangle$ occur with only the restriction $\Delta F=0,\pm 1$, which can produce sharp peaks in Doppler absorption signals (the upper curves in Figure 2.7(a1)) owing to enhanced atomic coherence in the EIT-like window, as well as the transitions from $|5P_{3/2}, F''=0,1,2\rangle$ to $|5S_{1/2}, F=1\rangle$. The distance between the right two sharp peaks in the absorption signals is at 157 MHz, according to $F''=1$ and $F''=2$, which is the same as the left two sharp peaks.

Figure 2.7(b2) shows the relevant $5P_{3/2}-5S_{1/2}-5P_{1/2}$ quasi-V-type four-level subsystem of ^{87}Rb in Figure 2.5(c) with $\boldsymbol{E}_1$ and $\boldsymbol{E}_4$ driving the transitions in a different manner from Figure 2.7(a2). Similarly, we can get pump transmission signals by versus Δ_4, as shown in Figure 2.7(b1). The pump transmission signal represents the Doppler absorption increased in the $Rb^{87}|5S_{1/2}, F=2\rangle$ dip (the upper curve in Figure 2.7(b1)), which has the contribution from $\boldsymbol{E}_1$ optical pumping of the population from the state$|5S_{1/2}, F=1\rangle$. The theoretical explanation is the same as in Figure 2.7(a1)).

Figure 2.8(a) to (c) presents the changes in the MWM signal, as the probing frequency detuning Δ_1 changes, in the N-type atomic system ($|0\rangle-|1\rangle-|2\rangle-|3\rangle$), as shown in Figure 2.5(b). When five laser fields $\boldsymbol{E}_1$, $\boldsymbol{E}_2$, $\boldsymbol{E}_3$, $\boldsymbol{E}'_3$, and $\boldsymbol{E}_4$ are turned on simultaneously, but $\boldsymbol{E}'_2$ is blocked, we measure the MWM signals versus the detuning Δ_1 in the five-level atomic system. We identify the MWM signals by selectively detuning different laser frequencies and blocking different laser beams to solve the problem of MWM diffraction in the same spatial direction. It is tedious to write explicitly analytical results for the MWM signals and the corresponding via, so we write out ones that are straightforward to understand in experimental results.

It is seen from Figure 2.8(a) that, under certain laser beam configurations, there is one broad FWM signal $\boldsymbol{E}_{F2}$ ($\mathbf{k}_{F2}=\mathbf{k}_1+\mathbf{k}_3-\mathbf{k}'_3$, labeled as F2) and two SWM signals which are a sharp SWM signal $\boldsymbol{E}_{S2}$ ($\mathbf{k}_{S2}=\mathbf{k}_1+\mathbf{k}_3-\mathbf{k}'_3+\mathbf{k}_2-\mathbf{k}_2$, labeled as S2) and a broad SWM signal $\boldsymbol{E}_{S3}$ (via $\rho_{00}^{(0)}\xrightarrow{\omega_1}\rho_{10}^{(1)}\xrightarrow{-\omega_3}\rho_{30}^{(2)}\xrightarrow{\omega_4}\rho_{10}^{(3)}\xrightarrow{-\omega_4}\rho_{20}^{(4)}\xrightarrow{\omega_3}\rho_{10}^{(5)}$ with $\mathbf{k}_{S3}=\mathbf{k}_1+\mathbf{k}_3-\mathbf{k}'_3+\mathbf{k}_4-\mathbf{k}_4$, labeled as S3). The two generated SWM signals are separating and the SWM signal $\boldsymbol{E}_{S2}$ falls in the EIT window ($\Delta_2=146$ MHz, the narrow peak in the left of the curve) which is obtained for the probe beam and satisfies $\Delta_1+\Delta_2=0$.

We set the large frequency detuning at the top and bottom traces. We can see that the two SWM signals begin to overlap and compete when the frequency difference between Δ_2 and Δ_1 is diminishing. The maximum suppression of both the SWM

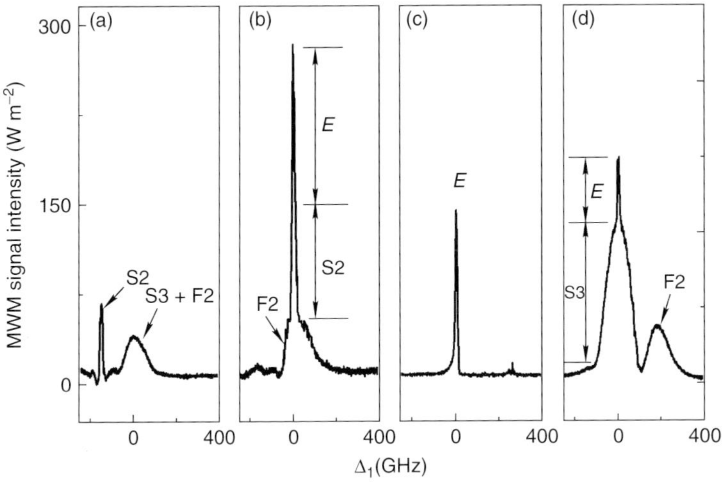

Figure 2.8 The different MWM signals versus Δ_1 for different Δ_2. (a) $\Delta_2 = 146$ MHz With SWM2, SWM3, and FWM2 signals. (b) $\Delta_2 = -7$ MHz With eight-wave mixing (EMW), FWM2, and SWM2 signals. (c) The conditions are the same as (a) with EMW signal only. (d) $\Delta_2 = -3$ MHz With EMW, SWM2, and FWM2 signal. The corresponding powers are $I_1 = 0.1 \times 10^4$ Wm^{-2}, $I_2 = 5.8 \times 10^4$ Wm^{-2}, $I_3 = I_3' = 0.7 \times 10^4$ Wm^{-2}, and $I_4 = 0.2 \times 10^4$ Wm^{-2}.

signals can be detected when $\Delta_1 = -\Delta_2 = 7$ MHz is satisfied; in this condition, we can get EWM (EWM, the sharp peak as shown in Figure 2.8(b)) satisfying the phase-matching condition $\mathbf{k}_{S3} = \mathbf{k}_1 + \mathbf{k}_3 - \mathbf{k}_3' + \mathbf{k}_2 - \mathbf{k}_2 + \mathbf{k}_4 - \mathbf{k}_4$, which are generated via $\rho_{00}^{(0)} \xrightarrow{\omega_1} \rho_{10}^{(1)} \xrightarrow{\omega_2} \rho_{20}^{(2)} \xrightarrow{-\omega_2} \rho_{10}^{(3)} \xrightarrow{-\omega_3} \rho_{30}^{(4)} \xrightarrow{\omega_4} \rho_{40}^{(5)} \xrightarrow{-\omega_4} \rho_{30}^{(6)} \xrightarrow{\omega_3} \rho_{10}^{(7)}$ (labeled as E). In Figure 2.8(b), we see that the EWM can coexist with a sharp SWM2 signal (labeled as S2) and a broad FWM2 signal (labeled as F2) which are the lower curves. The high-order atomic coherence plays a significant role in the enhancement of the MWM process.

More importantly, we describe our experimental observation of an efficient EWM process assisted by an EIT window and enhanced by optical pumping in which light is used to raise (or "pump") electrons from a lower energy level in an atom or molecule to a higher one, as shown in Figure 2.8(b), which is measured at a frequency detuning of $\Delta_2 = 7$ MHz. When pumping beams $\boldsymbol{E}_3$ ($\boldsymbol{E}_3'$) and $\boldsymbol{E}_4$ are added, the EWM signal is greatly enhanced and this is due to the contribution from $\boldsymbol{E}_3$ and $\boldsymbol{E}_4$ optical pumping of population from the ground state $|5S_{1/2}, F = 1\rangle$ and the state $|5P_{1/2}\rangle$, as shown in Figure 2.5(b). The main purpose of using such doubly optical pumping channels is to generate efficient EWM and, at the same time, to allow us to control the relative strengths of various wave-mixing processes, so such high-order nonlinear optical processes can be enhanced, manipulated, and studied in detail. On the other hand, one EIT that merges into the Doppler-broadened FWM peak satisfying $\Delta_1 + \Delta_2 = 0$ also enhances the EWM process. In other words, highly efficient EWM can be generated which are enhanced by one EIT window and double optical pumping channels simultaneously in this atomic system. Figure 2.8(c) has subtracted SWM2, SWM3, and FWM2 signals from Figure 2.8(a) and (b), in which the EWM signal is a sharp peak falling into an EIT window.

After studying the coexistent FWM, SWM, and EWM processes in an open five-level atomic system as shown in Figure 2.5(b), we now consider the processes in another open five-level atomic system as shown in Figure 2.5(c). When the probe laser beam $\boldsymbol{E}_1$ with wavelength near 780 nm, connecting transition $5S_{1/2}(F=2) - 5P_{3/2}$, the laser beam $\boldsymbol{E}_2$ with wavelength 776.16 nm connecting transition $5P_{3/2} - 5D_{3/2}$, the laser beams $\boldsymbol{E}_3$ and $\boldsymbol{E}'_3$ with wavelength 780.24 nm connecting transition between $5P_{1/2} - 5S_{1/2}(F=1)$, and the pumping laser beam $\boldsymbol{E}_4$ with wavelength 794.97 nm connecting transition between $5S_{1/2}(F=1) - 5P_{1/2}$, the N-type atomic system ($|0\rangle - |1\rangle - |2\rangle - |3\rangle - |4\rangle$) is as shown in Figure 2.5(c). Moreover, we make the intensity of the beam $\boldsymbol{E}_4$ larger than that in the former experiment and the intensity of $\boldsymbol{E}_2$ smaller than that in the former experiment. When $\boldsymbol{E}_2 \ll \boldsymbol{E}_4$ and $\Delta_2 = -3$ MHz, we can also get a coexistent EWM signal (the sharp top of the left peak, labeled as E), an SWM3 signal (lower curve of the left peak, labeled as S3), and an FWM2 signal (the right peak in the lower curve, labeled as F2) as shown in Figure 2.8(d), in which the other experimental condition and theoretical explanation are similar to that in the Figure 2.8(b). But as one can see that with the $\boldsymbol{E}_4$ beam enhanced and $\boldsymbol{E}_2$ weakened, the EWM signal strength changes significantly, showing a remarkable reduction in Figure 2.8(d) with reference to Figure 2.8(b). Such an EWM system is interesting and deserves further studies, both theoretically and experimentally.

2.2.3 Conclusion

We discussed that the FWM, SWM, and EWM processes can coexist in an open five-level atomic system with carefully arranged laser beams. Especially, we presented one EIT window and double optical pumping channels which can generate an EWM process under some conditions, and this has been clearly observed. By employing a weak coupling beam ($\boldsymbol{E}'_2$), the ratio of SWM to FWM signal strengths has been adjusted, and can be made to be desired values for application. Under certain laser beam configurations, the FWM, SWM, and EWM processes can be made to coexist with similar signal amplitudes and transmit through the same EIT window in such an open five-level atomic system.

2.3 Controlled MWM via Interacting Dark States

Investigations into interactions of doubly-dressed states and corresponding effects on atomic systems have also attracted many researchers. The interaction of double-dark states (nested scheme of doubly-dressing) and splitting of dark state (the secondarily dressed state) in a four-level atomic system with an EIT were studied theoretically by Lukin *et al.* [12]. Then, doubly-dressed states in cold atoms were observed, in which triple-photon absorption spectrum exhibits a constructive

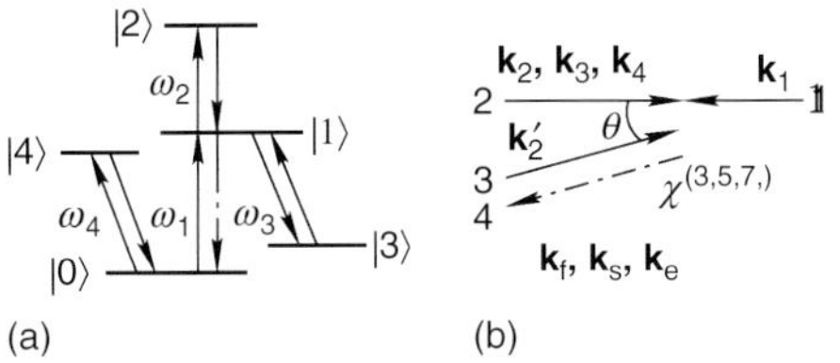

Figure 2.9 (a) Energy level diagram of a five-level system for EWM. (b) Phase-conjugation geometry.

interference between the excitation paths of two closely spaced, doubly-dressed states [28]. A similar result was obtained in the inverted-Y system [29–31] and double-Λ system [32, 33].

In this section, we go further to theoretically study, in a five-level system with three possible dressing fields, three types of dual-dressing (nested and sequential) schemes and precisely predict the splitting of dark states, resulting in five absorption peaks. Also, we present controlled MWM (FWM, SWM, and EWM) via interacting dark states in a five-level system (Figure 2.9(a)). Specifically, there exists the intermixture among three dressing schemes (nested, parallel, and sequential dressing schemes) in FWM, the quadruple nested scheme in SWM, and the parallel combination of two nested schemes in EWM.

Several features are distinctly different and gain an advantage over the previously studied MWM processes [22, 34, 35]. First, multi-dressed FWM, SWM, and EWM processes coexist and compete in a five-level system because FWM is suppressed while SWM and EWM are enhanced by controlling the offsets of fields. Second, various multi-dressed types in FWM, SWM, EWM, and EIT are viewed as the combination of three basic dual-dressing types (nested, parallel, and sequential dressing schemes), which is also separately investigated for a better explanation of multi-dressing. Third, we obtain triple AT splitting in the spectra of the FWM signal, quadruple AT splitting in SWM signal, and two triple splitting in EIT. The spectra of AT splitting, suppression, and enhancement of FWM and EWM signals are the superposition of two groups of the different AT splitting peaks. The interaction of multi-dressed states created by three dressing fields has been studied, which can result in the dramatic enhancement of MWM signals. Thus, by virtue of controlling the multi-dressed MWM signal, we can obtain the nonlinear susceptibility of the desired order. Investigations of such intermixing and interplay between different types of nonlinear wave-mixing processes will help us optimize the high-order multi-channel nonlinear optical signals, which have potential applications in achieving better nonlinear optical materials and optoelectronic devices (e.g., all optical-switch or controlled-NOT gate).

2.3.1
Basic Theory

With the basic system (in Figure 2.9(a)) of three energy levels ($|0\rangle - |1\rangle - |2\rangle$) and three laser fields (the pumping beams E_2 (with frequency ω_2, wave vector $\mathbf{k}_2$, and

Rabi frequency G_2), E_2' (ω_2, $\mathbf{k}_2'$, G_2') and the probe beam E_1 (ω_1, $\mathbf{k}_1$, G_1)), an FWM signal at frequency ω_1 will be generated and propagated in the $\mathbf{k}_f = \mathbf{k}_1 + \mathbf{k}_2 - \mathbf{k}_2'$ direction depending on the phase-matching condition. By adding another energy level ($|4\rangle$) and another laser field E_4 (ω_4, $\mathbf{k}_4$, G_4), system $|4\rangle - |0\rangle - |1\rangle - |2\rangle$ is formed and a small-angle grating in the atomic medium is established. E_4 interacts with the FWM process in the ladder system and generates SWM signals propagating along the $\mathbf{k}_s(= \mathbf{k}_1 + \mathbf{k}_2 - \mathbf{k}_2' + \mathbf{k}_4 - \mathbf{k}_4)$ direction owing to the phase-matching condition. When the pumping beam E_3 (ω_3, $\mathbf{k}_3$, G_3) is turned on, this five-level system also can generate an EWM signal, which propagates in the same direction owing to the phase-matching condition $\mathbf{k}_e = \mathbf{k}_1 + \mathbf{k}_2 - \mathbf{k}_2' + \mathbf{k}_3 - \mathbf{k}_3 + \mathbf{k}_4 - \mathbf{k}_4$. Simultaneously, there exists another SWM signal that satisfies the phase-matching condition $\mathbf{k}_{s1} = \mathbf{k}_1 + \mathbf{k}_2 - \mathbf{k}_2' + \mathbf{k}_3 - \mathbf{k}_3$. However, only SWM $\mathbf{k}_s(= \mathbf{k}_1 + \mathbf{k}_2 - \mathbf{k}_2' + \mathbf{k}_4 - \mathbf{k}_4)$ is discussed in the section.

The probe and pumping laser beams are aligned spatially in the pattern as shown in Figure 2.9(b), with four pumping beams (E_2, E_2', E_3, and E_4) transmitting through the atomic medium in the same direction with a small angle (0.3°) between them in phase-conjugation geometry. The probe beam (E_1) propagates in the opposite direction with an angle, as shown in Figure 2.9(b). By setting the propagation direction of the probe beam E_1, as indicated in Figure 2.9(b), the diffracted one FWM, two SWM, and one EWM signal beams will be in the same direction for the reason of equivalent phase-matching conditions ($\mathbf{k}_f$, $\mathbf{k}_s$, $\mathbf{k}_{s1}$, and $\mathbf{k}_e$) when the five laser beams are turned on.

According to the description given, the Hamiltonian of this system can be written as,

$$\begin{aligned} H_{\text{int}} = &-\hbar[\Delta_1|1\rangle\langle 1| + (\Delta_1 + \Delta_2)|2\rangle\langle 2| + (\Delta_1 - \Delta_3)|2\rangle\langle 2| + \Delta_4|2\rangle\langle 2|] \\ &-\hbar[G_1|1\rangle\langle 0| + G_4|4\rangle\langle 0| + (G_2 + G_2')|2\rangle\langle 1| + G_3|3\rangle\langle 1| + \text{HC}] \end{aligned}$$

Then, the density matrix equations that determine the evolution of the whole system can be obtained. As to a five-level system, there are 15 independent density matrix equations without considering that $\rho_{00} + \rho_{11} + \rho_{22} + \rho_{33} + \rho_{44} = 1$. However, if we consider the probe field being weak enough, the number of the required equations will be reduced because some weak terms will be omitted in further derivation. Moreover, the perturbation chain method, a suitable method for our special case and can remove density matrices that contribute little to MWM generation, is employed to calculate the density matrix related to the MWM process. Finally only six equations can be useful for derivation, as follows,

$$\begin{aligned} \dot{\rho}_{10}^{(r)} &= -d_{10}\rho_{10}^{(r)} - iG_4 e^{i\mathbf{k}_4\cdot\mathbf{r}}\rho_{14}^{(r)} + i(G_2')^* e^{-i\mathbf{k}_2'\cdot\mathbf{r}}\rho_{20}^{(r)} + iG_3 e^{i\mathbf{k}_3\cdot\mathbf{r}}\rho_{30}^{(r)} + iG_1 e^{i\mathbf{k}_1\cdot\mathbf{r}}\rho_{00}^{(r)} \\ \dot{\rho}_{20}^{(r)} &= -d_{20}\rho_{20}^{(r)} - iG_1 e^{i\mathbf{k}_1\cdot\mathbf{r}}\rho_{21}^{(r)} - iG_4 e^{i\mathbf{k}_4\cdot\mathbf{r}}\rho_{24}^{(r)} + iG_2 e^{i\mathbf{k}_2\cdot\mathbf{r}}\rho_{10}^{(r)} \\ \dot{\rho}_{30}^{(r)} &= -d_{30}\rho_{30}^{(r)} - iG_1 e^{i\mathbf{k}_1\cdot\mathbf{r}}\rho_{31}^{(r)} - iG_4 e^{i\mathbf{k}_4\cdot\mathbf{r}}\rho_{34}^{(r)} + iG_3{}^* e^{-i\mathbf{k}_3\cdot\mathbf{r}}\rho_{10}^{(r)} \\ \dot{\rho}_{14}^{(r)} &= -d_{14}\rho_{14}^{(r)} + iG_2{}^* e^{-i\mathbf{k}_2'\cdot\mathbf{r}}\rho_{24}^{(r)} + iG_3 e^{i\mathbf{k}_3\cdot\mathbf{r}}\rho_{34}^{(r)} - iG_4^* e^{-i\mathbf{k}_4\cdot\mathbf{r}}\rho_{10}^{(r)} \\ \dot{\rho}_{24}^{(r)} &= -d_{24}\rho_{24}^{(r)} + iG_2 e^{i\mathbf{k}_2\cdot\mathbf{r}}\rho_{14}^{(r)} - iG_4^* e^{-i\mathbf{k}_4\cdot\mathbf{r}}\rho_{20}^{(r)} \\ \dot{\rho}_{34}^{(r)} &= -d_{34}\rho_{34}^{(r)} + iG_3^* e^{-i\mathbf{k}_3\cdot\mathbf{r}}\rho_{14}^{(r)} - iG_4^* e^{-i\mathbf{k}_4\cdot\mathbf{r}}\rho_{30}^{(r)} \end{aligned} \tag{2.1}$$

where $d_{10} = \Gamma_{10} + i\Delta_1$, $d_{20} = \Gamma_{20} + i(\Delta_1 + \Delta_2)$, $d_{30} = \Gamma_{30} + i(\Delta_1 - \Delta_3)$, $d_{14} = \Gamma_{14} - i(\Delta_1 - \Delta_4)$, $d_{24} = \Gamma_{24} - i(\Delta_1 + \Delta_2 - \Delta_4)$, and $d_{34} = \Gamma_{34} - i(\Delta_1 - \Delta_3 - \Delta_4)$ with the frequency detuning $\Delta_i = \Omega_i - \omega_i$. Γ_{ij} is the transverse relaxation rate between states $|i\rangle$ and $|j\rangle$. Ω_i is the atomic resonance frequency. The Rabi frequencies are defined as $G_i = \varepsilon_i \mu_{ij}/\hbar$, where μ_{ij} are the transition dipole moments between level $|i\rangle$ and level $|j\rangle$. Next, we apply modified density matrix equations via perturbation chains to analyze the coexisting multi-dressed FWM, SWM, and EWM.

2.3.2 Multi-Wave Mixing (MWM)

To better understand the interplay among coexisting FWM, SWM, and EWM processes, we use the perturbation chain method to derive expressions of high-order density matrix elements standing for FWM, SWM, and EWM processes. Although such an approach makes significant approximations, it shows a simple but clear picture of those that provide leading contributions in the complicated nonlinear optical processes. Moreover, we approach the final complicated expressions by inserting the dressing field step by step. So the expression of the pure FWM without any dressing field of a ladder subsystem contained in the system we showed earlier are given a clear derivation process using the perturbation chain method in the following.

2.3.2.1 Four-Wave Mixing (FWM)

Considering a subsystem consisting of states $|0\rangle$, $|1\rangle$, $|2\rangle$ and three laser beams E_1, E_2, and E'_2, there is a simple FWM (ρ_{F1}) expressed as the FWM perturbation chain (F1) $\rho_{00}^{(0)} \xrightarrow{\omega_1} \rho_{10}^{(1)} \xrightarrow{\omega_2} \rho_{20}^{(2)} \xrightarrow{-\omega_2} \rho_{10}^{(3)}$. Density matrix elements in the chain denote the initial, intermediary, and final states of the transition process of generating FWM. According to the physics of the perturbation chain, as to a certain element denoting a state, only the higher-order element in the chain can contribute to it; the other weak term can be neglected. Consequently, simplified density matrix equations about $\rho_{10}^{(1)}$, $\rho_{20}^{(2)}$, and $\rho_{10}^{(3)}$ are

$$\begin{aligned}
\rho_{10}^{(1)} &= \frac{iG_1 e^{i\mathbf{k}_1 \cdot \mathbf{r}}}{d_{10}} \\
\rho_{20}^{(2)} &= \frac{iG_2 e^{i\mathbf{k}_2 \cdot \mathbf{r}} \rho_{10}^{(1)}}{d_{20}} \\
\rho_{10}^{(3)} &= \frac{iG_2'^{*} e^{-i\mathbf{k}'_2 \cdot \mathbf{r}} \rho_{20}^{(2)}}{d_{10}}
\end{aligned} \tag{2.2}$$

Then, we get:

$$\rho_{F1} = \rho_{10}^{(3)} = \frac{-iG_A e^{i\mathbf{k}_f \cdot \mathbf{r}}}{d_{10}^2 d_{20}} \tag{2.3}$$

where $G_A = G_1 G_2 G_2'^{*}$.

Let us talk about the case where an additional strong dressing field E_3 (connecting transition $|1\rangle$ and $|3\rangle$) is added in system $|0\rangle - |1\rangle - |2\rangle$. E_3 dresses level $|1\rangle$ and creates two new states $|G_3+\rangle$ and $|G_3-\rangle$, and then the dressed FWM is denoted as the dressed perturbation chain, (F2) $\rho_{00}^{(0)} \xrightarrow{\omega_1} \rho_{10}^{(1)} \xrightarrow{\omega_2} \rho_{20}^{(2)} \xrightarrow{-\omega_2} \rho_{G_3\pm 0}^{(3)}$, where the subscript "1" of element $\rho_{10}^{(3)}$ is replaced by "$G_3\pm$." In the following, we see that the dressed state is actually the interference between the polarizations $\rho_{10}^{(3)}$ and $\rho_{10}^{(5)}$ which is generated by E_3 combining with FWM (F1). This led us to expand $\rho_{G_3\pm 0}^{(3)}$ as subchain $\rho_{10}^{(3)} \xrightarrow{-\omega_3} \rho_{30}^{(4)} \xrightarrow{\omega_3} \rho_{10}^{(5)}$ and then the dressed FWM chain transformed into an SWM chain (S1) $\rho_{00}^{(0)} \xrightarrow{\omega_1} \rho_{10}^{(1)} \xrightarrow{\omega_2} \rho_{20}^{(2)} \xrightarrow{-\omega_2} \rho_{10}^{(3)} \xrightarrow{-\omega_3} \rho_{30}^{(4)} \xrightarrow{\omega_3} \rho_{10}^{(5)}$, which stands for the one that interfered with FWM (F1) and finally caused dressing states $|G_3\pm\rangle$. In the weak field approximation, terms containing weak field G_1 are all neglected in the density matrix equation, while terms containing strong field G_3 are saved. When doing this guided by subchain $\rho_{10}^{(3)} \xrightarrow{-\omega_3} \rho_{30}^{(4)} \xrightarrow{\omega_3} \rho_{10}^{(5)}$, the coupling equations are as follows:

$$\dot{\rho}_{10}^{(3)} = -d_{10}\rho_{10}^{(3)} + iG_2{}^{*}e^{i\mathbf{k}_2'\cdot\mathbf{r}}\rho_{20}^{(2)} + ie^{i\mathbf{k}_3\cdot\mathbf{r}}G_3\rho_{30}$$

$$\dot{\rho}_{30} = -d_{30}\rho_{30} + iG_3{}^{*}e^{-i\mathbf{k}_3\cdot\mathbf{r}}\rho_{10}^{(3)}$$

Solving these equations, we get

$$\rho_{G_3\pm 0}^{(3)} = \frac{iG_2'^{*}e^{i\mathbf{k}_2'\cdot\mathbf{r}}}{d_{10} + |G_3|^2/d_{30}}\rho_{20}^{(2)} \tag{2.4}$$

Combining Eq. (2.2) with Eq. (2.4), the dressed FWM is finally obtained as

$$\rho_{F1}' = \rho_{10}^{(3)} = \frac{-iG_A e^{ik_f\cdot r}}{d_{10}d_{20}}\frac{1}{d_{10} + |G_3|^2/d_{30}} \tag{2.5}$$

We can see the factor d_{10} in ρ_{F1}' is modified by the G_3 dressing term in this singly-dressed FWM. More interesting, under the weak field limit ($|G_3|^2 \ll \Gamma_{10}\Gamma_{30}$), Eq. (2.5) can be expanded to be $\rho_{F1}' = \rho_{F1} + iG_A|G_3|^2e^{i\mathbf{k}_s\cdot\mathbf{r}}/(d_{10}^2d_{20}d_{30}) = \rho_{F1} + \rho_{S1}$, here ρ_{S1} is the SWM expression that can be deduced via the chain S1. This means that the density matrix element of the singly-dressed FWM can be considered as a coherent superposition of one pure FWM (ρ_{F1}) process and one pure SWM (ρ_{S1}) process under the weak field limit. The enhancement and suppression of the FWM process consequently can be viewed as resulting from the constructive and destructive interference between these two processes, respectively. In addition, such a coherent superposition can be used to investigate the autoionization state of atoms by optics heterodyne between FWM and SWM. However, when the dressing field E_3 is strong enough, Eq. (2.5) cannot be series expanded. Here we believe the dressing field splits the level $|1\rangle$, and the enhancement and suppression of the FWM process results from the constructive and destructive interferences between two dressed paths ($|0\rangle \to |G_3+\rangle \to |1\rangle$ and $|0\rangle \to |G_3-\rangle \to |1\rangle$), respectively.

Next, let us discuss the dual-dressed scheme. It generally has three different types, namely, the nested, sequential, and parallel dressed schemes. Detailed analyses are as follows.

Here, besides the dressing field E_3, another strong field E_4 is turned on too. Then a dual-dressed FWM is generated in this five-level system. The dressing fields E_3 and E_4 dress the levels $|0\rangle$ and $|1\rangle$, respectively. So, they can act on two different transition processes, namely, two different matrix-density elements in the perturbation chain. For example, the dual-dressed four-wave mixing (DDFWM) expressed by (F3) $\rho_{00}^{(0)} \xrightarrow{\omega_1} \rho_{1G_4\pm}^{(1)} \xrightarrow{\omega_2} \rho_{20}^{(2)} \xrightarrow{-\omega_2} \rho_{G_3\pm0}^{(3)}$. Taking two subchains of dressing fields (D1) $\rho_{10}^{(1)} \xrightarrow{-\omega_4} \rho_{14}^{(2)} \xrightarrow{\omega_4} \rho_{10}^{(3)}$ and (D2) $\rho_{10}^{(5)} \xrightarrow{-\omega_3} \rho_{30}^{(6)} \xrightarrow{\omega_3} \rho_{10}^{(7)}$ to displace two matrix-density elements $\rho_{1G_4\pm}^{(1)}$ and $\rho_{G_3\pm0}^{(3)}$, the EWM perturbation chain (E1) $\rho_{00}^{(0)} \xrightarrow{\omega_1} \rho_{10}^{(1)} \xrightarrow{-\omega_4} \rho_{14}^{(2)} \xrightarrow{\omega_4} \rho_{10}^{(3)} \xrightarrow{\omega_2} \rho_{20}^{(4)} \xrightarrow{-\omega_2} \rho_{10}^{(5)} \xrightarrow{-\omega_3} \rho_{30}^{(6)} \xrightarrow{\omega_3} \rho_{10}^{(7)}$ is obtained. This doubly-dressing scheme is called a *parallel dressing scheme*, for the two dressing fields are influenced in this process parallely. According to the chain D1, we solve the coupling equations,

$$\dot{\rho}_{10}^{(1)} = -d_{10}\rho_{10}^{(1)} - iG_4 e^{i\mathbf{k}_4\cdot\mathbf{r}}\rho_{14} + iG_1 e^{i\mathbf{k}_1\cdot\mathbf{r}}\rho_{00}^{(0)}, \dot{\rho}_{14} = -d_{14}\rho_{14} - iG_4^* e^{-i\mathbf{k}_4\cdot\mathbf{r}}\rho_{10}^{(1)}$$

and obtain

$$\rho_{1G_4\pm}^{(1)} = \frac{iG_1 e^{i\mathbf{k}_1\cdot\mathbf{r}}}{d_{10} + |G_4|^2/d_{14}}\rho_{00}^{(0)} \tag{2.6}$$

According to the chain D2, we solve coupling equations

$$\dot{\rho}_{10}^{(3)} = -d_{10}\rho_{10}^{(3)} + iG_2{}^* e^{-i\mathbf{k}_2'\cdot\mathbf{r}}\rho_{20}^{(2)} + iG_3 e^{i\mathbf{k}_3\cdot\mathbf{r}}\rho_{30}, \dot{\rho}_{30} = -d_{30}\rho_{30} + iG_3{}^* e^{-i\mathbf{k}_3\cdot\mathbf{r}}\rho_{10}^{(3)}$$

and obtain

$$\rho_{G_3\pm0}^{(3)} = \frac{iG_2'^* e^{-i\mathbf{k}_2'\cdot\mathbf{r}}}{d_{10} + |G_3|^2/d_{30}}\rho_{20}^{(2)} \tag{2.7}$$

From chain (F3) and combining Eq. (2.2), Eq. (2.4), and Eq. (2.5), we obtain

$$\rho_{F1}'' = \rho_{10}^{(3)} = \frac{-iG_A e^{i\mathbf{k}_f\cdot\mathbf{r}}}{d_{20}} \frac{1}{d_{10} + |G_4|^2/d_{14}} \frac{1}{d_{10} + |G_3|^2/d_{30}} \tag{2.8}$$

where d_{10} in two terms are modified parallelly. Under the weak limit ($|G_4|^2 \ll \Gamma_{10}\Gamma_{14}$), Eq. (2.8) can be expanded as $\rho_{F1}' + G_A|G_4|^2/[d_{10}d_{14}d_{20}(d_{10} + |G_3|^2/d_{30})] = \rho_{F1}' + \rho_S'$. This means that the density matrix element of the DDFWM can be considered as a coherent superposition of one single-dressed FWM process and one single-dressed SWM process under the weak field limit. We can see that the DDFWM is actually the interference between the polarizations ρ_{F1}' and ρ_S'. In addition, under the weak field limit ($|G_3|^2 \ll \Gamma_{10}\Gamma_{30}$ and $|G_4|^2 \ll \Gamma_{10}\Gamma_{14}$), the parallel DDFWM can be considered as a coherent superposition of signals from one FWM, two SWM, and one EWM processes according to its expression expansion.

In this part, we talk about the dressing mechanism contained in five-dressing fields based on the three basic dual-dressing types. We focus on investigating the two portions of the expression, one is the two-dressing term, which is a simply nested dressing scheme, the other contains three dressing fields of which the outer dressing terms of the nested dressing scheme is a sequential dressing scheme.

At first, let us consider the nested dressing scheme, in which the dressing fields E_3 and E_4 perform their dressing effect in another manner denoted as subchain (D3) $\rho_{10}^{(3)} \xrightarrow{-\omega_3} \rho_{30} \xrightarrow{\omega_4} \rho_{34} \xrightarrow{-\omega_4} \rho_{30} \xrightarrow{\omega_3} \rho_{10}^{(3)}$, in which the subchain

"$\rho_{30} \xrightarrow{\omega_4} \rho_{34} \xrightarrow{-\omega_4} \rho_{30}$" induced by dressing field E_4 is nested into the subchains "$\rho_{10}^{(3)} \xrightarrow{-\omega_3} \rho_{30}$" and "$\rho_{30} \xrightarrow{\omega_3} \rho_{10}^{(3)}$" induced by dressing field E_3. That is why it is called the *nested dressing scheme* and $\rho_{10}^{(3)}$ in the chain (F1) can be modified as $\rho_{G_3\pm G_4\pm}^{(3)}$. Similarly, from chain D3, we choose the following equations as the coupling equations, $\dot{\rho}_{10}^{(3)} = -d_{10}\rho_{10}^{(3)} + iG_2{}^*e^{-i\mathbf{k}_2'\cdot\mathbf{r}}\rho_{20}^{(r)} + iG_3e^{i\mathbf{k}_3\cdot\mathbf{r}}\rho_{30}$, $\dot{\rho}_{30} = -d_{30}\rho_{30} - iG_4e^{i\mathbf{k}_4\cdot\mathbf{r}}\rho_{34} + iG_3{}^*e^{-i\mathbf{k}_3\cdot\mathbf{r}}\rho_{10}^{(3)}$, $\dot{\rho}_{34} = -d_{34}\rho_{34} + iG_4^*e^{-i\mathbf{k}_4\cdot\mathbf{r}}\rho_{30}$.

Solving these equations, we get

$$\rho_{G_3\pm G_4\pm}^{(3)} = \frac{iG_2'^*e^{-i\mathbf{k}_2'\cdot\mathbf{r}}}{d_{10} + |G_3|^2/(d_{30} + |G_4|^2/d_{34})}\rho_{20}^{(2)} \tag{2.9}$$

So, the single-photon term d_{10} is modified by G_3, which is called the *inner dressing field*, and then d_{30} is modified by the outer G_4 which is called the *outer dressing field*. From the expression, we can see that the G_4 term depends on the G_3 term; they interact tightly.

Next, we investigate the three dressing field terms. Here, we assume E_2 is also strong enough to be a dressing field. It can dress the same level $|1\rangle$ and act on element $\rho_{14}^{(2)}$ together with E_3, which is denoted as $\rho_{(G_2G_3\pm)4}^{(2)}$. It expands as the subchain (D4) $\rho_{14}^{(2)} \xrightarrow{\omega_2} \rho_{24} \xrightarrow{-\omega_2} \rho_{14} \xrightarrow{-\omega_3} \rho_{34} \xrightarrow{\omega_3} \rho_{14}^{(2)}$, which is the adhesion of two subchains of the dressing fields $\rho_{14}^{(2)} \xrightarrow{\omega_2} \rho_{24} \xrightarrow{-\omega_2} \rho_{14}$ and $\rho_{14} \xrightarrow{-\omega_3} \rho_{34} \xrightarrow{\omega_3} \rho_{14}^{(2)}$. Here, the subscript "1" of "$\rho_{14}^{(2)}$" is replaced by $G_2G_3\pm$. By virtue of the subchain D4, we take coupling equations, $\dot{\rho}_{14}^{(2)} = -d_{14}\rho_{14}^{(2)} + iG_2^*e^{-i\mathbf{k}_2'\cdot\mathbf{r}}\rho_{24} + iG_3e^{i\mathbf{k}_3\cdot\mathbf{r}}\rho_{34} - iG_4^*e^{i\mathbf{k}_4\cdot\mathbf{r}}\rho_{10}^{(1)}$, $\dot{\rho}_{24}^{(r)} = -d_{24}\rho_{24} + iG_2e^{i\mathbf{k}_2\cdot\mathbf{r}}\rho_{14}^{(2)}$, and $\dot{\rho}_{34} = -d_{34}\rho_{34} + iG_3^*e^{i\mathbf{k}_3\cdot\mathbf{r}}\rho_{14}^{(2)}$.

Under the dipole moment approximation, we have

$$\rho_{(G_2G_3\pm)4}^{(2)} = \frac{-iG_4^*e^{i\mathbf{k}_4\cdot\mathbf{r}}}{d_{14} + |G_2|^2/d_{24} + |G_3|^2/d_{34}}\rho_{10}^{(1)} \tag{2.10}$$

where the term d_{14} is modified by the G_2 and G_3 terms sequentially. From the expression, we can see that these two dressing terms can interact, but not firm as the nested scheme.

Here, we have known that the interaction between two dressing fields of the nested dressing scheme is strongest, while the parallel dressing scheme is weakest and the sequential dressing scheme is intermediate between them.

On the basis of this discussion, let us talk about the five-dressing scheme in the FWM process, which can be viewed as the multiplication of the two-dressing and three-dressing terms. When the dressing fields E_2, E_3, and E_4 are all turned on and strong enough, they dress the FWM process simultaneously and the created multi-dark states interact with each other. Actually, it is denoted as the five-dressing FWM chain (F4) $\rho_{00}^{(0)} \xrightarrow{\omega_1} \rho_{A}^{(1)} \xrightarrow{\omega_2} \rho_{20}^{(2)} \xrightarrow{-\omega_2} \rho_{G_3\pm G_4\pm}^{(3)}$, where $\rho_{A}^{(1)}$ represents the dressed subchain (D5) $\rho_{10}^{(1)} \xrightarrow{-\omega_4} \rho_{(G_2G_3\pm)4} \xrightarrow{\omega_4} \rho_{10}^{(1)}$. In D5, the subchain D4 (represented by $\rho_{(G_2G_3\pm)4}$) shows the sequential scheme, and then it is nested between subchains "$\rho_{10}^{(1)} \xrightarrow{-\omega_4} \rho_{14}$" and "$\rho_{14} \xrightarrow{\omega_4} \rho_{10}^{(1)}$". In addition, $\rho_{A}^{(1)}$ and $\rho_{G_3\pm G_4\pm}^{(3)}$ lie parallelly in the chain F4. We have the expressions of $\rho_{G_3\pm G_4\pm}^{(3)}$ and $\rho_{(G_2G_3\pm)4}$, and according to the subchain D5, we can obtain,

$$\rho_A^{(1)} = \frac{iG_1 e^{i\mathbf{k}_1\cdot\mathbf{r}}}{d_{10} + |G_4|^2/(d_{14} + |G_2|^2/d_{24} + |G_3|^2/d_{34})}\rho_{00}^{(0)} \tag{2.11}$$

So, the "ω_1" single-photon term d_{10} is modified directly by the G_4 term, while the G_2 and G_3 terms synchronously modify the "$\omega_1 + \omega_4$" two-photon term d_{14}. The outer dressing terms of the nested dressing scheme in Eq. (2.11) is a sequential combination of the G_2 and G_3 dressing terms. Three dressing fields E_2, E_3, and E_4 in this combining form can induce second or triple AT splitting in spectra numerical simulation.

The density matrix elements for the five-dressing FWM, based on the above-mentioned discussion, can be written as

$$\rho_{F1}''' = \rho_{10}^{(3)} = \frac{-iG_A e^{i\mathbf{k}_f\cdot\mathbf{r}}}{d_{20}[d_{10} + |G_3|^2/(d_{30} + |G_4|^2/d_{43})]} \frac{1}{d_{10} + [|G_4|^2/(d_{41} + |G_2|^2/d_{42} + |G_3|^2/d_{43})]} \tag{2.12}$$

Apparently, nested, parallel, and sequential dressing schemes coexist in this five-dressing FWM process. As the spectra of the parallel dressing scheme can be considered as a superposition of two groups of the different peaks, we separately investigate the two dressing terms, $\rho_{A1} = 1/[d_{10} + |G_3|^2/(d_{30} + |G_4|^2/d_{43})]$ (two dressing fields term) and $\rho_{A2} = 1/[d_{10} + |G_4|^2/(d_{41} + |G_2|^2/d_{42} + |G_3|^2/d_{43})]$ (three dressing fields term). After that, ρ_{F1}''', which is proportional to the multiplication of ρ_{A1} and ρ_{A2} terms, is investigated. More interesting, in the weak fields limit ($|G_3|^2 \ll \Gamma_{14}\Gamma_{34}$, $|G_4|^2 \ll \Gamma_{10}\Gamma_{14}$, and $|G_2|^2 \ll \Gamma_{14}\Gamma_{24}$), ρ_{F1}''' can be expanded as

$$\rho_{10}^{(3)} = \frac{-iG_A(1 - |G_4|^2/d_{10}d_{14} + |G_2|^2|G_4|^2/d_{10}d_{14}^2d_{24} + |G_3|^2|G_4|^2/d_{10}d_{14}^2d_{34})\rho_{A2}}{d_{10}d_{20}} \tag{2.13}$$

We can see that the density matrix element representing the five-dressing FWM is the summation of one FWM (first term), one SWM (second term), and two EWM (third and fourth terms). Hence, under the weak-dressing field condition, the polarization state of the generating five-dressing FWM can be viewed as the coherent superposition state of these states generating FWM, SWM, and EWM. The relative FWM signal intensity is given by $I_F \propto |\rho_{F1}'''|^2$.

2.3.2.2 Four-Dressing SWM

There also exists a four-dressing SWM in the five-level system, as shown in Figure 2.9(a). When E_3 is blocked, by means of the SWM perturbation chain (S) $\rho_{00}^{(0)} \xrightarrow{\omega_1} \rho_{10}^{(1)} \xrightarrow{-\omega_4} \rho_{14}^{(2)} \xrightarrow{\omega_4} \rho_{10}^{(3)} \xrightarrow{\omega_2} \rho_{20}^{(4)} \xrightarrow{-\omega_2} \rho_{10}^{(5)}$, we obtain the expression of pure SWM without any dressing field:

$$\rho_S = \rho_{10}^{(5)} = \frac{iG_B e^{i\mathbf{k}_s\cdot\mathbf{r}}}{d_{10}^3 d_{20} d_{14}} \tag{2.14}$$

where $G_B = G_A G_4 G_4^*$. When the strong dressing fields E_2, E_3, and E_4 dress the levels $|1\rangle$ and $|0\rangle$, they affect the atomic coherence $\rho_{10}^{(3)}$ in the form of subchain

(D6) $\rho_{10}^{(3)} \xrightarrow{-\omega_3} \rho_{30} \xrightarrow{\omega_4} \rho_{34} \xrightarrow{\omega_3} \rho_{G2\pm4} \xrightarrow{-\omega_3} \rho_{34} \xrightarrow{-\omega_4} \rho_{30} \xrightarrow{\omega_3} \rho_{10}^{(3)}$. In D6, $\rho_{G2\pm4}$, induced by the dressing field E_2, is nested in $\rho_{34} \xrightarrow{\omega_3} \rho_{G_2\pm4} \xrightarrow{-\omega_3} \rho_{34}$ induced by the dressing field E_3, which are then nested into "$\rho_{30} \xrightarrow{\omega_4} \rho_{34}$" and "$\rho_{34} \xrightarrow{-\omega_4} \rho_{30}$," and all are finally nested into "$\rho_{10}^{(3)} \xrightarrow{-\omega_3} \rho_{30}$" and "$\rho_{30} \xrightarrow{\omega_3} \rho_{10}^{(3)}$." Thus, we obtain

$$\rho_S'' = \rho_{10}^{(5)} = \frac{iG_B e^{ik_S \cdot r}}{d_{10}^2 d_{20} d_{14}} \frac{1}{d_{10} + \cfrac{|G_3|^2}{d_{30} + \cfrac{|G_4|^2}{d_{34} + \cfrac{|G_3|^2}{d_{14} + \cfrac{|G_2|^2}{d_{24}}}}}} \tag{2.15}$$

Here, the term d_{10}, which represents the ω_1 single-photon process, is dressed directly by the inner dressing field E_3, which involves a two-photon term d_{30}, and then it is modified by the first outer dressing field E_4, which shows that two dressing fields entangle with each other in such a nested scheme. Again, the second outer dressing field E_3 affects the $|G_4|^2$ term, and finally the $|G_3|^2$ term is modified by the third outer dressing field E_2. Hence, the chain D6 and the expression ρ_S'' show that four-dressing fields are entangled with each other in such a quadruple nested scheme. The inner dressing field E_3 controls the SWM signal directly, while three outer dressing fields E_4, E_3, and E_2 affect the SWM signal indirectly. In the weak-dressing field limit ($|G_3|^2 \ll \Gamma_{10}\Gamma_{30}$), ρ_S'' can be expanded to $\rho_S'' = \rho_{10}^{\prime(5)} + \rho_{10}^{(7)}$, where $\rho_{10}^{\prime(5)} = ie^{ik_s \cdot r} G_B/(d_{10}^2 d_{20} d_{14} B_1)$, $\rho_{10}^{(7)} = -ie^{ik_e \cdot r} G_B |G_3|^2/(d_{10}^4 d_{20} d_{14} B_1)$, and $B_1 = d_{30} + |G_4|^2/\{d_{34} + [|G_3|^2/(d_{14} + |G_2|^2/d_{24})]\}$. Under the weak field condition, the four-dressing SWM can be considered as a coherent superposition of the signals from one SWM process and one triple nested EWM process; in other words, the four-dressing SWM is actually the interference between the polarizations $\rho_{10}^{\prime(5)}$ and $\rho_{10}^{(7)}$. The relative SWM signal intensity is given by $I_S \propto |\rho_S''|^2$.

Therefore, four-dressing SWM is a quadruple nested dressing scheme that can induce quadruple AT splitting. To clearly analyze the spectra of the four-dressing SWM, we plot the figures by adding a dressing field into the pure SWM expression step by step.

2.3.2.3 Four-Dressing EWM

The four-dressing EWM process coexists with the multi-dressed FWM and SWM processes in the five-level system. When all five laser beams are turned on simultaneously, as shown in Figure 2.9(b), the EWM signal is generated, which is denoted as the EWM perturbation chain E1, and the pure EWM expression is

$$\rho_{E1} = \rho_{10}^{(7)} = \frac{-iG_C e^{ik_e \cdot r}}{d_{10}^4 d_{20} d_{30} d_{14}} \tag{2.16}$$

where $G_C = G_B G_3 G_3^*$. With strong dressing fields E_4, E_2 and E_3, $\rho_{10}^{(1)}$ and $\rho_{10}^{(3)}$ in chain E1 can be modified as $\rho_{G_2\pm G_4\pm}^{(1)}$ and $\rho_{G_3\pm G_4\pm}^{(3)}$, respectively, where $\rho_{G_2\pm G_4\pm}^{(1)}$ can be extended into the subchain of the nested scheme (D7) $\rho_{10}^{(1)} \xrightarrow{\omega_2} \rho_{20} \xrightarrow{\omega_4} \rho_{24} \xrightarrow{-\omega_4} \rho_{20} \xrightarrow{-\omega_2} \rho_{10}^{(1)}$. As a result, the four-dressing EWM chain can be written as (E2) $\rho_{00}^{(0)} \xrightarrow{\omega_1} \rho_{G_2\pm G_4\pm}^{(1)} \xrightarrow{-\omega_4} \rho_{14}^{(2)} \xrightarrow{\omega_4} \rho_{G_3\pm G_4\pm}^{(3)} \xrightarrow{\omega_2} \rho_{20}^{(4)} \xrightarrow{-\omega_2} \rho_{10}^{(5)} \xrightarrow{-\omega_3}$

$\rho_{30}^{(6)} \xrightarrow{\omega_3} \rho_{10}^{(7)}$. According to subchains D3, D7, and chain E2, we can easily obtain the four-dressing EWM expression,

$$\rho'_{E1} = \rho_{10}^{(7)} = \frac{-i\exp(i\mathbf{k}_e \cdot \mathbf{r})G_C}{d_{10}d_{20}d_{30}d_{41}} \frac{1}{d_{10} + [|G_2|^2/(d_{20} + |G_4|^2/d_{24})]}$$
$$\frac{1}{d_{10} + [|G_3|^2/(d_{30} + |G_4|^2/d_{34})]} \tag{2.17}$$

Both the multiplying terms of Eq. (2.17) contain nested dressing schemes (E_2 and E_4, E_3, and E_4). It means that the dressing effect occurs in two different transition processes and, therefore, they are independent of each other. In each dressing part, there are nested dressing terms; the inner dressing field G_2 or G_3 acts on the state of EWM transitions, while the outer dressing field G_4 acts on the states of the transition, respectively. In all, the understanding of these two basic dual-dressing schemes will enable us to easily analyze this complicated expression. The relative EWM signal intensity is given by $I_E \propto |\rho_{E_3}|^2$.

2.3.2.4 **Four-Dressing EIT**

The EIT displayed in the probe field absorption spectrum is a single-photon transition process which is dressed by the fields E_2, E_3, and E_4. Under the weak probe field approximation, the density matrix for probe filed absorption ρ_{10} is

$$\rho_{10} = \frac{iG_1}{d_{10} + |G_2|^2/(d_{20} + |G_4|^2/d_{24}) + |G_3|^2/(d_{30} + |G_4|^2/d_{34})} \tag{2.18}$$

Equation (2.18) gives the four-dressing EIT case, which shows the sequential combination of two nested schemes (E_2 and E_4, E_3 and E_4). Two sequentially dressing fields act on the state $|1\rangle$ together and then they are both dressed by field G_4. Three dressing fields (E_2, E_3, and E_4) can induce two triple EIT splitting. By changing the offsets and intensities of the dressing fields, we can obtain the different multi-EIT spectra.

When five laser beams are turned on, the five-dressing FWM, four-dressing SWM, four-dressing EWM, and four-dressing EIT as discussed in the preceding text, can exist simultaneously in the five-level system. Such FWM, SWM, and EWM signal beams are in the same direction for the reason of equivalent phase-matching conditions ($\mathbf{k}_f$, $\mathbf{k}_s$, and $\mathbf{k}_e$). By controlling the offsets and intensities of dressing fields, the coexisting MWMs fall into different EIT windows which can be clearly separated and distinguished.

2.3.3 Numerical Results and Discussion

The coexisting FWM and SWM processes have been observed in the Cd(S, Se) semiconductor-doped glasses. The advantages of solids include high density of atoms, compactness, and absence of atomic diffusion, except relatively broad optical line widths and fast decoherence rate. However, there is a narrow line width in an atomic vapor. In addition, FWM and SWM processes can coexist in the inverted-Y atomic subsystem ($|0\rangle = |5S_{1/2}\rangle$ (F = 2), $|1\rangle = |5P_{3/2}\rangle$, $|2\rangle = |5D_{3/2}\rangle$, $|3\rangle = |5S_{1/2}\rangle$; as shown in Figure 2.9(a)), there exists a ladder-type EIT window with a line width of 50 MHz in transmission Doppler broadening profile (about 500 MHz). In this section, we add an additional beam E_4 (connecting transition $|4\rangle = |5P_{1/2}\rangle$ (F = 3) to $|0\rangle = |5S_{1/2}\rangle$) into such an inverted-Y system, as shown in Figure 2.9(a), where the transverse relaxation rate Γ_{ij} between states $|i\rangle$ and $|j\rangle$ is given by $\Gamma_{ij} = (\Gamma_i + \Gamma_j)/2$ ($\Gamma_0 = \gamma_{30} = 0.1$ MHz, $\Gamma_1 = \gamma_{10} + \gamma_{13} = 5.9$ MHz, $\Gamma_2 = \gamma_{21} + \gamma_{24} = 1.77$ MHz, $\Gamma_3 = \gamma_{30} = 0.1$ MHz and $\Gamma_4 = \gamma_{40} + \gamma_{43} = 5.4$ MHz, where γ_{ij} describes decay rates of coherences between $|i\rangle$ and $|j\rangle$. In Figure 2.9(b), nonlinear susceptibility $\chi^{(3)}$ for the FWM signal, $\chi^{(5)}$ for the SWM signal, and $\chi^{(7)}$ for the EWM signal in this five-level system interact with up to five laser fields, and we assume that the intensity of the probe field G_1 is small, while G_2, G_2', G_3, and G_4 can be arbitrary magnitudes.

We have discussed multi-dressing schemes and the interaction of dressing fields in the coexisting multi-dressed MWM by their calculated expressions. In this section, we investigate the normalized MWM spectra based on these analytic expressions.

2.3.3.1 Five-Dressing FWM

Here, the spectra of five-dressing FWM ($|\rho_{F1}^{'''}|^2$) are investigated by plotting and analyzing the two portions ($|\rho_{A1}|^2$ and $|\rho_{A2}|^2$), respectively. The characteristic spectra of the AT splitting, suppression, and enhancement of the FWM signal are generally equivalent to the superposition of the two portions. This method is just a proof of the concept theory, and will help us understand the effects of each of the dressing terms and their interactions.

At first, let us consider the spectrum of $|\rho_{A2}|^2$ versus Δ_1, as Figure 2.10(a) to (d) shows, in which primary, secondary, and triple AT splitting corresponding to the dressed states in Figure 2.10(e1) to (e3) are presented, respectively. We approach the final spectrum by adding the dressing fields one by one. In the first step, as Figure 2.10(a) shows, the inner dressing field E_4 dresses and splits the single-photon resonant peak of pure FWM into two peaks located at $\Delta_1 = \pm\Delta_{G4}/2 \approx \pm G_4 = \pm 100$ MHz, where Δ_{G4} is the separation between two AT splitting peaks. It is the primary AT splitting corresponding to, in the dressed-states picture, primary dressed states $|G_4\pm\rangle$ (Figure 2.10(e1)) generated by E_4 dressing the ground level $|0\rangle$, which corresponds to the left and right peaks in Figure 2.10(a), respectively. The second step, by setting a proper offset, the outer dressing field E_2, can exactly hit on one of the primary dressed state, for example, $|G_4+\rangle$ (then $\Delta_2 = 100$ MHz), and create the secondary dressed states $|G_4+\pm\rangle$ (Figure 2.10(e2)). Reflecting on the spectra, the left peak in Figure 2.10(a) splits into two peaks, and then three peaks locate at $\Delta_1 = -\Delta_{G4}/2 \mp \Delta_{G_2}/2 = -140$ MHz and -60 MHz, and $\Delta_1 = \Delta_{G_4}/2 = 100$ MHz (where Δ_{G_2} is the separation induced by E_2) will appear in Figure 2.10(b). It is the so-called secondary splitting. The third step, the other outer dressing field E_3, is tuned close to $\Delta_3 = -140$ MHz and then the triply-dressed states $|G_4++\pm\rangle$ are generated around the secondarily dressed states $|G_4++\rangle$ (Figure 2.10(e3)). And in the spectra, the left peak of the secondary

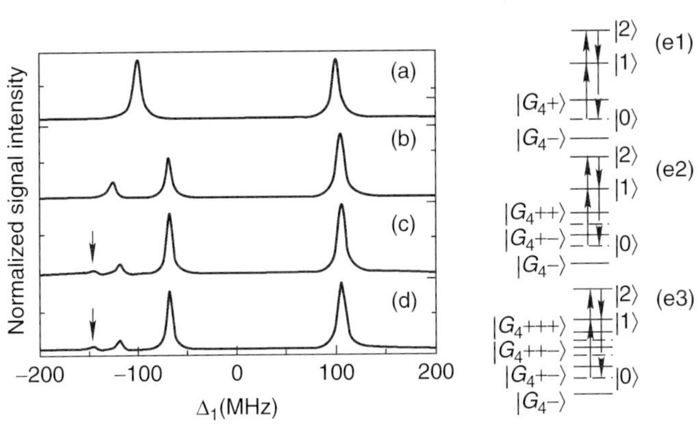

Figure 2.10 $|\rho_{A2}|^2$ in $|\rho'''_{F1}|^2$ spectra versus Δ_1 for $G_4 = 100$ MHz and $\Delta_4 = 0$ MHz, when (a) $G_2 = G_3 = 0$ MHz, (b) $G_2 = 40$ MHz, $\Delta_2 = 100$ MHz, $G_3 = 0$ MHz, (c) $G_2 = 40$ MHz, $\Delta_2 = 100$ MHz, $G_3 = 20$ MHz, and $\Delta_3 = -140$ MHz, and (d) $G_3 = 40$ MHz, $\Delta_3 = -100$ MHz, $G_2 = 20$ MHz, $\Delta_2 = 140$ MHz. The maximum of the intensity is normalized to be 1. (e1)–(e3) The dressed-state pictures. Source: Adapted from Ref. [20].

peaks will be split into two peaks, as Figure 2.10(c) shows. Finally, four peaks, which illustrate the triple-dressing effect, appear in Figure 2.10(c), which is located at $\Delta_1 = -\Delta_{G_4}/2 - \Delta_{G_2}/2 \mp \Delta_{G_3}/2 = -160\,\text{MHz}$ and -120 MHz, $\Delta_1 = -\Delta_{G_4}/2 + \Delta_{G_2}/2 = -60\,\text{MHz}$, and $\Delta_1 = \Delta_{G_4}/2 = 100\,\text{MHz}$.

Similarly, the dressed state $|G_4-\rangle$ can also be secondary dressed by the outer dressing field E_3 (by setting $\Delta_3 \approx -100$ MHz) and the secondary dressed state $|G_4 - \pm\rangle$ is created. Furthermore, the other outer dressing field E_2 ($\Delta_2 = 140$ MHz) creates the triple AT splitting, as shown in Figure 2.10(d). The triple AT splitting spectra in Figure 2.10(c) and (d) are identical, which prove that the outer dressing fields E_2 and E_3 are interchangeable because of their equivalent status in expression. The three-dressing term ($|\rho_{A2}|^2$) can obtain the secondary or triple AT splitting by changing the offsets of the dressing fields (E_2 and E_3) and the results show that the AT splitting of the sequential dressing scheme is similar to that of the nested dressing scheme.

On the other hand, the spectra of expression $|\rho_{A2}|^2$ in $|\rho'''_{F1}|^2$ can also present two secondary-dressed AT splitting. As Figure 2.11(a) shows, two peaks induced by the inner dressing field E_4 corresponding to $|G_4 \pm\rangle$ splits are split again by two outer dressing fields E_2 and E_3 ($\Delta_2 = \Delta_3 = G_4 = 50$ MHz) into two pairs of secondary-dressed AT splitting peaks (secondary splits $|G_4 + \pm\rangle$ and $|G_4 - \pm\rangle$ as is shown in Figure 2.11(d)). In addition, Figure 2.11(b) shows two peaks induced by $|\rho_{A1}|^2$ in $|\rho'''_{F1}|^2$.

So, we can see that the five-dressing FWM ($|\rho'''_{F1}|^2$) spectra (Figure 2.11(c)) is the superposition of the spectra of Figure 2.11(a) and (b). $|\rho'''_{F1}|^2$ is proportional to the multiplication of $|\rho_{A1}|^2$ and $|\rho_{A2}|^2$, and so its spectra of AT splitting caused by the five-dressing field is the superposition of these two portions ($|\rho_{A1}|^2$ and $|\rho_{A2}|^2$).

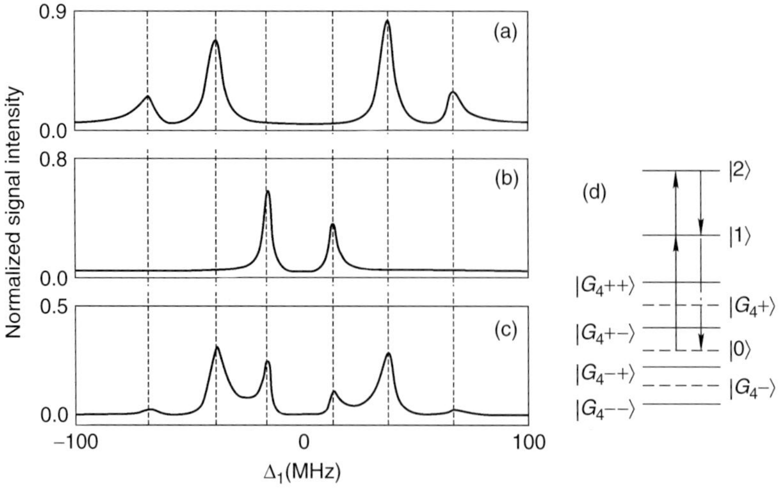

Figure 2.11 Five-dressing FWM signal intensity versus Δ_1 (a) for $|\rho_{A2}|^2$ in $|\rho_{F1}''''|^2$, (b) $|\rho_{A1}|^2$ in $|\rho_{F1}''''|^2$, (c) for Five-dressing FWM with $G_2 = G_3 = 20$ MHz, $G_4 = 50$ MHz, $\Delta_2 = \Delta_3 = 50$ MHz, $\Delta_4 = 0$, and (d) the dressed-state pictures.

Next, let us investigate the spectra of the suppression and enhancement of the five-dressing FWM signal. First, $|\rho_{A2}|^2$ intensity versus Δ_4 is discussed. Here, the dressed FWM signal intensity is normalized by rating pure FWM, which has no dressing fields. The intensity above or below "1" means enhancement or suppression of the FWM signal, respectively. We employ the same method as earlier to clearly understand the detailed process of the five-dressing FWM. First, the inner dressing field E_4 drives the transition from $|4\rangle$ to $|0\rangle$ and creates the dressed states $|G_4 \pm\rangle$, which leads to single-photon transition $|0\rangle \rightarrow |1\rangle$ off-resonance (Figure 2.12(b1)). So, at the exact single-photon resonance $\Delta_1 = 0$, the FWM signal intensity is greatly suppressed when scanning the dressing field E_4 across the resonance ($\Delta_4 = 0$), as Figure 2.12(a1) shows, one suppressed dip at the line center. Second, the outer dressing field E_2 is added and under the condition of $\Delta_1 + \Delta_2 - \Delta_4 = 0$, the single suppressed dip is split into two small suppressed dips, that is, $G_2 = 50$ MHz, as Figure 2.12(a2) shows. Further calculation shows that it has $\Delta_{G_2} \approx 2G_2$, under the condition of $G_2 \gg \Gamma_{10}$. It means that the outer dressing field E_2 dresses the state $|4\rangle$ to create two new states $|G_2 \pm\rangle$ (Figure 2.12(b2)); then the inner dressing field E_4 suppresses the resonant FWM signal directly from $|G_2 \pm\rangle$. Third, when the other outer dressing field E_3 is tuned close to $\Delta_3 = \Delta_1 - \Delta_{G_2}/2$, for example, $\Delta_3 = -50$ MHz, the left suppressed dip splits into two suppressed dips (Figure 2.12(a3)) owing to the secondary dressed states $|G_2 + \pm\rangle$ created by G_3. Therefore, a plot reflecting the outer dressing fields E_2 and E_3 comprehensive dressing effect are given in Figure 2.12(b3). However, it is not unique. Figure 2.12(a4) shows another two outer dressing fields E_2 and E_3 ($\Delta_2 = \Delta_3 = 0$ MHz) separate synchronously the suppressed dip in Figure 2.12(a1) into two suppressed dips at $\Delta_4 = \Delta_1 \pm \Delta_{G_2,G_3}/2 = \pm 70$ MHz. Under the condition

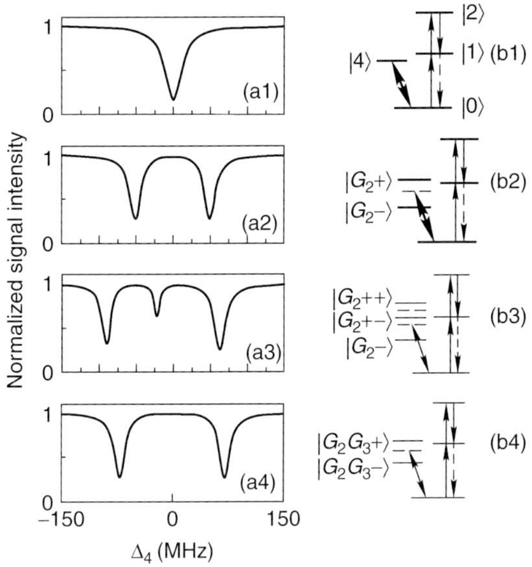

Figure 2.12 $|\rho_{A2}|^2$ spectra versus Δ_4 for $\Delta_1 = 0$, $G_4 = 5$ MHz, (a1) $G_2 = G_3 = 0$; (a2) $G_2 = 50$ MHz, $\Delta_2 = 0$, $G_3 = 0$; (a3) $G_2 = 50$ MHz, $\Delta_2 = 0$, $\Delta_3 = 50$ MHz, and $G_3 = 50$ MHz; and (a4) $\Delta_2 = \Delta_3 = 0$, $G_2 = G_3 = 50$ MHz. (b1–b4) The corresponding dressed-state pictures. Source: Adapted from Ref. [20].

of $\sqrt{G_2^2 + G_3^2} \gg \Gamma_{10}$, it has $\Delta_{G_2,G_3} \approx 2\sqrt{G_2^2 + G_3^2}$, where Δ_{G_2,G_3} is the separation between the dressed states $|G_2G_3\pm\rangle$ induced by two dressing fields together (Figure 2.12(b4)). Hence, the inner dressing field (E_4) suppresses FWM directly, while the outer dressing fields (E_2 and E_3) effect on the dressed states and influence the FWM indirectly.

Next, let us discuss $|\rho_{A2}|^2$ versus Δ_2 with $G_3 = 0$. Here the FWM signal intensity is also normalized for observing its suppression and enhancement. As Figure 2.13(a) shows, a suppressed dip, the height of which increases as G_2, increasingly appears. In fact, it induced by the outer dressing field E_2, which can largely weaken the suppression effect of the inner dressing field E_4 on the FWM signal, as is shown in Figure 2.13(b). Furthermore, we plot $|\rho_{A2}|^2$ versus Δ_2 and Δ_3 in Figure 2.13(c) and the results show that suppression of the sequential dressing scheme is like the one in the parallel dressing scheme. Specially, two independent suppressed peaks exist at $\Delta_2 = \Delta_3 = 0$, which means that these two outer dressing fields (E_2 and E_3) are indeed interchangeable.

Next, we investigate the spectra of the suppression and enhancement of the five-dressing FWM signal ($|\rho_{F1}'''|^2$) based on the analysis about $|\rho_{A2}|^2$ (Figure 2.12). Figure 2.14 shows the spectra of the suppression and enhancement of the five-dressing FWM signal ($|\rho_{F1}'''|^2$) versus Δ_4 and Δ_3. Under the resonance condition ($\Delta_1 = 0$), three dressing fields only can suppress the FWM signal in Figure 2.14(a). Figure 2.15(a) and (b) is the cross section of Figure 2.14(a) versus Δ_4 and Δ_3,

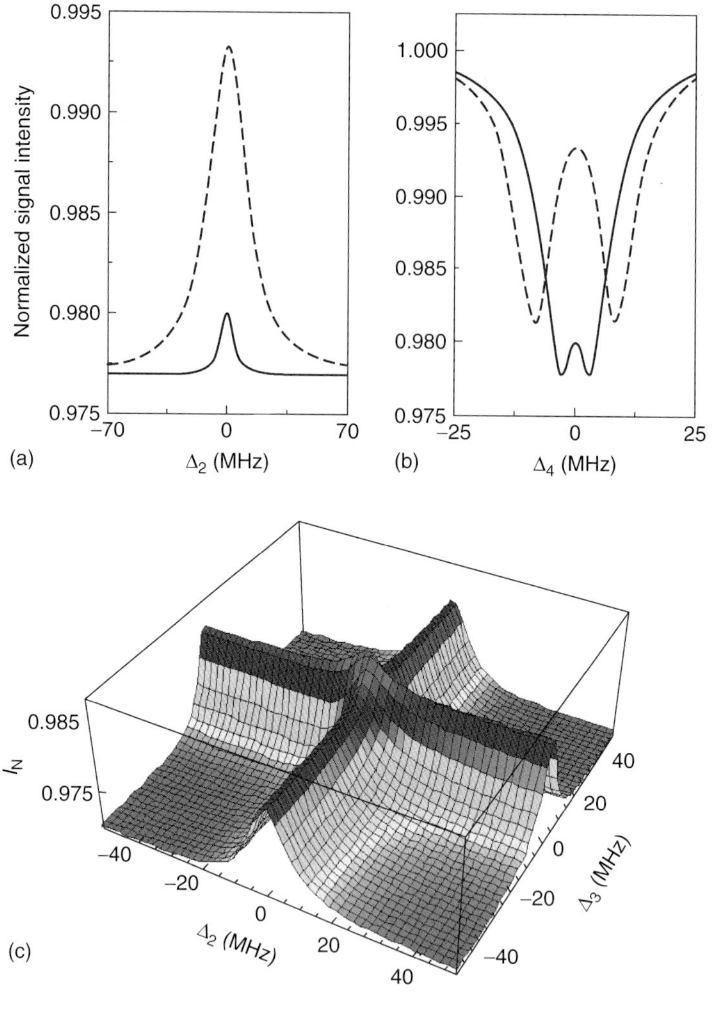

Figure 2.13 $|\rho_{A2}|^2$ spectra (a) versus Δ_2 for $\Delta_4 = 0$ and (b) versus Δ_4 for $\Delta_2 = 0$ with $G_2 = 2$ MHz (solid curve), $G_2 = 8$ MHz (dashed curve), $G_3 = 0$, $G_4 = 0.5$ MHz, and $\Delta_1 = \Delta_3 = 0$; and (c) $|\rho_{A2}|^2$ spectrum versus Δ_2 and Δ_3 for $\Delta_1 = \Delta_4 = 0$, $G_2 = G_3 = 2$ MHz, and $G_4 = 0.5$ MHz. The signal intensity with no dressing fields is normalized to be 1.

respectively. For Figure 2.15(a), when the outer dressing field is off-resonance ($\Delta_3 = 200$ MHz), one small enhanced suppression dip at $\Delta_4 = 200$ MHz via the "$\omega_3 + \omega_4$" two-photon resonance (see the inset plot in Figure 2.15(a)) and two big suppressed dips induced by the dressing fields E_4 and E_2 of the nested scheme in $|\rho_{A2}|^2$ appear simultaneously. Especially when $\Delta_3 = 50$ MHz, Figure 2.15(b) shows three suppressed dips that are caused by the same inducement as in Figure 2.11(c). Finally, when $\Delta_3 = 0$, $|\rho_{A1}|^2$ in $|\rho_{F1}'''|^2$ dominates in the FWM signal and the outer dressing field E_4 in $|\rho_{A1}|^2$ creates a suppressed peak at $\Delta_4 = 0$ (Figure 2.15(c)).

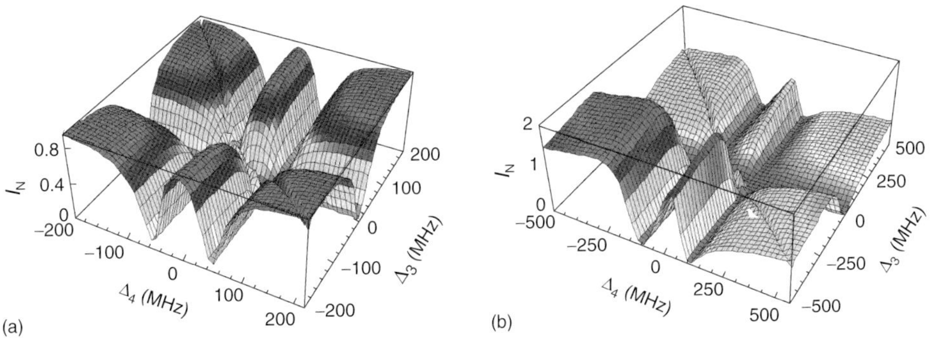

Figure 2.14 Five-dressing FWM signal intensity versus the dressing field detunings Δ_4 and Δ_3 for (a) $\Delta_1 = \Delta_2 = 0$, $G_2 = 50$ MHz, and $G_3 = G_4 = 10$ MHz. (b) $\Delta_1 = 2$ MHz, $\Delta_2 = 0$, $G_2 = 100$ MHz, and $G_3 = G_4 = 20$ MHz. The FWM signal intensity with no dressing fields is normalized to be 1.

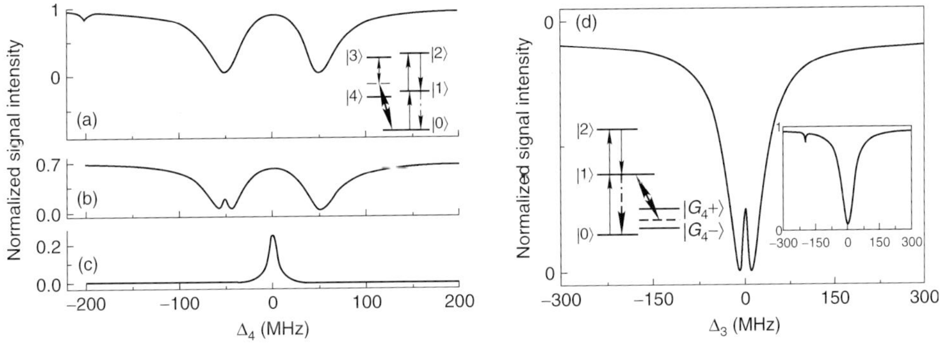

Figure 2.15 Five-dressing FWM signal intensity versus Δ_4 for (a) $\Delta_3 = 200$ MHz. The inserted plot, the corresponding dressed-state picture. (b) $\Delta_3 = 50$ MHz. (c) $\Delta_3 = 0$. (d) Five-dressing FWM signal intensity versus Δ_3 for $\Delta_4 = 0$, the right insert plot $\Delta_4 = 200$ MHz. The left insert plot is the corresponding dressed-state picture. The FWM signal intensity with no dressing fields is normalized to be 1. The other parameters are $\Delta_1 = \Delta_2 = 0$, $G_2 = 50$ MHz, and $G_3 = G_4 = 10$ MHz.

Figure 2.15(d) shows the five-dressing FWM spectra versus Δ_3. Two suppressed dips are induced by the dressing fields E_3 and E_4 in terms of the nested scheme (see the left insert plot of Figure 2.15(d)), which shows the $|\rho_{A1}|^2$ in $|\rho'''_{F1}|^2$ as $\Delta_4 = 0$. When $\Delta_4 = 200$ MHz, the dressing field E_3 creates one suppressed dip near $\Delta_3 = 0$ and the dressing field E_4 creates a small enhanced suppression dip via the "$\omega_3 + \omega_4$" two-photon resonance at $\Delta_3 = 200$ MHz (the right inset plot in Figure 2.15(d)).

Under the off-resonant condition ($\Delta_1 \neq 0$), three dressing fields can either suppress or enhance the FWM, as is shown in Figure 2.14(b). The enhancement effect is especially dramatic in the area of $\Delta_4 < 0$ and $\Delta_3 < 0$ because three dressing fields can reinforce their suppression and enhancement effect each other with the proper offsets.

Previous studies certify that the AT splitting of the sequential dressing scheme is like the nested dressing scheme, while suppression and enhancement spectra of the sequential dressing scheme is like the one in the parallel dressing scheme.

Therefore, in the AT splitting spectra of the five-dressing FWM, three dressing fields (E_2, E_3, and E_4) entangle tightly with each other. It generally is equivalent to the superposition of the two portions ($|\rho_{A1}|^2$ and $|\rho_{A2}|^2$). For the suppression and enhancement in the nested dressing scheme, the inner dressing field suppresses FWM directly, while the outer dressing fields create the dressed states to influence FWM indirectly. Moreover, the suppression and enhancement spectra of the sequential dressing scheme are like the ones in the parallel dressing scheme.

2.3.3.2 **Four-Dressing SWM**

The spectra of the quadruple nested, dressed SWM signal are investigated in this section. Also, we plot the spectra of the AT splitting by adding dressing fields into pure SWM expression one by one. At this stage this method is just a proof of concept theory.

Figure 2.16 presents the SWM signal intensity versus Δ_1 and the dressed-state picture for quadruple dressing. First, as Figure 2.16(b) shows, the inner dressing field E_3 splits the energy level $|1\rangle$ into two primarily dressed states $|G_3 \pm\rangle$ corresponding to two peaks from left to right in Figure 2.16(a1). Then, the outer dressing field E_4 (when $\Delta_4 = 200$ MHz) splits state $|G_3 -\rangle$ level into secondarily dressed levels $|G_3 - \pm\rangle$ corresponding to two right peaks from left to right in Figure 2.16(a2). After that, the outer dressing fields E_3 splits the secondary dressing state $|G_3 - +\rangle$ into triple dressed levels $|G_3 - + \pm\rangle$ corresponding to the second and the third peaks from left to right in Figure 2.16(a3). At last, as the outer dressing field E_2 is tuning close to one of the triply dressed state $|G_3 - + -\rangle$ ($\Delta_2 = -90$ MHz), it dresses to generate the separated quadruple-dressed states $|G_3 - + - \pm\rangle$ corresponding to the third and fourth peaks in Figure 2.16(a4). Hence, the quadruple AT splitting spectra of the SWM signal is obtained, where three dressing fields (E_2, E_3, and E_4) interact tightly.

Next, we investigate the normalized SWM spectra versus the dressing field's offset Δ_4. Under the resonant condition $\Delta_1 = \Delta_3 = 0$, as Figure 2.16(c1) shows, a suppressed peak created by E_4 appears at the line center. The outer dressing field E_3, however, splits such a suppressed peak into a pair of suppressed peaks, as is shown in Figure 2.16(c2). Under the condition $G_3 \gg \Gamma_{30}, \Gamma_{43}$, we can obtain $\Delta_{G_3} = 2G_3 = 100\,\text{MHz}$, where Δ_{G_3} is the separation induced by E_3. Finally, as is shown in Figure 2.16(c3), when the outer dressing field E_2 is tuned close to $\Delta_2 = \Delta_{G_3}/2$, the right suppressed peak generated by E_3 splits again into two new small suppressed peaks.

2.3.3.3 **Four-Dressing EWM**

In this section, we focus on the spectra of AT splitting, suppression, and enhancement of the four-dressing EWM signal.

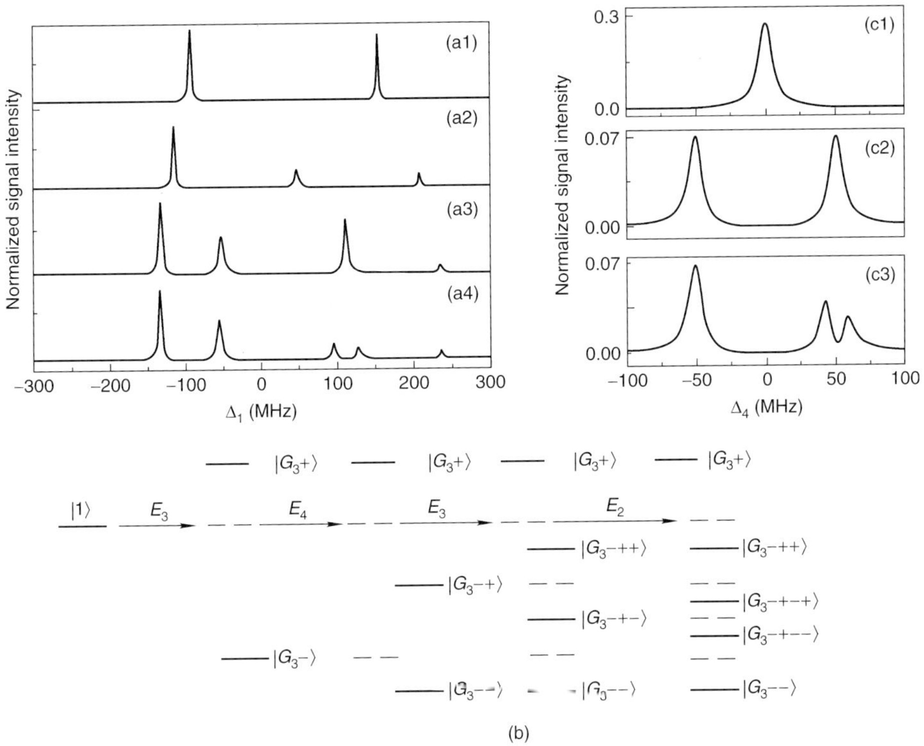

Figure 2.16 Four-dressing SWM signal intensity versus Δ_1 for (a1) $G_3 = 120$ MHz, $G_4 = 0$, $\Delta_3 = 60$ MHz, (a2) $G_4 = 100$ MHz, $\Delta_4 = 20$ MHz, we assume $G_3 = 0$ under the outer dressing field E_4, (a3) $G_3 = 120$ MHz, $G_2 = 0$, and (a4) $G_2 = 30$ MHz, $\Delta_2 = -90$ MHz. (b) The dressed-state picture. Four-dressing SWM signal intensity versus the outer dressing field detuning Δ_4 for (c1) $G_3 = G_4 = 50$ MHz, $\Delta_1 = \Delta_3 = 0$ we assume $G_3 = 0$ below the outer dressing field E_4, (c2) $G_3 = G_4 = 50$ MHz, $\Delta_1 = \Delta_3 = 0$ $G_2 = 0$, and (c3) $G_2 = 10$ MHz, $\Delta_2 = 50$ MHz. The SWM signal intensity with no dressing fields is normalized to be 1.

To clearly understand the AT splitting spectra of the EWM signal, we consider the spectra of two dressing terms separately, as the method applied in analyzing FWM spectra. Figure 2.17(a) and (b) shows the spectra of the second and third multiplying terms of Eq. (2.17), respectively. Figure 2.17(c) which is the superposition of the two portions (E_2, E_4 nested dressing term and E_3 and E_4 nested dressing term), shows the spectra of AT splitting of the EWM signal. So, as to the spectra of the parallel dressing scheme, it can be viewed as the superposition of the two portions.

In addition, Figure 2.18(a) shows the symmetrical full-suppression spectrum of the resonant ($\Delta_1 = 0$) EWM versus Δ_2 and Δ_3. Two suppressed dips in the Δ_2 side are induced by the inner dressing field E_2 and the outer dressing field E_4, while dips in the Δ_3 side are induced by the inner dressing field E_3 and the outer dressing field E_4. Under the off-resonant condition ($\Delta_1 \neq 0$), any dressing fields can either suppress or enhance the EWM signal intensity.

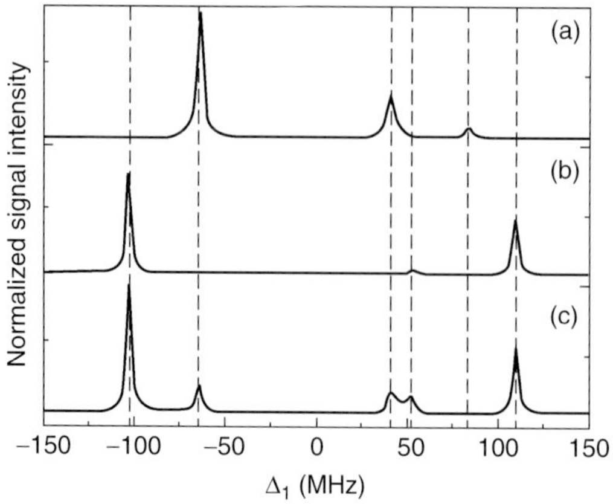

Figure 2.17 Four-dressing EWM signal intensity versus Δ_1 for (a) the spectra of the second multiplying term of Eq. (2.11), (b) the spectra of the third multiplying term of Eq. (2.17), and (c) four-dressing EWM signal spectra. When $G_2 = 60$ MHz, $G_3 = 100$ MHz, $G_4 = 30$ MHz, $\Delta_2 = \Delta_3 = 0$, $\Delta_4 = 60$ MHz.

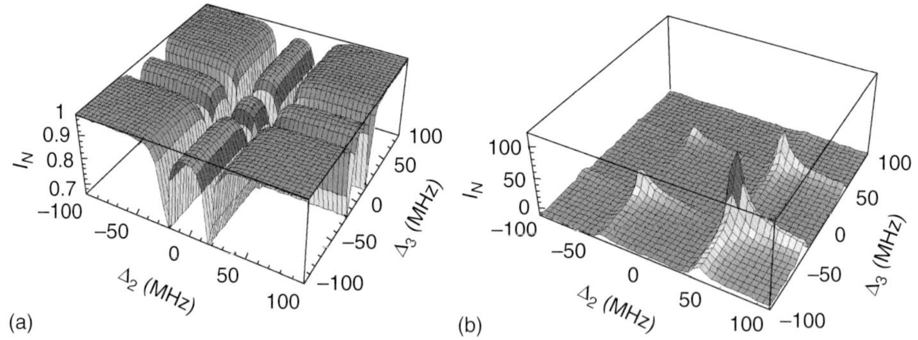

Figure 2.18 Four-dressing EMW suppression and enhancement (a) $G_2 = G_3 = 2$ MHz, $G_4 = 20$ MHz, $\Delta_1 = \Delta_4 = 0$ and (b) $G_2 = G_3 = 20$ MHz, $G_4 = 40$ MHz, $\Delta_1 = 15$ MHz. The EWM signal intensity with no dressing fields is normalized to be 1.

More interestingly, when the dressing fields are strong enough and $|\Delta_1| \gg 0$, a significant full enhancement with an amplitude of about 100 can be obtained, as is shown in Figure 2.18(b). The reason is that the parallel combination nested dressing scheme multiplies the enhanced effect. Further calculation based on the dressed-state theory gives the resonant enhanced peak's location, which is at $\Delta_2 = -\Delta_1 + \Delta_{G_2}/2 + \Delta_4 \pm \Delta_{G_4}/2 \approx \pm 50\,\text{MHz}$ and $\Delta_3 = \Delta_1 - \Delta_{G_3}/2 - \Delta_4 \pm \Delta_{G_4}/2 \approx \pm 50\,\text{MHz}$, respectively. It means that these dressing fields can enhance the EWM signal constructively with proper offsets. Therefore, whether the AT splits, or not, the suppression and enhancement of the EWM signal are generally equivalent to the superposition of the two portions (E_2, E_4 nested dressing term and E_3 and E_4 nested dressing term). Moreover, as with the nested dressed structure, the inner dressing fields (E_2 and E_3) suppress the EWM signal, while the outer dressing field (E_4) affects the EWM signal indirectly.

2.3.3.4 Absorption and Dispersion in the Four-Dressing EIT System

The four-dressing EIT windows induced by interacting dark states are used to transmit the coexisting FWM, SWM, and EWM signals. It is valuable to examine the coherence term ρ_{10} (Eq. (2.18)) in terms of its real and imaginary parts as a function of Δ_1 because the imaginary (real) part of ρ_{10} versus Δ_1 represents the probe absorption (dispersion) spectrum.

Also, we plot the spectra of EIT by adding dressing fields one by one. First, Figure 2.19(a1) shows the spectra of the primary EIT splitting, which is induced by the dressing field E_2. Second, Figure 2.19(a2) presents the spectra of the secondary EIT splitting, which is induced by the dressing field E_3. Third, Figure 2.19(a3) and (a4) shows the spectra of two triple EIT splitting, the former is induced by dressing

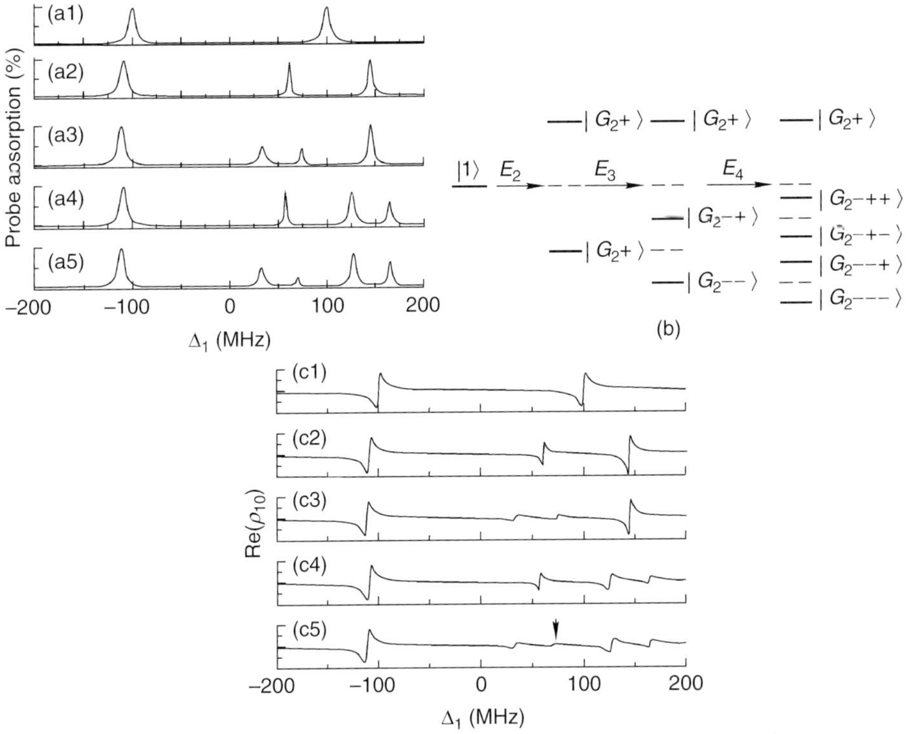

Figure 2.19 Im(ρ_{10}) as a function of Δ_1 (a1) $G_2 = 100$ MHz and $\Delta_2 = 0$ MHz, when $G_3 = 0$ MHz, $G_4 = 0$ MHz, (a2) $G_3 = 50$ MHz, $\Delta_3 = 100$ MHz, when $G_4 = 0$; (a3) $G_4 = 30$ MHz (G_4 under G_2) and $\Delta_4 = 50$ MHz, when the other $G_4 = 0$; (a4) $G_4 = 30$ MHz (G_4 under G_3), $\Delta_4 = 50$ MHz, when $G_4 = 0$ under G_2; and (a5) $G_4 = 30$ MHz, $\Delta_4 = 0$ MHz. (b) The dressed-state picture for EIT system. Re (ρ_{10}) as a function of Δ_1 show in (c1)–(c5) with all other conditions same as in Figures 2.19 (a1)–(a5), respectively.

field E_4 under E_2, and the latter is induced by dressing field E_4 under E_3. When all the dressing fields are acted upon, we can obtain the probe absorption spectra as shown in Figure 2.19(a5) and the corresponding dressed states' picture is shown in Figure 2.19(b). Figure 2.19(a5) shows two triple EIT splitting and totally gives four EIT windows.

As to the dispersion curve, as Figure 2.19(c1) to (c5) shows, the spectra of EIT dispersion (Re (ρ_{10})) with the same parameters in Figure 2.19(a1) to (a5). Figure 2.19(c1) shows the typical EIT dispersion curve in the $|0\rangle - |1\rangle - |2\rangle$ subsystem and Figure 2.19(c2) represents the EIT dispersion in the four-level system with the inverted-Y configuration. It displays a secondary EIT splitting. In Figure 2.19(c3) and (c4), which shows the dispersion spectra of triple EIT splitting, the most prominent change appears because of three EIT conditions. We can see that the absorption spectrum exhibits four absorptive peaks in Figure 2.19(a3) and (a4), while the corresponding dispersion profile exhibits three steep pure slopes, which should be useful for supporting slow light pulses with different frequencies. Figure 2.19(c5) presents the dispersion spectra of two triple EIT splitting and exhibits four steep pure slopes.

Next, we discuss the EIT absorption and dispersion curves by changing the offsets of the dressing fields. First, as is shown in Figure 2.20(a), as setting $\Delta_4 = 150$ MHz, the secondarily dressed absorptive peak (the right peak in Figure 2.19(a2)) is separated into two triple dressed peaks (the third and the forth peaks in Figure 2.20(a) from left to right) by one dressing field E_4, while the other dressing field E_4 (under G_3 in Eq. (2.17)) induces the three-photon absorption peak appearing at $\Delta_1 = \Delta_3 + \Delta_4 = 250$ MHz (Figure 2.20(a)). Second, when

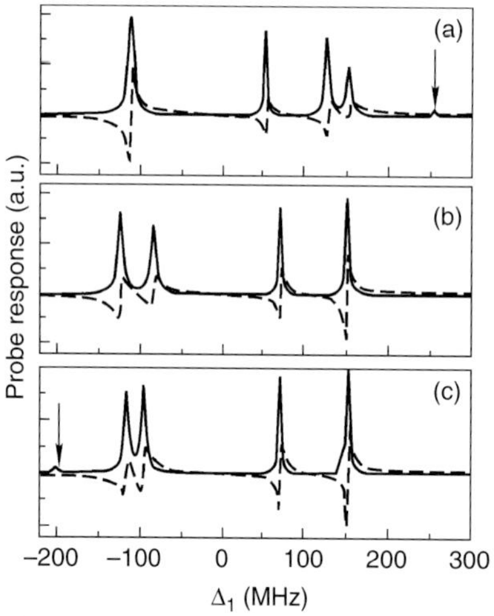

Figure 2.20 Im (ρ_{10}) as a function of the probe frequency detuning (a) $\Delta_4 = 150$ MHz, (b) $\Delta_4 = -100$ MHz, and (c) $\Delta_4 = -200$ MHz. Other parameters are $G_2 = 100$ MHz, $G_3 = 50$ MHz, $G_4 = 50$ MHz, $\Delta_2 = 0$, and $\Delta_3 = 100$ MHz. The dashed lines are dispersion profiles.

$\Delta_4 = \Delta_{G_2}/2 = -100\,\text{MHz}$, the left absorptive peak in Figure 2.20(a) is separated into two peaks in Figure 2.20(b), while the dressing field E_4 (under G_3 in Eq. (2.17)) generates a very small three-photon absorption peak at $\Delta_1 = \Delta_3 + \Delta_4 = 0$ (Figure 2.20(b)). Finally, as Figure 2.20(c) shows, when $\Delta_4 = -200$ MHz, the left absorption peak in Figure 2.20(a) splits into two peaks by the dressing field E_4 (under G_3 in Eq. (2.17)), and the three-photon absorption peak is generated by the dressing field E_4 (under G_2 in Eq. (2.17)) at $\Delta_1 = -\Delta_2 + \Delta_4 = -200$ MHz. The dashed curves in Figure 2.20 also show the quadruple-dressed dispersion corresponding to the solid absorption. So, a different multi-EIT phenomenon is obtained by changing the offsets of the dressing fields, and, moreover, the locations can be controlled as well as the shape/height of the EIT profile of $\text{Im}(\rho_{10})$ and $\text{Re}(\rho_{10})$ in other ways in this four-dressing EIT system.

2.3.4
Discussion and Conclusion

We investigate the multi-dressed FWM, SWM, and EWM in the section, which can coexist in different EIT windows. In the five-level system we investigated, when the five laser beams (E_1, E_2', E_2, E_3, and E_4) are on, there exist three different EIT windows, that is, $|0\rangle \to |1\rangle \to |2\rangle$ (ladder-type) with two counter-propagation beams E_2 (and E_2') and E_1, $|0\rangle \to |1\rangle \to |3\rangle$ (Λ-type) with two counter-propagation beams E_3 and E_1, and $|0\rangle \to |1\rangle \to |4\rangle$ (V-type) with two counter-propagation beams E_4 and E_1. Moreover, there also exist one multi-dressed FWM ($\mathbf{k}_f = \mathbf{k}_1 + \mathbf{k}_2 - \mathbf{k}_2'$), two multi-dressed SWM ($\mathbf{k}_s = \mathbf{k}_1 + \mathbf{k}_2 - \mathbf{k}_2' + \mathbf{k}_4 - \mathbf{k}_4$ and $\mathbf{k}_{s1} = \mathbf{k}_1 + \mathbf{k}_2 - \mathbf{k}_2' + \mathbf{k}_3 - \mathbf{k}_3$), and one multi-dressed EWM ($\mathbf{k}_e = \mathbf{k}_1 + \mathbf{k}_2 - \mathbf{k}_2' + \mathbf{k}_3 - \mathbf{k}_3 + \mathbf{k}_4 - \mathbf{k}_4$) in the five-level system. They are in the same direction determined by the phase-matching conditions, as shown in Figure 2.9(b), which enable us to investigate the coexisting, intermixing, and temporal and spatial interference between FWM and SWM, or even to EWM. We accomplish this by controlling the offsets and intensities of dressing fields for the aim of MWM coexisting in different EIT windows, which could be individually controlled and the generated MWM signals can be clearly separated and distinguished or pulled together (by frequency offsets) to observe interferences and competitions between them. The relative strengths of the FWM, SWM, and EWM can be adjusted freely by controlling the intensities of the dressed fields (via dressed states). Therefore, the SWM or EWM signal can be enhanced to the same order as the FWM signal. To investigate the efficient energy transfer between FWM ($\mathbf{k}_f' = \mathbf{k}_1 + \mathbf{k}_2 - \mathbf{k}_2'$) and SWM ($\mathbf{k}_s' = \mathbf{k}_1 + \mathbf{k}_2 - \mathbf{k}_2' + \mathbf{k}_3 - \mathbf{k}_3'$), or EWM ($\mathbf{k}_e' = \mathbf{k}_1 + \mathbf{k}_2 - \mathbf{k}_2' + \mathbf{k}_3 - \mathbf{k}_3' + \mathbf{k}_4 - \mathbf{k}_4'$), we can set the different spatial beam geometry, and then the MWM signal propagates different directions because of the phase-matching condition.

In this section, we present the controlled MWM via interacting dark states in a five-level system. The spectra of the AT splitting, suppression, and enhancement of the MWM signal, which are plotted by adding dressing fields one by one into pure MWM expression without any dressing terms, are well described. This method is just a proof of concept theory, and will help us understand the effects of each of the dressing terms and their interactions. As a result, we obtain triple AT splitting in the spectra of the FWM signal, quadruple AT splitting in the SWM signal, and two triple splitting in EIT. The spectra of AT splitting, suppression, and enhancement of the FWM and EWM signals are the superposition of two groups of the different AT splitting peaks. By controlling intensities and frequency offsets of laser fields, these higher-order nonlinear wave-mixing processes can be enhanced and suppressed to obtain the desired magnitude of nonlinearities. Understanding the higher order, multi-channel nonlinear optical processes can help us optimize these higher order, nonlinear optical processes, which have potential applications in optical communication and quantum information processing.

2.4
Observation of Dressed Odd-Order MWM

In this section, we show seven coexisting distinguishable MWM signals (including three FWM and four SWM signals) by selectively blocking different laser beams in a K-type five-level atomic system. Also, by blocking certain laser beams, the interactions among six MWM signals were studied. In addition, when scanning the frequency detuning of external-dressing, self-dressing, and probe fields respectively in the dressed FWM process, we analyze the corresponding relationship and differentiate between the experimental results of different scanning methods, and demonstrate that scanning the dressing field can be used as a technique to directly observe the dressing effects of the FWM process. Also, we discuss the enhancement and suppression of the SWM signal at large detuning, due to the atomic velocity component and optical pumping effect. Moreover, we demonstrate the temporal and spatial interferences between two FWM signals.

2.4.1
Basic Theory and Experimental Scheme

The experiments are performed in a five-level atomic system as shown in Figure 2.21(b) where the five energy levels are $5S_{1/2}(F=3)$ ($|0\rangle$), $5P_{3/2}$ ($|1\rangle$), $5D_{5/2}$ ($|2\rangle$), $5S_{1/2}(F=2)$ ($|3\rangle$), and $5D_{3/2}$ ($|4\rangle$) in ^{85}Rb. The resonant frequencies are Ω_1, Ω_2, Ω_3, and Ω_4 for transitions $|0\rangle$ to $|1\rangle$, $|1\rangle$ to $|2\rangle$, $|1\rangle$ to $|3\rangle$, and $|1\rangle$ to $|4\rangle$ respectively. The lower transition $|0\rangle$ to $|1\rangle$ is driven and probed by a weak laser beam E_1, and the two strong coupling beams E_2 and E'_2 connect with the transition $|1\rangle$ to $|2\rangle$. Two pumping beams E_3 and E'_3 are applied to drive the transition $|1\rangle$ to $|3\rangle$ and two additional strong coupling beams E_4 and E'_4 drive the transition $|1\rangle$ to $|4\rangle$. In the experimental setup, the two coupling beams E_2 (frequency ω_2, wave vector $\mathbf{k}_2$, Rabi frequency G_2, and frequency detuning Δ_2, where $\Delta_i = \Omega_i - \omega_i$) and E'_2 (ω_2, $\mathbf{k}'_2$, G'_2, and Δ_2) with vertical polarization (wavelength 775.978 nm) are from the same ECDL. The other two vertically polarized coupling beams E_4 (ω_4, $\mathbf{k}_4$, G_4, and Δ_4) and E'_4 (ω_4, $\mathbf{k}'_4$, G'_4, and Δ_4) are from a tapered-amplifier diode laser with the same wavelength of 776.157 nm. Two pumping beams E_3(ω_3, $\mathbf{k}_3$, G_3, and Δ_3) and E'_3 (ω_3,

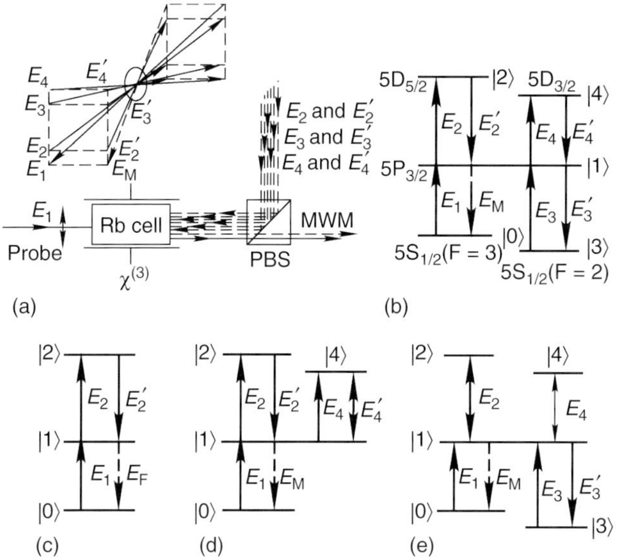

Figure 2.21 (a) Spatial beam geometry used in the experiment. (b) The diagram of the K-type five-level atomic system. (c) The diagram of the ladder-type three-level atomic subsystem when the fields E_3, E_3', E_4, and E_4' are blocked and E_1, E_2, and E_2' are turned on. (d) The diagram of the Y-type four-level atomic subsystem when the fields E_3 and E_3' are blocked and other beams are turned on. (e) The diagram of the K-type five-level atomic system when only the fields E_2' and E_4' are blocked only.

$\mathbf{k}_3'$, G_3', and Δ_3) with equal power are split from a laser diode (LD) beam and have polarizations vertical to each other via a polarization beam splitter. The probe beam E_1 (ω_1, $\mathbf{k}_1$, G_1, and Δ_1) is generated by an ECDL (Toptica DL100L) with horizontal polarization. These laser beams are spatially designed in a square-box pattern as shown in Figure 2.21(a), in which the laser beams E_2, E_2', E_3, E_3', E_4, and E_4' propagate through the Rubidium vapor cell with a temperature of 60 °C in the same direction with small angles (about 0.3°) between one another, and the probe beam E_1 propagates in the opposite direction with a small angle from the other beams.

In such a beam geometric configuration, the two-photon Doppler-free conditions will be satisfied for the two ladder-type subsystems both $|0\rangle \to |1\rangle \to |2\rangle$ and $|0\rangle \to |1\rangle \to |4\rangle$, thus, two EIT windows appear. When all the seven laser beams (E_1, E_2, E_2', E_3, E_3', E_4, E_4') are turned on, three FWM processes E_{F2} (satisfying the phase-matching condition $\mathbf{k}_{F2} = \mathbf{k}_1 + \mathbf{k}_2 - \mathbf{k}_2'$), E_{F4} (satisfying $\mathbf{k}_{F4} = \mathbf{k}_1 + \mathbf{k}_4 - \mathbf{k}_4'$) and E_{F3} (satisfying $\mathbf{k}_{F3} = \mathbf{k}_1 - \mathbf{k}_3' + \mathbf{k}_3$), and four SWM processes E_{S2} (satisfying $\mathbf{k}_{S2} = \mathbf{k}_1 - \mathbf{k}_3' + \mathbf{k}_3 + \mathbf{k}_2 - \mathbf{k}_2$), E_{S2}' (satisfying $\mathbf{k}_{S2}' = \mathbf{k}_1 - \mathbf{k}_3' + \mathbf{k}_3 + \mathbf{k}_2' - \mathbf{k}_2'$), E_{S4} (satisfying $\mathbf{k}_{S4} = \mathbf{k}_1 - \mathbf{k}_3' + \mathbf{k}_3 + \mathbf{k}_4 - \mathbf{k}_4$), and E_{S4}' (satisfying $\mathbf{k}_{S4}' = \mathbf{k}_1 - \mathbf{k}_3' + \mathbf{k}_3 + \mathbf{k}_4' - \mathbf{k}_4'$) can occur simultaneously. The propagation direction of all the generated signals with horizontal polarization is determined by the phase-matching conditions, so all the signals propagate along the same direction deviated from the probe beam at an angle θ, as shown in Figure 2.21(a). The wave-mixing signals are detected by an avalanche photodiode detector, and the probe beam transmission is simultaneously detected by a silicon photodiode.

Generally, the expression of the density matrix element related to the MWM signals can be obtained by solving the density matrix equations. For the simple FWM process of E_{F2}, via the perturbation chain $\rho_{00}^{(0)} \xrightarrow{\omega_1} \rho_{10}^{(1)} \xrightarrow{\omega_2} \rho_{20}^{(2)} \xrightarrow{-\omega_2} \rho_{10}^{(3)}$ we can obtain the third-order density element $\rho_{F2}^{(3)} = G_{F2}/(d_1^2 d_2)$, the amplitude of which determines the intensity of the simple FWM process, where $G_{F2} = -iG_1G_2(G_2')^* \exp(i\mathbf{k}_{F2} \cdot \mathbf{r})$, $d_1 = \Gamma_{10} + i\Delta_1$, $d_2 = \Gamma_{20} + i(\Delta_1 + \Delta_2)$, and Γ_{ij} is the transverse relaxation rate between states $|i\rangle$ and $|j\rangle$. Similarly, for the simple FWM process of E_{F4}, we can obtain $\rho_{F4}^{(3)} = G_{F4}/(d_1^2 d_4)$ via $\rho_{00}^{(0)} \xrightarrow{\omega_1} \rho_{10}^{(1)} \xrightarrow{\omega_4} \rho_{40}^{(2)} \xrightarrow{-\omega_4} \rho_{10}^{(3)}$, where $G_{F4} = -iG_1G_4(G_4')^* \exp(i\mathbf{k}_{F4} \cdot \mathbf{r})$, $d_4 = \Gamma_{40} + i(\Delta_1 + \Delta_4)$. And for the simple FWM process of E_{F3}, we can obtain $\rho_{F3}^{(3)} = G_{F3}/(d_1^2 d_4)$ via $\rho_{00}^{(0)} \xrightarrow{\omega_1} \rho_{10}^{(1)} \xrightarrow{-\omega_3} \rho_{30}^{(2)} \xrightarrow{\omega_3} \rho_{10}^{(3)}$, where $G_{F3} = -iG_1G_3(G_3')^* \exp(i\mathbf{k}_{F3} \cdot \mathbf{r})$, $d_3 = \Gamma_{30} + i(\Delta_1 - \Delta_3)$. In addition, via perturbation chain $\rho_{00}^{(0)} \xrightarrow{\omega_1} \rho_{10}^{(1)} \xrightarrow{-\omega_3} \rho_{30}^{(2)} \xrightarrow{\omega_3} \rho_{10}^{(3)} \xrightarrow{\omega_2} \rho_{20}^{(4)} \xrightarrow{-\omega_2} \rho_{10}^{(5)}$ we obtain $\rho_{S2}^{(5)} = G_{S2}/(d_1^3 d_2 d_3)$ for the simple SWM process of E_{S2}, where $G_{S2} = iG_1G_2G_2^*G_3'^*G_3 \exp(i\mathbf{k}_{S2} \cdot \mathbf{r})$ (or $\rho_{S2}^{\prime(5)} = G_{S2}'/(d_1^3 d_2 d_3)$ for E_{S2}', where $G_{S2}' = iG_1G_2'G_2'^*G_3'^*G_3 \exp(i\mathbf{k}_{S2}' \cdot \mathbf{r})$). And via $\rho_{00}^{(0)} \xrightarrow{\omega_1} \rho_{10}^{(1)} \xrightarrow{-\omega_3} \rho_{30}^{(2)} \xrightarrow{\omega_3} \rho_{10}^{(3)} \xrightarrow{\omega_4} \rho_{40}^{(4)} \xrightarrow{-\omega_4} \rho_{10}^{(5)}$ we obtain $\rho_{S4}^{(5)} = G_{S4}/(d_1^3 d_4 d_3)$ for the simple SWM process of E_{S4}, where $G_{S4} = iG_1G_4G_4^*G_3'^*G_3 \exp(i\mathbf{k}_{S4} \cdot \mathbf{r})$ (or $\rho_{S4}^{\prime(5)} = G_{S4}'/(d_1^3 d_4 d_3)$ for E_{S4}', where $G_{S4}' = iG_1G_4'G_4'^*G_3'^*G_3 \exp(i\mathbf{k}_{S4}' \cdot \mathbf{r})$). Further, the dressing effect on these MWM signals and the interaction among them will be researched in the following section.

In the experiments, these MWM signals are researched by selectively blocking different laser beams. When the beams E_3, E_3', E_4, and E_4' are blocked and E_1, E_2, and E_2' are turned on (as shown in Figure 2.21(c)), only the E_{F2} signal with self-dressing effect could be generated in the ladder-type three-level subsystem $|0\rangle \to |1\rangle \to |2\rangle$. When the beams E_3 and E_3' are blocked and other beams on (Figure 2.21(d)), two FWM signals E_{F2} and E_{F4} will coexist in the Y-type four-level subsystem; both of them will be perturbed by the self-dressing effect and external-dressing effect, and spatiotemporal coherent interference and interaction between them will be observed. When the beams E_2' and E_4' are blocked only (Figure 2.21(e)), two SWM signals E_{S2} and E_{S4} will coexist and interact with each other.

2.4.2 Dressed Odd-Order MWM

By individually adjusting the frequency detuning Δ_2 and Δ_4, these generated wave-mixing signals can be separated in spectra for the identification, or be overlapped for investigating the interplay among them. First, by detuning the frequency of the participating laser beams and blocking one or two participating laser beams, we can successfully separate two EIT windows and these MWM signals can be identified. Figure 2.22(a) to (e) present the measured signals versus the probe detuning Δ_1 with different laser beams blocked, in which the lower curves are

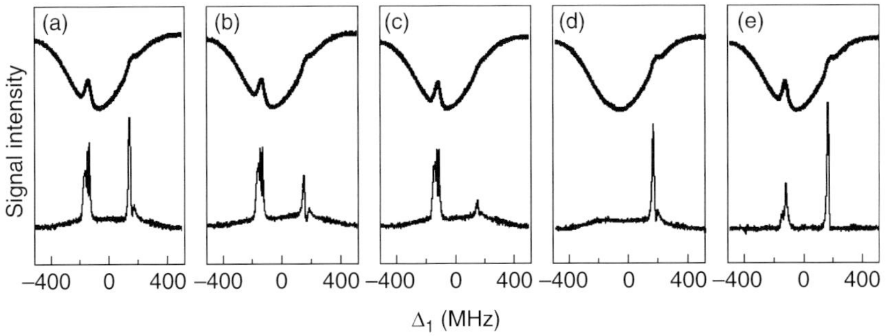

Figure 2.22 The probe transmission (the upper curves) and measured MWM signals (the bottom curves) with different ways of blocking laser beams. The left peaks show the EIT window and MWM signals related to $E_4(E_4')$ and the right peaks are related to $E_2(E_2')$. (a) Measured total MWM signals with all beams turned on. (b) Measured MWM signals related to $E_4(E_4')$ and SWM signal E_{S2}, with E_2' blocked. (c) Measured MWM signals related to $E_4(E_4')$ and SWM signal E_{S2}', with E_2 blocked. (d) Measured MWM signals related to $E_2(E_2')$ with E_4 and E_4' blocked. (e) Measured FWM signal E_{F2} and E_{F4} with E_3 and E_3' blocked. The experimental parameters are $\Delta_2 = -150$ MHz, $\Delta_4 = 140$ MHz, $\Delta_3 = 0$ MHz, and the powers of all laser beams are 5.3 mW (E_1), 40.9 mW (E_2), 5 mW (E_2'), 44 mW (E_3 and E_3'), 21 mW (E_4 and E_4'). Source: Adapted from Ref. [36].

the measured MWM signals while the corresponding probe transmission signals versus Δ_1 are shown in the upper curves. Figure 2.22(a) depicts the case when all the participating laser beams are turned on, and Figure 2.22(b) to (e) show the measured signals when the laser beams or beam combinations E_2', E_2, E_4 and E_4', and E_3 and E_3' are blocked. In the upper curves, the left EIT window is created by $E_4(E_4')$ at $\Delta_1 = -140$ MHz, satisfying $\Delta_1 + \Delta_4 = 0$, and the right EIT window is created by $E_2(E_2')$ at $\Delta_1 = 150$ MHz, satisfying $\Delta_1 + \Delta_2 = 0$. The generated FWM and SWM signals, except for E_{F3}, all fall into these two separate EIT windows, so the linear absorptions of the generated signals are greatly suppressed. Specifically, the MWM signals related to $E_4(E_4')$ (i.e., E_{F4}, E_{S4}, and E_{S4}') fall into the $|0\rangle \rightarrow |1\rangle \rightarrow |4\rangle$ EIT window, while the MWM signals related to $E_2(E_2')$ (i.e., E_{F2}, E_{S2}, and E_{S2}') fall into the $|0\rangle \rightarrow |1\rangle \rightarrow |2\rangle$ EIT window. The FWM signal E_{F3} can be obtained in Figure 2.22(a)–(d), and it appears as a Doppler-broadened background signal because the Doppler-free condition cannot be satisfied in the FWM process generating E_{F3} with opposite direction between E_1 and $E_3(E_3')$. By comparing Figure 2.22(a) with (d), we find the interaction between MWM signals related to $E_4(E_4')$ and those related to $E_2(E_2')$ does not exist when the two EIT widows are separated from each other, as the intensity of the MWM signals related to $E_2(E_2')$ behaves identically in both curves.

Moreover, the interaction between the FWM and SWM processes in the same EIT window of $E_2(E_2')$ can be observed, by comparing the total MWM signal related to $E_2(E_2')$ (Figure 2.22(a)) with the sum of FWM signal E_{F2} (Figure 2.22(e)), SWM signals E_{S2} (Figure 2.22(b)), and E_{S2}' (Figure 2.22(c)) in amplitude. We can find that the MWM signal is suppressed by 40%, which shows the interaction and competition between FWM and SWM when they coexist. This phenomenon can be explained by the dressing effect of $E_3(E_3')$ on FWM signal E_{F2}. The dressed FWM process can be described by $\rho_{00}^{(0)} \xrightarrow{\omega_1} \rho_{(G_3\pm)0}^{(1)} \xrightarrow{\omega_2} \rho_{20}^{(2)} \xrightarrow{-\omega_2} \rho_{(G_3\pm)0}^{(3)}$, and we obtain $\rho_{F2}^{(3)} = G_{F2}/[d_2(d_1 + |G_3^b|^2/d_3)^2]$, where $G_3^b = G_3 + G_3'$. And the density matrix element related to the SWM processes can be obtained as $\rho_{S2}^{(5)} = G_{S2}/(d_1^3 d_2 d_3)$ and $\rho_{S2}'^{(5)} = G_{S2}'/(d_1^3 d_2 d_3)$ (for simplicity, the self-dressing effects on E_{F2}, E_{S2}, and E_{S2}' are not considered here). As these MWM processes exist at the same time in the experiment, and the signals are copropagating in the same direction, the total detected MWM signal (Figure 2.22(a)) will be proportional to the mod square of ρ_M, where $\rho_M = \rho_{S2}^{(5)} + \rho_{S2}'^{(5)} + \rho_{F2}^{(3)}$.

Next, we investigate the singly-dressed FWM process in the ladder-type three-level subsystem $|0\rangle \rightarrow |1\rangle \rightarrow |2\rangle$ (shown as Figure 2.21(c)) when only the laser beams E_1, E_2, and E_2' are turned on. In this three-level subsystem, only the wave-mixing signal E_{F2} is generated, with a self-dressing effect of $E_2(E_2')$. According to the perturbation chain of the self-dressed FWM process $\rho_{00}^{(0)} \xrightarrow{\omega_1} \rho_{(G_2\pm)0}^{(1)} \xrightarrow{\omega_2} \rho_{20}^{(2)} \xrightarrow{-\omega_2} \rho_{(G_2\pm)0}^{(3)}$, we obtain $\rho_{F2}^{(3)} = G_{F2}/[d_2(d_1 + |G_2^b|^2/d_2)^2]$ for this singly-dressed FWM process, where $G_2^b = G_2 + G_2'$.

The spectra of the singly-dressed FWM process are shown in Figure 2.23. Figure 2.23(a1) and (a2), respectively, presents the intensities of the probe transmission (Figure 2.23(a1)) and E_{F2} (Figure 2.23(a2)) versus Δ_2 at discrete Δ_1 values. Figure 2.23(a3) and (a4), respectively, depicts the intensities of probe transmission (Figure 2.23(a3)) and E_{F2} (Figure 2.23(a4)) versus Δ_1 at discrete Δ_2 values, and the Doppler broadening of the probe transmission signal in Figure 2.23(a3) has been subtracted. Figure 2.23(b1), (b3), (b4), and (b5) are the theoretical calculations corresponding to Figure 2.23(a1) to (a4), and Figure 2.23(b2) represents the theoretical enhancement and suppression of E_{F2}, which is depicted as the peak and dip on each baseline of the curves, by the self-dressing effect. Notice the experimentally obtained E_{F2} signal when scanning Δ_2 (Figure 2.23(a2)) includes two components, the pure FWM signal when not considering the dressing effect, and the modification (enhancement and suppression) of the FWM process that is theoretically shown in Figure 2.24(b2). Figure 2.23(c) shows the singly-dressed energy level diagrams corresponding to the curves at discrete frequency detuning in Figure 2.23(a).

When scanning Δ_2, the FWM signal shows the evolution from pure enhancement ($\Delta_1 = -80$ MHz), to first enhancement and next suppression ($\Delta_1 = -20$ MHz), to pure suppression ($\Delta_1 = 0$), to first suppression and next enhancement ($\Delta_1 = 20$ MHz), to pure enhancement ($\Delta_1 = 80$ MHz), as shown in Figure 2.23(b2). And the corresponding probe transmission shows the evolution from pure EIA, to first EIA and next EIT, to pure EIT, to first EIT and next EIA, finally to pure EIA in series as shown in Figure 2.23(a1). The height of each baseline of the curves represents the probe transmission without the dressing effect of $E_2(E_2')$ versus probe detuning Δ_1, while the peak and dip on each baseline represent EIT and EIA, respectively. We can see that every enhancement

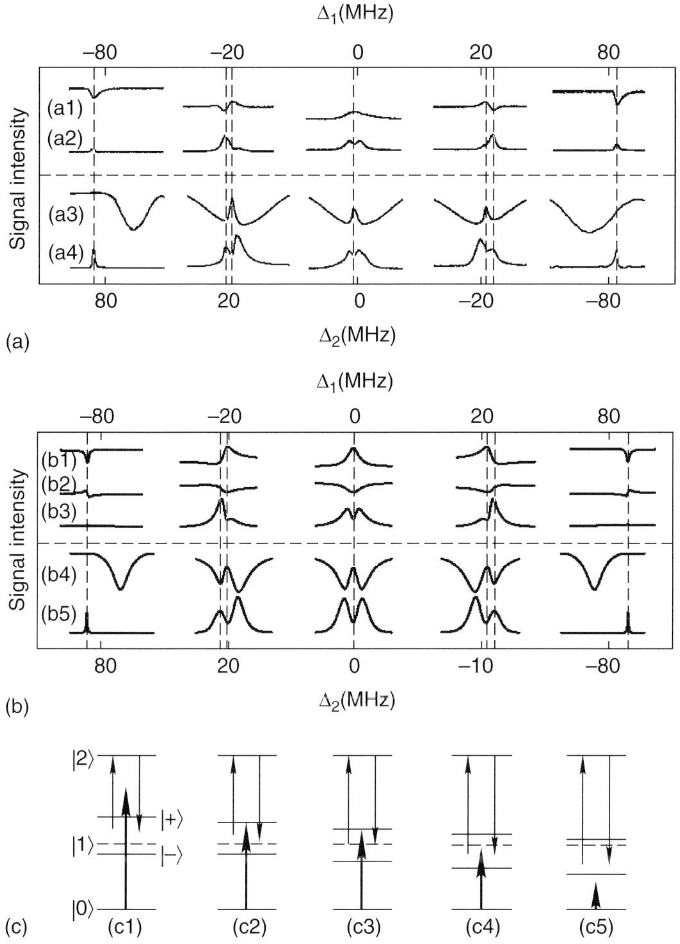

Figure 2.23 (a) The measured intensity of (a1) the probe transmission and (a2) the singly-dressed FWM signal E_{F2} versus Δ_2 at discrete probe detunings $\Delta_1 = -80$, -20, 0, 20, and 80 MHz, and the measured intensity of (a3) the probe transmission and (a4) the singly-dressed FWM signal E_{F2} versus Δ_1 at discrete dressing detuning $\Delta_2 = 80$, 20, 0, -20, and -80 MHz. (b1), (b3), (b4), (b5) are theoretical calculations corresponding to (a1)–(a4). (b2) is the theoretical calculations of enhancement and suppression of singly-dressed E_{F2}. (c) The dressed energy level diagrams corresponding to (a). Powers of participating laser beams are 4 mW (E_1), 34.5 mW (E_2), and 8.7 mW (E_2').

and suppression correspond to EIA and EIT, respectively, and the curves show symmetric behavior.

In order to understand the above-mentioned phenomena, we resort to the singly-dressed energy level diagrams in Figure 2.23(c). With the self-dressing effect of $E_2(E_2')$, the energy level $|1\rangle$ will be split into two dressed states $|G_2\pm\rangle$, as shown in Figure 2.23(c1)–(c5). When Δ_2 is scanned at $\Delta_1 = 0$, on the one hand,

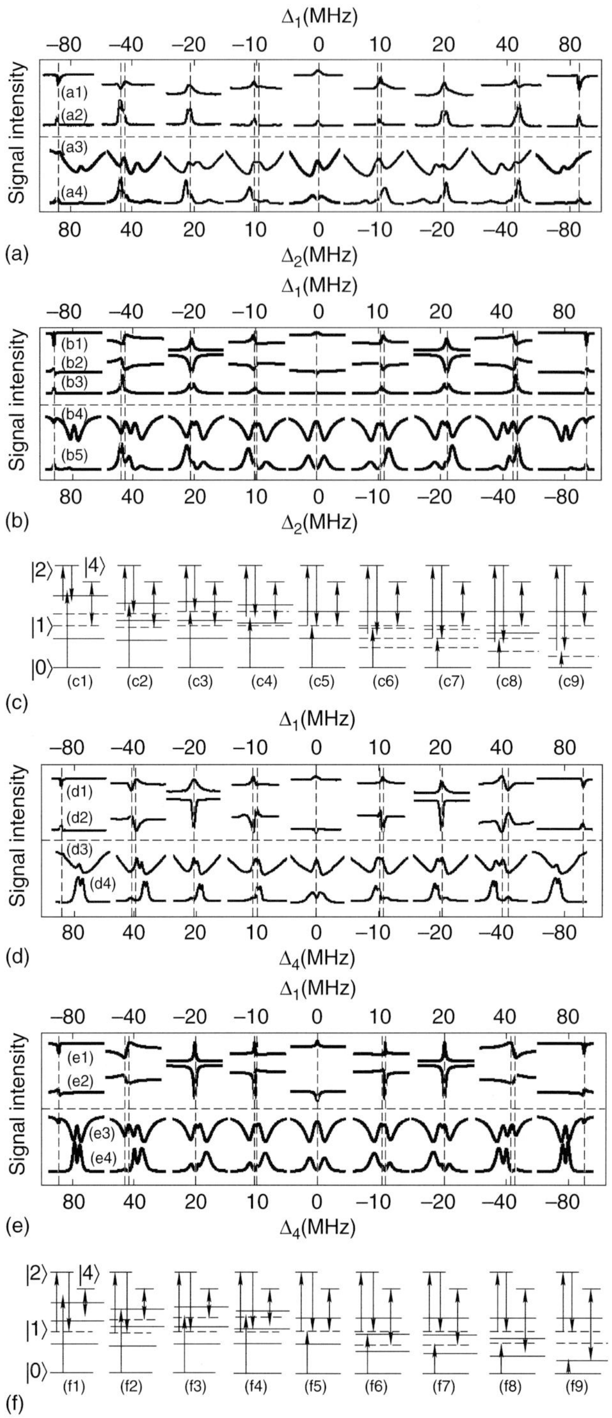

Δ1(MHz)
−80 −40 −20 −10 0 10 20 40 80
Signal intensity
(a1)
(a2)
(a3)
(a4)
80 40 20 10 0 −10 −20 −40 −80
(a)
Δ2(MHz)
Δ1(MHz)
−80 −40 −20 −10 0 10 20 40 80
Signal intensity
(b1)
(b2)
(b3)
(b4)
(b5)
80 40 20 10 0 −10 −20 −40 −80
(b)
Δ2(MHz)
|2⟩
|4⟩
|1⟩
|0⟩
(c1) (c2) (c3) (c4) (c5) (c6) (c7) (c8) (c9)
(c)
Δ1(MHz)
−80 −40 −20 −10 0 10 20 40 80
Signal intensity
(d1)
(d2)
(d3)
(d4)
80 40 20 10 0 −10 −20 −40 −80
(d)
Δ4(MHz)
Δ1(MHz)
−80 −40 −20 −10 0 10 20 40 80
Signal intensity
(e1)
(e2)
(e3)
(e4)
80 40 20 10 0 −10 −20 −40 −80
(e)
Δ4(MHz)
|2⟩
|4⟩
|1⟩
|0⟩
(f1) (f2) (f3) (f4) (f5) (f6) (f7) (f8) (f9)
(f)

EIT is obtained in Figure 2.23(a1) at the point $\Delta_2 = 0$ where the suppression condition $\Delta_1 + \Delta_2 = 0$ is satisfied. On the other hand, a pure suppression of E_{F2} is obtained in Figure 2.23(b2) because the probe field E_1 could not resonate with either of the two dressed energy levels $|G_2 \pm\rangle$, as shown in Figure 2.23(c3). In the region with $\Delta_1 < 0$, when Δ_2 is scanned, the probe transmission shows EIA first and EIT afterwards in Figure 2.23(a1) at $\Delta_1 = -20$ MHz. Correspondingly, E_{F2} is first enhanced when the EIA is obtained and next suppressed when the EIT is obtained, shown in Figure 2.23(b2) at $\Delta_1 = -20$ MHz. The reason for the first EIA and the corresponding enhancement of E_{F2} is that the probe field E_1 resonates with the dressed state $|G_2+\rangle$ at first, thus the enhancement condition $\Delta_1 + (\Delta_2 + \sqrt{\Delta_2^2 + 4|G_2^b|^2})/2 = 0$ is satisfied. While the reason for the next EIT and the corresponding suppression of E_{F2} is that two-photon resonance occurs so as to satisfy the suppression condition $\Delta_1 + \Delta_2 = 0$ (Figure 2.23(c2)). When Δ_1 changes to be positive, the curves at $\Delta_1 = 20$ MHz show symmetric evolution behavior with the curves at $\Delta_1 = -20$ MHz, that is, EIT as well as a suppression of E_{F2} are obtained owing to the two-photon resonance that matched the suppression condition $\Delta_1 + \Delta_2 = 0$ first; and then EIA as well as an enhancement of E_{F2} are obtained when E_1 is in resonance with $|G_2-\rangle$ satisfying the enhancement condition $\Delta_1 + (\Delta_2 - \sqrt{\Delta_2^2 + 4|G_2^b|^2})/2 = 0$, as depicted in Figure 2.23(c4). When Δ_1 is far away from the resonance point ($\Delta_1 = \pm 80$ MHz), the pure EIA as well as the pure enhancement of E_{F2} are obtained because the probe field can only resonate with one of the two dressed states $|G_2 \pm\rangle$ (as shown in Figure 2.23(c1) and (c5)).

On the other hand, when Δ_2 is set at discrete values orderly from positive to negative and Δ_1 is scanned, the probe transmission shows an EIT window on each curve in Figure 2.23(a3) satisfying $\Delta_1 + \Delta_2 = 0$. Also, the FWM signal E_{F2} presents double peaks (Figure 2.23(a4)), due to AT splitting. The two peaks are obtained when E_1 resonates with $|G_4+\rangle$ and $|G_4-\rangle$, respectively. The theoretical calculations (Figure 2.23(b)) are in good agreement with the experimental results (Figure 2.23(a)).

Figure 2.24 (a) The measured intensity of (a1) the probe transmission and (a2) the doubly-dressed FWM signal E_{F2} versus Δ_2 at discrete probe detunings $\Delta_1 = -80$, -40, -20, -10, 0, 10, 20, 40, and 80 MHz, and the measured intensity of (a3) the probe transmission, and (a4) the doubly-dressed FWM signal E_{F2} versus Δ_1 at discrete self-dressing detunings $\Delta_2 = 80$, 40, 20, 10, 0, -10, -20, -40, and -80 MHz (Δ_4 is fixed at $\Delta_4 = 0$). (b1), (b3), (b4), and (b5) are theoretical calculations corresponding to (a1)–(a4). (b2) is the theoretical calculations of enhancement and suppression of doubly-dressed E_{F2}. (c) The energy level diagrams corresponding to (a). (d) The measured intensity of (d1) the probe transmission and (d2) the doubly-dressed FWM signal E_{F2} versus Δ_4 at discrete probe detunings $\Delta_1 = -80$, -40, -20, -10, 0, 10, 20, 40, and 80 MHz, and the measured intensity of (d3) the probe transmission, and (d4) the doubly-dressed FWM signal E_{F2} versus Δ_1 at discrete external-dressing detunings $\Delta_4 = 80$, 40, 20, 10, 0, -10, -20, -40, and -80 MHz (Δ_2 is fixed at $\Delta_2 = 0$). (e1)–(e4) Theoretical calculations corresponding to (d1)–(d4). (f) The energy level diagrams corresponding to (d). Powers of participating laser beams are 4.6 mW (E_1), 33 mW (E_2), 8.4 mW (E_2'), 39 mW (E_4). Source: Adapted from Ref. [36].

Moreover, when we compare the results of these two kinds of scanning methods (i.e., scanning Δ_2 at discrete Δ_1 values and scanning Δ_1 at discrete Δ_2 values), an interesting corresponding relationship between them could be discovered, as expressed by the dashed lines in Figure 2.23(a) and (b). Referring to the dressed energy level diagrams in Figure 2.23(c), one can easily find that the curves in the same column, which are connected by dashed lines, correspond to the same dressed energy level diagram in Figure 2.23(c), although these curves are obtained by scanning different fields. In other words, the positions of enhancement points and suppression points of the FWM signal in the probe frequency detuning (Δ_1) domain correspond to the positions in the dressing frequency detuning (Δ_2) domain, satisfying the same enhancement/suppression conditions. Take the curves obtained when scanning Δ_2 at $\Delta_1 = -20$ MHz in Figure 2.23(b1) and (b2), and the curves obtained when scanning Δ_1 at $\Delta_2 = 20$ MHz in Figure 2.23(b4) and (b5), for example. These four curves, although obtained by scanning different fields, correspond to the same energy state depicted in Figure 2.23(c2) and reveal some common features. When E_1 resonates with the dressed state $|G_2+\rangle$, a dip of probe transmission (EIA) is obtained both in Figure 2.23(b1) and (b4), and a peak (enhancement point) of E_{F2} appears correspondingly both in Figure 2.23(b2) and (b5), as the left dashed line expresses; when two-photon resonance occurs at the point $\Delta_1 + \Delta_2 = 0$, a peak (EIT) is obtained both in Figure 2.23(b1) and (b4), and a dip (suppression point) of E_{F2} appears correspondingly both in Figure 2.23(b2) and (b5), as the right dashed line expresses. In addition, we notice that when scanning the probe detuning, two enhancement points (i.e., the two peaks of AT splitting) and one suppression point could be obtained, whereas when scanning the dressing detuning, only one enhancement point and one suppression point could be obtained at most. The reason is that the two splitting states $|G_2\pm\rangle$ could not move across the original position of $|1\rangle$ and therefore only one of them could resonate with E_1 when scanning the dressing detuning.

Furthermore, the spectra of the doubly-dressed FWM process of E_{F2} in the Y-type four-level subsystem are investigated as shown in Figure 2.24, with the laser beams E_3, E'_3, and E'_4 blocked and E_1, E_2, E'_2, and E_4 turned on. As E_4 is turned on, the FWM signal E_{F2} is dressed by E_4 (external-dressing effect) as well as $E_2(E'_2)$ (self-dressing effect). According to the perturbation chain of the doubly-dressed FWM process $\rho_{00}^{(0)} \xrightarrow{\omega_1} \rho_{(G_2\pm G_4\pm)0}^{(1)} \xrightarrow{\omega_2} \rho_{20}^{(2)} \xrightarrow{-\omega_2} \rho_{(G_2\pm G_4\pm)0}^{(3)}$, we can obtain $\rho_{F2}^{(3)} = G_{F2}/[(d_1 + |G_2^b|^2/d_2 + |G_4|^2/d_4)^2 d_2]$ for the doubly-dressed FWM process.

First, the probe transmission and the FWM signal E_{F2} versus probe detuning Δ_1 and self-dressing detuning Δ_2 are investigated (Figure 2.24(a) to (c)). Figure 2.24(a1) and (a2) present the intensities of the probe transmission (Figure 2.24(a1)) and E_{F2} (Figure 2.24(a2)), respectively, versus Δ_2 at discrete Δ_1 values. Figure 2.24(a3) and (a4) depict the intensities of probe transmission (Figure 2.24(a3)) and E_{F2} (Figure 2.24(a4)) versus Δ_1, respectively, at discrete Δ_2 values (with fixed Δ_4 at $\Delta_4 = 0$). Notice the Doppler broadening of the probe transmission signal in Figure 2.23(a3) has been subtracted. Figure 2.24(b1), (b3), (b4), and (b5) are the theoretical calculations corresponding to Figure 2.24(a1) to (a4), while Figure 2.24(b2) represents the theoretical enhancement and suppression of E_{F2}, respectively, expressed by the peak and dip on each baseline of the curves. Figure 2.24(c) shows the doubly-dressed energy level diagrams corresponding to the curves at discrete detuning values in Figure 2.24(a).

When Δ_1 is set at discrete values orderly from negative to positive and Δ_2 is scanned, the experimentally obtained E_{F2} signal is shown in Figure 2.24(a2), including two components, the pure FWM signal when not considering the dressing effect, and the modification (enhancement and suppression) of the FWM process which is theoretically shown in Figure 2.24(b2). The profile of all the baselines in Figure 2.24(b2) reveals AT splitting of E_4, and the transition of enhancement and suppression in each curve is induced by the interaction between $E_2(E_2')$ and E_4, showing the evolution from pure enhancement ($\Delta_1 = -80$ MHz), to first enhancement and next suppression ($\Delta_1 = -40$ MHz), to pure suppression ($\Delta_1 = -20$ MHz), to first suppression and next enhancement ($\Delta_1 = -10$ MHz), to pure suppression ($\Delta_1 = 0$), to first enhancement and next suppression ($\Delta_1 = 10$ MHz), to pure suppression ($\Delta_1 = 20$ MHz), to first suppression and next enhancement ($\Delta_1 = 40$ MHz), and finally to pure enhancement ($\Delta_1 = 80$ MHz). Correspondingly, the probe transmission shows the evolution from pure EIA, to first EIA and next EIT, to pure EIT, to first EIT and next EIA, to pure EIT, to first EIA and next EIT, to pure EIT, to first EIT, and next EIA, and finally to pure EIA in series as shown in Figure 2.24(a1). The height of the baseline of each curve represents the probe transmission without the dressing field $E_2(E_2')$ versus probe detuning Δ_1. The profile of these baselines reveals an EIT window induced by external-dressing field E_4 at $\Delta_1 = -\Delta_4$. While the peak and dip on each baseline of the curves represent EIT and EIA induced by self-dressing fields E_2 and E_2', we can see that every enhancement and suppression correspond to EIA and EIT, respectively, which is similar to the singly-dressing case observed in Figure 2.23.

Such variations in the probe transmission and the transition of enhancement and suppression of E_{F2} are caused by the interaction of the dressing fields $E_2(E_2')$ and E_4. Because of the doubly-dressing effect, the energy level $|1\rangle$ is totally split into three dressed states (shown in Figure 2.24(c)). First under the external-dressing effect of E_4, the energy level $|1\rangle$ will be broken into two primarily dressed states $|G_4 \pm\rangle$. Then, in the region with $\Delta_1 < 0$, when Δ_2 is scanned around $|G_4 +\rangle$, two secondarily dressed states $|G_4 + G_2 \pm\rangle$ could be created from $|G_4 +\rangle$ by the self-dressing effect of $E_2(E_2')$, as shown in Figure 2.24(c1) to (c4). Symmetrically, in the region with $\Delta_1 > 0$, when Δ_2 is scanned around $|G_4 -\rangle$, two secondarily dressed states $|G_4 - G_2 \pm\rangle$ could be created from $|G_4 -\rangle$, as shown in Figure 2.24(c6) to (c9). As the phenomena and analysis method are similar to those in the singly-dressing case, here we only give the enhancement and suppression conditions as $\Delta_1 + (\Delta_2' \pm \sqrt{\Delta_2'^2 + 4|G_2^b|^2})/2 + G_4 = 0$ and $\Delta_1 + \Delta_2 = 0$ for $\Delta_1 < 0$, where $\Delta_2' = \Delta_2 - G_4$ represents the detuning of $E_2(E_2')$ from $|G_4 +\rangle$, and $\Delta_1 + (\Delta_2' \pm \sqrt{\Delta_2'^2 + 4|G_2^b|^2})/2 - G_4 = 0$, and $\Delta_1 + \Delta_2 = 0$ for $\Delta_1 > 0$, where $\Delta_2' = \Delta_2 + G_4$ represents the detuning of $E_2(E_2')$ from $|G_4 -\rangle$. When the enhancement condition is satisfied, the probe field E_1 resonates with one of the secondarily dressed states, leading to an EIA of probe transmission and enhancement of E_{F2}. When the suppression condition is satisfied, two-photon resonance occurs, leading to an EIT and suppression of E_{F2}. Especially, we notice the enhancement and suppression of E_{F2} in Figure 2.24(b2) shows symmetric behavior with three symmetric centers at $\Delta_1 = 0$, -20, and 20 MHz, all of which are pure suppression. The pure suppression at $\Delta_1 = 0$ MHz is induced by the primary dressing effect of E_4, while pure suppressions at $\Delta_1 = \pm 20$ MHz are caused by the secondary dressing effect of $E_2(E_2')$.

On the other hand, when Δ_2 is set at discrete values orderly from positive to negative and Δ_1 is scanned, the intensity of the probe transmission shows double EIT windows on each curve in Figure 2.24(a3), which are EIT windows $|0\rangle \to |1\rangle \to |2\rangle$ (appearing at $\Delta_1 = -\Delta_2$) and $|0\rangle \to |1\rangle \to |4\rangle$ (appearing at $\Delta_1 = -\Delta_4$), respectively. As Δ_4 is fixed at $\Delta_4 = 0$ and Δ_2 is set at discrete values from positive to negative, the EIT window $|0\rangle \to |1\rangle \to |4\rangle$ is fixed at $\Delta_1 = -\Delta_4 = 0$ and the EIT window $|0\rangle \to |1\rangle \to |2\rangle$ moves from negative to positive. Especially, when Δ_2 is set at $\Delta_2 = 0$, the two EIT windows overlap as shown in Figure 2.24(a3), and a double-peak FWM signal is obtained because both $E_2(E_2')$ and E_4 dress the energy level $|1\rangle$ simultaneously into two dressed states $|+\rangle$ and $|-\rangle$, as shown in Figure 2.24(a4). When Δ_2 is set at $\Delta_2 \neq 0$, in the process of scanning Δ_1, the FWM signal E_{F2} presents three peaks, corresponding to the three-dressed state respectively. First, E_4 dresses $|1\rangle$ into two primary dressed state $|G_4+\rangle$ and $|G_4-\rangle$, corresponding to the primary AT splitting. Then, when the frequency of E_2 and E_2' is tuned so as to move the $|0\rangle \to |1\rangle \to |2\rangle$ EIT window into the left FWM peak ($\Delta_2 > 0$), secondary AT splitting occurs and the left peak splits into two peaks, respectively, corresponding to the secondarily dressed states $|G_4 + G_2+\rangle$ and $|G_4 + G_2-\rangle$. Symmetrically, in the region with $\Delta_2 < 0$, the three peaks corresponding to $|G_4+\rangle$, $|G_4 - G_2+\rangle$, and $|G_4 - G_2-\rangle$ respectively. The theoretical calculations (Figure 2.24(b)) are in good agreement with the experimental results (Figure 2.24(a)).

When comparing the results of these two kinds of scanning method (i.e., scanning Δ_2 at discrete Δ_1 values and scanning Δ_1 at discrete Δ_2 values), the corresponding relationship could also be discovered, as expressed by the dashed lines in Figure 2.24(a) and (b). By referring to the energy level diagrams in Figure 2.24(c), one can easily find that positions of enhancement points and suppression points of FWM signal in the probe frequency detuning (Δ_1) domain correspond with the positions in the self-dressing frequency detuning (Δ_2) domain, satisfying the same enhancement/suppression conditions. Take the curves at $\Delta_1 = -40$ MHz in Figure 2.24(b1) and (b2) and the curves at $\Delta_2 = 40$ MHz in Figure 2.24(b4) and (b5), for example. These four curves, although obtained by scanning different fields, correspond to the same energy state depicted in Figure 2.24(c2) and reveal some common features. When E_1 resonates with the dressed state $|G_4 + G_2+\rangle$, a dip of probe transmission (EIA) is gotten both in Figure 2.24(b1) and (b4), and a peak (enhancement point) of E_{F2} signal appears both in Figure 2.24(b2) and (b5), as the left dashed line expresses; when two-photon resonance ($\Delta_1 + \Delta_2 = 0$) occurs, a peak (EIT) is obtained both in Figure 2.24(b1) and (b4), and a dip (suppression point) appears both in Figure 2.24(b2) and (b5), as the right dashed line expresses. Especially, the position of the pure suppression at $\Delta_1 = 0$ in Figure 2.24(b2) corresponds to the center of primary AT splitting at $\Delta_2 = 0$

in Figure 2.24(b5); and the positions of the pure suppression at $\Delta_1 = \pm 20$ MHz in Figure 2.24(b2) correspond to the center of secondary AT splitting at $\Delta_2 = \pm 20$ MHz in Figure 2.24(b5). We also notice that when scanning Δ_1, three enhancement points and two suppression points could be obtained, whereas when scanning Δ_2 only one enhancement point and one suppression point could be obtained at most.

Next, we investigate the probe transmission and the enhancement and suppression of E_{F2} versus the probe detuning Δ_1 and external-dressing detuning Δ_4 (Figure 2.24(d) and (e)). Figure 2.24(d1) and (d2), respectively, present the intensities of the probe transmission (Figure 2.24(d1)) and the enhancement and suppression of E_{F2} (Figure 2.24(d2)) versus Δ_4 at discrete Δ_1 values. While Figure 2.24(d3) and (d4) depict the intensities of probe transmission (Figure 2.24(d3)) and E_{F2} (Figure 2.24(d4)) versus Δ_1 at discrete Δ_4 values (with fixed Δ_2 at $\Delta_2 = 0$). Figure 2.24(e1) to (e4) are the theoretical calculations corresponding to Figure 2.24(d1) to (d4). Figure 2.24(f) shows the doubly-dressed energy level diagrams corresponding to the curves at discrete detuning values in Figure 2.24(d). Similar to the above discussion of Figure 2.24(a) to (c), the signal E_{F2} is dressed by both $E_2(E_2')$ and E_4. First, under the self-dressing effect of $E_2(E_2')$, the state $|1\rangle$ will be broken into two primarily dressed states $|G_2 \pm\rangle$. Then in the region with $\Delta_1 < 0$, when Δ_4 is scanned around $|G_2 +\rangle$, two secondarily dressed states $|G_2 + G_4 \pm\rangle$ could be created from $|G_2 +\rangle$ by the external-dressing field E_4, as shown in Figure 2.24(f1)–(f4). Symmetrically, in the region with $\Delta_1 > 0$, when Δ_4 is scanned around $|G_2 -\rangle$, two secondarily dressed states $|G_2 - G_4 \pm\rangle$ could be created from $|G_2 -\rangle$, as shown in Figure 2.24(f6) to (f9). Here we only give the enhancement and suppression conditions as $\Delta_1 + (\Delta_4' \pm \sqrt{\Delta_4'^2 + 4|G_4|^2})/2 + G_2 = 0$ and $\Delta_1 + \Delta_4 = 0$ for $\Delta_1 < 0$, where $\Delta_4' = \Delta_4 - G_2$ represents the detuning of E_4 from $|G_2 +\rangle$, and $\Delta_1 + (\Delta_4' \pm \sqrt{\Delta_4'^2 + 4|G_4|^2})/2 - G_2 = 0$ and $\Delta_1 + \Delta_4 = 0$ for $\Delta_1 > 0$, where $\Delta_4' = \Delta_4 + G_2$ represents the detuning of E_4 from $|G_2 -\rangle$. Unlike scanning self-dressing detuning (Figure 2.24(a2)), by scanning the external-dressing detuning the enhancement and suppression of E_{F2} could be detected directly, excluding the pure FWM component (Figure 2.24(d2)). We can see that the enhancement and suppression of E_{F2} in Figure 2.24(d2) shows a similar evolution with that in Figure 2.24(b2). The profile of all the baselines, which has two peaks, reveals AT splitting of $E_2(E_2')$, and the transition of enhancement and suppression in each curve is induced by the interaction between $E_2(E_2')$ and E_4 with three symmetric centers at $\Delta_1 = 0$, -20, and 20 MHz, all of which are pure suppression. The pure suppression at $\Delta_1 = 0$ MHz is induced by the primary dressing effect of $E_2(E_2')$, while pure suppressions at $\Delta_1 = \pm 20$ MHz are caused by the secondary dressing effect of E_4. On the other hand, when Δ_1 is scanned, the FWM signal E_{F2} in Figure 2.24(d4) also presents three peaks, corresponding to the three-dressed state. Moreover, the corresponding relationship between scanning probe detuning and scanning external-dressing detuning is similar to that mentioned earlier, as expressed by the dashed lines in Figure 2.24(d) and (e). It is obvious that the theoretical calculations (Figure 2.24(e)) are in good agreement with the experimental results (Figure 2.24(d)).

Comparing with the singly-dressed FWM process in Figure 2.23, we notice that the doubly-dressed FWM process, although derived from the former, shows more complexities as one more dressing field is considered. When scanning probe detuning, the doubly-dressed FWM signal shows three peaks resulting from two orders of AT splitting (Figure 2.24(a4) and (d4)), whereas the singly-dressed FWM signal shows only two peaks resulting from the AT splitting of the self-dressing effect (Figure 2.23(a4)). When scanning the dressing detuning, only one symmetric center appears in the singly-dressing case (Figure 2.23(b2)), whereas three symmetric centers appear in the doubly-dressing case respectively at $\Delta_1 = 0, -20$, and 20 MHz (Figure 2.24(b2) and (e2)), all of which reveal pure suppression. The symmetric center at $\Delta_1 = 0$ is caused by the primary dressing effect, while the two symmetric centers at $\Delta_1 = \pm 20$ MHz are due to the secondary dressing effect.

Synthetically, on the basis of the above-mentioned analysis, we find the methods of scanning the probe detuning (Figure 2.23(a3) and (a4) and Figure 2.24(a3) and (a4), (d3) and (d4)), scanning self-dressing detuning (Figure 2.23(a1) and (a2) and Figure 2.24(a1) and (a2)), and scanning external-dressing detuning (Figure 2.24(d1) and (d2)) individually show some different features and advantages on researching the FWM process. When scanning the probe detuning, the obtained FWM signal includes two components, the pure FWM signal when not considering dressing effects and the modification (revealing AT splitting) of the FWM process. When scanning self-dressing detuning, the obtained signal also includes two components, the pure FWM signal and the modification (revealing the transition between enhancement and suppression) of the FWM process, whereas by scanning external-dressing detuning, the enhancement and suppression could be detected directly, excluding the pure FWM component. On the other hand, by scanning the probe detuning, all enhancement points and suppression points could be observed corresponding to the peaks and dips of AT splitting. In the singly-dressing case, there are two enhancement points and one suppression point (Figure 2.23(a3) and (a4)), and in the doubly-dressing case, three enhancement points and two suppression points (Figure 2.24(a3) and (a4)), and so on. In contrast, by scanning dressing detuning, at most one enhancement point and one suppression point could be obtained in the spectra. Furthermore, the positions of the enhancement and suppression points when scanning dressing detuning match with the positions of corresponding points when scanning probe detuning, as the dashed lines express in Figure 2.23 and Figure 2.24.

After that, we demonstrate a new type of phase-controlled, spatiotemporal coherent interference between two FWM processes (E_{F2} and E_{F4}) in a four-level, Y-type subsystem when the laser beams E_3 and E'_3 are turned off (shown in Figure 2.21(d)). With a specially designed spatial configuration for the laser beams with phase matching and an appropriate optical delay introduced in one of the coupling laser beams, we can have a controllable phase difference between the two FWM processes in the subsystem. When this relative phase is varied, temporal and spatial interferences can be observed. The interference in the time domain is in the femtosecond timescale, corresponding to the optical transition frequency excited by the delayed laser beam. In the experiment, the beam E'_2 is delayed by an amount τ using a computer-controlled stage. The CCD and an avalanche photo diode (APD) are set on the propagation path of the two FWM signals to measure them. By changing the frequency detuning Δ_4, the $|0\rangle - |1\rangle - |4\rangle$ EIT window can be shifted toward the $|0\rangle - |1\rangle - |2\rangle$ EIT window. When the two EIT windows overlap, temporal and spatial interferences of two FWM signals E_{F2} and E_{F4} can be observed, as shown in Figure 2.25.

The coexisting E_{F1} and E_{F2} signals give the total detected intensity as, $I(\tau, r) \propto |\chi_{F2}^{(3)}|^2 + |\chi_{F4}^{(3)}|^2 + 2|\chi_{F2}^{(3)}||\chi_{F4}^{(3)}| \cos(\varphi_2 - \varphi_4 + \varphi)$, where

$$\chi_{F2}^{(3)} = \frac{N\mu_1^2}{\hbar\varepsilon_0 G_1}\rho_{F2}^{(3)}\frac{\mu_2^2}{\hbar^2 G_2 G_2'^*} = \frac{-i\mu_1^2\mu_2^2 N}{\varepsilon_0\hbar^3(d_1 + G_2^2/d_2 + G_4^2/d_4)^2 d_2} = |\chi_{F2}^{(3)}| \exp(i\varphi_2)$$

$$\chi_{F4}^{(3)} = \frac{N\mu_1^2}{\hbar\varepsilon_0 G_1}\rho_{F4}^{(3)}\frac{\mu_4^2}{\hbar^2 G_4 G_4'^*} = \frac{-i\mu_1^2\mu_4^2 N}{\varepsilon_0\hbar^3(d_1 + G_2^2/d_2 + G_4^2/d_4)^2 d_4} = |\chi_{F4}^{(3)}| \exp(i\varphi_4)$$

and $\varphi = \Delta\mathbf{k}\cdot\mathbf{r} - \omega_2\tau$ with the frequency of spatial interference $\Delta\mathbf{k} = \mathbf{k}_{F2} - \mathbf{k}_{F4} = (\mathbf{k}_2 - \mathbf{k}_2') - (\mathbf{k}_4 - \mathbf{k}_4')$. From the expression of $I(\tau,r)$, we can see that the total

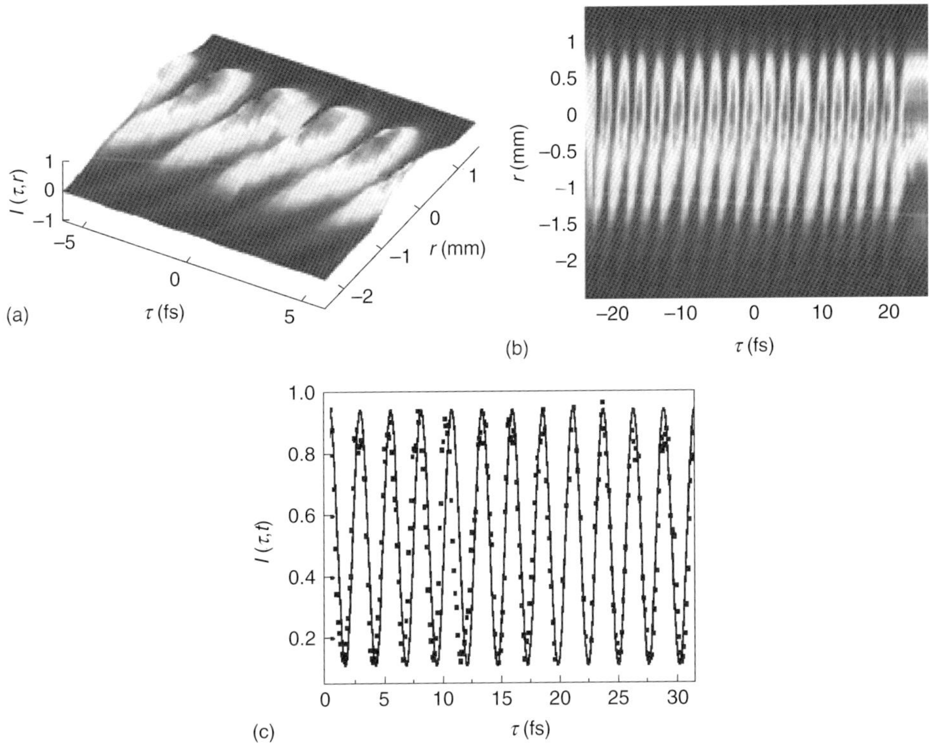

Figure 2.25 The spatiotemporal interferograms of E_{F2} and E_{F4} in the Y-type atomic subsystem. (a) A three-dimensional spatiotemporal interferogram of the total FWM signal intensity $I(\tau,r)$ versus time delay τ of beam E_2' and transverse position r. (b) The temporal interference with a much longer time delay of beam E_2'. (c) Measured beat signal intensity $I(\tau,r)$ versus time delay τ together with the theoretically simulated result (solid curve).

signal has an ultrafast time oscillation with a period of $2\pi/\omega_2$ and spatial interference with a period of $2\pi/\Delta k$, which forms a spatiotemporal interferogram. Figure 2.25(a) shows a three-dimensional interferogram pattern, and Figure 2.25(b) shows the temporal interference with a much longer time delay in beam E_2', while Figure 2.25(c) shows its projections on time. Figure 2.25(c) depicts a typical temporal interferogram with the temporal oscillation period of $2\pi/\omega_2 = 2.6$ fs corresponding to the $|1\rangle$ to $|2\rangle$ transition frequency of $\Omega_2 = 2.4\ \text{fs}^{-1}$ in ^{85}Rb. This gives a technique for precision measurement of atomic transition frequency in optical wavelength range. The solid curve in Figure 2.25(c) is the theoretical calculation from the full density matrix equations. It is easy to see that the theoretical results fit well with the experimentally measured results.

Now, we concentrate on the SWM process when Δ_3 is at large detuning, with E_2' and E_4' blocked (shown in Figure 2.21(e)). When we take the atomic velocity component and the dressing effect of $E_3(E_3')$ into consideration, the enhancement and suppression of the SWM signal would be shifted far away from the resonance, as shown in Figure 2.26.

When considering the Doppler effect, with the atom moving toward the probe laser beam with velocity v, the frequency of E_1 is changed to $\omega_1 + \omega_1 v/c$, and the frequencies of E_2, E_3, and E_4 are changed to $\omega_2 - \omega_2 v/c$, $\omega_3 - \omega_3 v/c$, and $\omega_4 - \omega_4 v/c$ under our experimental geometry configuration; therefore, their detunings are changed to $\Delta_1 - \omega_1 v/c$, $\Delta_2 + \omega_2 v/c$, $\Delta_3 + \omega_3 v/c$, $\Delta_4 + \omega_4 v/c$. Noticing that in such beam geometric configuration, the two-photon Doppler-free condition will not be satisfied for the Λ-type, three-level subsystem $|0\rangle \to |1\rangle \to |3\rangle$, the atomic velocity component $\omega_3 v/c$ will behave dominant.

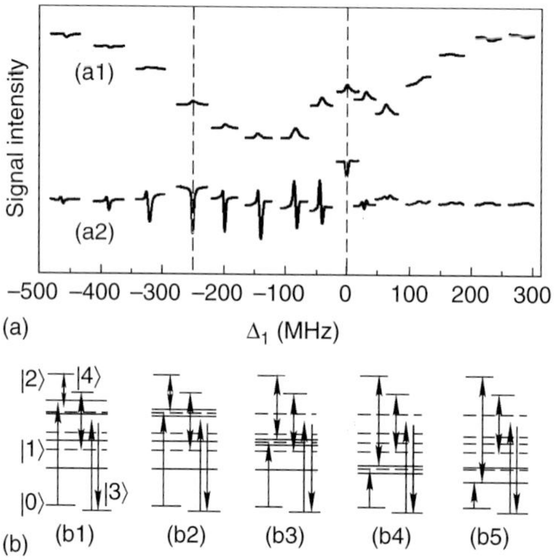

Figure 2.26 (a1) The probe transmission signal and (a2) the SWM signal with the enhancement and suppression effect versus Δ_2 for different Δ_1 with the laser beams E_2' and E_4' blocked when Δ_3 is at large detuning. (b) The doubly-dressed state diagram of the SWM signal. Powers of participating laser beams are 3 mW (E_1), 4.6 mW (E_2), 44 mW (E_3 and E_3'), and 65 mW (E_4).

The density matrix element of the SWM signal E_{S4} can be obtained as $\rho_{S4}^{(5)} = G_{S4}/[d_1'd_3'd_4'(d_1' + |G_4|^2/d_4' + |G_3^b|^2/d_3')^2]$ via the self-dressed perturbation chain $\rho_{00}^{(0)} \xrightarrow{\omega_1} \rho_{(G_4\pm G_3\pm)0}^{(1)} \xrightarrow{-\omega_3} \rho_{30}^{(2)} \xrightarrow{\omega_3} \rho_{(G_4\pm G_3\pm)0}^{(3)} \xrightarrow{\omega_4} \rho_{40}^{(4)} \xrightarrow{-\omega_4} \rho_{10}^{(5)}$, where $d_1' = d_1 - i\omega_1 v/c + \gamma_1$, $d_3' = d_3 - i\omega_3 v/c - i\omega_1 v/c + \gamma_1 - \gamma_3$, $d_4' = d_4 + i\omega_4 v/c - i\omega_1 v/c + \gamma_1 + \gamma_4$, and γ_1, γ_3, and γ_4 are the half line width of laser beams E_1, E_3, and E_4, respectively. When the SWM signal E_{S4} is externally dressed by E_2 (defined as E_{S4}^D), the solved expression is $\rho_{S4}^{(5)} = G_{S4}/[d_3'd_4'(d_1' + |G_2|^2/d_2')(d_1' + |G_2|^2/d_2' + |G_4|^2/d_4' + |G_3^b|^2/d_3')^2]$ via the dressed perturbation chain, $\rho_{00}^{(0)} \xrightarrow{\omega_1} \rho_{(G_4\pm G_3\pm G_2\pm)0}^{(1)} \xrightarrow{-\omega_3} \rho_{30}^{(2)} \xrightarrow{\omega_3} \rho_{(G_4\pm G_3\pm G_2\pm)0}^{(3)} \xrightarrow{\omega_4} \rho_{40}^{(4)} \xrightarrow{-\omega_4} \rho_{(G_2\pm)0}^{(5)}$, where $d_2' = d_2 + i\omega_2 v/c - i\omega_1 v/c + \gamma_1 + \gamma_2$ and γ_2 are the half line width of laser beam E_2. Similarly, the density matrix element of the SWM signal E_{S2} can be obtained as $\rho_{S2}^{(5)} = G_{S2}/[d_1'd_2'd_3'(d_1' + |G_2|^2/d_2' + |G_3^b|^2/d_3')^2]$ via the self-dressed perturbation chain, $\rho_{00}^{(0)} \xrightarrow{\omega_1} \rho_{(G_2\pm G_3\pm)0}^{(1)} \xrightarrow{-\omega_3} \rho_{30}^{(2)} \xrightarrow{\omega_3} \rho_{(G_2\pm G_3\pm)0}^{(3)} \xrightarrow{\omega_2} \rho_{20}^{(4)} \xrightarrow{-\omega_2} \rho_{10}^{(5)}$. When the SWM signal E_{S2} is externally dressed by E_4 (defined as E_{S2}^D), the solved expression is $\rho_{S2}^{(5)} = G_{S2}/[d_2'd_3'(d_1' + |G_4|^2/d_4')(d_1' + |G_2|^2/d_2' + |G_4|^2/d_4' + |G_3^b|^2/d_3')^2]$ via the dressed perturbation chain, $\rho_{00}^{(0)} \xrightarrow{\omega_1} \rho_{(G_2\pm G_3\pm G_4\pm)0}^{(1)} \xrightarrow{-\omega_3} \rho_{30}^{(2)} \xrightarrow{\omega_3} \rho_{(G_2\pm G_3\pm G_4\pm)0}^{(3)} \xrightarrow{\omega_2} \rho_{20}^{(4)} \xrightarrow{-\omega_2} \rho_{(G_4\pm)0}^{(5)}$. From the expressions of E_{S4}^D and E_{S2}^D signals, one can see that the two SWM processes are closely connected by mutual dressing effect.

In Figure 2.26(a), we present the probe transmission (Figure 2.26(a1)) and the measured SWM signal (Figure 2.26(a2)) by scanning Δ_2 at different designated Δ_1 values, with $G_2 \ll G_4$. In Figure 2.26(a1), The profile of each baseline represents the probe transmission without dressing field E_2 versus probe detuning Δ_1, which reveals an EIT window (-80 MHz $< \Delta_1 < 80$ MHz) induced by E_4, and the peak on each baseline is the EIT induced by E_2. In Figure 2.26(a2), the profile of each baseline represents the intensity variation of the triple-peak SWM signal E_{S4} versus Δ_1. The peak and dip on each baseline include the dressed SWM signal E_{S2}^D and the enhancement and suppression of E_{S4} induced by E_2. Considering $G_2 < G_4$, we can deduce the signal of E_{S2}^D is quite small. Therefore, the peak and dip on each baseline mainly represent the enhancement and suppression of the SWM signal E_{S4} induced by E_2.

One can see that the curves in Figure 2.26(a2) shows pure suppression at $\Delta_1 = 0$ MHz and $\Delta_1 = -250$ MHz. These two pure suppressions can be explained by the triple-dressing effect of E_2, $E_3(E_3')$, and E_4. The enhancement and suppression of the SWM is caused by the triply-dressing fields. First, owing to the self-dressing effect of E_4, the state $|1\rangle$ would be split into two dressed states $|G_4 \pm\rangle$. Next, the dressing field $E_3(E_3')$ splits the state $|G_4+\rangle$ into $|G_4 + G_3 \pm\rangle$. Finally, when Δ_2 is scanned, E_2 will further split $|G_4 + G_3 \pm\rangle$ into two dressed states $|G_4 + G_3 + G_2 \pm\rangle$ or $|G_4 + G_3 - G_2 \pm\rangle$; or split $|G_4 -\rangle$ into two dressed states $|G_4 - G_2 \pm\rangle$, as shown in Figure 2.26(b). When two-photon resonance occurs at the original states $|G_4 + G_3 +\rangle$ (Figure 2.26(b2)), $|G_4 + G_3 -\rangle$ (Figure 2.26(b3)) or $|G_4 -\rangle$ (Figure 2.26(b4)), the pure suppressions of SWM can be obtained. In the Doppler effect being considered, the dominant atomic velocity component $\omega_3 v/c$ moving the $|G_4 + G_3 +\rangle$ state far away from the resonance, the pure suppression on the left induced by the triply-dressing effect will be shifted to large detuning. In the experiment, only two pure suppressions can be obtained, of which the left one is caused by two-photon resonance at the original $|G_4 + G_3 +\rangle$ state, and the right one is related to the original $|G_4 + G_3 -\rangle$ state. This inconsistence is because when the frequency of $E_3(E_3')$ is at large detuning $(\Delta_3 \gg 0)$, the enhancement and suppression of the SWM signal is no more symmetrical. Specifically, owing to the optical pumping effect corresponding to the transition from $|3\rangle$ to $|1\rangle$ by E_3 and E_3', the suppression will be intensified with $\Delta_1 < 0$ (especially when $\Delta_1 + \Delta_3 = 0$) as shown in Figure 2.26(b); but with $\Delta_1 > 0$, the inexistence of such an effect makes the suppression caused by the two-photon resonance at the original $|G_4 -\rangle$ state cannot be obtained.

In this way, we first demonstrate that the enhancement and suppression signal can be observed out of the EIT window (-80 MHz $< \Delta_1 < +80$ MHz) through the Doppler frequency shift led by atomic velocity component and optical pumping. From the figures one can see that even Δ_1 is at large detuning (Figure 2.26(b1) and (b5)), the enhancement and suppression of SWM will still exist in the region with $\Delta_1 < 0$.

Finally, the interaction of the SWM signals is studied. When all seven laser beams are turned on, two FWM signals (E_{F2} and E_{F4}) and four SWM signals (E_{S2}, E'_{S2}, E_{S4}, and E'_{S4}) can be generated simultaneously and interact with each other (considering the FWM signal E_{F3} is so weak that it could be negligible). When the $|0\rangle \to |1\rangle \to |4\rangle$ EIT window and the $|0\rangle \to |1\rangle \to |2\rangle$ EIT window are tuned separate, the interaction of the MWM processes related to the same EIT window is displayed in Figure 2.22, by scanning probe detuning under different blocking conditions. Here, we overlap the two EIT windows experimentally; therefore, the interaction between these two groups of wave-mixing signals (those related to $E_2(E_2')$ and those related to $E_4(E_4')$) will be studied, by scanning the dressing field detuning Δ_2 at discrete probe detuning Δ_1 values (as shown in Figure 2.27).

Generally, owing to the mutual dressings of the two ladder subsystems, we can obtain the density elements $\rho_{F2}^{(3)} = G_{F2}/[d_2(d_1 + |G_4^b|^2/d_4)^2]$ (where $G_4^b = G_4 + G_4'$) for the external-dressed FWM process of E_{F2}, $\rho_{S2}^{(5)} = G_{S2}/[d_3 d_2(d_1 + |G_4^b|^2/d_4)^3]$ (or $\rho_{S2}'^{(5)} = G'_{S2}/[d_3 d_2(d_1 + |G_4^b|^2/d_4)^3]$) for the external-dressed SWM process of E_{S2} (or E'_{S2}); and $\rho_{F4}^{(3)} = G_{F4}/[d_4(d_1 + |G_2^b|^2/d_2)^2]$ for the external-dressed FWM process E_{F4}, $\rho_{S4}^{(5)} = G_{S4}/[d_3 d_4(d_1 + |G_2^b|^2/d_2)^3]$ (or $\rho_{S4}'^{(5)} = G'_{S4}/[d_3 d_4(d_1 + |G_2^b|^2/d_2)^3]$) for the external-dressed SWM process of E_{S4} (or E'_{S4}), when not considering the self-dressing effects. The total detected MWM signal (Figure 2.27(a)) will be proportional to the mod square of ρ'_M, where $\rho'_M = (\rho_{S2}^{(5)} + \rho_{S2}'^{(5)} + \rho_{F2}^{(3)}) + (\rho_{S4}^{(5)} + \rho_{S4}'^{(5)} + \rho_{F4}^{(3)})$.

The measured total MWM signal when all the laser beams are turned on is depicted in Figure 2.27(a). The global profile of the baselines of each curve, which mainly includes the self-dressed E_{F4}, E_{S4}, and E'_{S4} signals, exhibits AT splitting induced by $E_4(E_4')$. The peak on each profile is mainly composed of the doubly-dressed E_{F2}, E_{S2}, and E'_{S2} signals and the enhancement and suppression of E_{F4}, E_{S4}, and E'_{S4} induced by $E_2(E_2')$. To understand the interaction of these six generated signals in depth, we divide them into two parts, the interaction of FWM signals E_{F2} and E_{F4} (shown in Figure 2.27(b)), and the interaction of SWM signals E_{S2}, E'_{S2}

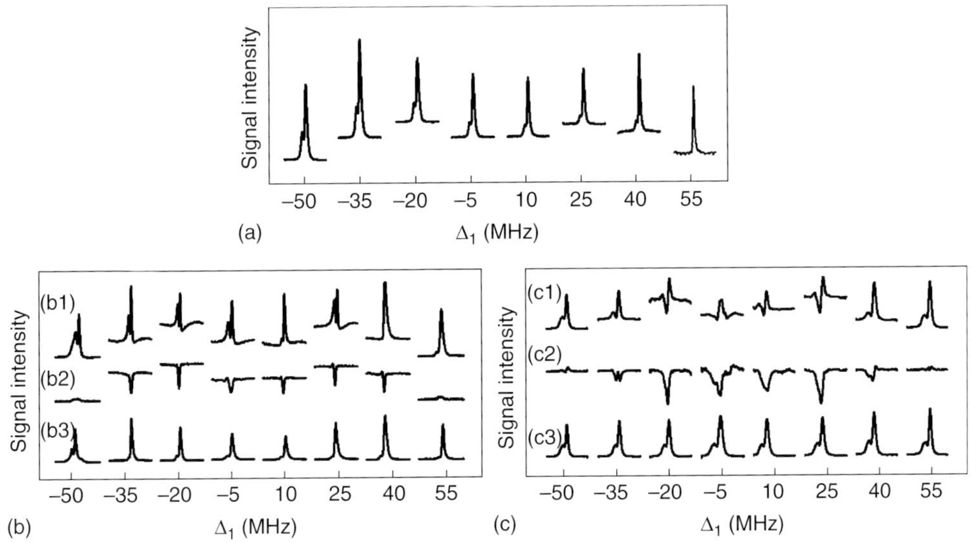

Figure 2.27 (a) Measured total MWM signal versus Δ_2 at discrete Δ_1 when all seven laser beams are on. (b) Measured FWM signals versus Δ_2 at discrete Δ_1. (b1) Signal obtained with the laser beams E_3 and E'_3 blocked and others on. (b2) The enhancement and suppression of the FWM signal E_{F4} when the laser beams E_3, E'_3, and E'_2 are blocked. (b3) The FWM signal when the laser beams E_3, E'_3, and E'_4 are blocked. (c) Measured SWM signals versus Δ_2 at discrete Δ_1. (c1) Signal obtained with laser beams E'_2, E'_4 blocked and others on. (c3) Signal obtained with the laser beams E'_2, E'_4, and E_4 blocked. (c2) The sum of $E^D_{S4} - E_{S4}$ and $E^D_{S2} - E_{S2}$. Powers of all laser beams are 3.7 mW (E_1), 55 mW (E_2), 5.3 mW (E'_2), 44 mW (E_3 and E'_3), 85 mW (E_4), and 8.6 mW (E'_4).

and E_{S4}, and E'_{S4}. As E'_{S2} and E_{S2} share similar characteristics (so do E_{S4} and E'_{S4}), the interaction between E'_{S2}, E_{S2} and E_{S4}, E'_{S4} can be studied by investigating the interaction between E_{S2} and E_{S4} (shown in Figure 2.27(c)) . Therefore, by blocking different laser beams and scanning Δ_2 at discrete Δ_1 values, the interaction of these SWM signals can be observed directly, separated into the interplay between two FWM signals, two SWM signals, and the interplay between FWM and SWM signals.

First, we investigate the interplay between the two FWM signals E_{F2} and E_{F4} in the Y-type subsystem (Figure 2.21(d)) by blocking the laser beams E_3 and E'_3. The interplay between these two FWM signals will occur when we overlap the two separated EIT windows, as shown in Figure 2.27(b). Figure 2.27(b1) shows the measured FWM signal versus Δ_2 at discrete $\Delta_1$values, which include the information of both E_{F2} and E_{F4}, with mutual dressings. In Figure 2.27(b1), the global profile of baselines of all the curves represents the intensity variation of E_{F4} at designated probe detuning values, and the peak and dip on each baseline include two components, the doubly-dressed E_{F2} signal and the enhancement and suppression of E_{F4} induced by $E_2(E'_2)$. These two components could be individually detected by additionally blocking E_2' or E_4', as shown in Figure 2.27(b2) and (b3) separately. When blocking E_2', the information related to E_{F4} could be extracted as E_{F2} is turned off (Figure 2.27(b2)). The global profile of all the baselines in Figure 2.27(b2) reveals AT splitting of E_{F4}, and the peak and dip of each curve represent the enhancement and suppression of E_{F4} induced by E_2, which show an evolution similar to the curves in Figure 2.24(d2). On the other hand, when turning on E_2' and blocking E_4', the doubly-dressed E_{F2} signal could be obtained in Figure 2.27(b3), which is similar to Figure 2.24(a2). It is quite obvious that measured total FWM signal (Figure 2.27(b1)) is approximate to the sum of the enhancement and suppression of E_{F4} which mainly behave as dips (Figure 2.27(b2)), and the dressed FWM signal E_{F2}, which mainly behave as peaks (Figure 2.27(b3)).

Next, we investigate the interplay between two SWM signals E_{S2} and E_{S4} in Figure 2.27(c), with E_2' and E_4' blocked (shown as Figure 2.21(e)). When all the five laser beams (E_1, E_2, E_4, E_3, and E_3') are turned on, two external-dressed SWM signals E_{S2}^D and E_{S4}^D will form simultaneously, as shown in Figure 2.27(c1). The global baseline variation profile shows the intensity variation of the SWM signal E_{S4} revealing AT splitting. The peak and dip on each baseline include the SWM signal E_{S2}^D and $E_{S4}^D - E_{S4}$ which represents the enhancement and suppression of E_{S4} caused by E_2. When the beam E_4 is also blocked, only the measured SWM signal E_{S2} remains, as shown in Figure 2.27(c3). By subtracting the SWM signal E_{S2} (Figure 2.27(c3)) and the height of each baseline from the total signal (Figure 2.27(c1)), the sum of the signals $E_{S4}^D - E_{S4}$ and $E_{S2}^D - E_{S2}$ revealing the pure dressing effect can be obtained, as shown in Figure 2.27(c2). Here, $E_{S2}^D - E_{S2}$ expresses the enhancement or suppression of the E_{S2} caused by the E_4. On the one hand, we can see from the curve (c2) that when two-photon resonance occurs at $\Delta_1 = -20$ MHz and $\Delta_1 = 25$ MHz, the depth of the dip is approximately maximum, meaning that the suppression is most significant. On the other hand, when E_1 resonates with $|G_4 + G_2 +\rangle$ and $|G_4 - G_2 -\rangle$, the generated SWM signals are enhanced as shown by the small peaks.

When the FWM signals and SWM signals coexist in Figure 2.27(a) with all seven beams on, the interaction of these generated wave-mixing signals can be obtained. Theoretically, the intensity of the measured total MWM signal in Figure 2.27(a) can be described as sum of the FWM signal intensity (Figure 2.27(b1)), the SWM signal intensity (Figure 2.27(c1)) and the intensity of the SWM signals relate to E_{S2}' and E_{S4}' which is similar to Figure 2.27(c1)). From the experimental result, one can see that the generated signal with all laser beams tuned on in (a) is approximate to the sum of the FWM intensity in (b1) and the SWM intensity in (c1), but behaves FWM dominant because the SWM signals are too weak to be distinguished when compared with the FWM. We also notice that when the FWM signals and SWM signals coexist and interplay with each other, the enhancement and suppression effect of the FWM will be weaken by the interaction of the six MWM signal.

2.4.3
Conclusion

In summary, we distinguish seven coexisting MWM signals in a K-type five-level atomic system by selectively blocking different laser beams. And the interactions among these MWM signals have been studied by investigating the interaction between two FWM signals, between two SWM signals, and the interaction between FWM and SWM signals. We also discuss the dressed FWM process by scanning the frequency detuning of the probe field, the self-dressing field, and the external-dressing field, proving the corresponding relationship between different scanning methods. Especially, by scanning external-dressing detuning, the enhancements and suppressions of FWM can be detected directly. In addition, we successfully demonstrate the temporal interference between two FWM signals with a femtosecond time scale. Moreover, when Δ_3 is far away from resonance, we discussed the enhancement and suppression of SWM signal at large detuning, which is moved out of the EIT window through the Doppler frequency shift led by atomic velocity component.

Problems

2.1 Present how to get MWM signals via EIT channels.

2.2 Summarize the conditions with which different odd-order MWM signals can coexist and why such a phenomenon occurs.

2.3 Find the differences among the three types of dual-dressed schemes (nested, sequential, and parallel schemes) in MWM signals.

References

1. Hemmer, P., Katz, D., Donoghue, J., Cronin-Golomb, M., Shahriar, M., and Kumar, P. (1995) Efficient low-intensity optical phase conjugation based on coherent population trapping in sodium. *Opt. Lett.*, **20**, 982–984.
2. Grove, T.T., Shahriar, M., Hemmer, P., Kumar, P., Sudarshanam, V., and Cronin-Golomb, M. (1997) Distortion-free gain and noise correlation in sodium vapor with four-wave mixing and coherent population trapping. *Opt. Lett.*, **22**, 769–771.
3. Korsunsky, E., Maichen, W., and Windholz, L. (1997) Dynamics of coherent optical pumping in a sodium atomic beam. *Phys. Rev. A*, **56**, 3908.
4. Lü, B., Burkett, W., and Xiao, M. (1998) Nondegenerate four-wave mixing in a double-Lambda system under the influence of coherent population trapping. *Opt. Lett.*, **23**, 804–806.
5. Boller, K.J., Imamolu, A., and Harris, S. (1991) Observation of electromagnetically induced transparency. *Phys. Rev. Lett.*, **66**, 2593–2596.
6. Li, Y. and Xiao, M. (1995) Observation of quantum interference between dressed states in an electromagnetically induced transparency. *Phys. Rev. A*, **51**, 4959.
7. Mitsunaga, M., Yamashita, M., and Inoue, H. (2000) Absorption imaging

of electromagnetically induced transparency in cold sodium atoms. *Phys. Rev. A*, **62**, 013817.

8. Hau, L.V., Harris, S.E., Dutton, Z., and Behroozi, C.H. (1999) Light speed reduction to 17 metres per second in an ultracold atomic gas. *Nature*, **397**, 594–598.
9. Fleischhauer, M. and Lukin, M. (2000) Dark-state polaritons in electromagnetically induced transparency. *Phys. Rev. Lett.*, **84**, 5094–5097.
10. Liu, C., Dutton, Z., Behroozi, C.H., and Hau, L.V. (2001) Observation of coherent optical information storage in an atomic medium using halted light pulses. *Nature*, **409**, 490–493.
11. Phillips, D., Fleischhauer, A., Mair, A., Walsworth, R., and Lukin, M. (2001) Storage of light in atomic vapor. *Phys. Rev. Lett.*, **86**, 783–786.
12. Lukin, M., Matsko, A., Fleischhauer, M., and Scully, M. (1998) Quantum noise and correlations in resonantly enhanced wave mixing based on atomic coherence. *Phys. Rev. Lett.*, **82**, 1847–1850 Arxiv preprint quant-ph/9811028.
13. van der Wal, C.H., Eisaman, M.D., André, A., Walsworth, R.L., Phillips, D.F., Zibrov, A.S., and Lukin, M.D. (2003) Atomic memory for correlated photon states. *Science*, **301**, 196–200.
14. Li, Y. and Xiao, M. (1996) Enhancement of nondegenerate four-wave mixing based on electromagnetically induced transparency in rubidium atoms. *Opt. Lett.*, **21**, 1064–1066.
15. Zibrov, A., Matsko, A., Kocharovskaya, O., Rostovtsev, Y., Welch, G., and Scully, M. (2002) Transporting and time reversing light via atomic coherence. *Phys. Rev. Lett.*, **88**, 103601.
16. Rostovtsev, Y.V., Sariyanni, Z.E., and Scully, M.O. (2006) Electromagnetically induced coherent backscattering. *Phys. Rev. Lett.*, **97**, 113001.
17. Balić, V., Braje, D.A., Kolchin, P., Yin, G., and Harris, S. (2005) Generation of paired photons with controllable waveforms. *Phys. Rev. Lett.*, **94**, 183601.
18. Kolchin, P., Du, S., Belthangady, C., Yin, G., and Harris, S. (2006) Generation of narrow-bandwidth paired photons: use of a single driving laser. *Phys. Rev. Lett.*, **97**, 113602.
19. Du, S., Wen, J., Rubin, M.H., and Yin, G. (2007) Four-wave mixing and biphoton generation in a two-level system. *Phys. Rev. Lett.*, **98**, 53601.
20. Li, C., Zhang, Y., Nie, Z., Zheng, H., Zuo, C., Du, Y., Song, J., Lu, K., and Gan, C. (2010) Controlled multi-wave mixing via interacting dark states in a five-level system. *Opt. Commun.*, **283**, 2918–2928.
21. Zhang, Y., Nie, Z., Wang, Z., Li, C., Wen, F., and Xiao, M. (2010) Evidence of Autler–Townes splitting in high-order nonlinear processes. *Opt. Lett.*, **35**, 3420–3422.
22. Zhang, Y., Anderson, B., and Xiao, M. (2008) Coexistence of four-wave, six-wave and eight-wave mixing processes in multi-dressed atomic systems. *J. Phys. B: At. Mol. Opt. Phys.*, **41**, 045502.
23. Zhang, Y., Khadka, U., Anderson, B., and Xiao, M. (2009) Temporal and spatial interference between four-wave mixing and six-wave mixing channels. *Phys. Rev. Lett.*, **102**, 13601.
24. Zhang, Y., Brown, A.W., and Xiao, M. (2007) Opening four-wave mixing and six-wave mixing channels via dual electromagnetically induced transparency windows. *Phys. Rev. Lett.*, **99**, 123603.
25. Zhang, Y., Anderson, B., Brown, A.W., and Xiao, M. (2007) Competition between two four-wave mixing channels via atomic coherence. *Appl. Phys. Lett.*, **91**, 061113.
26. Wang, Z., Li, P., Zheng, H., Sang, S., Zhang, R., Zhang, Y., and Xiao, M. (2011) Interference of three multiwave mixings via electromagnetically induced transparency. *J. Opt. Soc. Am. B*, **28**, 1922–1927.
27. Wu, Z., Yuan, C., Zhang, Z., Zheng, H., Huo, S., Zhang, R., Wang, R., and Zhang, Y. (2011) Observation of eight-wave mixing via electromagnetically induced transparency. *EPL (Europhys. Lett.)*, **94**, 64005.
28. Yan, M., Rickey, E.G., and Zhu, Y. (2001) Observation of doubly dressed states in cold atoms. *Phys. Rev. A*, **64**, 013412.

29. Drampyan, R., Pustelny, S., and Gawlik, W. (2009) Electromagnetically induced transparency versus nonlinear faraday effect: coherent control of the light beam polarization. *Phys. Rev. A.*, **80**, 033815 (Arxiv preprint arXiv:0906.0571).

30. Joshi, A. and Xiao, M. (2005) Phase gate with a four-level inverted-Y system. *Phys. Rev. A*, **72**, 062319.

31. Joshi, A. and Xiao, M. (2005) Generalized dark-state polaritons for photon memory in multilevel atomic media. *Phys. Rev. A*, **71**, 041801.

32. Han, Y., Xiao, J., Liu, Y., Zhang, C., Wang, H., Xiao, M., and Peng, K. (2008) Interacting dark states with enhanced nonlinearity in an ideal four-level tripod atomic system. *Phys. Rev. A*, **77** 023824.

33. Rebić, S., Vitali, D., Ottaviani, C., Tombesi, P., Artoni, M., Cataliotti, F., and Corbalan, R. (2004) Polarization phase gate with a tripod atomic system. *Phys. Rev. A*, **70** 032317.

34. Nie, Z., Zheng, H., Li, P., Yang, Y., Zhang, Y., and Xiao, M. (2008) Interacting multiwave mixing in a five-level atomic system. *Phys. Rev. A*, **77**, 063829.

35. Zhang, Y. and Xiao, M. (2007) Generalized dressed and doubly-dressed multi-wave mixing, *Opt. Express*, **15**, 7182–7189.

36. Li, N., Zhao, Z., Chen, H., Li, P., Li, Y., Zhao, Y., Zhou, G., Jia, S., and Zhang, Y. (2012) Observation of dressed odd-order multi-wave mixing in five-level atomic medium. *Opt. Express*, **20**, 1912–1929.

3
Controllable Autler–Townes Splitting of MWM Process via Dark State

Highlights

Ac-Stark shift and Autler–Townes splitting in high-order nonlinear optical processes is experimentally observed. Such controlled multi-channel shift and splitting signals can have potential applications in optical communication and quantum information processing.

By EIT windows, and induced atomic coherence, the FWM and SWM signals can be made very efficient and they can pass through the dense atomic medium nearly transparently. Besides generating various MWM processes, the multi-level atomic system can also be used to generate coexisting FWM (SWM) processes with specially designed spatial patterns and phase-matching conditions for laser beams. In this chapter, the ac-Stark effect and AT splitting of MWM and their various effects are introduced. In the first, we discuss the two-photon ac-Stark effect in the FWM process, which can be used to determine the energy-level shift of the atom under dressing fields. Then, the primary and secondary AT splitting of FWM and SWM are introduced. Such controlled multi-channel splitting of nonlinear optical signals can have potential applications in optical communication and quantum information processing.

3.1
Measurement of ac-Stark Shift via FWM

When a strong laser beam is applied to a two-level atomic system, the energy levels experience a frequency shift, known as *ac-Stark effect* [1, 2]. Such ac-Stark effects were well studied in various atomic [3], molecular [4], and solid [5] systems. When multiple laser beams interact with a three-level system, more interesting phenomena, such as ac-Stark splitting (i.e., AT splitting), can be observed by using a pump–probe laser configuration, with one atomic transition coupled by a strong fixed-frequency beam and another probed by a weak beam [6]. Multi-peak ac-Stark splitting was studied by applying two strong pump fields in the three-level atoms [7–12]. In addition, the ac-Stark effect in a four-level system was investigated,

Quantum Control of Multi-Wave Mixing, First Edition. Yanpeng Zhang, Feng Wen, and Min Xiao.

where the effect of the two strong pump fields connecting the cascade three-level subsystem was measured by probing the frequency shift of one of the sublevels with transition to a fourth level [3]. Generally, these ac-Stark shifts are caused by single-beam (or one photon) transitions and can be interpreted easily by the standard dressed-state theory.

Also, nonlinear FWM spectroscopies have been well developed as powerful tools to detect material properties [13]. For example, the dipole moment of an atomic transition can be measured by using such an FWM process. In multi-level EIT systems [14], the FWM processes can be generated very efficiently even with low-power cw-laser beams due to induced atomic coherence and the two-photon Doppler-free EIT window [15]. We have previously studied the effects of dressing fields on the enhancement and suppression of FWM processes [16], but their effects on frequency shifts have not been addressed.

In this section, we present a new method to measure the ac-Stark shift due to the two-photon cascade dressing scheme in an inverted Y-type four-level atomic system via the frequency shift of the FWM signal in an improved experimental system. In the current experiment, the laser beams are arranged to satisfy two-photon Doppler-free configurations for both the cascade- and Λ-subsystems, as shown in Figure 3.1(a). The FWM signal is generated from the Λ-subsystem. To have a better understanding of this technique, we also analyze the distribution of the FWM efficiencies within the Doppler width, which relates to the *hf* structures of the atomic energy levels. In comparison with the previous pump–probe method [3], this technique gives a more direct measurement of the energy-level shift.

3.1.1
Experiment and Basic Theory

The relevant atomic system is shown in Figure 3.1(b). Four energy levels from ^{85}Rb atoms in vapor cell (60 °C) are involved in the experiment. Four continuous

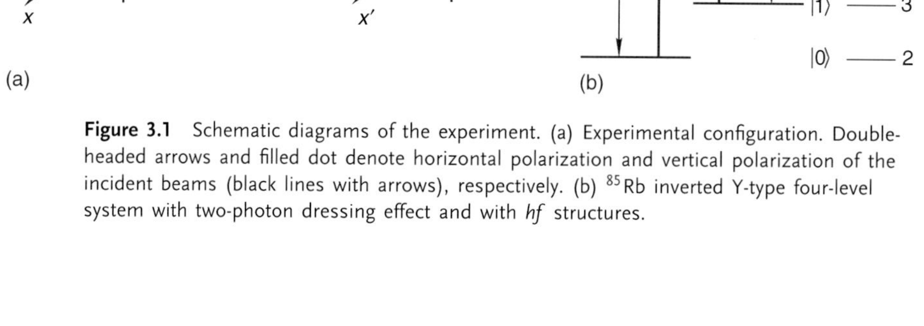

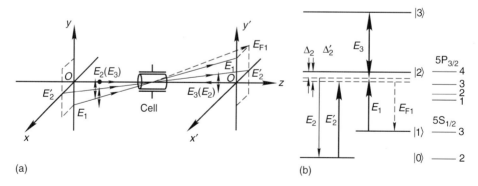

Figure 3.1 Schematic diagrams of the experiment. (a) Experimental configuration. Double-headed arrows and filled dot denote horizontal polarization and vertical polarization of the incident beams (black lines with arrows), respectively. (b) ^{85}Rb inverted Y-type four-level system with two-photon dressing effect and with *hf* structures.

wave (cw)-laser beams in the experiment are spatially aligned as shown in Figure 3.1(a).

First, the laser beam $\boldsymbol{E}_1$ from a diode laser (as scanning beam) connects the transition between energy levels $|1\rangle$ and $|2\rangle$. Another two laser beams $\boldsymbol{E}_2$ and $\boldsymbol{E}'_2$ with frequency detuning Δ_2 and Δ'_2 from a second diode laser are coupled to the energy levels $|0\rangle$ and $|2\rangle$, which propagate in the same direction with a small angle from the beam $\boldsymbol{E}_1$ (see Figure 3.1(a)). $\boldsymbol{E}_2$ and $\boldsymbol{E}'_2$ have an 80 MHz frequency difference ($\omega_2 = \omega'_2 + 80\,\text{MHz}$) introduced by an acoustic-optical modulator (AOM). The polarization of each beam is set by a half-wave plate and a polarization beam splitter (PBS) cube. On the basis of the arrangement of the laser beams, the FWM signal $\boldsymbol{E}_{F1}$ from the three-level Λ-type subsystem ($|0\rangle-|1\rangle-|2\rangle$) will be generated with EIT window [15], satisfying the phase-matching condition $\mathbf{k}_{F1} = \mathbf{k}_1 + \mathbf{k}'_2 - \mathbf{k}_2$, where $\mathbf{k}_i$ is the wave vector. This FWM signal $\boldsymbol{E}_{F1}$ (shown in Figure 3.2) is detected by an avalanche photodiode detector (APD).

We investigate the FWM efficiency as a function of transition frequency, which relates to the *hf* structures of the atom. Figure 3.2 shows the experimentally measured intensities of the FWM signal $\boldsymbol{E}_{F1}$ from the three-level Λ-type ($|0\rangle-|1\rangle-|2\rangle$) atomic system via the detuning Δ_2 for different laser powers. The horizontal axis is calibrated by choosing the lowest point of the Doppler absorption profile as zero point, corresponding to the highest *hf* transition probability, that is, $F = 3 \rightarrow F' = 4$ [18]. The Doppler width of rubidium atoms at 60 °C is about 523 MHz. The powers of the $\boldsymbol{E}_2$ and $\boldsymbol{E}'_2$ beams are, respectively, 0.9 mW (diameter 0.6 mm) and 7.3 mW (diameter 0.9 mm) in Figure 3.2(a), and 7.3 and 1.3 mW in Figure 3.2(b). The power of beam $\boldsymbol{E}_1$ is fixed at 8.5 mW (diameter 1.0 mm). The FWM signals are always generated in the negative part of the Doppler absorption spectrum, $\Delta_2 < 0$ and hardly observed in the positive part regardless of the powers of the laser beams.

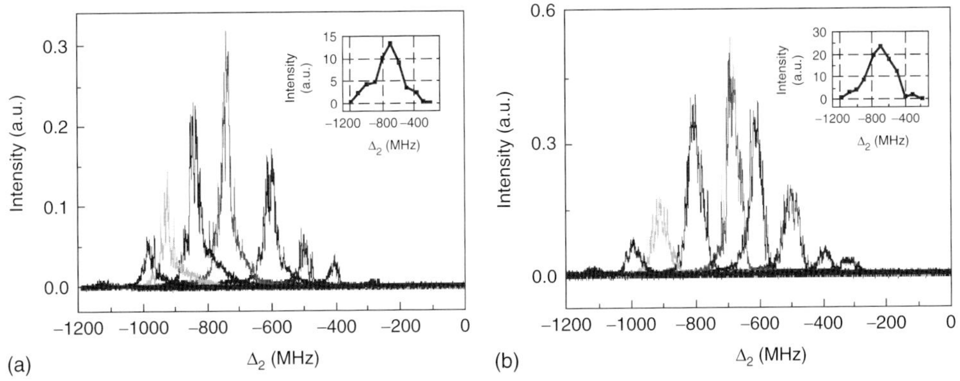

Figure 3.2 Spectra of the FWM process via the frequency detuning Δ_2. The power of beam $\boldsymbol{E}_1$ is 8.5 mW (diameter 1.0 mm). The powers of beams $\boldsymbol{E}_2$ and $\boldsymbol{E}'_2$ are (a) 0.9 mW (diameter 0.6 mm) and 7.3 mW (diameter 0.9 mm), (b) 7.3 mW (diameter 0.9 mm), and 1.3 mW (diameter 0.6 mm), respectively. Insets are the FWM efficiencies versus frequency detuning Δ_2 by calculating the area of each FWM signal $\boldsymbol{E}_{F1}$. Source: Adapted from Ref. [17].

Such an effect can be attributed to the transition rule for the *hf* structures (see Figure 3.1(b)), that is, the transition between energy levels $|0\rangle$ (F = 2) and $|2\rangle$ (F′ = 4) is forbidden for beams $\boldsymbol{E}_2$ and $\boldsymbol{E}'_2$. In addition, the transition between energy levels $|1\rangle$ (F = 3) and $|2\rangle$ (F′ = 1) is also forbidden for beam $\boldsymbol{E}_1$. There is a frequency range for an efficient FWM process as can be seen in the insets of Figure 3.2, where each point is calculated by the area of the FWM signal. Note that because all incident beams are nondegenerate in frequency, interference phenomenon between incident beams is greatly reduced or even cancelled, which is similar to the coherent anti-Stokes Raman spectroscopy (CARS) [19].

Next, we apply a coupling beam $\boldsymbol{E}_3$ with frequency detuning Δ_3, which comes from a Ti:Sapphire ring laser, to drive the transition between the upper energy levels $|2\rangle$ and $|3\rangle$. This beam propagates in the direction opposite to the $\boldsymbol{E}_2$ beam and creates (together with $\boldsymbol{E}_2$) atomic coherence between the dipole forbidden transitions of $|1\rangle$ to $|3\rangle$ via two-photon process with $\Delta_1 + \Delta_3 = 0$. In the cascade three-level atomic subsystem ($|1\rangle-|2\rangle-|3\rangle$) coupled by two strong lights with frequencies ω_1 and ω_3 as shown in Figure 3.1(b), each energy level is affected by the one-photon ac-Stark effect that has been investigated before. Here, we consider the situation in which the ground state $|1\rangle$ is shifted by the two-photon process connecting transition between energy levels $|1\rangle$ and $|3\rangle$. This can be understood in terms of the dressed-state theory. In the following, we present a brief description of the two-photon dressing situation [3].

The total Hamiltonian $\widehat{H}$ (consisting of an atomic part, a field part, and an interaction part) of the cascade three-level system can be written as [3]

$$\widehat{H} = \widehat{H}_{\mathrm{A}} + \widehat{H}_{\mathrm{F}} + \widehat{H}_{\mathrm{I}} \tag{3.1}$$

$$\text{with}\quad \widehat{H}_{\mathrm{A}} = \hbar\omega_{21}|2\rangle\langle 2| + \hbar\omega_{31}|3\rangle\langle 3|, \quad \widehat{H}_{\mathrm{F}} = \hbar\omega_1 a_1^{\dagger} a_1 + \hbar\omega_3 a_2^{\dagger} a_2,$$

$$\hat{H}_{\mathrm{I}} = \left(\frac{\hbar\Omega_1}{2}\right)[a_1|2\rangle\langle 1| + a_1^{\dagger}|1\rangle\langle 2|] + \left(\frac{\hbar\Omega_2}{2}\right)[a_2|3\rangle\langle 2| + a_2^{\dagger}|2\rangle\langle 3|] \tag{3.2}$$

where a_i and $a_i^{\dagger}$ are the annihilation and creation operators for the field modes and Ω_i is the Rabi frequency. The ground-state energy is set to zero. We use the notation in Ref. [3] to denote the three energy levels as $|1,n,m\rangle$, $|2, n-1, m\rangle$, and $|3, n-1, m-1\rangle$, where $|i,n,m\rangle$ indicates that the atom is in level $|i\rangle$ with n photons in field $\boldsymbol{E}_1$ and m photons in field $\boldsymbol{E}_3$. These states are the eigenstates of the uncoupled atom plus field Hamiltonian, that is, $\widehat{H} = \widehat{H}_{\mathrm{A}} + \widehat{H}_{\mathrm{F}}$. Thus, we can rewrite Eq. (3.1) as

$$\widehat{H} - \hat{I}(n\hbar\omega_1 + m\hbar\omega_2) = \begin{pmatrix} 0 & \hbar\Omega_1/2 & 0 \\ \hbar\Omega_1/2 & 0 & \hbar\Omega_2/2 \\ 0 & \hbar\Omega_2/2 & -\hbar\Delta_3 \end{pmatrix} \tag{3.3}$$

where $\hat{I}$ is the identity matrix. The dressed-state wave functions by two-photon dressing relate to the unperturbed states $|1,n,m\rangle$, $|2, n-1, m\rangle$, and $|3, n-1, m-1\rangle$ are

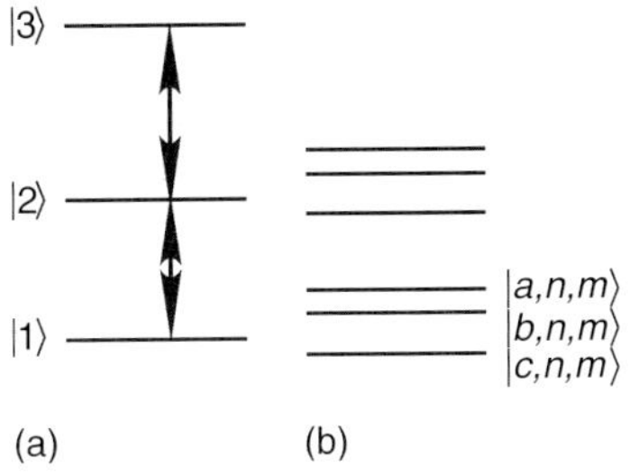

Figure 3.3 Energy-level scheme illustrating the two-photon dressed states. (a) Bare states. (b) Dressed states.

$$
\begin{aligned}
|a,n,m\rangle &= \frac{1}{\sqrt{2}}\left(\frac{\Omega_1}{\Omega}|1,n,m\rangle + |2,n-1,m\rangle + \frac{\Omega_2}{\Omega}|3,n-1,m-1\rangle\right)\\
|b,n,m\rangle &= -\frac{\Omega_2}{\Omega}|1,n,m\rangle + \frac{\Omega_1}{\Omega}|3,n-1,m-1\rangle\\
|c,n,m\rangle &= \frac{1}{\sqrt{2}}\left(\frac{\Omega_1}{\Omega}|1,n,m\rangle - |2,n-1,m\rangle + \frac{\Omega_2}{\Omega}|3,n-1,m-1\rangle\right)
\end{aligned}
\quad (3.4)
$$

where $\Omega = \sqrt{(\Omega_1)^2 + (\Omega_2)^2}$, and their corresponding energies are

$$
E_{a,n,m} = \frac{\hbar\Omega}{2},\quad E_{b,n,m} = 0,\quad E_{c,n,m} = \frac{-\hbar\Omega}{2} \quad (3.5)
$$

Under the two-photon resonant condition for fields $\boldsymbol{E}_1$ and $\boldsymbol{E}_3$, levels $|1,n,m\rangle$, $|2,n-1,m\rangle$, and $|3,n-1,m-1\rangle$ can be considered as a quasi-degenerate triplet as shown in Figure 3.3. In addition, from Eq. (3.5), the energy-level spacing of the dressed state is closely related to the intensity of coupling fields (Ω_1 and Ω_2). However, energy level $|0\rangle$ is not involved in such a dressed-state picture due to $\Delta_2 + \Delta_3 \neq 0$.

3.1.2 Experiment and Result

Figure 3.4 shows the observations of the dressed effect on an atom by the two-photon process. The upper curves of each panel are the FWM signals from Λ-type subsystem ($|0\rangle - |1\rangle - |2\rangle$), while the lower curves are the transmission spectra of the scanning $\boldsymbol{E}_1$ beam. Beams $\boldsymbol{E}_1$ and $\boldsymbol{E}_3$ counter-propagate through the atomic medium and satisfy the EIT condition (see lower curves of Figure 3.4). As the power of the $\boldsymbol{E}_3$ beam increases, the EIT peak (from down to up) varies from one peak (not shown in Figure 3.4) to two peaks shown as red and green curves in Figure 3.4. The reason is that the energy level $|2\rangle$ is split by the strong coupling beam $\boldsymbol{E}_3$ (Figure 3.3).

Next, we set the detuning Δ_2 at off-resonance to study the dressed effect of the FWM signal. Two cases setting the detuning Δ_3 at resonance ($\Delta_3 = 0$ MHz) (Figure 3.4(a)) and near-resonance ($\Delta_3 = -150$ MHz) (Figure 3.4(b)) are measured, respectively. They have similar effects, that is, the frequency of the FWM signal $\boldsymbol{E}_{F1}$

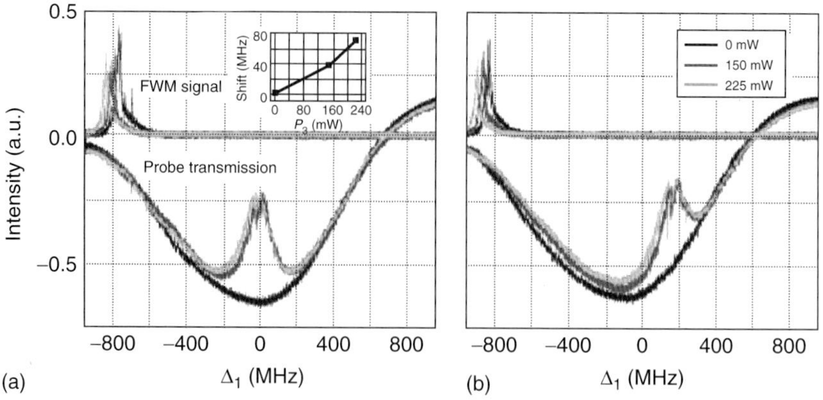

Figure 3.4 Frequency shift of the FWM signal by increasing the power of $\boldsymbol{E}_3$ beam. The upper curves of each panel are the FWM signals from Λ-type subsystem ($|0\rangle-|1\rangle-|2\rangle$) while the lower curves are the transmission spectra of the scanning $\boldsymbol{E}_1$ beam. The powers of beams $\boldsymbol{E}_1$, $\boldsymbol{E}_2$, and $\boldsymbol{E}_2'$ are 8.5, 0.9, and 7.3 mW, respectively. From right curve to left curve of the FWM signal, the power of $\boldsymbol{E}_3$ beam with diameter 1.2 mm varies from 0 mW (bottom curve) to 150 mW (middle curve), then 225 mW (top curve). The detuning are (a) $\Delta_3 = 0$ and (b) $\Delta_3 = -150$ MHz, respectively. Inset is the power dependence of the FWM signal. Source: Adapted from Ref. [17].

is shifted (from right curve to left curve) as the power of the $\boldsymbol{E}_3$ beam increases. Note that this FWM signal is emitted from energy level $|2\rangle$ to level $|1\rangle$. However, besides the phase-matching condition, the FWM process must also satisfy the energy conservation $\omega_{F1} = \omega_1 + \omega_2' - \omega_2$. Thus, the $\boldsymbol{E}_2'$ beam and the FWM signal $\boldsymbol{E}_{F1}$ share the common upper level determined by the frequency of the $\boldsymbol{E}_2'$ beam. The shift of the energy level $|1\rangle$ becomes the only factor to influence the frequency of the FWM signal. Meanwhile, the frequency shift of the FWM signal $\boldsymbol{E}_{F1}$ depends on the power of the beam $\boldsymbol{E}_3$ (Figure 3.4) which does not act on the energy level $|1\rangle$ directly. So the energy level $|1\rangle$ is shifted by coaction of the beam $\boldsymbol{E}_1$ and the coupling beam $\boldsymbol{E}_3$, that is, via the two-photon dressing effect, which creates atomic coherence between the forbidden transition energy levels $|1\rangle$ and $|3\rangle$. From Figure 3.4, the frequency of the FWM signal $\boldsymbol{E}_{F1}$ shifts about −70 MHz as the power of the upper coupling beam $\boldsymbol{E}_3$ varies from 0 to 225 mW (inset of Figure 3.4). We believe that the frequency shift will be larger, if the power of the beam $\boldsymbol{E}_3$ gets higher. In addition, owing to the increase in the two-photon absorption and the population distribution, the intensity of the FWM signal $\boldsymbol{E}_{F1}$ is reduced as frequency shifts further. So, this method provides a more direct way to measure the ac-Stark shifts due to dressing beams than the previous pump–probe method [3].

3.1.3 Conclusion

We have both experimentally and theoretically studied the two-photon ac-Stark effect in the FWM signal, which can be used to determine the energy-level shifts of the atoms due to dressing fields even in a vapor cell. To better understand the relationship between the FWM process and the *hf* structures of the atomic energy levels, we also analyzed the distribution of the FWM efficiencies within the Doppler profile from the Λ-type system related to the *hf* structures and transition rules. This two-photon dressing technique can also be used to directly determine the energy shifts due to dressing fields in inhomogeneously broadened atomic media.

3.2 Evidence of AT Splitting in FWM

Enhanced MWM processes due to atomic coherence have been experimentally demonstrated in several multi-level atomic systems [20, 21]. The keys in such enhanced nonlinear optical processes include enhanced nonlinear susceptibilities due to the induced atomic coherence and slowed laser beam propagation in the atomic medium, as well as greatly reduced linear absorption of the generated optical fields due to EIT [22–26]. Atomic AT splitting was first observed on a radio frequency transition [27], and then in calcium atoms [28]. Such an AT splitting effect was also investigated in the lithium molecule using cw triple-resonant spectroscopy [29, 30] and in semiconductor material with ultrashort intense laser pulses [31]. Recently, an antiblockade effect due to the AT-split Rydberg population was studied theoretically [32] and experimentally [33, 34] with two-photon excitation in a three-level atomic system.

In this section, we discuss the primary and secondary AT splitting of the dressed FWM process in an EIT window of a four-level Y-type atomic vapor system. Theoretical calculations are carried out to explain the observed results, giving a full physical understanding of the interesting multiple AT splitting in the high-order nonlinear optical processes. Although primary AT splitting in molecular lithium was reported previously by detecting fluorescence [29], the current method is a coherent phenomenon in high-order nonlinear processes, making use of the unique spatial phase-matching conditions and laser-induced atomic coherence in the multi-level atomic system, so it can be used to control the direction of optical signals.

3.2.1 Basic Theory

The experiment is carried out in atomic vapor of ^{85}Rb. The energy levels of $5S_{1/2}(F=3)$ ($|0\rangle$), $5P_{3/2}(F=3)$ ($|1\rangle$), $5D_{3/2}$ ($|3\rangle$), and $5D_{5/2}$ ($|2\rangle$) form the four-level Y-type system, as shown in Figure 3.5(a). The vapor cell temperature is set at 60 °C. Two vertically polarized pump laser fields (wavelength 775.98 nm), split from a CW Ti:Sapphire laser with equal power ($P_2 \approx P_2'$) with E_2 (ω_2, $\mathbf{k}_2$, and Rabi frequency G_2) and E_2' (ω_2, $\mathbf{k}_2'$, and G_2'), drive the upper transition ($|1\rangle$ to $|2\rangle$). A strong coupling field E_3 (ω_3, $\mathbf{k}_3$, and G_3, with wavelength 776.16 nm) from a tapered-amplifier diode

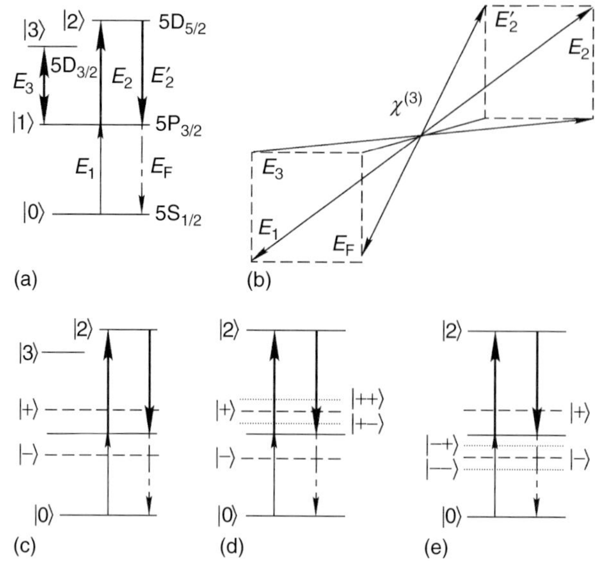

Figure 3.5 (a) Four-level Y-type atomic system. (b) Spatial phase-matching beam geometry used in the experiment. (c)–(e) Corresponding dressed-state pictures of (a).

laser with vertical polarization, drives the other upper transition ($|1\rangle$ to $|3\rangle$). The probe field E_1 (ω_1, $\mathbf{k}_1$, and G_1, with wavelength 780 nm) from an ECDL probes the lower transition $|0\rangle$ to $|1\rangle$. The pump and coupling laser beams (E_2, E_2', and E_3) are spatially aligned in a square-box pattern as shown in Figure 3.5(b), which propagate through the atomic medium in the same direction with small angles ($\sim 0.3^\circ$) between them. The probe beam (E_1) propagates in the opposite direction with a small angle. The diameters of the pump, coupling, and probe beams are about 0.5, 0.5, and 0.3 mm at the vapor cell center. This configuration satisfies the two-photon Doppler-free conditions for the two ladder-type EIT subsystems [20]. Under this configuration, the diffracted FWM (E_F) signal with a horizontal polarization is in the direction determined by the phase-matching condition $\mathbf{k}_F = \mathbf{k}_1 + \mathbf{k}_2 - \mathbf{k}_2'$ and is detected by an avalanche photodiode. The transmitted probe beam is simultaneously detected by a silicon photodiode.

The two ladder-type EIT subsystems form two EIT windows [20]. These two EIT windows can either overlap or be separated by changing the frequency detuning of the pump and coupling laser beams. First, without the strong coupling field E_3, a simple FWM process (with E_1, E_2, and E_2') will generate a signal field E_F with frequency ω_1. The density matrix element $\rho_{10}^{(3)}$ of the FWM signal can be got via the perturbation chain (I) $\rho_{00}^{(0)} \xrightarrow{\omega_1} \rho_{10}^{(1)} \xrightarrow{\omega_2} \rho_{20}^{(2)} \xrightarrow{-\omega_2} \rho_{10}^{(3)}$ [20]. When the powers of E_2 and E_2' are strong enough, they will start to dress the energy level $|1\rangle$ to create the primary dress states $|+\rangle$ and $|-\rangle$, as shown in Figure 3.5(c), which can be described via the perturbation chain (II) $\rho_{00}^{(0)} \xrightarrow{\omega_1} \rho_{\pm0}^{(1)} \xrightarrow{\omega_2} \rho_{20}^{(2)} \xrightarrow{-\omega_2} \rho_{\pm0}^{(3)}$ [20]. Similarly, a stronger probe field E_1 can also modify such an FWM process. Such a self-dressing effect, that is, the participating FWM fields dress the involved energy level, which then affects the FWM process itself, is unique for such MWM processes in multi-level systems and has not been systematically studied before. Next, when the coupling field E_3 is added, these fields (E_2 (E'_2) and E_3) can dress the energy level $|1\rangle$ together. E_2 (E'_2) first creates the primary-dressed states $|\pm\rangle$, then E_3 creates the secondary-dressed states $|\pm\pm\rangle$ at a proper frequency detuning (tuned to near either the upper or lower dressed state $|+\rangle$ or $|-\rangle$), as shown in Figure 3.5(d) and (e), via the perturbation chain (III) $\rho_{00}^{(0)} \xrightarrow{\omega_1} \rho_{\pm\pm0}^{(1)} \xrightarrow{\omega_2} \rho_{20}^{(2)} \xrightarrow{-\omega_2} \rho_{\pm\pm0}^{(3)}$, which generates the secondary AT splitting for the FWM signal. The two primary-dressed states induced by E_2 and E'_2 can be written as $|\pm\rangle = \sin\theta_1|1\rangle + \cos\theta_1|2\rangle$ (Figure 3.5(c)). When E_3 only couples the dressed state $|+\rangle$, the secondary-dressed states are given by $|+\pm\rangle = \sin\theta_2|+\rangle + \cos\theta_2|3\rangle$ (Figure 3.5(d)), where $\sin\theta_1 = -a_1/a_2$, $\cos\theta_1 = G_2^b/a_2$, $\sin\theta_2 = -a_3/a_4$, $\cos\theta_2 = G_3/a_4$, $a_1 = \Delta_2 - \lambda_\pm$, $a_2 = \sqrt{a_1^2 + |G_2^b|^2}$, $a_3 = \Delta_3 - \lambda_+ - \lambda_{+\pm}$, $a_4 = \sqrt{a_3^2 + |G_3|^2}$, $\Delta_i = \Omega_i - \omega_i$ (Ω_i is the atomic resonance frequency), and $G_2^b = G_2 + G'_2$. The eigenvalues are $\lambda_\pm = (\Delta_2 \pm \sqrt{\Delta_2^2 + 4|G_2|^2})/2$ for $|\pm\rangle$, and $\lambda_{+\pm} = \left(\Delta'_3 \pm \sqrt{\Delta_3'^2 + 4|G_3|^2}\right)/2$ for $|+\pm\rangle$, where $\Delta'_3 = \Delta_3 - \lambda_+$.

In order to see the AT splitting of the FWM signal in the EIT window, we first calculate these nonlinear susceptibilities via appropriate perturbation chains here for simplicity. When the coupling beam E_3 is blocked, the simple FWM (with E_1, E_2, and E'_2) process via chain (I) gives $\rho_{10}^{(3)} = G_a/(d_1^2 d_2)$, where $G_a = -iG_1G_2(G'_2)^* \exp(i\mathbf{k}_F \cdot \mathbf{r})$, $d_1 = \Gamma_{10} + i\Delta_1$, $d_2 = \Gamma_{20} + i(\Delta_1 + \Delta_2)$, and Γ_{ij} is the transverse relaxation rate between states $|i\rangle$ and $|j\rangle$. When E_3 is turned on, the above-mentioned simple FWM process will be dressed by fields E_2, E'_2, E_3 and even E_1 (if it is not too weak), and the multi-dressed FWM process is

$$\rho_{10}^{(3)} = \frac{G_a}{[(d_2 + |G_1|^2/d_4)(d_1 + |G_2^b|^2/d_2 + |G_3|^2/d_3)(d_1 + |G_1|^2/\Gamma_0 + |G_3|^2/d_3)]} \tag{3.6}$$

where $d_3 = \Gamma_{30} + i(\Delta_1 + \Delta_3)$ and $d_4 = \Gamma_{21} + i\Delta_2$. As the probe field is weak ($G_1 \ll G_2^b$), the primary AT separation Δ_a is determined mainly by the fields E_2 and E'_2 (i.e., $\Delta_a = \lambda_+ - \lambda_- \approx 2G_2^b$, as shown in Figure 3.6(a)). The secondary AT separation Δ_c is caused by the dressing field E_3 (i.e., $\Delta_c = \lambda_{++} - \lambda_{+-} = \lambda_{-+} - \lambda_{--} \approx 2G_3$, as shown in Figure 3.6(b) and (c)). From the expansions $\rho_{10}^{(3)} \approx G_a[1 - |G_2^b|^2/(d_1 d_2)]/(d_1^2 d_2)$ and $\rho_{10}^{(3)} \approx G_a[1 - |G_3|^2/(d_1 d_3)]^2/(d_1^2 d_2)$, one can see that the first and second terms stand for three-photon and five-photon processes, respectively, which show that the AT splitting results from the destructive interference between the three-photon and five-photon processes.

3.2.2 Experimental Results

Figure 3.6(a) to (c) presents the FWM signal intensity versus the probe field detuning Δ_1 for different dressing field detuning $\Delta_3 = \lambda_\pm \approx \pm G_2^b$. The upper curve in each figure is the probe transmission with two ladder-type EIT windows, and the lower curve is the FWM signal. In Figure 3.6(a), the right ($|0\rangle - |1\rangle - |2\rangle$

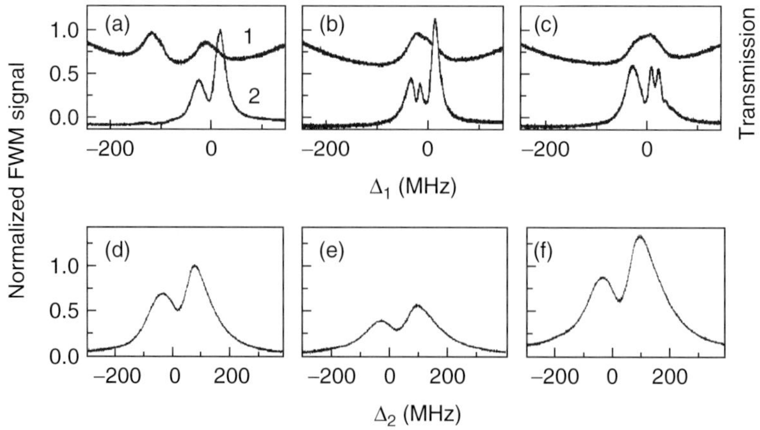

Figure 3.6 Measured multi-peak FWM signals (lower curves) and the corresponding EIT (upper curves) induced by the fields $E_2 + E_2'$ and E_3 versus Δ_1 for $\Delta_2 = 0$, $\Delta_3 = 125$ (a), $\Delta_3 = 20$ (b), $\Delta_3 = -20$ MHz (c), and versus Δ_2 for $\Delta_1 = -20$, $\Delta_3 = 125$ (d), $\Delta_3 = 20$ (e), $\Delta_3 = -20$ MHz (f). The other parameters are $P_1 = 1.3$ mW, $P_2 = P'_2 = 16$ mW, and $P_3 = 146$ mW. Source: Adapted from Ref. [35].

satisfying $\Delta_1 + \Delta_2 = 0$) EIT window is induced by the pump fields $E_2 + E_2'$ and the left ($|0\rangle - |1\rangle - |3\rangle$ satisfying $\Delta_1 + \Delta_3 = 0$) one is induced by the coupling field E_3. The right EIT window contains the double-peak FWM signal E_F, with the two peaks created by $E_2 + E_2'$ (i.e., the primary AT splitting). The left and right peaks of the FWM signal correspond to the dressed states $|+\rangle$ and $|-\rangle$, respectively (Figure 3.5(c)). As the left EIT window is quite far from the right EIT window, the coupling field E_3 basically has no effect on the two peaks in the FWM signal (Figure 3.6(a)).

However, when the frequency of E_3 is tuned to move the left EIT window into the left FWM peak, secondary AT splitting occurs and the left FWM signal peak splits into two peaks (Figure 3.6(b)). Moreover, the right FWM peak is enhanced simultaneously, as the coupling field E_3 dresses the state $|+\rangle$ and separates it into two secondary-dressed states $|++\rangle$ and $|+-\rangle$ (satisfying $\Delta_3 = \lambda_+$). The three peaks in the triple-peak FWM signal (Figure 3.6(b)), from left to right, correspond to the secondary-dressed states $|++\rangle$ and $|+-\rangle$, and the primary-dressed state $|-\rangle$, respectively (Figure 3.5(d)). Similarly, the right FWM peak is separated into two peaks, while the left FWM peak is enhanced when the coupling beam is tuned to the $|-\rangle$ state, as shown in Figure 3.6(c). The three peaks, from left to right, correspond to the primary-dressed state $|+\rangle$ and the secondary-dressed states $|-+\rangle$ and $|--\rangle$ (satisfying $\Delta_3 = \lambda_-$), respectively (Figure 3.5(e)). Figure 3.6(d) shows the AT splitting of the FWM signal versus the pump field detuning Δ_2, when the $|0\rangle - |1\rangle - |3\rangle$ EIT window is tuned quite far from the $|0\rangle - |1\rangle - |2\rangle$ EIT window. In this case, the primary AT splitting is mainly caused by $E_2 + E_2'$. When the two EIT windows get close and overlap, Figure 3.6(e) and (f) depicts the suppressed and enhanced FWM signal intensities versus Δ_2 for different coupling field detuning Δ_3, respectively. Compared with Figure 3.6(d), the FWM signals are suppressed (Figure 3.6(e)) and enhanced (Figure 3.6(f)) at $\Delta_1 + \Delta_3 = 0$ and $\Delta_1 + \Delta_3 = |G_3|^2/\Delta_1$, respectively. Such an FWM signal is enhanced via a single-photon resonance.

With the coupling field E_3 blocked and the probe field E_1 weak in Figure 3.7(a) and (e), the AT splitting separations Δ_a and Δ_b mainly result from the fields E_2 and E_2' (i.e., $\Delta_a, \Delta_b \approx 2G_2^b$). Such AT splitting separations Δ_a (Figure 3.7(b)) and Δ_b (Figure 3.7(f)) get larger with increasing E_2 and E_2'. However, when the probe field E_1 is strong enough (in Figure 3.7(c) and (g)), the AT splitting separations Δ_a and Δ_b are then determined by both the pump fields (E_2, E_2') and the probe field E_1. As Δ_a and Δ_b get larger with increasing E_1, there exists a turning point at $P_1 = 10$ mW (Figure 3.7(d) and (h)), beyond which both AT splitting separations reach their respective saturation values at high probe power (Figure 3.7(d) and (h)). More interestingly, Eq. (3.6) for the dressed FWM with $G_3 = 0$ can be

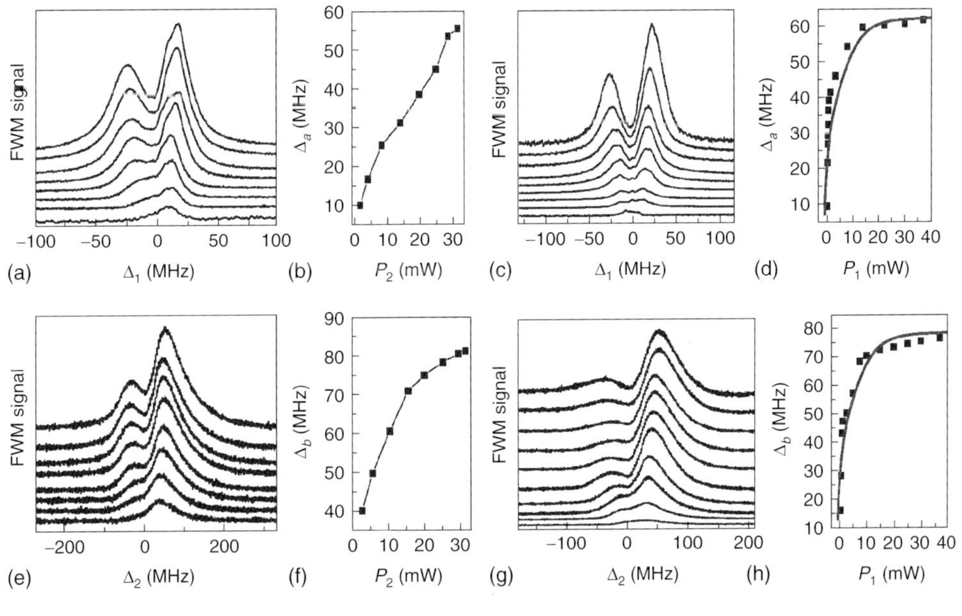

Figure 3.7 Measured double-peak FWM signals versus Δ_1 with $\Delta_2 = 0$ for (a) increasing $P_2 = 1.6$, 4, 8.3, 14, 20, 25, 29, and 32 mW from bottom to top and (b) the power dependence of Δ_a when $P_1 = 1.3$ mW, and for (c) increasing $P_1 = 0.24$, 0.48, 0.58, 0.68, 0.83, 0.93, 1.1, 3.4, and 8 mW from bottom to top and (d) the power dependence of Δ_a when $P_2 = 17$ mW. Measured double-peak FWM signal versus Δ_2 with $\Delta_1 = 0$ for (e) increasing $P_2 = 2.6$, 5.5, 10.3, 15.5, 20, 25, 29, and 31.5 mW from bottom to top and (f) the power dependence of Δ_b when $P_1 = 1.3$ mW, and for (g) increasing $P_1 = 0.5$, 0.75, 1, 2.5, 5, 7.5, 10, 15, 20, and 25 mW from bottom to top and (h) the power dependence of Δ_a when $P_2 = 7.6$ mW. Source: Adapted from Ref. [35].

expanded as

$$\rho_{10}^{(3)} \approx G_a' \left[1 - |G_1|^2 \left(\frac{1}{d_2 d_4} + \frac{1}{d_1 \Gamma_0} \right) + \frac{|G_1|^4}{(\Gamma_0 d_1 d_2 d_4)} \right] \tag{3.7}$$

where $G_a' = G_a / \{[1 + |G_2^b|^2/(d_1 d_2)] d_1^2 d_2\}$.

The first term of Eq. (3.7) represents a three-photon process in the probe and pump fields, and the second term is the five-photon process that interferes destructively with the first term and is responsible for the AT splitting in Figure 3.7(d) and (h). Finally, the third term gives the seven-photon process, which has the same sign as the first term and hence tends to increase the FWM signal. In fact, the second term is dominant at lower probe power, while the third term is dominant at higher probe power. Physically, such a saturation behavior in Figure 3.7(d) and (h) is induced by the balanced interactions between the destructive and constructive interferences of these multi-photon transition pathways [25].

Figure 3.8 presents the FWM signal intensity versus the probe field detuning Δ_1 for different frequency detuning Δ_3 and coupling field power P_3. As the probe E_1 is weak, the primary AT splitting Δ_a mainly results from E_2 and E_2' (i.e., $\Delta_a = \lambda_+ - \lambda_- \approx 2G_2^b$). The secondary AT splitting Δ_c (i.e., $\Delta_c = \lambda_{++} - \lambda_{+-} \approx 2G_3$ in Figure 3.8(a) and $\Delta_c = \lambda_{-+} - \lambda_{--} \approx 2G_3$ in Figure 3.8(c)) gets larger with increasing E_3 (Figure 3.8(b) and (d)). Moreover, the experimentally measured results in Figure 3.8(a) and (c) are in good agreement with our theoretical calculations shown in Figure 3.8(e) and (f), respectively.

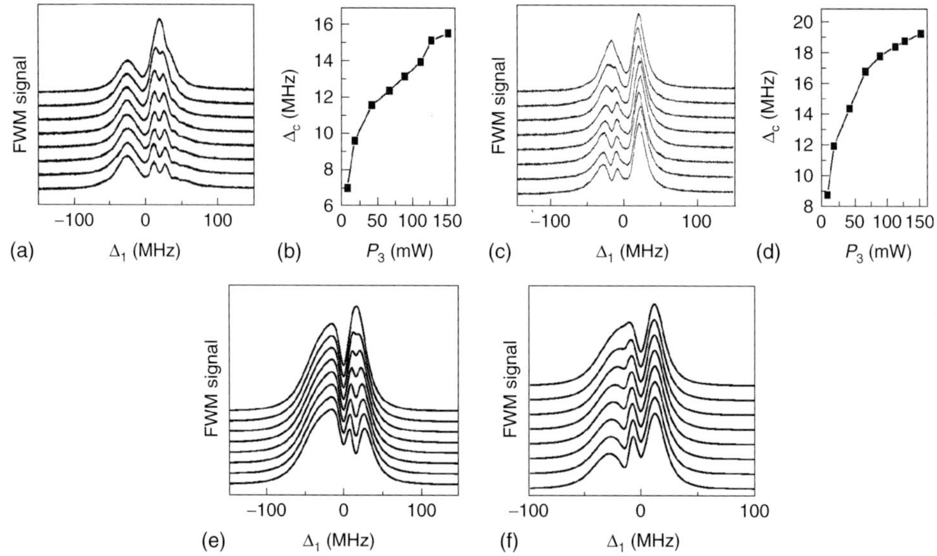

Figure 3.8 Measured multi-peak FWM signals versus Δ_1 with (a) $\Delta_3 = -20$ MHz and (c) $\Delta_3 = 20$ MHz for increasing $P_3 = 8$, 18, 42, 66, 88, 111, 126, and 150 mW from bottom to top when $P_1 = 1.3$ mW and $\Delta_2 = 0$. (b) and (d) Power dependence of Δ_a versus P_3 for the cases of (a) and (c), respectively. (e) and (f) Theoretically calculated results correspond to (a) and (c), respectively.

3.3 Observation of AT Splitting in SWM

The interaction of double-dark states (nested scheme of doubly-dressing) and splitting of dark state (the secondarily-dressed state) in a four-level atomic system with EIT were studied theoretically by Lukin *et al.* [36]. Then, the doubly-dressed states in cold atoms were observed, in which the triple-photon absorption spectrum exhibits a constructive interference between the excitation pathways of two closely spaced, doubly-dressed states [37]. A similar result was obtained in the inverted-Y [38] and double-Λ [39] atomic systems.

In this section, we analyzed the self-, doubly-, and triply-dressed AT splitting states of the SWM process within the EIT window in a five-level atomic system. Theoretical calculations are carried out and used to explain the observed results, giving a full physical understanding of the interesting multiple AT splitting in the high-order nonlinear optical processes. On the basis of the last section, the AT splitting in the FWM process [35], we go further to investigate the complex AT splitting phenomena in the SWM process.

3.3.1 Theoretical Model and Experimental Scheme

The experimental demonstration of the AT splitting of SWM within the EIT window is carried out in the atomic system of ^{85}Rb. The energy levels of $5S_{1/2}(F=3)$, $5S_{1/2}(F=2)$, $5P_{3/2}(F=3)$, $5D_{3/2}$, and $5D_{5/2}$ form the five-level atomic system, as shown in Figure 3.9(b). The atomic vapor cell temperature is set at 60 °C. The probe laser beam E_1 (with frequency ω_1, wave vector $\mathbf{k}_1$, Rabi frequency G_1, and wavelength of 780.245 nm, connecting the transition $5S_{1/2}-5P_{3/2}$) is from an external cavity diode laser (Toptica DL100L), which is horizontally polarized and has a power of about $P_1 \approx 1.3$ mW. The coupling laser beams E_2 (ω_2, $\mathbf{k}_2$, G_2, and wavelength 775.978 nm, connecting the transition $5P_{3/2}-5D_{5/2}$) and E_4 (ω_4, $\mathbf{k}_4$, G_4, and wavelength 776.157 nm, connecting the transition $5P_{3/2}-5D_{3/2}$) are from two external cavity diode lasers (Hawkeye Optoquantum and UQEL100), respectively. The pump laser beams E_3 (ω_3, $\mathbf{k}_3$, and G_3) and E_3' (ω_3, $\mathbf{k}_3'$, and G_3'), which are split from a tapered-amplifier diode laser (Thorlabs TCLDM9) with equal power ($P_3 \approx P_3'$) and vertical polarization, drive the transition $5S_{1/2}-5P_{3/2}$. The diameters of the probe pump and coupling beams are about 0.3, 0.5, and 0.5 mm at the cell center. The pump and coupling laser beams (E_3, E_3', E_2, and E_4) are spatially aligned in a square-box pattern as shown in Figure 3.9(a), which propagate through the atomic medium in the same direction with small angles ($\sim 0.3^\circ$) between them (the angles are exaggerated in the figure). The probe beam (E_1) propagates in the opposite direction with a small angle from the other beams. Under this configuration, the diffracted FWM signal (E_F) and two SWM signals (E_{S1} and E_{S2}) with the same horizontal polarization are in the directions determined by the phase-matching conditions $\mathbf{k}_F = \mathbf{k}_1 + \mathbf{k}_3 - \mathbf{k}_3'$, $\mathbf{k}_{S1} = \mathbf{k}_1 + \mathbf{k}_2 - \mathbf{k}_2 + \mathbf{k}_3 - \mathbf{k}_3'$, and $\mathbf{k}_{S2} = \mathbf{k}_1 + \mathbf{k}_4 - \mathbf{k}_4 + \mathbf{k}_3 - \mathbf{k}_3'$. These signals are in the same direction as E_F (at the

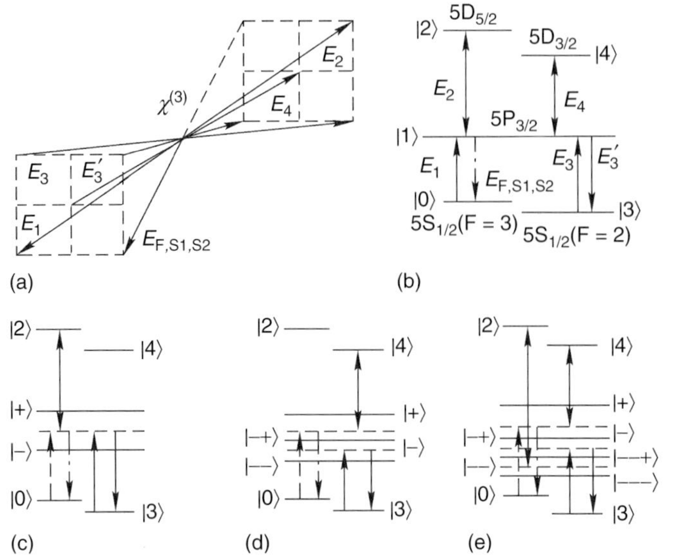

Figure 3.9 (a) Phase-matching spatial beam geometry used in the experiment. (b) Five-level atomic system with one probe field E_1, two pump fields E_3 and E_3', and two coupling (dressing) fields E_2 and E_4. E_F is the generated FWM signal. E_{S1} and E_{S2} are the two generated SWM signals. (c)–(e) Corresponding dressed-state pictures of (b).

lower right corner of Figure 3.9(a)), and are detected by an avalanche photodiode detector. The transmitted probe beam is simultaneously detected by a silicon photodiode.

For the five-level atomic system as shown in Figure 3.9(b), if two strong coupling laser fields (E_2 and E_4) drive two separate upper transitions ($|1\rangle \to |2\rangle$ and $|1\rangle \to |4\rangle$), respectively, and a weak laser field (E_1) probes the lower transition ($|0\rangle$ to $|1\rangle$), two ladder-type EIT subsystems will form with two-photon Doppler-free configuration [23] and two EIT windows appear. Depending on the frequency detuning of the two coupling laser beams, these two EIT windows can either overlap or be separated in frequency on the probe beam transmission signal. On the other hand, if the probe field E_1 drives the transition ($|0\rangle$ to $|1\rangle$) and the two pump fields (E_3 and E_3') drive another transition ($|3\rangle$ to $|1\rangle$) in the three-level Ξ-type subsystem, as shown in Figure 3.9(b), there will be a corresponding FWM signal generated at frequency ω_1 (satisfying $\mathbf{k}_F = \mathbf{k}_1 + \mathbf{k}_3 - \mathbf{k}_3'$). However, the FWM signal without the EIT window (not satisfying the two-photon Doppler-free configuration [23]) can be neglected. When the two coupling laser fields E_2 (connecting transition $|1\rangle$ to $|2\rangle$) and E_4 (connecting the transition $|1\rangle$ to $|4\rangle$) are added, two SWM processes will occur [20]. First, without the strong coupling field E_4, a simple SWM1 process (E_{S1}) at frequency ω_1 is generated from the probe beam (E_1), the coupling field (E_2), and two pump fields (E_3 and E_3'), via the perturbation chain (I): $\rho_{00}^{(0)} \xrightarrow{G_1} \rho_{10}^{(1)} \xrightarrow{G_2} \rho_{20}^{(2)} \xrightarrow{G_2^*} \rho_{10}^{(3)} \xrightarrow{(G_3')^*} \rho_{30}^{(4)} \xrightarrow{G_3} \rho_{10}^{(5)}$ (satisfying $\mathbf{k}_{S1} = \mathbf{k}_1 + \mathbf{k}_2 - \mathbf{k}_2 + \mathbf{k}_3 - \mathbf{k}_3'$) [20]. When the power of E_2 is strong enough, it will start to dress the energy level $|1\rangle$ to create the primarily-dressed states $|+\rangle$ and $|-\rangle$, as shown in Figure 3.9(c). This dressed SWM1 process can be described via the perturbation chain (II): $\rho_{00}^{(0)} \xrightarrow{\omega_1} \rho_{\pm 0}^{(1)} \xrightarrow{\omega_2} \rho_{20}^{(2)} \xrightarrow{-\omega_2} \rho_{\pm 0}^{(3)} \xrightarrow{-\omega_3} \rho_{30}^{(4)} \xrightarrow{\omega_3} \rho_{\pm 0}^{(5)}$. Such a self-dressing effect, that is, one of the participating fields for generating SWM dresses the involved energy level $|1\rangle$, which then modifies the SWM process itself, is unique for such MWM processes in multi-level systems. Similarly, for another SWM process (with fields E_1, E_4, E_3, and E_3'), without the strong coupling field E_2, it will generate a signal field E_{S2} at frequency ω_1 via the perturbation chain (III): $\rho_{00}^{(0)} \xrightarrow{G_1} \rho_{10}^{(1)} \xrightarrow{G_4} \rho_{40}^{(2)} \xrightarrow{G_4^*} \rho_{10}^{(3)} \xrightarrow{(G_3')^*} \rho_{30}^{(4)} \xrightarrow{G_3} \rho_{10}^{(5)}$ (satisfying $\mathbf{k}_{S2} = \mathbf{k}_1 + \mathbf{k}_4 - \mathbf{k}_4 + \mathbf{k}_3 - \mathbf{k}_3'$). When the power of E_4 is strong enough, it will start to dress the energy level $|1\rangle$ to create the primarily-dressed states $|+\rangle$ and $|-\rangle$, as shown in Figure 3.9(d). This dressed SWM2 process can be described via the perturbation chain (IV): $\rho_{00}^{(0)} \xrightarrow{\omega_1} \rho_{\pm 0}^{(1)} \xrightarrow{\omega_4} \rho_{40}^{(2)} \xrightarrow{-\omega_4} \rho_{\pm 0}^{(3)} \xrightarrow{-\omega_3} \rho_{30}^{(4)} \xrightarrow{\omega_3} \rho_{\pm 0}^{(5)}$.

Next, when both coupling fields (E_2 and E_4) are on at the same time, they can dress the energy level $|1\rangle$ together. For the SWM1 process (E_{S1}), E_2 first produces the primarily-dressed states $|\pm\rangle$, then E_4 produces the secondarily-dressed states $|\pm\pm\rangle$ at a proper frequency detuning (i.e., either tuned to the upper or lower dressed state, $|+\rangle$ or $|-\rangle$), as shown in Figure 3.9(e) via the perturbation chain (V): $\rho_{00}^{(0)} \xrightarrow{\omega_1} \rho_{\pm\pm 0}^{(1)} \xrightarrow{\omega_2} \rho_{20}^{(2)} \xrightarrow{-\omega_2} \rho_{\pm\pm 0}^{(3)} \xrightarrow{-\omega_3} \rho_{30}^{(4)} \xrightarrow{\omega_3} \rho_{\pm\pm 0}^{(5)}$. This generates the secondary AT splitting for the SWM1 signal. The situation for the SWM2 (E_{S2}) process is similar.

The two primarily-dressed states induced by E_2 can be written as $|\pm\rangle = \sin\theta_{1\pm}|1\rangle + \cos\theta_{1\pm}|2\rangle$ (Figure 3.9(c)). When E_4 only couples to the dressed state $|+\rangle$, the secondarily-dressed states are then given by $|+\pm\rangle = \sin\theta_{2\pm}|+\rangle + \cos\theta_{2\pm}|4\rangle$ (Figure 3.9), where $\tan\theta_{1\pm} = -a_{1\pm}/G_2$, $\tan\theta_{2\pm} = -a_{2\pm}/G_4$, $a_{1\pm} = \Delta_2 - \lambda_\pm^{(1)}$, and $a_{2\pm} = \Delta_4 - \lambda_+^{(1)} - \lambda_{+\pm}^{(1)}$. One can obtain the eigenvalues $\lambda_\pm^{(1)} = (\Delta_2 \pm \sqrt{\Delta_2^2 + 4|G_2|^2})/2$ (measured from level $|1\rangle$) of $|\pm\rangle$, and $\lambda_{+\pm}^{(1)} = (\Delta_4' \pm \sqrt{\Delta_4'^2 + 4|G_4|^2})/2$ (measured from level $|+\rangle$) of $|+\pm\rangle$, where $\Delta_4' = \Delta_4 - \lambda_+^{(1)}$. When E_4 only couples to the dressed state $|-\rangle$, the secondarily-dressed states are given by $|-\pm\rangle = \sin\theta_{2\pm}|-\rangle + \cos\theta_{2\pm}|4\rangle$, where $a_{2\pm} = \Delta_4 - \lambda_-^{(1)} - \lambda_{-\pm}^{(1)}$, and other parameters are the same as before. Then, we obtain the same eigenvalues $\lambda_\pm^{(1)}$and $\lambda_{-\pm}^{(1)} = (\Delta_4' \pm \sqrt{\Delta_4'^2 + 4|G_4|^2})/2$ (measured from level $|-\rangle$) of $|-\pm\rangle$, where $\Delta_4' = \Delta_4 - \lambda_-^{(1)}$. Similarly, when E_4 induces the two primarily-dressed states, and E_2 acts as the external-dressing field, one can get the following corresponding eigenvalues: $\lambda_\pm^{(2)} = (\Delta_4 \pm \sqrt{\Delta_4^2 + 4|G_4|^2})/2$, $\lambda_{+\pm}^{(2)} = (\Delta_2' \pm \sqrt{\Delta_2'^2 + 4|G_2|^2})/2$ ($\Delta_2' = \Delta_2 - \lambda_+^{(2)}$), and $\lambda_{-\pm}^{(2)} = (\Delta_2' \pm \sqrt{\Delta_2'^2 + 4|G_2|^2})/2$ ($\Delta_2' = \Delta_2 - \lambda_-^{(2)}$).

In general, for arbitrary strengths of the fields E_1, E_3, E_3', E_2, and E_4, one needs to solve the coupled density matrix equations to obtain $\rho_{10}^{(5)}$ for the SWM processes, which we have done in simulating the experimental results later on. For simplicity, we have solved the coupled equations with perturbation chain (I) to obtain the nonlinear density matrix element for the multi-dressed SWM processes (including self-dressing and external-dressing) as $\rho_{10}^{(5)} = iG_{S1}/(ABCDE)$, where $G_{S1} =$

$G_1G_2G_2^*G_3'^*G_3$, $A = d_1 + G_1^2/\Gamma_1 + G_1^2/\Gamma_0 + G_2^2/d_2 + G_4^2/d_4$, $B = d_2 + G_1^2/d_2'$, $C = d_1 + G_1^2/\Gamma_1 + G_2^2/d_2 + G_3^2/d_3 + G_4^2/d_4$, $D = d_3 + G_1^2/d_3'$, $E = d_1 + G_1^2/\Gamma_1 + G_1^2/\Gamma_0 + G_3^2/d_3 + G_4^2/d_4$, $d_1 = \Gamma_{10} + i\Delta_1$, $d_2 = \Gamma_{20} + i(\Delta_1 + \Delta_2)$, $d_2' = i\Delta_2 + \Gamma_{21}$, $d_3 = \Gamma_{30} + i(\Delta_1 - \Delta_3)$, $d_3' = \Gamma_{31} - i\Delta_3$, $d_4 = \Gamma_{40} + i(\Delta_1 + \Delta_4)$ with $\Delta_i = \Omega_i - \omega_i$ and Γ_{ij} being the transverse relaxation rate between states $|i\rangle$ and $|j\rangle$. For the SWM1 signal (due to the weak probe field), the expression can be simplified as $\rho_{S1}^{(5)} = iG_{S1}/[(d_1 + G_2^2/d_2)d_2(d_1 + G_2^2/d_2 + G_3^2/d_3)d_3(d_1 + G_3^2/d_3)]$. Similarly, for the SWM2 signal, the expression is simplified as $\rho_{S2}^{(5)} = iG_{S2}/[(d_1 + G_4^2/d_4)d_4(d_1 + G_3^2/d_3 + G_4^2/d_4)d_3(d_1 + G_3^2/d_3)]$, where $G_{S2} = G_1G_3'^*G_3G_4G_4^*$.

There exist two ladder-type EIT windows in Figure 3.9(b), that is, the $|0\rangle - |1\rangle - |2\rangle$ EIT1 window satisfying $\Delta_1 + \Delta_2 = 0$ (induced by the coupling field E_2) and the $|0\rangle - |1\rangle - |4\rangle$ EIT2 window satisfying $\Delta_1 + \Delta_4 = 0$ (induced by the coupling field E_4). The EIT1 and EIT2 windows contain the SWM1 signal (E_{S1}) and the SWM2 signal (E_{S2}), respectively. Next, we consider the AT splitting of the SWM signals within the EIT windows.

3.3.2
Experiment and Result

When the external-dressing field E_4 is blocked, we get the SWM1 signal within the EIT1 window (which is an inverted-Y system) [38]. Figure 3.10(a1), (b1), and (c1) presents the SWM1 signal intensity versus the probe field detuning (Δ_1) for different field powers of P_1, P_2, and P_3 with the same frequency detuning of $\Delta_2 = -50$ MHz. Obviously, the SWM1 signal shows two peaks due to multi-dressing effects (Figure 3.9(c)). When the power increases, the intensity of the SWM1 signals increases accordingly, while the left peak height is always greater than the height of the right peak. Meanwhile, the increments of the AT splitting separations $\Delta_i = \lambda_+^{(1)} - \lambda_-^{(1)} \approx 2\sqrt{|G_j|^2 + |G_{i0}|^2}$ (i equals; a, b, c corresponds to j equals; 1, 2, 3, respectively, and $|G_{a0}|^2 = |G_{20}|^2 + |G_{30}|^2$, $|G_{b0}|^2 = |G_{10}|^2 + |G_{30}|^2$, $|G_{c0}|^2 = |G_{10}|^2 + |G_{20}|^2$), increase obviously with increased powers (Rabi frequencies) P_1 (G_1), P_2 (G_2), and P_3 (G_3), respectively, and fixed P_2 and P_3 (G_{20} and G_{30}), P_1 and P_3 (G_{10} and G_{30}), and P_1 and P_2 (G_{10} and G_{20}), respectively.

The two peaks of the double-peak SWM1 signal (Figure 3.10(a1), (b1), and (c1)) correspond, from left to right, to the primarily-dressed states $|+\rangle$ and $|-\rangle$, respectively (Figure 3.9(c)). Moreover, the experimentally measured (peak separation) results in Figure 3.10(a1), (b1), and (c1) are in good agreement with our theoretical calculations (solid curves), as shown in Figure 3.10(a2), (b2), and (c2).

When the external-dressing field E_2 is blocked, we get the SWM2 signal in the EIT2 window. Figure 3.11 presents the SWM2 signal intensity versus the probe field detuning (Δ_1) for different powers of P_1, P_3, and P_4 with the same frequency detuning of $\Delta_4 = 0$ MHz. Figure 3.11(a) depicts the measured EIT windows induced by the self-dressing field E_4 versus Δ_1 (satisfying $\Delta_1 + \Delta_4 = 0$). Such an EIT window increases with P_4 power increasing. The SWM2 signal has three peaks. In general, when the power increases, the intensity of the SWM2 signal also

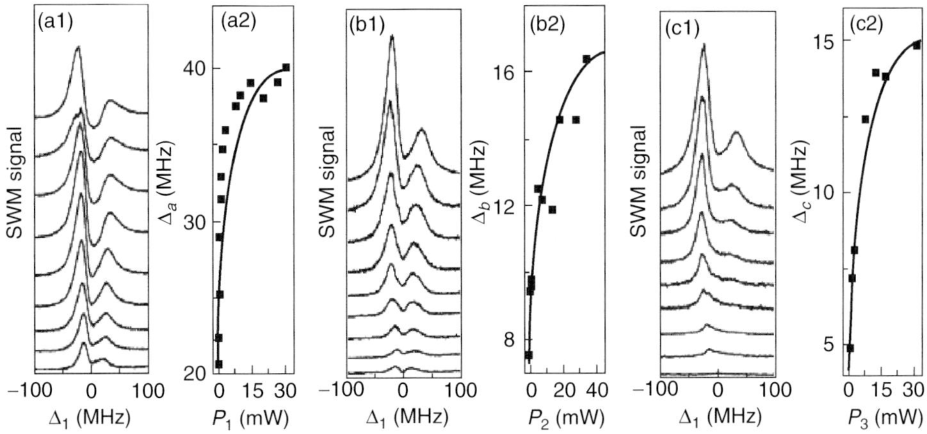

Figure 3.10 (a1), (b1), and (c1) are the measured SWM1 self-dressing AT splitting signals versus Δ_1 for $\Delta_2 = -50$ MHz under different P_1, P_2, and P_3 powers, respectively. (a1) is increasing $P_1 = 0.30$, 0.36, 0.43, 0.90, 1.37, 1.85, 2.32, 3.66, 5.33, 8.18, 10.3, 13.61, 15.46, 22.5, 24.2, 25.3, 28.8, and 29.5 mW from bottom to top. (b1) is increasing $P_2 = 0.3$, 0.6, 0.9, 1.2, 1.5, 2.1, 4.5, 7.5, 12.5, 17.2, 25.5, and 31.4 mW from bottom to top. (c1) is increasing $P_3 = 1.8$, 3.9, 6.3, 7.7, 11.4, 16.7, 22.2, 33.1, and 52.2 mW from bottom to top. (a2), (b2), and (c2) are the corresponding power dependences of (a1), (b1), and (c1), respectively. Here Δ_a, Δ_b, and Δ_c are the increments of distance between the two AT splitting peaks when P_1, P_2, and P_3 are increased, respectively, and the squares are the experimental results, while the solid lines in (a2, b2, and c2) are the theoretical calculations. The fixed powers in (a1, a2), (b1, b2), and (c1, c2) are ($P_2 = 32.0$ mW and $P_3 = 55.0$ mW), ($P_1 = 13.0$ mW and $P_3 = 55.0$ mW), and ($P_1 = 13.0$ mW and $P_2 = 32.0$ mW), respectively. Source: Adapted from Ref. [40].

increases accordingly. However, the states of AT splitting change differently for different power changes. If only P_1 power increases, the right peak first increases and then decreases, while the height of the middle peak is always larger than the height of either the left or right peak. If only P_3 power increases, the right peak always increases, while the height of the middle peak is always larger than the height of either the left or right peak. If only P_4 power increases, the height of the right peak increases monotonously to approach the height of the middle peak. The two primarily-dressed states $|+\rangle$ and $|-\rangle$ are induced by E_4. When E_1 and E_3 couple to the dressed state $|-\rangle$, the secondarily-dressed states $|-+\rangle$ and $|--\rangle$ appear. The three peaks in the SWM2 signal (Figure 3.11(b1), (c1), and (d1)) correspond, from left to right, to the primarily-dressed state $|+\rangle$, the secondarily-dressed states $|-+\rangle$ and $|--\rangle$, respectively (Figure 3.9(d)). On the basis of our theoretical analysis, the two primarily-dressed states dressed by E_4 can be written as $|\pm\rangle = \sin\alpha_{1\pm}|1\rangle + \cos\alpha_{1\pm}|4\rangle$, and the secondarily-dressed states dressed by E_3 and E_1 are given by $|-\pm\rangle = \sin\alpha_{2\pm}|-\rangle + \cos\alpha_{2\pm}|3\rangle$, where $\tan\alpha_{1\pm} = -b_{1\pm}/G_4$, $\tan\alpha_{2\pm} = -b_{2\pm}/\sqrt{|G_1|^2 + |G_3|^2}$, $b_{1\pm} = \Delta_4 - \lambda_\pm^{(3)}$, and $b_{2\pm} = \Delta_3 - \lambda_-^{(3)} - \lambda_{-\pm}^{(3)}$. One can obtain the eigenvalues $\lambda_\pm^{(3)} = (\Delta_4 \pm \sqrt{\Delta_4^2 + 4|G_4|^2})/2$ (measured from level $|1\rangle$)

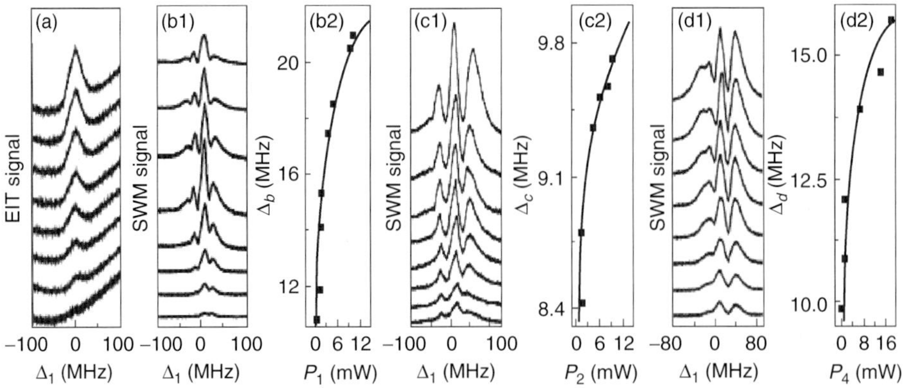

Figure 3.11 (a) is the measured EIT2 induced by the field E_4 versus Δ_1. (b1), (c1), and (d1) are the measured SWM2 self-dressing AT splitting signals versus Δ_1 for $\Delta_4 = 0$ under different P_1, P_3, and P_4 powers, respectively. (b1) is increasing P_1 = 0.4, 0.9, 1.4, 1.8, 3.7, 5.3, 9.3, 10.3, 13.6, 15.5, 24.7, and 29.5 mW from bottom to top. (c1) is increasing P_3 = 1.2, 1.4, 2.9, 4.6, 6.6, 8.2, 9.5, and 27.2 mW from bottom to top. (d1) is increasing P_4 = 0.12, 0.48, 0.6, 1.36, 2.47, 3.1, 5.4, 7.7, 14.0, 23.0, 39.0, 71.0, 116.0, 142.0, 184.0, 205.0, 225.0, 242.0, and 258.0 mW from bottom to top. (b2), (c2), and (d2) are the corresponding power dependences of (b1), (c1), and (d1), respectively. Here Δ_b, Δ_c, and Δ_d are the increments of the distance between the right (for Δ_b and Δ_c) or left (for Δ_d) two AT splitting peaks, respectively, when P_1, P_3, and P_4 are increased, respectively, and the squares are the experimental results, while the solid lines in (b2, c2, and d2) are the theoretical calculations. The fixed powers in (b1, b2), (c1, c2), and (d1, d2) are (P_3 = 30.0 mW and P_4 = 150.0 mW), (P_1 = 1.0 mW and P_4 = 150.0 mW), and (P_1 = 1.0 mW and P_3 = 30.0 mW), respectively.

of $|\pm\rangle$, and $\lambda_{-\pm}^{(3)} = (\Delta_3' \pm \sqrt{\Delta_3'^2 + 4(|G_1|^2 + |G_3|^2)})/2$ (measured from level $|-\rangle$) of $|-\pm\rangle$, where $\Delta_3' = -\Delta_3 - \lambda_-^{(3)}$.

Meanwhile, the increments of the AT splitting separations $\Delta_b = \lambda_{-+}^{(3)} - \lambda_{--}^{(3)} = 2\sqrt{|G_1|^2 + |G_{30}|^2}$, $\Delta_c = \lambda_{-+}^{(3)} - \lambda_{--}^{(3)} = 2\sqrt{|G_3|^2 + |G_{10}|^2}$, and $\Delta_d = \lambda_{+}^{(3)} - \lambda_{-+}^{(3)} = 2G_4 - \sqrt{|G_{10}|^2 + |G_{30}|^2} \approx 2G_4$ ($G_4 \gg G_{10,30}$), increase obviously with increased powers (Rabi frequencies) of P_1 (G_1), P_3 (G_3), and P_4 (G_4), respectively, and fixed P_3 and P_4 (G_{30} and G_{40}), P_1 and P_3 (G_{10} and G_{30}), and P_1 and P_4 (G_{10} and G_{40}), respectively. When the self-dressing P_4 power changes, the state of the AT splitting obviously has an essential distinction from the former two states as the power of P_1 and P_3 change. The experimentally measured results in Figure 3.11(b1), (c1), and (d1) are in good agreement with our theoretical calculations shown in Figure 3.11(b2), (c2), and (d2), respectively.

After studying the self-dressing AT splitting of the individual SWM1 or SWM2 signal, we now consider the cross-dressing AT splitting between the two SWM signals. Figure 3.12(a1) to (a3) presents the interplay between the two SWM signals versus the probe field detuning (Δ_1) for different external-dressing field detuning (Δ_2) with $\Delta_4 = 0$ MHz. Here, we consider the case with $G_2 < G_4$. The upper curve in each figure is the probe transmission with two ladder-type EIT windows and

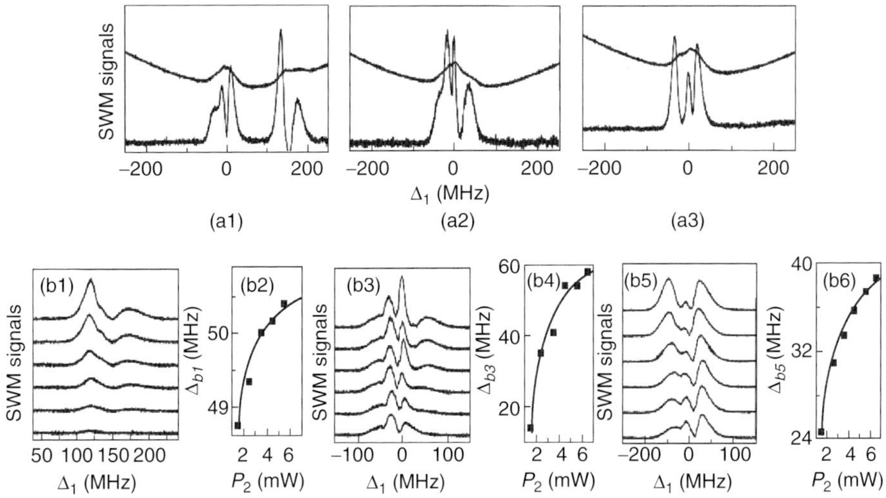

Figure 3.12 Measured SWM1 moving toward SWM2 signal (lower curves) and the corresponding EIT (upper curves) versus Δ_1 for $\Delta_4 = 0$, $\Delta_2 = -150$ (a1), $\Delta_2 = -15$ (a2), $\Delta_2 = 15$ MHz and (a3). (b3) is the SWM1 signal splitting the right peak of the SWM2 signal versus Δ_1 for increasing $P_2 = 0.5$, 2.5, 3.5, 4.5, 5.5, and 6.5 mW from bottom to top when $P_1 = 6.5$ mW, $P_3 = 38.5$ mW, and $P_4 = 22.6$ mW. (b5) is the SWM1 splitting the left peak of SWM2 signals versus Δ_1 under the same power of (b3). (b1) is the double-peaked SWM1 signals versus Δ_1 with $\Delta_2 = -150$ MHz. (b2), (b4), and (b6) are the corresponding power dependences. Here Δ_{b1}, Δ_{b3}, and Δ_{b5} are the increments of the distances between the two large peaks in (b1), right two peaks in (b3), left two peaks in (b5), respectively, when P_2 is increased, and the squares are the experimental results, while the solid lines in (b2, b4, and b6) are the theoretical calculations. Source: Adapted from Ref. [40].

the lower curve gives the measured SWM signals. In Figure 3.12(a1), the left EIT window ($|0\rangle - |1\rangle - |4\rangle$ satisfying $\Delta_1 + \Delta_4 = 0$) is induced by the coupling field E_4,which contains the SWM2 signal (E_{S2}), and the right one ($|0\rangle - |1\rangle - |2\rangle$ satisfying $\Delta_1 + \Delta_2 = 0$, $\Delta_2 = -150$ MHz) is induced by the coupling field E_2, which contains the SWM1 signal (E_{S1}). As the right EIT window ($\Delta_2 = -150$ MHz) is quite far from the left EIT window, these two SWM signals have little effect on each other (Figure 3.12(a1)). When the frequency of E_2 is tuned to move the right EIT window ($|0\rangle - |1\rangle - |2\rangle$) toward the left one, the two EIT windows overlap at $\Delta_2 = -15$ MHz, which leads to the triple AT splitting for SWM2, that is, the right peak of SWM2 signal is further split into two peaks (Figure 3.12(a2), satisfying $\Delta_2 = \lambda_{--}^{(3)}$). It is the coupling field E_2 that couples the secondarily-dressed state $|--\rangle$ dressed by E_3 and splits it into two triply-dressed states $|--+\rangle$ and $|---\rangle$. The four peaks of the SWM2 signal in Figure 3.12(a2) correspond, from left to right, to the primarily-dressed state $|+\rangle$, the secondarily-dressed states $|-+\rangle$, the triply-dressed states $|--+\rangle$ and $|---\rangle$, respectively (Figure 3.9(e)). We can write the triply-dressed states as $|--+\rangle = \sin\alpha_{3\pm}|--\rangle + \cos\alpha_{3\pm}|2\rangle$, where $\tan\alpha_{3\pm} = -b_{3\pm}/G_2$, $b_{3\pm} = \Delta_2 - \lambda_{-}^{(3)} - \lambda_{--}^{(3)} - \lambda_{--\pm}^{(3)}$, and the other parameters are the same as before. One can obtain the eigenvalues $\lambda^{(3)}_{--\pm} = (\Delta'_2 \pm \sqrt{\Delta'^2_2 + 4|G_2|^2})/2$ (measured from level $|--\rangle$) for the dressed states $|--\pm\rangle$, where $\Delta'_2 = \Delta_2 - \lambda^{(3)}_{-} - \lambda^{(3)}_{--}$. With the right EIT window ($|0\rangle-|1\rangle-|2\rangle$) continuously moving to the left, the SWM1 signal splits the left peak of the SWM2 signal when $\Delta_2 = 15$ MHz, as shown in Figure 3.12(a3). Corresponding to the moving SWM1 signal in Figure 3.12(a1) and (b1) presents the measured self-dressing double-peak SWM1 signal versus Δ_1 with $\Delta_2 = 0$ MHz. Corresponding to the fixed SWM2 signal in Figure 3.12(a2), (a3), (b3), and (b5) shows the measured SWM2 signals, the right and left peaks of which are split by the SWM1 signal, versus Δ_1 with an increasing external-dressing P_2 power, respectively. Figure 3.12(b2), (b4), and (b6) gives the corresponding power dependences of the AT splitting separations. The corresponding separations are determined by $\Delta_{b1} = \lambda^{(1)}_{+} - \lambda^{(1)}_{-} \approx 2G_2$, $\Delta_{b3} = \lambda^{(3)}_{--+} - \lambda^{(3)}_{---} \approx 2G_2$, and $\Delta_{b5} \approx 2G_2$, respectively.

3.3.3 Conclusion

We have observed the self- and externally-dressed AT splitting of the SWM processes within EIT windows in a five-level atomic system. Such an AT splitting demonstrates the interactions between the two coexisting SWM processes. The controlled multi-channel splitting signals in nonlinear optical processes can find potential applications in optical communication and quantum information processing, such as wavelength-demultiplexer [41–44].

Problems

3.1 Compare the pump–probe method and two-photon FWM technique in measuring the ac-Stark shift, and answer the question why the two-photon FWM technique is a more direct measurement of the energy level shift.

3.2 Point out the difference and relationship between ac-Stark shift and AT splitting of MWM.

3.3 Try to use the dressing state theory to explain the primary and secondary AT of the MWM signal, and then give the corresponding relationship between the AT splitting and EIT/EIA of probe field signal.

References

1. Happer, W. and Mathur, B.S. (1967) Effective operator formalism in optical pumping. *Phys. Rev.*, **163**, 12–25.
2. Mathur, B.S., Tang, H., and Happer, W. (1968) Light shifts in the Alkali atoms. *Phys. Rev.*, **171**, 11–19.
3. Wei, C., Suter, D., Windsor, A.S.M., and Manson, N.B. (1998) AC Stark effect in a doubly driven three-level atom. *Phys. Rev. A*, **58**, 2310–2318.
4. Bakos, J. (1977) AC Stark effect and multiphoton processes in atoms. *Phys. Rep.*, **31**, 209–235.
5. Otis, C.E. and Johnson, P.M. (1981) The ac stark effect in molecular multiphoton

ionization spectroscopy. *Chem. Phys. Lett.*, **83**, 73–77.

6. Schuster, D.I., Wallraff, A., Blais, A., Frunzio, L., Huang, R.S., Majer, J., Girvin, S.M., and Schoelkopf, R.J. (2005) AC Stark shift and dephasing of a superconducting qubit strongly coupled to a cavity field. *Phys. Rev. Lett.*, **94**, 123602.
7. Papademetriou, S., Van Leeuwen, M., and Stroud, C. Jr., (1996) Autler-Townes effect for an atom in a 100% amplitude-modulated laser field. II. Experimental results. *Phys. Rev. A*, **53**, 997.
8. Van Leeuwen, M., Papademetriou, S., and Stroud, C. (1996) Autler-Townes effect for an atom in a 100% amplitude-modulated laser field. I. A dressed-atom approach. *Phys. Rev. A*, **53**, 990–996.
9. Ficek, Z. and Freedhoff, H. (1993) Resonance-fluorescence and absorption spectra of a two-level atom driven by a strong bichromatic field. *Phys. Rev. A*, **48**, 3092.
10. Manson, N.B., Wei, C., Holmstrom, S.A., and Martin, J.P.D. (1995) Transitions between dressed states. *Laser Phys.*, **5**, 486–489.
11. Narducci, L., Scully, M., Oppo, G.L., Ru, P., and Tredicce, J. (1990) Spontaneous emission and absorption properties of a driven three-level system. *Phys. Rev. A*, **42**, 1630.
12. Manka, A., Doss, H., Narducci, L., Ru, P., and Oppo, G.L. (1991) Spontaneous emission and absorption properties of a driven three-level system. II. The Λ and cascade models. *Phys. Rev. A*, **43**, 3748.
13. Zheng, H., Zhang, Y., Nie, Z., Li, C., Chang, H., Song, J., and Xiao, M. (2008) Interplay among multidressed four-wave mixing processes. *Appl. Phys. Lett.*, **93**, 241101.
14. Li, Y. and Xiao, M. (1995) Electromagnetically induced transparency in a three-level Λ-type system in rubidium atoms. *Phys. Rev. A*, **51**, 2703–2706.
15. Zhang, Y., Brown, A.W., and Xiao, M. (2007) Observation of interference between four-wave mixing and six-wave mixing. *Opt. Lett.*, **32**, 1120–1122.
16. Li, C., Zheng, H., Zhang, Y., Nie, Z., Song, J., and Xiao, M. (2009) Observation of enhancement and suppression in four-wave mixing processes. *Appl. Phys. Lett.*, **95**, 041103.
17. Zheng, H., Khadka, U., Song, J., Zhang, Y., and Xiao, M. (2011) Measurement of ac-Stark shift by a two-photon dressing process via four-wave mixing. *Europhys. Lett.*, **93**, 23002.
18. Alzetta, G., Cartaleva, S., Dancheva, Y., Andreeva, C., Gozzini, S., Botti, L., and Rossi, A. (2001) Coherent effects on the Zeeman sublevels of hyperfine states at the D1 and D2 lines of Rb. *J. Opt. B: Quantum Semiclassical Opt.*, **3**, 181.
19. Zheltikov, A. (2000) Coherent anti-Stokes Raman scattering: from proof-of-the-principle experiments to femtosecond CARS and higher order wave-mixing generalizations. *J. Raman Spectrosc.*, **31**, 653–667.
20. Zhang, Y., Brown, A.W., and Xiao, M. (2007) Opening four-wave mixing and six-wave mixing channels via dual electromagnetically induced transparency windows. *Phys. Rev. Lett.*, **99**, 123603.
21. Zuo, Z., Sun, J., Liu, X., Jiang, Q., Fu, G., Wu, L.-A., and Fu, P. (2006) Generalized n-photon resonant 2n-wave mixing in an $(n+1)$-level system with phase-conjugate geometry. *Phys. Rev. Lett.*, **97**, 193904.
22. Harris, S. and Yamamoto, Y. (1998) Photon switching by quantum interference. *Phys. Rev. Lett.*, **81**, 3611–3614.
23. Xiao, M., Li, Y.-q., Jin, S.-z., and Gea-Banacloche, J. (1995) Measurement of dispersive properties of electromagnetically induced transparency in rubidium atoms. *Phys. Rev. Lett.*, **74**, 666–669.
24. Moseley, R.R., Shepherd, S., Fulton, D.J., Sinclair, B.D., and Dunn, M.H. (1995) Spatial consequences of electromagnetically induced transparency: observation of electromagnetically induced focusing. *Phys. Rev. Lett.*, **74**, 670–673.
25. Wielandy, S. and Gaeta, A.L. (1998) Investigation of electromagnetically induced transparency in the strong probe regime. *Phys. Rev. A*, **58**, 2500–2505.
26. Harris, S.E. (1997) Electromagnetically induced transparency. *Phys. Today*, **50**, 36.

27. Autler, S.H. and Townes, C.H. (1955) Stark effect in rapidly varying fields. *Phys. Rev.*, **100**, 703–722.
28. Walker, B., Kaluža, M., Sheehy, B., Agostini, P., and DiMauro, L. (1995) Observation of continuum-continuum Autler-Townes splitting. *Phys. Rev. Lett.*, **75**, 633–636.
29. Qi, J., Lazarov, G., Wang, X., Li, L., Narducci, L.M., Lyyra, A.M., and Spano, F.C. (1999) Autler-Townes splitting in molecular lithium: prospects for all-optical alignment of nonpolar molecules. *Phys. Rev. Lett.*, **83**, 288–291.
30. Qi, J., Spano, F.C., Kirova, T., Lazoudis, A., Magnes, J., Li, L., Narducci, L.M., Field, R.W., and Lyyra, A.M. (2002) Measurement of transition dipole moments in lithium dimers using electromagnetically induced transparency. *Phys. Rev. Lett.*, **88**, 173003.
31. Mücke, O.D., Tritschler, T., Wegener, M., Morgner, U., and Kärtner, F.X. (2002) Role of the carrier-envelope offset phase of few-cycle pulses in nonperturbative resonant nonlinear optics. *Phys. Rev. Lett.*, **89**, 127401.
32. Ates, C., Pohl, T., Pattard, T., and Rost, J.M. (2007) Many-body theory of excitation dynamics in an ultracold Rydberg gas. *Phys. Rev. A*, **76**, 013413.
33. Schempp, H., Günter, G., Hofmann, C.S., Giese, C., Saliba, S.D., DePaola, B.D., Amthor, T., Weidemüller, M., Sevinçli, S., and Pohl, T. (2010) Coherent population trapping with controlled interparticle interactions. *Phys. Rev. Lett.*, **104**, 173602.
34. Amthor, T., Giese, C., Hofmann, C.S., and Weidemüller, M. (2010) Evidence of antiblockade in an ultracold Rydberg gas. *Phys. Rev. Lett.*, **104**, 13001.
35. Zhang, Y., Nie, Z., Wang, Z., Li, C., Wen, F., and Xiao, M. (2010) Evidence of Autler-Townes splitting in high-order nonlinear processes. *Opt. Lett.*, **35**, 3420–3422.
36. Lukin, M., Yelin, S., Fleischhauer, M., and Scully, M. (1999) Quantum interference effects induced by interacting dark resonances. *Phys. Rev. A*, **60**, 3225–3228.
37. Yan, M., Rickey, E.G., and Zhu, Y. (2001) Observation of doubly dressed states in cold atoms. *Phys. Rev. A*, **64**, 013412.
38. Drampyan, R., Pustelny, S., and Gawlik, W. (2009) Electromagnetically induced transparency versus nonlinear Faraday effect. Coherent control of the light beam polarization. *Phys. Rev. A*, **80**, 033815 Arxiv preprint arXiv:0906.0571.
39. Wasik, G., Gawlik, W., Zachorowski, J., and Kowal, Z. (2001) Competition of dark states: optical resonances with anomalous magnetic field dependence. *Phys. Rev. A*, **64**, 051802.
40. Zhang, Y., Li, P., Zheng, H., Wang, Z., Chen, H., Li, C., Zhang, R., and Xiao, M. (2011) Observation of Autler-Townes splitting in six-wave mixing. *Opt. Express*, **19**, 7769–7777.
41. Boyer, V., Marino, A.M., Pooser, R.C., and Lett, P.D. (2008) Entangled images from four-wave mixing. *Science*, **321**, 544–547.
42. Camacho, R.M., Vudyasetu, P.K., and Howell, J.C. (2009) Four-wave-mixing stopped light in hot atomic rubidium vapour. *Nat. Photonics*, **3**, 103–106.
43. Dolgaleva, K., Shin, H., and Boyd, R.W. (2009) Observation of a microscopic cascaded contribution to the fifth-order nonlinear susceptibility. *Phys. Rev. Lett.*, **103**, 113902.
44. Marino, A., Pooser, R., Boyer, V., and Lett, P. (2009) Tunable delay of Einstein–Podolsky–Rosen entanglement. *Nature*, **457**, 859–862.

4
Controllable Enhancement and Suppression of MWM Process via Dark State

Highlights

The generated MWM signal can be selectively enhanced/suppressed by changing powers, polarizations, and detuning's of the dressing fields or the probe field. Such studies can find applications in optical switch, optical communication and quantum information processing.

The switch between EIT and EIA and some other methods for realizing optical switching at low intensities have been widely studied, in which the double-dark resonance for quantum switching has been performed in various media. The constructive and destructive interference can be selected to control the absorptive photon switching by changing the phase of a weak laser. In this chapter, the switching between enhancement and suppression of MWM sin both frequency and spatial domains and their various effects will be introduced. First of all, two measurements for controlling enhanced/suppressed two-photon FWM in EIT window, namely, changing the probe detuning as well as changing the powers of the dressing or probe fields, will be introduced. Then we will focus on the spatial effect of the enhancement/suppression of FWM as well as the spatial splitting of FWM images when additional dressing state scheme is changed. And finally, controlling the enhancement/suppression of two coexisting SWM processes is introduced.

4.1
Enhancing and Suppressing FWM in EIT Window

The generated MWM signals in multi-level atomic systems can transmit through the resonant medium with little absorption under the EIT conditions [1, 2]. Several experiments indicate FWM processes enhanced by atomic coherence in multi-level atomic systems [3–6]. Also, efficient SWM has been observed in a four-level close-cycled atomic system [7]. Furthermore, EWM signal has been experimentally demonstrated in a folded five-level atomic system [8]. Afterward, it is demonstrated that the generated FWM and SWM can coexist with two ladder-type EIT windows [9] in an open-cycled atomic system. Recently, the enhancement and suppression of

Quantum Control of Multi-Wave Mixing, First Edition. Yanpeng Zhang, Feng Wen, and Min Xiao.

FWM was experimentally studied and the generated FWM signals can be selectively enhanced and suppressed [10, 11]. The AT splitting, enhancement, and suppression in the doubly-dressed FWM process have also been theoretically studied [12], where the dressing fields are all considered as external-dressing fields. In addition, the AT splitting in the FWM and SWM processes has been demonstrated [13, 14] in the experiments. On the other hand, when EIT and FWM processes are modulated by the different polarization of the strong coupling fields, it has been demonstrated that selective transitions among polarization dark states of degenerate Zeeman sublevels can be obtained [11, 15–20].

In this section, we show the enhancement and suppression of the FWM signal in an EIT window for different probe beam detuning and polarizations in Y-type ^{85}Rb atomic system. The generated FWM signal can be selectively enhanced and suppressed via an EIT window.

4.1.1
Theory and Experimental Results

The laser beams are aligned spatially as shown in Figure 4.1(a). A weak probe beam E_1 (ω_1, $\mathbf{k}_1$, and frequency detuning Δ_1) is modulated by a quarter wave plate (QWP) and propagates through the atomic medium, and two pump beams E_2 (ω_2, $\mathbf{k}_2$, and Δ_2) and E_2'(ω_2, $\mathbf{k}_2'$, and Δ_2) propagate in the opposite direction with small angle (0.3°) between them to generate one FWM signal beam E_F with phase-matching condition $\mathbf{k}_\text{F} = \mathbf{k}_1 + \mathbf{k}_2 - \mathbf{k}_2'$. At the same time, the strong dressing laser beam E_3 (ω_3, $\mathbf{k}_3$, and Δ_3) propagates in the same direction as beam E_2 to influence on this FWM signal. Here, we define detuning $\Delta_i = \omega_i - \Omega_i$. For a simple four-level Y-type atomic system, as shown in Figure 4.1(b1), E_2 and E_2' drive the upper transition $|1\rangle$ to $|2\rangle$ and E_3 drives the transition $|1\rangle$ to $|3\rangle$. The laser E_1 probes the lower transition $|0\rangle$ to $|1\rangle$.

In a ^{85}Rb vapor cell, the energy levels $5S_{1/2}$ (F = 3), $5P_{3/2}$ (F = 3), $5D_{3/2}$, and $5D_{5/2}$ (F = 2, 3, 4) form such a Y-type system. The four laser beams are aligned spatially as in Figure 4.1(a). The probe laser beam E_1 with a wavelength of 12 821 cm^{-1} is from an ECDL of Co. Toptica Photonics AG, connecting transition $5S_{1/2} - 5P_{3/2}$. It has a power of 3 mW (corresponding to Rabi frequency $G_1 = 0.0025$ cm^{-1}). The other laser beams E_2, E_2' (wavelength of 12 887 cm^{-1}, connecting transition $5P_{3/2} - 5D_{5/2}$) are from the second ECDL split, with equal power of 16 mW corresponding to $G_2 = G_2' = 0.0075$ cm^{-1}, and E_3 (wavelength of 12 884 cm^{-1}, connecting transition $5P_{3/2} - 5D_{3/2}$) is from the third ECDL with power of 108 mW corresponding to $G_3 = 0.0509$ cm^{-1}. The generated FWM signal I_P in the direction $\mathbf{k}_\text{F}$ emerging from the "P" polarization direction and the transmitted probe beam are detected by an avalanche photodiode detector and a photodiode, respectively.

With the dressing field E_3, the suppression peak satisfies the condition $\Delta_1 + \Delta_3 = 0$ and the enhancement condition is $\Delta_1 + \Delta_3 \pm G_3 = 0$ [8, 11]. Figure 4.2(a1) to (a3) presents evolution of the FWM signal intensity versus the probe field detuning Δ_1 for different Δ_3 values. In Figure 4.2(a1) to (a3), the up-curve is the probe transmission with two ladder-type EIT windows [21] and the down-curve is the measured FWM signal. With E_3 dressing, the perturbation

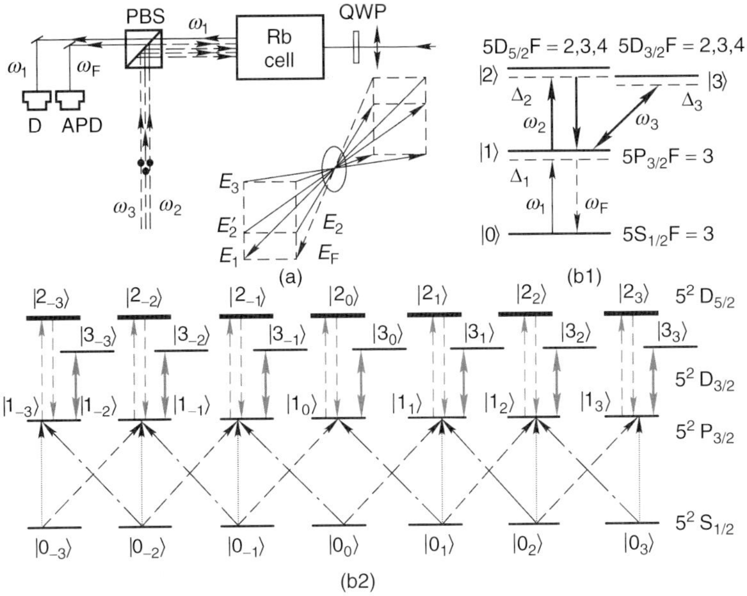

Figure 4.1 (a) The schematic diagram of the experiment. D denotes the photodiode, APD denotes the avalanche photo diode and PBS denotes the polarization beam splitter. Inset: the spatial alignment of the laser beams. (b1) Relevant ^{85}Rb energy levels and (b2) the corresponding Zeeman sublevels with various transition pathways. Solid line: dressing field E_3; linearly (dot lines, $q=0$), left (dash-dot lines, $q=-1$), and right (dash lines, $q=+1$) circularly polarized probe fields; and long-dash lines: the pump fields E_2 and E_2'.

chain is $\rho_{00}^{(0)} \xrightarrow{G_1} \rho_{G_3\pm 0}^{(1)} \xrightarrow{G_2} \rho_{20}^{(2)} \xrightarrow{(G_2')^*} \rho_{G_3\pm 0}^{(3)}$ [9, 10] and the modified third-order nonlinear susceptibility is $\rho_{10}^{(3)} = g/d_2[d_1 + G_3^2/d_3]^2$, where $d_1 = i\Delta_1 + \Gamma_{10}$, $d_2 = i(\Delta_1 + \Delta_2) + \Gamma_{20}$, $d_3 = i(\Delta_1 + \Delta_3) + \Gamma_{30}$, $g = -iG_1G_2G_2'^*$. In Figure 4.2(a1), the left $|0\rangle - |1\rangle - |2\rangle$ (satisfying $\Delta_1 + \Delta_2 = 0$) EIT window created by the pump fields E_2 and E_2' corresponds to the double-peak FWM signal and the right one is $|0\rangle - |1\rangle - |3\rangle$ (satisfying $\Delta_1 + \Delta_3 = 0$) EIT window created by the dressing field E_3. This right EIT window moves to the left when the dressing field detuning Δ_3 gets larger. In Figure 4.2(a2) and (a3), we can see that when two EIT windows get close and overlap, as a result of satisfying the enhancement or suppression condition, the right peak of FWM is suppressed and separated into two peaks; while the left FWM peak is enhanced, then the left peak of FWM is suppressed and separated into two peaks while the right FWM peak is enhanced.

4.1.2
Experiment and Result

In Figure 4.2(b) we experimentally study the enhancement and suppression of FWM spectra versus the dressing field detuning Δ_3 for different probe detuning

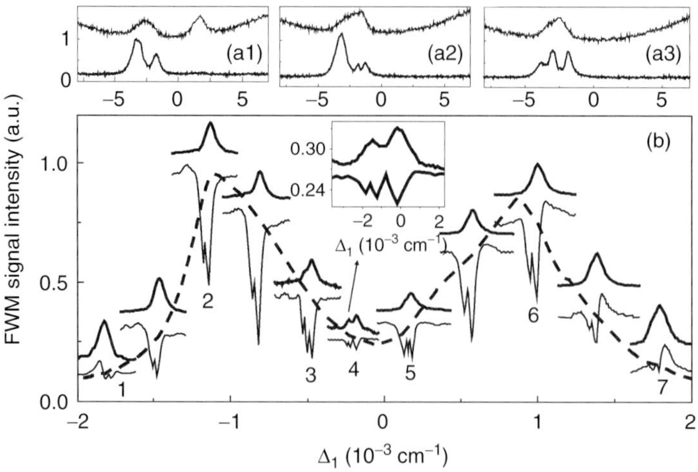

Figure 4.2 (a1)–(a3) Measured double-peak FWM signal and the corresponding EIT window induced by the dressing field E_3 versus Δ_1 for $\Delta_2 = 0$, $\Delta_3 = -0.0027$ cm^{-1} (a1), $\Delta_3 = 0$ (a2), and $\Delta_3 = 0.0023$ cm^{-1} (a3). (b) The enhanced and suppressed FWM signal (down curves) and the corresponding EIT windows (up curves) versus Δ_3 for different Δ_1 increasing from $\Delta_1 = -0.002$ to 0.002 cm^{-1} with the step of 0.0004 cm^{-1}. The dash curve is the double-peak FWM signal versus Δ_1. The inset is a zoom onto curve 4. Source: Adapted from Ref. [22].

Δ_1. The up-curves in Figure 4.2(b) are the probe transmission versus Δ_3. The down-curves are the enhancement and suppression of FWM signals. The constant background (two sides of down-curves) represents the signal intensity of the FWM without the dressing field, the dips lower than the background (we call them as suppression peaks, e.g., the curves 2–6) and the peaks higher than the background (enhancement peaks, e.g., the peaks of curves 1 and 7) represent that the FWM signal is suppressed and enhanced, respectively. The dashed curve in Figure 4.2(b) is the profile of the double-peak FWM signal versus Δ_1.

Specifically, we can see that all the curves in Figure 4.2(b) almost show axial symmetric evolution behavior by $\Delta_1 = 0$. Specifically, in the $\Delta_1 < 0$ region, with Δ_1 increasing, the FWM signals change from all-enhanced to half-suppressed and half-enhanced, and finally all-suppressed around resonant point. By contrast, in the $\Delta_1 > 0$ region, the FWM signals change oppositely. Moreover, we find that the FWM signals show left-enhancement and right-suppression shapes in the half suppression and half enhancement region (from $\Delta_1 = -0.00067$ to -0.0017 cm^{-1}), but left suppression and right enhancement in $\Delta_1 > 0$ region.

There exist a typical structure of two enhancement and two suppression peaks in the FWM spectra versus Δ_3 (e.g., curves 2 and 6 in Figure 4.2(b)), which result from the modification of the double-peak FWM signal. Moreover, the dressing field E_3 and the pump field E_2 interact with each other strongly in the all-enhancement and all-suppression region, and create triple peaks of the enhancement (curves 1 and 7) and suppression (curves 3–5) spectra. In the inset of Figure 4.2(b), two left-suppression peaks might be induced by the sequential-cascade double-dressing effect [12] of E_3 and E_2 described by the perturbation chain $\rho_{00}^{(0)} \xrightarrow{G_1} \rho_{(G_2 \pm G_3 \pm)0}^{(1)} \xrightarrow{G_2} \rho_{20}^{(2)} \xrightarrow{(G_2')^*} \rho_{G_3 \pm 0}^{(3)}$ and the modified third-order nonlinear susceptibility. By solving the following coupled equations: $\partial \rho_{10}/\partial t = -d_1\rho_{10} + iG_2^*\rho_{20} + iG_3^*\rho_{30}$, $\partial \rho_{20}/\partial t = -d_2\rho_{20} + iG_2\rho_{10}$, and $\partial \rho_{30}/\partial t = -d_3\rho_{30} + iG_3\rho_{10}$ with a weak probe field, we can finally obtain $\rho_{10}^{(3)} = g/[d_2(d_1 + |G_2|^2/d_2 + |G_3|^2/d_3)(d_1 + |G_3|^2/d_3)]$.

Similarly, we can consider the probe transmission versus Δ_3. The height of the up-curve represents the transparent degree of probe field E_1 in Figure 4.2(b). When the pump field E_2 is blocked, the pure E_3 EIT ($|0\rangle - |1\rangle - |3\rangle$) peak is higher at $\Delta_1 = 0$ than that at large detuning $|\Delta_1|$. However, with E_2 on, as the E_2 field destroys the E_3 EIT condition when $\Delta_1 = -\Delta_3 = -\Delta_2 \approx 0$, we can see that the E_3 EIT peak is strongly suppressed at $\Delta_1 = 0$ in Figure 4.2(b).

We have shown the enhancement and suppression of the FWM signal in EIT window for different probe beam detuning. Next, we will consider the influence of different probe laser polarization configurations on the enhancement and suppression of FWM spectra. We use one QWP with a rotation angle θ to change E_1 polarization state to decompose it into linearly or circularly polarized components. Thus, the Zeeman sublevels in Figure 4.1(b2) need to be considered in the FWM process. Figure 4.3(a) and (b) arranges the FWM spectra when scanning the dressing field from $\theta = 0^\circ$ to 90°, with the variation step of 5° when $\Delta_1 = 0.000723\ \text{cm}^{-1}$ and $\Delta_1 = 0.0023\ \text{cm}^{-1}$, respectively. The background represents the signal intensity of the FWM with no dressing field, while the dips and the peaks represent that the signal was suppressed or enhanced, respectively.

Square points in Figure 4.3(c) shows the variation in the FWM with no dressing fields (the background in Figure 4.3(a)), which are in accordance with the classical polarization spectroscopy (sinusoidal law). Triangle points give the dependence of the maximum depth of suppression peaks versus θ, which denotes that the dressing strength is increasing as E_1 changes from the linearly to circularly polarized state. It is because the polarization variation transforms the energy distribution among various transition paths, for example, linearly polarized transitions to circular ones. As different FWM transition paths are dressed by different transition dressing fields, the total dressing strength is generally different (Figure 4.1(b2)). The FWM transition paths and their expressions for linearly and circularly polarized subsystems in Figure 4.1(b2) can be written as

$$\rho_{0_M 0_M}^{(0)} \xrightarrow{G_{1_{M+q}}^q} \rho_{1_{M+q}0_M}^{(1)} \xrightarrow{G_{2_{M+q}}} \rho_{2_{M+q}0_M}^{(2)} \xrightarrow{(G'_{2_{M+q}})^*} \rho_{1_{M+q}0_M}^{(3)}$$

$$\rho_{1_{M+q}0_M}^q = \frac{\sum_{M=-3,-2,\ldots,3} g_{M+q}}{d_{2_{M+q}}(d_{1_{M+q}} + G_{3_{M+q}}^2/d_{3_{M+q}})^2}$$

where $g_{M+q} = -iG_{2_{M+q}}G_{2_{M+q}}^*G_{1_{M+q}}^q$, $d_{1_{M+q}} = i\Delta_1 + \Gamma_{1_{M+q}0_M}$, $d_{2_{M+q}} = i(\Delta_1 + \Delta_2) + \Gamma_{2_{M+q}0_M}$, $d_{3_{M+q}} = i(\Delta_1 + \Delta_3) + \Gamma_{3_{M+q}0_M}$, and $q = 0, \pm 1$. The theoretical curves are fit well with the experimentally measured results (Figure 4.3(c)).

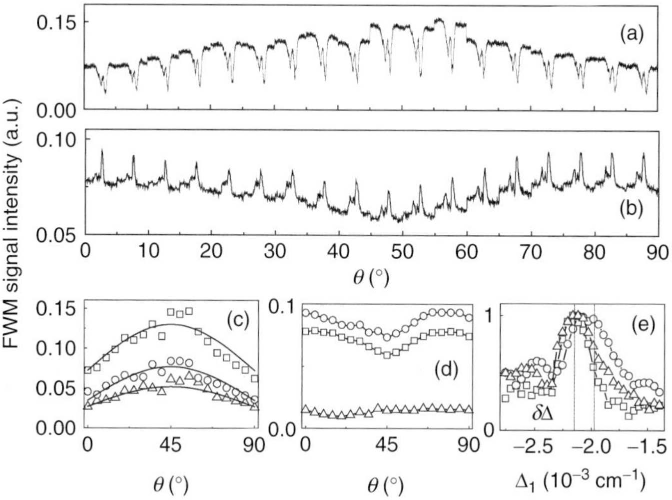

Figure 4.3 Polarization dependence of (a) the suppression (with $\Delta_1 = 0.00072\ \text{cm}^{-1}$) and (b) the enhancement (with $\Delta_1 = 0.0023\ \text{cm}^{-1}$) of FWM signal I_p (by scanning Δ_3) versus rotation angle θ (increases from 0° to 90° with the variation step of 5°) at $\Delta_2 = 0$. Variation of the (c,d) background (square), (c) minimums of the suppression peaks (circle) and suppression depths (triangle) and (d) maximums of the peaks (circle) and enhanced height (triangle) of the dependence curves. The solid curves in (c) are the corresponding theory results. (e) Normalized I_p versus Δ_3 for $\theta = 0°$ (square), 45° (circle), and 90° (triangle). Source: Adapted from Ref. [22].

Triangle points in Figure 4.3(d) give the dependence of the maximum height of enhancement peaks versus θ, which denotes that the dressing strength is decreasing as E_1 changes from the linearly to circularly polarized state. We can see that the variation is fairly stable as rotating QWP compared with Figure 4.3(c). It discloses that the far detuning condition homogenizes the difference of the transition strengths induced by the CG (Clebsch–Gordan) coefficients. Square points show the variation of the FWM with no dressing fields (the background in Figure 4.3(b)), which are in accordance with the traditional polarization rules. Moreover, the enhancement peaks are shifting as rotating QWP. The shift distance $\delta\Delta$ between the enhancement peaks is about 5 MHz, as shown in Figure 4.3(e). It is because the enhancement conditions are different for different Zeeman sublevels ($\Delta_1 + \Delta_3 \pm G_{3,M} = 0$).

When the frequencies of the probe field are set at a proper position ($\Delta_1 = 0.00181\ \text{cm}^{-1}$), which is not too far from the resonant position, both suppression and enhancement appear simultaneously, as shown in Figure 4.4(a). The polarization rules for suppression (Figure 4.4(b)) and enhancement Figure 4.4(c) are both similar to the above-mentioned results when they appear singly (Figure 4.3). Also, the positions of the enhanced peaks are shifting with rotating QWP, as shown in Figure 4.4(d). More importantly, we find that in Figure 4.4(a) the enhanced triple peaks transform into two enhancement and two

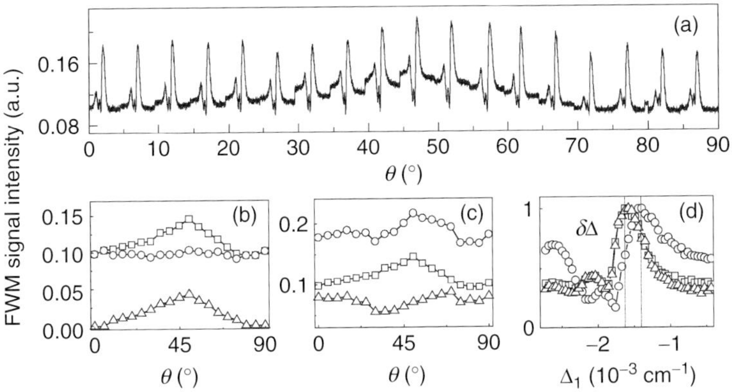

Figure 4.4 (a) Polarization dependence of the FWM signal I_p (by scanning Δ_3) versus rotation angle θ (increases from 0° to 90° with the variation step of 5°). (b) and (c) Variation of the background (square), (b) minimums of the suppression peaks (circle) and suppressed depth (triangle) and (c) maximums of the peaks (circle) and enhanced height (triangle) of the dependence curves. (d) I_p versus Δ_3 for $\theta = 0°$ (square), 45° (circle), and 90° (triangle). The other parameters are $\Delta_1 = 0.0018\ \text{cm}^{-1}$ and $\Delta_2 = 0$.

suppression peaks as θ changes from 0° to 45°. It results from the frequency shift $\delta\Delta$ (Figure 4.4(d)) due to the different polarization states (i.e., the enhancement condition $\Delta_1 + \Delta_3 \pm (G_3 + \delta\Delta) = 0$).

4.1.3
Conclusion

In summary, we discuss the enhancing and suppressing FWM in EIT windows. When the dressing laser EIT window moves to the EIT window of FWM, the generated FWM signal can be selectively enhanced and suppressed. Moreover, the sequential-cascade double dressing modifies the EIT of the probe and the enhancement and suppression of FWM signals. Finally, we observe polarization dependence of the enhancement and suppression of the FWM signal. The enhanced peaks can also be shifted by the different polarization states.

4.2
Cascade Dressing Interaction of FWM Image

Efficient FWM processes enhanced by atomic coherence in multi-level atomic systems [1, 6, 23] are of great interest currently. Recently, destructive and constructive interferences in a two-level atomic system and competition via atomic coherence in a four-level atomic system [10] with two coexisting FWM processes were studied. Also, the interactions of doubly-dressed states and the corresponding effects of atomic systems have attracted many researchers in recent years [24, 25].

The interaction of double-dark state and splitting of a dark state in a four-level atomic system were studied theoretically in an EIT system by Lukin *et al.* [25]. The triple-peak absorption spectrum, which was observed later in the N-type cold atomic system by Zhu *et al.* [24], verified the existence of the secondarily dressed states. Recently, we had theoretically investigated three types of doubly-dressed schemes in a five-level atomic system [12] and observed three-peak AT splitting of the secondary dressing FWM signal [14]. In addition, we reported the evolution of suppression and enhancement of the FWM signal by controlling an additional laser field [11].

As two or more laser beams pass through an atomic medium, the cross-phase modulation (XPM), as well as modified self-phase modulation (SPM), can potentially affect the propagation and spatial patterns of the incident laser beams. Laser beam self-focusing [26] and pattern formation [27] have been extensively investigated, with two laser beams propagating in atomic vapors. Recently, we have observed spatial shift [28] and spatial splitting [29–31] of the FWM beams generated in multi-level atomic systems, which can be well controlled by an additional dressing laser beam via XPM. Studies on such spatial shift and splitting of the laser beams can be useful in understanding the formation and interactions of spatial solitons [31] in the Kerr nonlinear systems and signal processing applications, such as spatial image storage [32], entangled spatial images [33], soliton pair generation [34], and influences of higher order (such as fifth-order) nonlinearities [35].

In this section, we discuss the interaction of four coexisting FWM processes in a two-level atomic system by blocking different laser beams. Next, we investigate the various suppression/enhancement of the degenerate four-wave mixing (DFWM) signals and two dispersion centers, which are caused by the cascade dressing interaction of two dressing fields. The experimental results clearly show the evolutions of the enhancement and suppression, from pure enhancement to partial enhancement/suppression, then to pure suppression, further to partial enhancement/suppression, and finally to enhancement, which are in good agreement with the theoretical calculations. In addition, we also observe the spatial splitting in the x- and y-directions of the DFWM signal due to different spatial alignment of the probe and coupling beams.

4.2.1 Theoretical Model and Experimental Scheme

The two relevant experimental systems are shown in Figure 4.5(a) and (b). Three energy levels from the sodium atom in a heat-pipe oven are involved in the experimental schemes. The pulse laser beams are aligned spatially as shown in Figure 4.5(c). In Figure 4.5(a), energy levels $|0\rangle$ ($3S_{1/2}$) and $|1\rangle$ ($3P_{3/2}$) form a two-level atomic system. Coupling field E_1 (with wave vector $\mathbf{k}_1$ and the Rabi frequency G_1), together with E_1' ($\mathbf{k}_1'$ and G_1') (connecting the transition between $|0\rangle$ and $|1\rangle$) having a small angle ($\sim 0.3^\circ$), propagates in the direction opposite to the probe field E_3 ($\mathbf{k}_3$ and G_3) (also connecting the transition between $|0\rangle$ and $|1\rangle$). These three

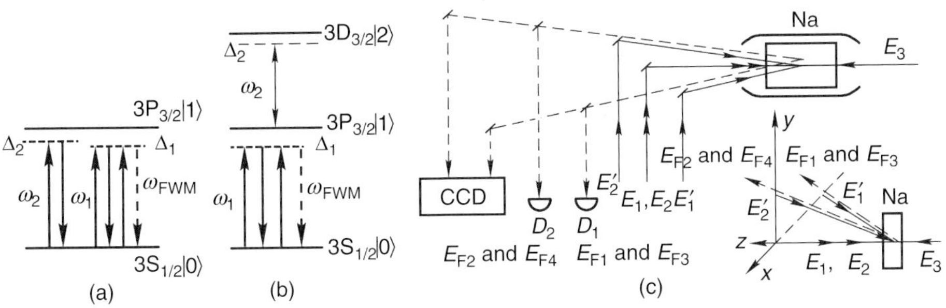

Figure 4.5 (a) and (b) The diagram of relevant Na energy levels. (c) The scheme of the experiment. Inset gives the spatial alignments of the incident beams.

laser beams come from the same near-transform-limited dye laser (with a 10 Hz repetition rate, 5 ns pulse width, and 0.04 cm^{-1} line width) with the same frequency detuning $\Delta_1 = \omega_{10} - \omega_1$, where ω_{10} is the transition frequency between $|0\rangle$ and $|1\rangle$. The coupling fields E_1 and E_1' induce a population grating between states $|0\rangle$ and $|1\rangle$, which is probed by E_3. This generates a DFWM process (Figure 4.5(a)) satisfying the phase-matching condition of $\mathbf{k}_{F1} = \mathbf{k}_1 - \mathbf{k}_1' + \mathbf{k}_3$. Then, two additional coupling fields E_2 ($\mathbf{k}_2$, G_2) and E_2' ($\mathbf{k}_2'$, G_2') are applied as scanning fields connecting the transition between $|0\rangle$ and $|1\rangle$, with the same frequency detuning $\Delta_2 = \omega_{10} - \omega_2$, which are from another similar dye laser set at ω_2 to dress the energy level $|1\rangle$. The fields E_2, E_2', and E_3 produce a nondegenerate four-wave mixing (NDFWM) signal $\mathbf{k}_{F2}$ (satisfying $\mathbf{k}_{F2} = \mathbf{k}_2 - \mathbf{k}_2' + \mathbf{k}_3$). When the five laser beams are all on, there also exist two other FWM processes $\mathbf{k}_{F3}$ (satisfying $\mathbf{k}_{F3} = \mathbf{k}_2 - \mathbf{k}_1' + \mathbf{k}_3$) and $\boldsymbol{E}_{F1}$ (satisfying $\mathbf{k}_{F4} = \mathbf{k}_1 - \mathbf{k}_2' + \mathbf{k}_3$) in the same directions as $\boldsymbol{E}_{F1}$ and $\boldsymbol{E}_{F2}$, respectively.

Under the experimental condition, E_1 (or E_1') with detuning Δ_1 depletes two groups of atoms with different velocities at the same time, such as the negative velocities group and positive velocities group. At $\Delta_1 < 0$, the positive velocities group will see E_1 (or E_1') with detuning $\Delta_1 + k_1 v$ and E_3 with detuning E_3. The frequency of the DFWM $\boldsymbol{E}_{F1}$ in this case will be $\omega_f = (\omega_1 - kv) - (\omega_1 - kv) + (\omega_1 + k_3 v) = \omega_1 + k_3 v$ due to the conservation of energy. Correspondingly, at $\Delta_1 > 0$, the negative velocities group will see E_1 (or E_1') with detuning $\Delta_1 - k_1 v$ and E_3 with detuning $\Delta_1 + k_3 v$. The frequency of $\boldsymbol{E}_{F1}$ will be $\omega_f = (\omega_1 + kv) - (\omega_1 + kv) + (\omega_1 - k_3 v) = \omega_1 - k_3 v$. Such a change implies that a group of atoms with certain velocities can satisfy the condition E_2', where Δ_1' is the detuning of E_1 (or E_1') based on both saturation excitation and the atomic coherence effect. As a result, the self-dressing field E_1 (or E_1') can be considered as the outer dressing field and separate the level $|0\rangle$ into two dressing states $|G_1 \pm\rangle$, as shown in Figure 4.8(a). In addition, the Doppler effect and the power broadening effect on the weak FWM signals need to be considered.

When E_1, E_1', E_2, E_2', and E_3 are open, the DFWM process $\boldsymbol{E}_{F1}$ and NDFWM processes $\boldsymbol{E}_{F2}$, $\boldsymbol{E}_{F3}$, and $\boldsymbol{E}_{F4}$ are generated simultaneously and there exists interplay among these four FWM signals in the two-level atomic system. These FWM signals generated have the frequencies $\omega_{F1} = \omega_{F2} = \omega_1$, $\omega_{F3} = \omega_2$, and $\omega_{F4} = 2\omega_1 - \omega_2$. They are split into two equal components by a 50% beam splitter before being detected. One is captured by the CCD camera and the other is detected by photomultiplier tubes (D_1 or D_2) and a fast gated integrator (gate width of 50 ns). Also, they are monitored by the digital acquisition card.

In order to interpret the following experimental results, we perform the theoretical calculation on the four coexisting FWM processes. First, we consider four FWM processes to be or perturbed by corresponding laser beams. In the two-level configuration, there exist the transition paths to generate FWM signals. They can be described by perturbation chains, (F1) $\rho_{00}^{(0)} \xrightarrow{(E_1)^*} \rho_{10}^{(1)} \xrightarrow{(E_1')^*} \rho_{00}^{(2)} \xrightarrow{E_3} \rho_{10}^{(3)}$, (F2) $\rho_{00}^{(0)} \xrightarrow{(E_2')^*} \rho_{10}^{(1)} \xrightarrow{E_2} \rho_{00}^{(2)} \xrightarrow{E_3} \rho_{10}^{(3)}$, (F3) $\rho_{00}^{(0)} \xrightarrow{E_2} \rho_{10}^{(1)} \xrightarrow{(E_1')^*} \rho_{00}^{(2)} \xrightarrow{E_3} \rho_{10}^{(3)}$, and (F4) $\rho_{00}^{(0)} \xrightarrow{E_1} \rho_{10}^{(1)} \xrightarrow{(E_2')^*} \rho_{00}^{(2)} \xrightarrow{E_3} \rho_{10}^{(3)}$, respectively. For the DFWM signal $\boldsymbol{E}_{F1}$, in fact, this DFWM generation process can be viewed as a series of transitions: the first step is from $|0\rangle$ to $|1\rangle$ with absorption of a coupling photon E_1 and the final state of this process can be dressed by the dressing field E_2 (or E_2'). The second step is the transition from $|1\rangle$ to $|0\rangle$ and the final state cannot be dressed by any field. The third step is the transition from $|0\rangle$ to $|1\rangle$ with the emission of a probe photon E_3 and the final state of this process can be dressed by E_2 (or E_2'). Then, the last transition is from $|1\rangle$ to $|0\rangle$, which emits an FWM photon at frequency ω_1. Thus, we can obtain the dressed perturbation chain $\rho_{00}^{(0)} \xrightarrow{(E_1)^*} \rho_{G2\pm0}^{(1)} \xrightarrow{(E_1')^*} \rho_{00}^{(2)} \xrightarrow{E_3} \rho_{G2\pm0}^{(3)}$. Similarly, we can obtain the other dressed perturbation chains as follows: (DF2) $\rho_{00}^{(0)} \xrightarrow{E_2} \rho_{G1\pm0}^{(1)} \xrightarrow{(E_2')^*} \rho_{00}^{(2)} \xrightarrow{E_3} \rho_{G1\pm0}^{(3)}$, (DF3) $\rho_{00}^{(0)} \xrightarrow{E_2} \rho_{G2\pm0}^{(1)} \xrightarrow{(E_1')^*} \rho_{00}^{(2)} \xrightarrow{E_3} \rho_{G2\pm0}^{(3)}$, and (DF4) $\rho_{00}^{(0)} \xrightarrow{E_1} \rho_{G1\pm0}^{(1)} \xrightarrow{(E_2')^*} \rho_{00}^{(2)} \xrightarrow{E_3} \rho_{G1\pm0}^{(3)}$, respectively. The expressions of the corresponding density matrix elements related to the four FWM processes are $\rho_{F1}^{(3)} = -iG_3G_1(G_1')^*/(\Gamma_{00}B_1^2)$, $\rho_{F2}^{(3)} = -iG_3G_2(G_2')^*/(\Gamma_{00}d_3B_2)$, $\rho_{F3}^{(3)} = -iG_2G_3(G_1')^*/(d_4B_3^2)$, and $\rho_{F4}^{(3)} = -iG_1G_3(G_2')^*/d_5d_6B_1$, respectively, where $d_1 = \Gamma_{10} + i\Delta_1$, $d_2 = \Gamma_{00} + i(\Delta_1/m - \Delta_2)$, $d_3 = \Gamma_{10} + i\Delta_2$, $d_4 = \Gamma_{00} + i(\Delta_2 - \Delta_1)$, $d_5 = \Gamma_{10} + i(2\Delta_1 - \Delta_2)$, $d_6 = [\Gamma_{00} + i(\Delta_1 - \Delta_2)]$, $A_1 = |G_2|^2/d_2$, $A_2 = G_1^2/\Gamma_{00}$, $A_3 = |G_2|^2/\Gamma_{00}$, $B_1 = d_1 + A_1$, $B_2 = d_1 + A_2$, and $B_3 = d_3 + A_3$. Here $G_i = -\mu_i E_i/\hbar$ ($i = 1, 2, 3$) is the Rabi frequency. Γ_{10}, Γ_{20}, and Γ_{00} are the transverse relaxation rates; Δ_i ($i = 1, 2$) is the detuning factor.

The experiments are carried out in a vapor cell containing sodium. The cell, 18 cm long, is heated up to a temperature of about 230 °C and crossed by linearly polarized laser beams that interact with the atoms. In the two-level atomic system, the coupling fields E_1 and E_1' (with diameter of 0.8 mm and power of 9 μW), and the probe field E_3 (with diameter of 0.8 mm and power of 3 μW) are tuned to the line center (589.0 nm) of the $|0\rangle$ to $|1\rangle$ transition, which generate the DFWM signal $\boldsymbol{E}_{F1}$ at frequency ω_1. The coupling fields E_2 and E_1' (with diameter of 1.1 mm and powers of 20 and 100 μW, respectively) are scanned simultaneously around the $|0\rangle$ to $|1\rangle$ transition to dress the DFWM process $\boldsymbol{E}_{F1}$.

4.2.2
Cascade Dressing Interaction

We first investigate the interaction of four coexisting FWM signals in the two-level atomic system by blocking different laser beams. First, by blocking E_2 (or E_1), the DFWM signal ϕ (or the FWM signal $\boldsymbol{E}_{F3}$) is suppressed by the coupling field E_2', as can be seen from the upper triangle points (or the right triangle points in Figure 4.6(a1)), compared to the pure DFWM signal E_{F1} (or the FWM signal $n_2 \propto \text{Re}(-i\mu G_{F1} F_a^2/h)$). Next, when laser beams E_1, E_1', E_2, and E_3 are turned on, two coexisting FWM processes (Δ_2 and $\boldsymbol{E}_{F3}$) couple to each other (the lower triangle points), and the intensities of total FWM signals are increased, as can be attributed to the combination of two FWM signal processes ($\boldsymbol{E}_{F1}$ and $\boldsymbol{E}_{F3}$). Finally, when all the five laser beams are turned on, the DFWM signal I_{F1}' and the FWM signal $\boldsymbol{E}_{F3}$

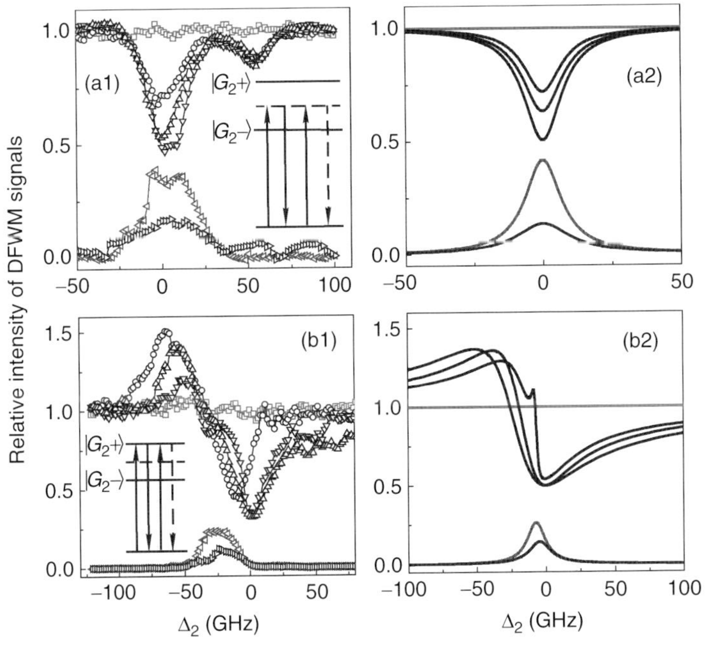

Figure 4.6 The interplay and mutual suppression/enhancement between two coexisting FWM signals ($\boldsymbol{E}_{F1}$ and $\boldsymbol{E}_{F3}$). (a1) The upper curves: pure DFWM signal $\boldsymbol{E}_{F1}$ (with both E_2 and E_2' blocked) (squares), singly-dressed DFWM signal $\boldsymbol{E}_{F1}$ (with E_2 blocked) (triangles), coexisting singly-dressed DFWM signal $\boldsymbol{E}_{F1}$ and FWM signal $\boldsymbol{E}_{F3}$ (with E_2' blocked) (reverse triangles), and coexisting dressed DFWM signal $\boldsymbol{E}_{F1}$ and FWM signal $\boldsymbol{E}_{F3}$ (circles); lower curves: pure FWM signal $\boldsymbol{E}_{F3}$ (with both E_1 and E_2' blocked) (left triangle), singly-dressed FWM signal (with E_1 blocked) (right triangle); $\Delta_1 = 0$. The inserted plot corresponding to the dressed-state picture. (b1) The condition are the same to that in (a1) except E_1. The inserted plot corresponding to the dressed-state picture. (a2) and (b2) Theoretical plots corresponding to the experimental parameters of (a1) and (b1), respectively. Source: Adapted from Ref. [36].

are both greatly suppressed by corresponding dressing fields. So, the intensities of total FWM signals are extremely decreased, as shown in the circle points in Figure 4.6(a1).

These effects can be explained effectively by the dressed-state picture. The dressing field E_2' couples the transition $|0\rangle$ to $|1\rangle$ and creates the dressed state E_2', which leads to single-photon transition $|0\rangle \rightarrow |1\rangle$ off-resonance (the inserted plot in Figure 4.6(a1)). At exact single-photon resonance with $\Delta_1 = 0$, the DFWM signal $\boldsymbol{E}_{\text{F1}}$ intensity is greatly suppressed by means of scanning the dressing field $4D_{3/2,5/2}$ across the resonance ($I''_{E_{\text{F1}}} \propto |\rho''^{(3)}_{E_{\text{F1}}}|^2$), as the upper triangle points in Figure 4.6(a1) shows. At the same time, the FWM signal $\boldsymbol{E}_{\text{F3}}$ experiences a similar process (the right triangle points in Figure 4.6(a1)).

Furthermore, an appropriate Δ_1 value at which $\boldsymbol{E}_{\text{F1}}$ is either enhanced or suppressed is chosen in the investigation. In this case, compared to the pure DFWM signal $\boldsymbol{E}_{\text{F1}}$ (square points in Figure 4.6(b1)), the dressed DFWM signal $\boldsymbol{E}_{\text{F1}}$ is enhanced (the upper and low triangle points in Figure 4.6(b1)). However, the dressed FWM signal $\boldsymbol{E}_{\text{F3}}$ is suppressed because of the destructive interference (right triangle points in Figure 4.6(b1)) compared to the pure signal (left triangle points in Figure 4.6(b1)). The upper triangle in Figure 4.6(b1) combines the two FWM processes ($\boldsymbol{E}_{\text{F1}}$ and $\boldsymbol{E}_{\text{F3}}$), which are dressed by laser beams E_1' and E_2', respectively. After calculating $\rho^{(3)}_{\text{F1}}$ and $\rho^{(3)}_{\text{F3}}$ under the above-mentioned experiment conditions, good agreements are obtained between the theoretical calculations and the experimental results as shown in Figure 4.6(a2) and (b2), respectively.

After that, we investigate the evolutions of the interaction between these two coexisting FWM signals by means of setting different frequency detuning Δ_1 values, where the fixed spectra corresponds to the suppression and enhancement of the DFWM signal $\boldsymbol{E}_{\text{F1}}$, and the shifting spectra corresponds to the FWM signal $\boldsymbol{E}_{\text{F3}}$ as shown in Figure 4.7(a1) to (a7). It is obvious in Figure 4.7(a1) to (a3) that, as the frequency detuning Δ_1 varies from $\Delta_1 < 0$ to zero from up to down, the DFWM signal $\boldsymbol{E}_{\text{F1}}$ shows the evolution from enhancement to partial enhancement/suppression, and then to suppression. At the same time, the FWM signal $\boldsymbol{E}_{\text{F3}}$ varies from intense to weak (when two FWM signals $\boldsymbol{E}_{\text{F1}}$ and $\boldsymbol{E}_{\text{F3}}$ overlap) and shifts from left to right, which satisfies the two-photon resonant condition ($\Delta_1 - \Delta_2 = 0$). When Δ_1 changes further to be positive, a symmetric process is observed (i.e., suppression in Figure 4.7(a5), partial suppression/enhancement in Figure 4.7(a6), and pure enhancement in Figure 4.7(a7)). It should be noted here that the FWM signal $\boldsymbol{E}_{\text{F3}}$ still shifts from left to right. Figure 4.7(a4) shows the weakened FWM signal due to the strong effect of the Doppler absorption. Specially, the DFWM signal $\boldsymbol{E}_{\text{F1}}$ at a large one-photon detuning is extremely weak when $G_2 = 0$. However, the strong dressing field can cause the resonant excitation of one of the dressed states if the enhanced condition $\Delta_1 - \Delta_2 \pm G_2 = 0$ is satisfied. In such a case, the DFWM signal $\boldsymbol{E}_{\text{F1}}$ is strongly enhanced (Figure 4.7(a1)), mainly due to the one-photon resonance ($|0\rangle \rightarrow |G_2+\rangle$) (the insert plot in Figure 4.6(b1)). As $\Delta_1 = 0$, the intensity of the DFWM signal $\boldsymbol{E}_{\text{F1}}$ is greatly suppressed (Figure 4.7(a3)), similar to the case of the upper triangle curves in Figure 4.6(a1). Also, we can observe that the FWM signal $\boldsymbol{E}_{\text{F3}}$ is suppressed because of the destructive interference.

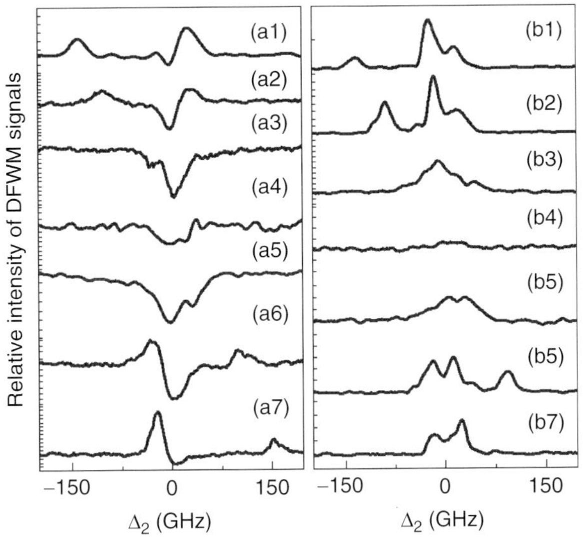

Figure 4.7 (a) and (b) Measured evolution of the four FWM signals (($\boldsymbol{E}_{\mathrm{F1}}$ and $\boldsymbol{E}_{\mathrm{F3}}$) and ($\boldsymbol{E}_{\mathrm{F2}}$ and $\boldsymbol{E}_{\mathrm{F4}}$), respectively) via Δ_2 for different Δ_1 values. (a1)–(a7) and (b1)–(b7): $\Delta_1 = -139.1$, -103.87, -29.5, 0, 29.8, 100.1, 155.7 GHz, respectively.

In addition, Figure 4.7(b1) to (b7) shows the interaction of another two coexisting the FWM processes ($\boldsymbol{E}_{\mathrm{F2}}$ and $\boldsymbol{E}_{\mathrm{F4}}$), in which the fixed spectra corresponds to the FWM signal $\boldsymbol{E}_{\mathrm{F2}}$, and the shifting spectra corresponds to the FWM signal $\boldsymbol{E}_{\mathrm{F4}}$ for different frequency detuning Δ_1 values.

Now, we concentrate on the cascade dressing interaction and the two dispersion centers of FWM images with two dressing fields E_1' and E_2' in the two-level atomic system. In order to investigate the cascade dressing interaction, the power of the coupling field E_1' is set at 80 μW. So, the DFWM signal $\boldsymbol{E}_{\mathrm{F1}}$ shows a spectrum of the AT splitting due to the self-dressed effect [14] induced by beam $\mathbf{k}_1'$ when Δ_1 is scanned and the dressing field E_2' is off, as shown in the dashed curve of Figure 4.8(a). When the beam $\mathbf{k}_2'$ is on, the DFWM signal $\boldsymbol{E}_{\mathrm{F1}}$ is dressed by both E_1' and E_2', and therefore shows the cascade dressing interaction, as shown in Figure 4.8(a) and (c). Specifically, by discretely choosing different detuning values within $\Delta_1 < 0$ and scanning Δ_2, the DFWM signal $\boldsymbol{E}_{\mathrm{F1}}$ shows the evolution of the successively occurring pure enhancement, partial suppression/enhancement, pure suppression, partial enhancement/suppression, and enhancement processes, as shown in the left side of Figure 4.8(a). When Δ_1 is changed to positive, a symmetric process occurs, in the right side of Figure 4.8(a), which is well described by the theoretical curves (Figure 4.8(b)).

In order to explain this phenomenon, the dressed-state picture is adopted, as shown in Figure 4.8(d). First, the DFWM signal $\boldsymbol{E}_{\mathrm{F1}}$ is dressed by both fields E_2' and E_1'. The corresponding expression of the modified density matrix element of DFWM $\boldsymbol{E}_{\mathrm{F1}}$ process is $\rho_{E_{\mathrm{F1}}}^{\prime(3)} = -iG_3G_1(G_1')^*/(B_4B_5^2)$, where $d_6 = \Gamma_{00} + A_5$, $d_7 = \Gamma_{10} - i\Delta_2$, $A_4 = G_1^2/B_1$, $A_5 = G_2^2/d_7$, $A_6 = G_1^2/d_6$, $B_4 = \Gamma_{00} + A_4$, and $B_5 = B_1 + A_6$. Next, the

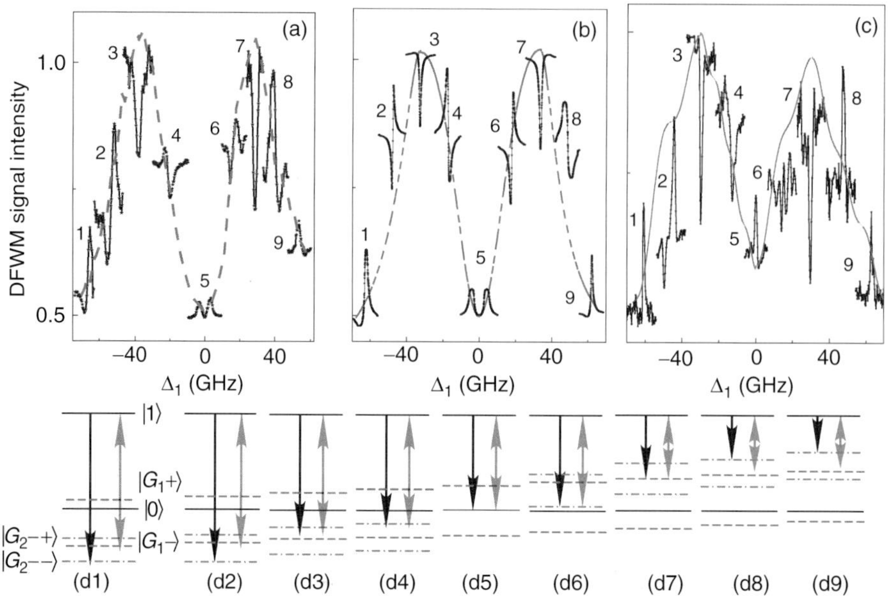

Figure 4.8 (a) Measured suppression and enhancement of DFWM signal $\boldsymbol{E}_{F1}$ via Δ_2 for different Δ_1 values in the two-level system. $\Delta_1 = -69.1$, -55.5, -38.7, -19.2, 0, 14.7, 28.8, 42.2, and 57.3 GHz, respectively. The dashed curve is the double-peak DFWM signal $\boldsymbol{E}_{F1}$ via Δ_1. (b) Theoretical plots corresponding to the experimental parameters in (a). (c) The same measures to (a) with the same condition, except that the laser beams E_1, E_1' overlap in the middle of heat oven. (d1)–(d9) The dressed-state pictures of the suppression or enhancement of the DFWM signal. The states $|G_1\pm\rangle$ (dashed lines) and the states $|G_1+\pm\rangle$ or $|G_1-\pm\rangle$ (dot-dashed lines), respectively. Source: Adapted from Ref. [36].

inner dressing field E_1' dresses the state $|0>$ to create two new dressing states $|G_1\pm\rangle$, and then the strong dressing field E_2' creates new states $|G_1+\pm\rangle$ or $|G_1-\pm\rangle$ around states $|G_1\pm\rangle$, by scanning the frequency detuning Δ_2. As a result of this dressing scheme, the DFWM signal $\boldsymbol{E}_{F1}$ is extremely small when $G_2 = 0$ and Δ_1 is set far away from the resonance point at both $\Delta_1 < 0$ and $\Delta_1 > 0$, respectively. Another result is that the strong fields can cause resonant excitation of one of the dressed states ((i.e., $|G_1--\rangle$ or $|G_1++\rangle$ Figure 4.8(d1) and (d9), respectively), which can lead to the enhancement of FWM signals. Specifically, if the condition $\Delta_1 - \Delta_2 = -G_2$ and $\Delta_1 - \Delta_2 = G_2$ (corresponding to the dressed states shown in Figure 4.8(d1) and (d9)) is satisfied, the DFWM signal $\boldsymbol{E}_{F1}$ is obviously enhanced, as shown in the curves of Figure 4.8(a1) and (a9), respectively. As Δ_1 changes to be near the resonance point, we can get a partial enhancement/suppression of the DFWM signal $\boldsymbol{E}_{F1}$. The first and second transition states satisfy suppression condition $\Delta_1 - \Delta_2 = 0$ and enhancement condition $\Delta_1 - \Delta_2 = -G_2$ (Figure 4.8(d2)) (enhancement condition $\Delta_1 - \Delta_2 = G_2$ and suppression condition $\Delta_1 - \Delta_2 = 0$, as shown in Figure 4.8(d4)), and lead to the first suppression and next enhancement (the curve of Figure 4.8(a2)) (or the first enhancement and next suppression, as shown in the curve of Figure 4.8(a4)). When Δ_1 reaches the point $\Delta_1 - \Delta_2 = 0$, the suppression effect gets dominant owing to the dressed states $|G_1 --\rangle$ (Figure 4.8(d3)); therefore, the DFWM signal $\boldsymbol{E}_{\mathrm{F1}}$ is purely suppressed, as shown in the curve of Figure 4.8(a3). For the point $\Delta_1 = \Delta_2 = 0$ between the two resonance points, the curve of Figure 4.8(a5) shows a pair of suppressed peaks. In fact, they are induced by the outer dressing field E_2', which can largely weaken the suppression effect of the inner dressing field E_1' on the DFWM signal. Furthermore, the other cascade dressing field E_2' splits such a suppressed peak into a pair of suppressed peaks, as shown in the curve of Figure 4.8(a5). Figure 4.8(a) shows the various suppressions/enhancements of the DFWM signal $\boldsymbol{E}_{\mathrm{F1}}$ and its two dispersion centers, which are caused by the cascade dressing interaction of the two dressing fields E_1 and E_2'.

In addition, the spatial splitting in the x-direction of the FWM signal beams induced by additional dressing laser beams is observed simultaneously, as shown in Figure 4.9(a). It is observed that the number of the splitting spots increases when the FWM intensity is suppressed. To understand these phenomena, we need to consider the XPM on the FWM signals. As described in our previous investigation [29], the spatial splitting of the FWM beam can be controlled by the intensities of the involved laser beams, the cross-Kerr nonlinear coefficients, and the atomic density, according to the nonlinear phase shift $\phi = 2k_{\mathrm{F1}} n_2 z I_1 \mathrm{e}^{-r^2/2}/(n_0 I_{\mathrm{F1}}')$. Here, the additional transverse propagation wave vector is $\delta k_r = \partial \phi / \partial r$. The change of phase ($\phi$) distribution in the laser propagating equations determines the spatial splitting of the laser beams. In theoretical calculation, we can obtain the intensity of the $\boldsymbol{E}_{\mathrm{F1}}$ beam by $I_{\mathrm{F1}}' \propto |\rho_{E\mathrm{F1}}^{\prime(3)}|^2 = |-iG_3 F_a|^2$ [29], with $F_a = G_1 (G_1')^* / (B_4 B_5^2)$ and the nonlinear cross-Kerr refractive index $n_2 \propto \mathrm{Re}(-i\mu G_{\mathrm{F1}} F_a^2 / h)$. When Δ_2 is scanned in the experiment, the intensity I_1 of the laser beam E_1' and n_2 almost stays constant for different detuning Δ_2. So, ϕ is primarily determined by I_{F1}'. When the suppression condition ($\Delta_1 - \Delta_2 = 0$) is satisfied, and the intensity of the FWM signal I_{F1}' reaches its minimum, the spatial splitting will become stronger as shown in Figure 4.9(a) (the suppression positions located at $\Delta_2 \approx -12$ GHz). While in the enhancement condition with the larger I_{F1}', ϕ is decreased, and therefore the splitting is weakened correspondingly, as shown in Figure 4.9(a), where the enhancement condition is located at $\Delta_2 \approx -28.3$ GHz.

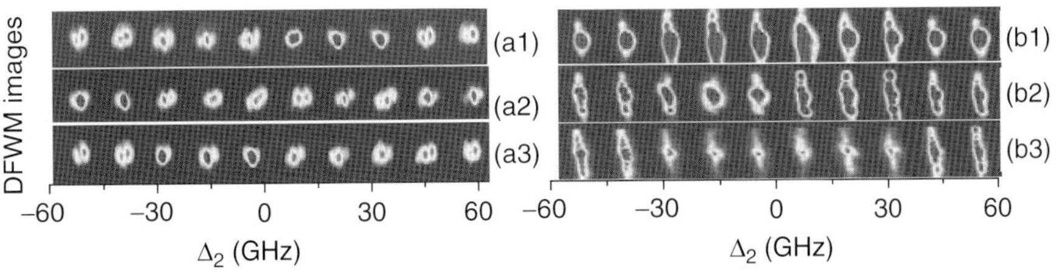

Figure 4.9 (a) DFWM signal $\boldsymbol{E}_{\mathrm{F1}}$ images when $\Delta_1 = -69.1, -55.5, -38.7$ GHz. (b) DFWM signal images when $\Delta_1 = -55.5, -38.7, -19.2$ GHz.

Specially, we observed the y-direction spatial splitting images of the DFWM signal $\boldsymbol{E}_{F1}$ (Figure 4.9(b)) by carefully arranging laser beams $\mathbf{k}_1$ and $\mathbf{k}_2'$. In the experiment, the beams E_1 and E_1' are deliberately aligned in the y–z-plane with an angle θ ($\sim 0.05^\circ$) to induce a grating in the same plane with the fringe spacing $\Lambda = \lambda_1/\theta$. Because θ is far less than the angle of E_1 and E_1' in the x–z-plane, Λ is big enough for observing the splitting caused by the induced grating. Furthermore, when E_1 and E_1' are set in the middle of the oven, $\boldsymbol{E}_{F1}$ and E_1' overlap in y-direction because of the phase-matching condition. As a result, the splitting of $\boldsymbol{E}_{F1}$ in the x-direction due to the nonlinear cross-Kerr effect from E_1' disappears simultaneously. Because Λ remains nearly the same for the changeless θ and λ_1, a larger spot of the $\boldsymbol{E}_{F1}$ beam with larger intensity will be split to more parts. In Figure 4.9(b), the field E_3 is stronger than that in Figure 4.9(a), which leads to stronger FWM signals passing through the grating in the y-direction. So, we can easily obtain the splitting in the y-direction. Moreover, in the enhanced position, the profile of the FWM signal becomes larger, and more split parts induced by the grating can be obtained. Here, Figure 4.9(b1) to (b3) shows the experimental spots corresponding to the curves in Figure 4.8(c1) to (c3). However, the effects of suppression and enhancement of the DFWM signal $\boldsymbol{E}_{F1}$ are much worse because of the special spatial alignment of the laser beams, as shown in Figure 4.8(c) compared to that in Figure 4.8(a). In Figure 4.8(a), the E_1, E_1' and E_2, E_2' are all set at the back of the heat oven. But in Figure 4.8(c), only E_1 and E_1' are deliberately moved to the middle of the oven to demonstrate the splitting of $\boldsymbol{E}_{F1}$ in the y-direction. So, the dressing effect on $\boldsymbol{E}_{F1}$ by E_2' in Figure 4.8(c) appears worse than that in Figure 4.8(a).

In order to verify the cascade dressing interaction and two dispersion centers of the FWM image, the dressing field E_2' is tuned to the line center (568.8 nm) of the $|1\rangle$ to $|2\rangle$ ($4D_{3/2,5/2}$) transition, and a ladder-type three-level atomic system forms, as shown in Figure 4.5(b). With the dressed perturbation chains, we can obtain $I''_{EF1} \propto |\rho''^{(3)}_{EF1}|^2$, where $\rho''^{(3)}_{EF1} = -iG_3G_1(G_1')^*/(B_6B_7^2)$, with $d_8 = \Gamma_{00} + i(\Delta_1 + \Delta_2)$, $d_9 = \Gamma_{21} + i\Delta_2$, $d_{10} = \Gamma_{11} + A_9$, $d_{11} = d_1 + A_7$, $A_7 = G_2^2/d_8$, $A_8 = G_1^2/d_{11}$, $A_9 = G_2^2/d_9$, $A_{10} = G_1^2/d_{10}$, $B_6 = \Gamma_{00} + A_8$, and $B_7 = d_1 + A_7 + A_{10}$.

We repeated the above-mentioned experiment with the same experimental conditions (the data points in Figure 4.8(a)), and obtained the results as shown in Figure 4.10(a). Comparing the results in Figure 4.10(a) with those in Figure 4.8(a), we can obtain the similar observations of suppression and enhancement of the DFWM signal $\boldsymbol{E}_{F1}$, except the shapes of partial suppressions/enhancement. For instance, the curve of Figure 4.8(a2) shows first suppression, and next enhancement, is different from the curve of Figure 4.10(a2), which shows first enhancement and next suppression. The reason for this contrast is the difference of levels structure. As Δ_1 is set near the resonance point ($\Delta_1 = -55.8$ GHz), the new dressed state $|G_1 --\rangle$ moves from upper to lower as Δ_2 scans from positive to zero (Figure 4.8(d2)). So, the DFWM signal $\boldsymbol{E}_{F1}$ is first enhanced as the condition $\Delta_1 + \Delta_2 = -G_2$ is satisfied, and then suppressed (the suppression condition $\Delta_1 + \Delta_2 = 0$), as shown in the curve of Figure 4.10(a2). Figure 4.10(b) shows theoretical plots corresponding to the experimental parameters in Figure 4.10(a). Simultaneously, we obtain the corresponding suppressions and

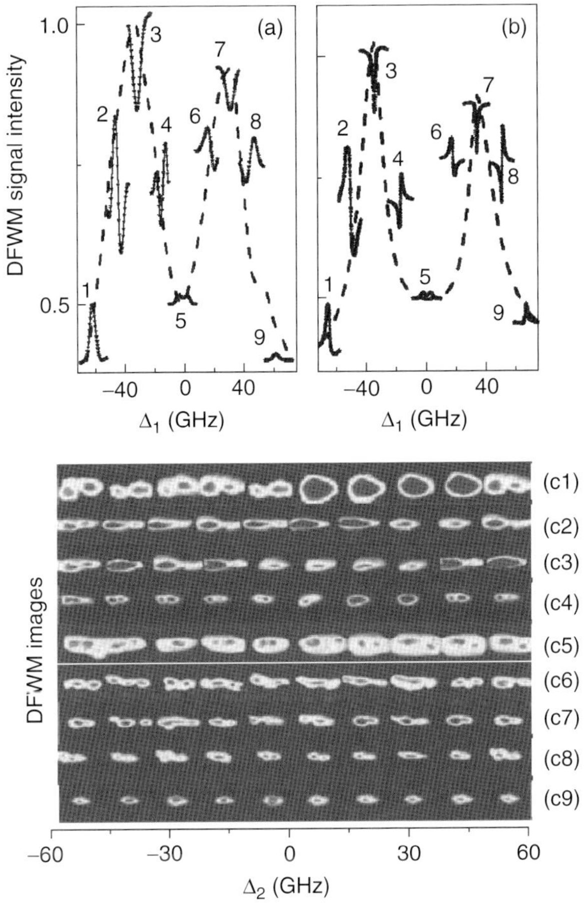

Figure 4.10 (a) Measured suppression and enhancement of DFWM signal $\boldsymbol{E}_{F1}$ via Δ_2 for different Δ_1 values. $\Delta_1 = -70.5$, -55.8, -32.7, -15.4, 0, 16.5, 32.5, 48.2, and 72.1 GHz, respectively. The dashed curve is the double-peak DFWM signal $\boldsymbol{E}_{F1}$ via Δ_1. (b) Theoretical plots corresponding to the experimental parameters in (a). (c1)–(c9) DFWM signal $\boldsymbol{E}_{F1}$ images. The condition here is same to that in (a).

enhancements of the x-direction spatial splitting of DFWM signal $\boldsymbol{E}_{F1}$ images, as shown in Figure 4.10(c). Here, Figure 4.10(c1) to (c9) shows the experimental spots corresponding to the curves in Figure 4.10(a1) to (a9).

4.2.3
Conclusion

We have discussed the suppression and enhancement of the spatial FWM signal by the controlled cascade interaction of additional dressing fields, and the corresponding controlled spatial splitting of the FWM signal caused by the enhanced cross-Kerr nonlinearity owing to the atomic coherence in two- and three-level atomic systems. In addition, we discuss the interplay between the two coexisting FWM signals, which can be tuned to overlap or separate by varying frequency detuning. Such controllable FWM processes can have important applications in wavelength conversion for spatial signal processing and optical communication.

4.3 Multi-Dressing Interaction of FWM

The basis for electromagnetically induced focusing (EIF), which is caused by spatial variations in the coupling laser strength via atomic coherence, has been studied [37]. It has been shown that the self- and cross-Kerr nonlinearities can be significantly enhanced and modified in three-level atomic systems owing to atomic coherence [38]. In particular, the enhanced self- and cross-Kerr nonlinearities in the system are essential in generating efficient FWM processes [30]. In addition, some studies have shown that atomic coherence and FWM processes can be effectively controlled by the polarization states of the involved laser beams [15]. Furthermore, investigations about the interactions of doubly-dressed states and corresponding effects on atomic systems have also attracted the attention of many researchers. The interaction of double-dark states and splitting of the dark state in a four-level atomic system via atomic coherence has been studied theoretically by Lukin and colleagues [39]. Then the doubly-dressed states in cold atoms were observed, in which the triple-photon absorption spectrum exhibits a constructive interference between the transition paths of two closely spaced, doubly-dressed-states [24]. In recent years, spatial shifting and splitting of the weak laser beam by stronger beam in Kerr nonlinear optical media due to the XPM [26] or EIG [40] were predicted and experimentally demonstrated. In our recent works, three schemes (nested-, parallel-, and sequential-cascade) for doubly-dressed FWM processes in an open five-level atomic system were presented [22, 41–43]. We also have studied the two-photon resonant NDFWM in a dressed cascade four-level system. In the presence of a strong dressing field, the two-photon resonant NDFWM spectrum exhibits AT splitting, accompanied with either suppression or enhancement of the NDFWM signal by scanning the detuning of dressing field [11].

In this section, we discuss the multi-dressing interaction of FWM in a ladder-type three-level system. First, we observe the evolution of the dressing processes of FWM from singly dressing to doubly dressing, and then to the multi-dressing case by changing the powers of the probe and pump fields. Second, the suppression of FWM signals can be enlarged as the power of the probe field increases compared with the pump fields that strengthens the enhancement of the FWM. So, by controlling the powers of the dressing, pump, and probe fields, the enhancement and suppression of FWM signal can be switched flexibly. Third, by scanning the frequency detuning of the probe and dressing fields, respectively, we compare the dressing effects and interaction between the dressing fields, which can be directly obtained by scanning the frequency detuning of the dressing field. Finally, we observe that the spatial splitting of the FWM in the y-direction is induced by XPM or EIG by arranging laser beams in a certain spatial configuration. Such switching of the enhancement and suppression of the FWM signal has potential applications in optical switch, optical communication, and quantum information processing.

4.3.1
Theoretical Model

The energy level diagram and the experiment systems are shown in Figure 4.11(a). A ladder-type three-level atomic system ($|0\rangle$ ($3S_{1/2}$), $|1\rangle$ ($3P_{1/2}$), $|2\rangle$ ($3P_{3/2}$)) in a sodium atom (in heat-pipe oven) is involved in the experiment. The pulse laser beams are aligned spatially, as shown in Figure 4.11(b), including two pump fields $\boldsymbol{E}_1$ (frequency ω_1, wave vector $\mathbf{k}_1$) and $\boldsymbol{E}_1'$ (ω_1, $\boldsymbol{k}_1'$), a probe field $\boldsymbol{E}_3$ (ω_1, $\mathbf{k}_3$), and two dressing fields $\boldsymbol{E}_2$ (ω_2, $\mathbf{k}_2$) and E_2' (ω_2, $\boldsymbol{k}_2'$). Driving the transition $|0\rangle - |1\rangle$, the laser fields $\boldsymbol{E}_1$, $\boldsymbol{k}_1'$, and $\boldsymbol{E}_3$ are all from a near-transform-limited dye laser (10 Hz repetition rate, 5 ns pulse width, and 0.04 cm^{-1} line width) and therefore have the same frequency detuning $\Delta_1 = \Omega_{10} - \omega_1$, where Ω_{10} is the transition frequency between $|0\rangle$ and $|1\rangle$. Driving the transition $|1\rangle - |2\rangle$, the dressing fields $\boldsymbol{E}_2$ and $\boldsymbol{E}_2'$ (G_2 and G_2') with the detuning $\Delta_2 = \Omega_{21} - \omega_2$, where Ω_{21} is the transition frequency between $|1\rangle$ and $|2\rangle$, are all from another dye laser. The Rabi frequencies of such beams are given as $G_i = \mu_{mn} E_i/\hbar$ ($i = 1, 2, 3$) or $G_i' = \mu_{mn} E_i'/\hbar$ ($i = 1, 2$) in which μ_{mn} is the transition dipole moments of the transition $|m\rangle - |n\rangle$ and E_i (E_i') is the electric field intensity of ($\boldsymbol{E}_i'$) $\boldsymbol{E}_i$. The pump (dressing) beams $\boldsymbol{E}_1$ and $\boldsymbol{E}_1'$ ($\boldsymbol{E}_2$ and $\boldsymbol{E}_2'$) propagate in the same direction with a small angle (0.3°) between them, while the probe beam propagates in the opposite direction along the z-axis in Figure 4.11(b). $\boldsymbol{E}_1$ and $\boldsymbol{E}_2$ propagate collinearly and the plane spanned by $\boldsymbol{E}_1$ and $\boldsymbol{E}_1'$ tends to be perpendicular to that spanned by $\boldsymbol{E}_2$ and $\boldsymbol{E}_2'$ (Figure 4.11(b)). In such a configuration, a DFWM $\boldsymbol{E}_{F1}$ (with Δ_{F1}, $\Delta_{F1} = \Omega_{10} - \omega_{F1}$, and $G_{F1} = \mu_{10} E_{F1}/\hbar$) satisfying the phase-matching condition $\mathbf{k}_{F1} = \mathbf{k}_1 - \mathbf{k}_1' + \mathbf{k}_3$ is obtained. Another NDFWM field $\boldsymbol{E}_{F2}$ (with ω_{F2}, $\Delta_{F2} = \Omega_{10} - \omega_{F2}$, and G_{F2}) satisfying the phase-matching condition $\mathbf{k}_{F2} = \mathbf{k}_2 - \mathbf{k}_2' + \mathbf{k}_3$ is also obtained at the same time. In experiment, only $\boldsymbol{E}_{F1}$ is detected and it is split into two equal components by a 50% beam splitter before being detected. One component is captured by a CCD camera,

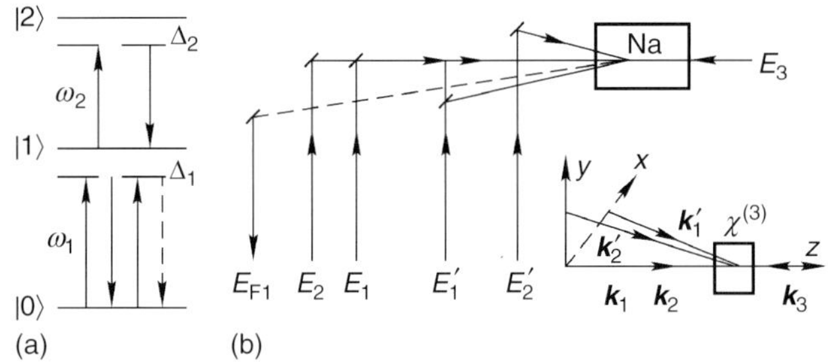

Figure 4.11 (a) The diagram of the ladder-type three-level atomic system. (b) Spatial beam geometry used in the experiment.

and the other is detected by a photomultiplier tube and fast gated integrator (gate width of 50 ns). Also, they are monitored by the digital acquisition card.

In order to interpret the following experimental results, we perform the theoretical calculation on the $\boldsymbol{E}_{F1}$ process. For the DFWM signal $\boldsymbol{E}_{F1}$, its generation process can be viewed as a series of transition steps: the first step is a rising transition from $|0\rangle$ to $|1\rangle$ with absorption of a pump photon $\boldsymbol{E}_1$, the second step is a falling transition from $|1\rangle$ to $|0\rangle$, and the third step is another transition from $|0\rangle$ to $|1\rangle$ with the absorption of a probe photon $\boldsymbol{E}_3$. Then, the last transition is from $|1\rangle$ to $|0\rangle$, which emits an FWM photon at frequency ω_1. According to the Liouville pathway of the pure DFWM $\rho_{00}^{(0)} \xrightarrow{G_1} \rho_{10}^{(1)} \xrightarrow{(G_1')^*} \rho_{00}^{(2)} \xrightarrow{G_3} \rho_{10}^{(3)}$, we can obtain the third-order nonlinear density matrix element $\rho_{10}^{(3)} = g/(\Gamma_{00} d_1^2)$, the absolute value of the square of which determines the intensity of the DFWM signal. Here $g = -iG_3 G_1 G_1'^*$, $d_1 = i\Delta_1 + \Gamma_{10}$, and Γ_{ij} is the transverse relaxation rate between the states $|i\rangle$ and $|j\rangle$.

In addition, the Doppler effect and the power broadening effect on the weak FWM signals need to be considered for the hot atom in the experiment. Under the experimental condition, for an atom with velocity v along the z-axis, the frequency of $\boldsymbol{E}_1$ ($\boldsymbol{E}_1'$) will shift to $\omega_1 - k_1 v$, and that of $\boldsymbol{E}_3$ will shift to $\omega_1 + k_1 v$; so, the frequency of the DFWM $\boldsymbol{E}_{F1}$ is $\omega_{F1} = (\omega_1 - k_1 v) - (\omega_1 - k_1 v) + (\omega_1 + k_1 v) = \omega_1 + k_1 v$ due to the conservation of energy. In such a case, if the intensity of $\boldsymbol{E}_1$ is nearly equal to that of $\boldsymbol{E}_3$, the fields $\boldsymbol{E}_1$ and $\boldsymbol{E}_3$ should be depleted into a pair of traveling wave components along the z-axis, and their frequencies are $\omega_1 - k_1 v$ and $\omega_1 + k_1 v$, respectively. So, the configuration in which atoms with a nonzero velocity interact with fields with the same frequency but different wave-vectors can be equal to the situation that the stationary atom are pumped by a bichromatic wave, in which the two waves have a frequency difference $2\delta = 2k_1 v$. In particular, if the time inference between these two fields in certain positions is considered, the total intensities of the dressing fields in the medium can obtain a time-periodic expression $|E_1 e^{-i(\omega_1 - k_1 v)t} + E_3 e^{-i(\omega_1 + k_1 v)t}|^2 = E_1{}^2 + E_3^2 + 2E_1 E_3 \cos(2k_1 vt)$ and the corresponding dressed n-th order atomic coherences can be expanded into a complex Fourier series $\rho_{10}^{(n)}(t) = \sum_{m=-\infty}^{m=\infty} \rho_{10}^{(n,m)} e^{imk_1 vt}$ with the fundamental frequency $k_1 v$ under bichromatic dressing fields [44]. All these Fourier coefficients can affect the height and location of the suppression and enhancement of the DFWM signal $\boldsymbol{E}_{F1}$, yet only some items are dominant in the experiment. In the same way, when the pump fields $\boldsymbol{E}_1$ and $\boldsymbol{E}_1'$ are both large compared with the probe field, $\boldsymbol{E}_3$ is relatively weak, and they can be considered as a set of bichromatic waves with a frequency difference $2\delta = 2k_1 v \sin^2(\theta/2)$ due to the small angle θ between their propagation directions along the vertical direction of the angular bisector of their propagating directions. So, a time-periodic oscillating dressing effect is induced and the nth order atomic coherence under such a bichromatic dressing effect can be expanded as $\rho_{10}^{(n)}(t) = \sum_{m=-\infty}^{m=\infty} \rho_{10}^{(n,m)} e^{i(mk_1 v \sin^2(\theta/2))t}$, with the fundamental frequency $k_1 v \sin^2(\theta/2)$. But if the powers of $\boldsymbol{E}_1$ and $\boldsymbol{E}_1'$ are weak, the dressing effects of the bichromatic fields can be ignored and the frequency detuning of the DFWM $\boldsymbol{E}_{F1}$ (Δ_{F1}) is equal to the frequency detuning of the probe field (Δ_1).

4.3.2
Experimental Result

4.3.2.1 Single-Dressed DFWM

First, the single-dressed system is investigated. When the powers of $\boldsymbol{E}_3$, $\boldsymbol{E}_1$ ($\boldsymbol{E}_1'$), and $\boldsymbol{E}_2$ ($\boldsymbol{E}_2'$) are 8, 8, and 22 μW, the DFWM signal is single dressed by $\boldsymbol{E}_2$ ($\boldsymbol{E}_2'$). According to the modified Liouville pathway of the $\boldsymbol{E}_2$ dressed DFWM $\rho_{00}^{(0)} \xrightarrow{G_1} \rho_{G_2\pm0}^{(1)} \xrightarrow{(G_1')^*} \rho_{00}^{(2)} \xrightarrow{G_3} \rho_{G_2\pm0}^{(3)}$, we can obtain the dressing-modified matrix element $\rho_{10}^{(3)} = g/[\Gamma_{00}(d_1 + |G_2|^2/d_2)^2]$, with $d_2 = \Gamma_{20} + i(\Delta_1 + \Delta_2)$. In physical meaning, the expression implies that the state $|1\rangle$ will be split into two dressed states $|G_2\pm\rangle$ by $\boldsymbol{E}_2$ ($\boldsymbol{E}_2'$).

When Δ_1 is scanned with $G_2 = 0$, that is, no dressing effect, the pure DFWM signal which has one emission peak at $\Delta_1 = 0$ is obtained, and is shown as the global profile in Figure 4.12(a) (experimental results) and (b) (theoretical simulation). When $\boldsymbol{E}_2$ ($\boldsymbol{E}_2'$) is turned on and Δ_2 is scanned at discrete Δ_1 value, the intensity of the DFWM signal shows the evolution from pure enhancement (peak higher than baseline which represents the signal strength of the undressed DFWM and $\Delta_1 = -60.0$ GHz) to first enhancement and next suppression (dip lower than baseline and $\Delta_1 = -30.1$ GHz), next to pure suppression ($\Delta_1 = 0$ GHz),

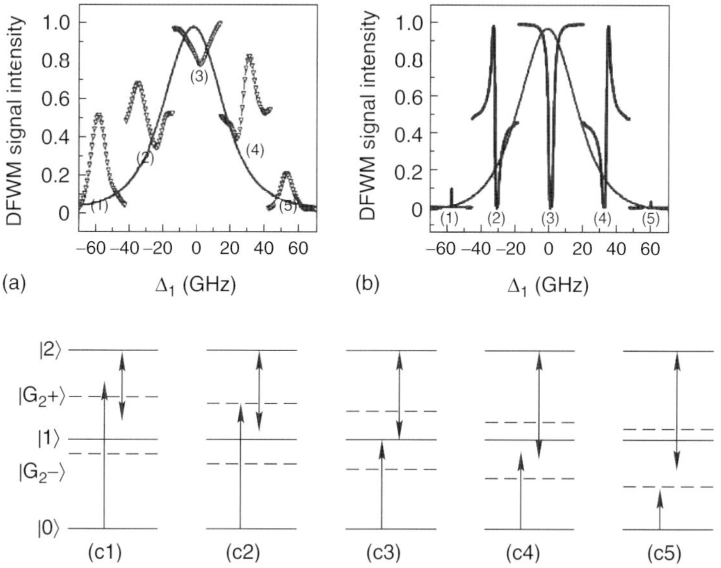

Figure 4.12 (a) The experimentally measured intensities of the single-dressed DFWM $\boldsymbol{E}_{F1}$, versus Δ_2 at different Δ_1 and the involved powers are set at $P_1 = 8$ μW, $P_2 = 22$ μW, and $P_3 = 8$ μW. (1) $\Delta_1 = -60.0$ GHz, (2) $\Delta_1 = -30.1$ GHz, (3) $\Delta_1 = 0$ GHz, (4) $\Delta_1 = 32.0$ GHz, and (5) $\Delta_1 = 55.3$ GHz, respectively. The background profile is the single-peak DFWM signal $\boldsymbol{E}_{F1}$ versus Δ_1. (b) The theoretical results of the single-dressed DFWM. The condition is same as that in (a). (c1)–(c5) The dressed-state pictures of the DFWM signal. The states $|G_2\pm\rangle$ are dashed lines.

further to first suppression and next enhancement ($\Delta_1 = 32.0$ GHz), and finally to pure enhancement ($\Delta_1 = 55.3$ GHz) sequentially as shown in Figure 4.12(a1) and (a5) and (b1) to (b5). At $\Delta_1 = -60.0$ GHz, where Δ_1 is set far away from the resonance point, because the DFWM signal approaches to $|G_2+\rangle$ (Figure 4.12(c1)) in the scanning of Δ_2, where the enhancement condition $\Delta_{F1} + \Delta_2 + G_2 = 0$ is matched, the pure enhancement is acquired as shown in Figure 4.12(a1) and (b1). At $\Delta_1 = -30.1$ GHz, the DFWM signal is first enhanced and next suppressed as shown in Figure 4.12(a2) and (b2), because it firstly resonates with the state $|G_2+\rangle$ satisfying the enhancement condition $\Delta_{F1} + \Delta_2 + G_2 = 0$ and next has two-photon resonance where the suppression condition $\Delta_{F1} + \Delta_2 = 0$ is satisfied (Figure 4.12(c2)). At $\Delta_1 = 0$, the pure suppression (Figure 4.12(a3) and (b3)) is obtained because the DFWM signal cannot resonate with $|G_2\pm\rangle$ as shown in Figure 4.12(c3) and only the two-photon resonance can be obtained. At $\Delta_1 > 0$, the DFWM signal shows symmetric evolving behavior with respect to $\Delta_1 = 0$. The experimental results (Figure 4.12(a)) agree well with the theoretical results (Figure 4.12(b)).

Two kinds of scanning methods (i.e., the method where the detuning of the dressing field (Δ_2) is scanned at discrete detuning of the probe field (Δ_1), and the method where Δ_1 is scanned at certain Δ_2 values) are used in the experiment. On the basis of the above-mentioned analysis, by scanning Δ_2 the enhancements and suppressions of the FWM can be detected directly. In contrast, the enhancements and suppressions of the FWM can hardly be demonstrated clearly by scanning Δ_1 because the enhancements and suppressions are relatively small against the background of the generated FWM signal. However, because the effective scanning ranges for Δ_1 can be relatively extensive, the entire derivation of the generated FWM signal could be revealed.

4.3.2.2 **Doubly-Dressed DFWM**

Next, we study the two schemes for generating doubly-dressed DFWM. The first is sequential-cascade mode. When the powers of $\boldsymbol{E}_1$ and $\boldsymbol{E}_1'$ are moderate compared to larger $\boldsymbol{E}_2$ ($\boldsymbol{E}_2'$) and lower $\boldsymbol{E}_3$, they can be considered as bichromatic fields. So, one dressing effect on the weak DFWM signal is induced by the bichromatic fields composed of two strong fields $\boldsymbol{E}_1$ and $\boldsymbol{E}_1'$ with different Doppler shifts due to the small angle θ between their propagating directions. Another dressing effect is induced by the outer dressing fields $\boldsymbol{E}_2$ ($\boldsymbol{E}_2'$). With the modified Liouville pathway of the doubly-dressed DFWM $\rho_{00}^{(0)} \xrightarrow{G_1} \rho_{G_2\pm0}^{(1)} \xrightarrow{(G_1')^*} \rho_{00}^{(2)} \xrightarrow{G_3} \rho_{((G_1G_1')\pm G_2\pm)0}^{(3)}$, we could get the second-order Fourier coefficient of the third-order atomic coherence as

$$\rho_{10}^{(n=3,m=2)} = \frac{g}{((d_1 - d_4 + |G_2|^2/d_2)(\Gamma_{00} - d_3 + |G_1'|^2/(\Gamma_{00} - d_4)))}$$
$$\times \frac{1}{(d_1+|G_1|^2/(\Gamma_{00}-d_3+|G_2|^2/(\Gamma_{00}-d_4))+|G_1'|^2/(\Gamma_{00}+d_3+|G_1|^2/(\Gamma_{00}+d_4))+|G_2|^2/d_2)}$$

the absolute value of the square, which determines the intensity of the doubly-dressed DFWM. Here, $d_3 = ik_1 v \sin^2(\theta/2)$ and $d_4 = i2k_1 v \sin^2(\theta/2)$. In such a case, the doubly-dressed DFWM signal under the bichromatic field dressing effects and

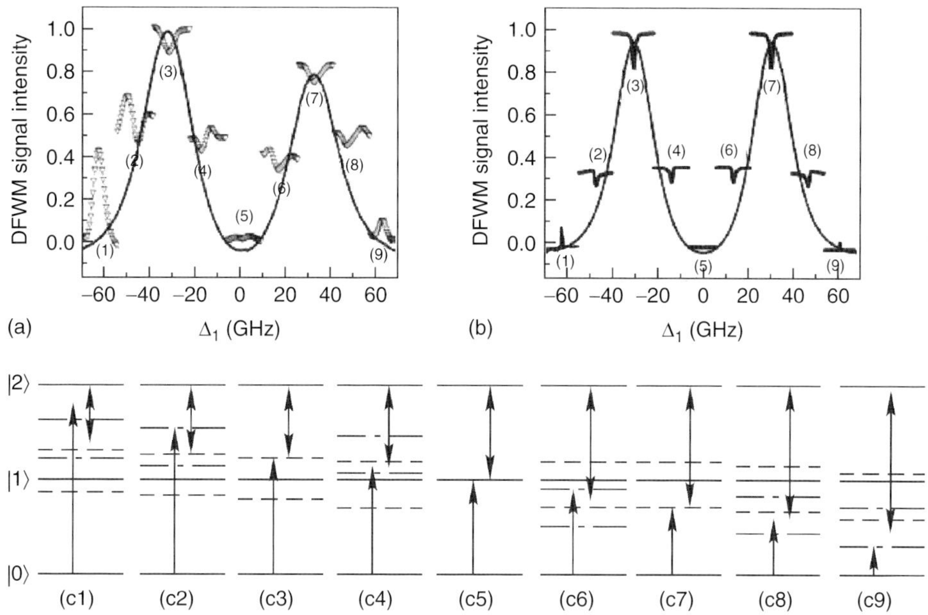

Figure 4.13 (a) The experimentally measured intensities of the DFWM sequentially doubly-dressed by the dressing fields and the pump fields, versus Δ_2 at different Δ_1 and the involved powers are set at $P_1 = 20$ μW, $P_2 = 25$ μW, and $P_3 = 10$ μW. Measured suppression and enhancement of DFWM signal I_0 versus I_1 at different I_0 values. (1) $\Delta_1 = -62.0$ GHz, (2) $\Delta_1 = -50.5$ GHz, (3) $\Delta_1 = -31.7$ GHz, (4) $\Delta_1 = -16.5$ GHz, (5) $\Delta_1 = 0$ GHz, (6) $\Delta_1 = 15.8$ GHz, (7) $\Delta_1 = 35.5$ GHz, (8) $\Delta_1 = 50.2$ GHz, and (9) $\Delta_1 = 65.8$ GHz respectively. The background profile is the double-peak DFWM signal $\boldsymbol{E}_{F1}$ versus Δ_1. (b) The theoretical results of the sequential doubly-dressed DFWM. The condition is same as that in (a). (c1)–(c9) The dressed-state energy level diagrams of the DFWM signal. The states $|\pm\rangle$ are dashed lines and the states $|+G_2\pm\rangle$ or $|-G_2\pm\rangle$ are dot-dashed lines respectively.

$\boldsymbol{E}_2$ ($\boldsymbol{E}_2'$) are studied theoretically, as shown in Figure 4.13(b), and the corresponding dressed-states diagrams are shown in Figure 4.13(c). The second doubly-dressed scheme is the nested-cascade mode in which $\boldsymbol{E}_3$ and $\boldsymbol{E}_2$ ($\boldsymbol{E}_2'$) are dressing fields, which is appropriate for the case that the intensity of $\boldsymbol{E}_1$ ($\boldsymbol{E}_1'$) is relatively weak and the dichromatic field dressing effects can be ignored in theoretical calculation, as shown in Figure 4.14(b). Correspondingly, the modified Liouville pathway of this doubly-dressed DFWM is $\boldsymbol{E}_1$. In the nesting doubly-dressing picture, we can get $\rho_{10}^{(3)} = g/\{\Gamma_{00}[d_1 + |G_3|^2/(\Gamma_{11} + |G_2|^2/d_5)]^2\}$, where $d_5 = \Gamma_{12} - i\Delta_2$. In this case, the state $|0\rangle$ will be split into two dressed states $|G_3 \pm\rangle$ by $\boldsymbol{E}_3$, and the state $|1\rangle$ will be split into two dressed states $|G_2 \pm\rangle$ by $\boldsymbol{E}_2$ ($\boldsymbol{E}_2'$) at the same time. The effect that split the state $|1\rangle$ into the states $|G_2 \pm\rangle$ can be equivalent to a step in the doubly-dressing effect in which the dressed states $|G_3 \pm\rangle$ are further split into the final dressed states $|G_3 \pm G_2 \pm\rangle$.

When the powers of $\boldsymbol{E}_3$, $\boldsymbol{E}_1$ ($\boldsymbol{E}_1'$), and $\boldsymbol{E}_2$ ($\boldsymbol{E}_2'$) are set to be 10, 20, and 25 μW, we can obtain the sequential doubly-dressed DFWM signal with two peaks by

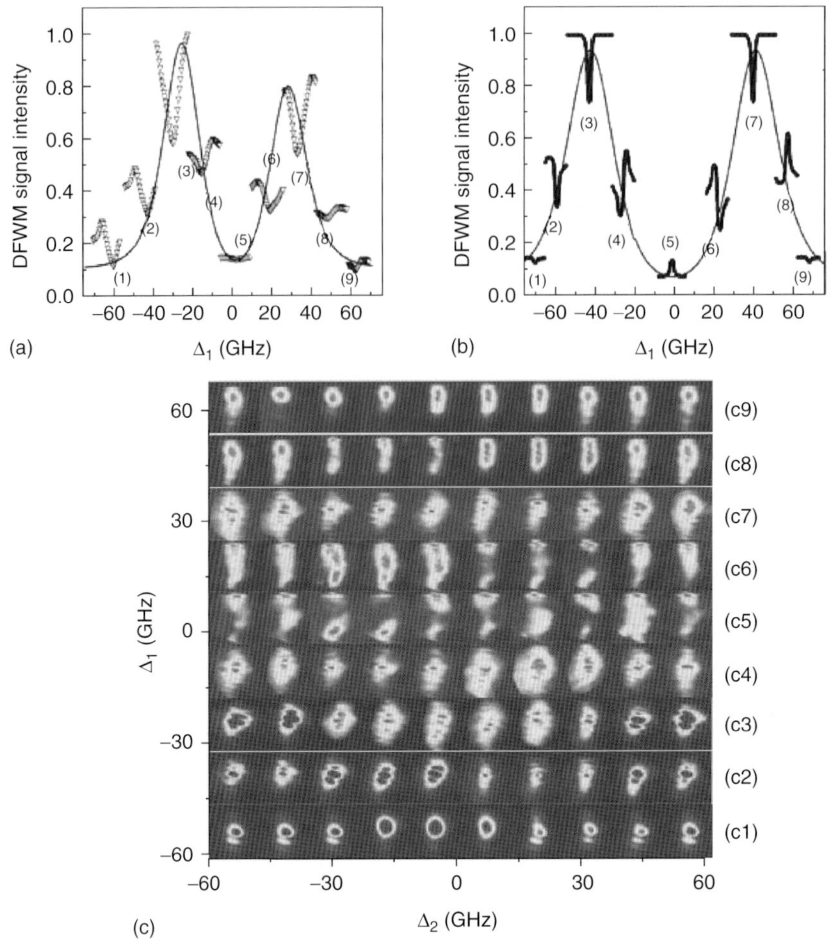

Figure 4.14 (a) The experimentally measured intensities of the DFWM doubly-dressed in nesting scheme by the dressing fields and probe field, versus Δ_2 at different Δ_1 value with the powers $P_1 = 10$ μW, $P_2 = 25$ μW, and $P_3 = 20$ μW. (1) $\Delta_1 = -63.3$ GHz, (2) $\Delta_1 = -50.7$ GHz, (3) $\Delta_1 = -32.5$ GHz, (4) $\Delta_1 = -17.8$ GHz, (5) $\Delta_1 = 0$ GHz, (6) $\Delta_1 = 16.4$ GHz, (7) $\Delta_1 = 35.2$ GHz, (8) $\Delta_1 = 50.5$ GHz, and (9) $\Delta_1 = 62.8$ GHz respectively. The background profile is the double-peak DFWM signal $\boldsymbol{E}_{F1}$ versus Δ_1. (b) The theoretical results of the nesting doubly-dressed DFWM in (a). The condition is same as that in (a). (c) The DFWM signal images when $\boldsymbol{E}_1$ and E_1' is set in the middle of the heat-pipe oven. The condition is same as that in (a).

scanning Δ_1 in Figure 4.13(a) (experimental results) and Figure 4.13(b) (theoretical simulation). If Δ_1 is set at discrete values in order from negative to positive and I_1 is scanned, the evolution of the intensity of the DFWM shows pure enhancement, first enhancement and next suppression; pure suppression, first suppression and next enhancement; dual enhancement, first enhancement and next suppression; pure suppression, first suppression and next enhancement;

and pure enhancement, successively, as shown in Figure 4.13(a1) to (a9). The various suppressions and enhancements in Figure 4.13(a) or (b) are caused by the interaction between these two dressing fields. First, we obtain two dressed states $|\pm\rangle$ (Figure 4.13(c)) while the bichromatic field dressing effects are considered. When Δ_1 is scanned with $G_2=0$, the DFWM field can resonate with $|\pm\rangle$, which leads to two maximal values of DFWM intensity corresponding to the two global peaks, respectively, as shown in Figure 4.13(a) and (b). Then, in the region with $\Delta_1<0$ when Δ_2 is scanned around the first dressed state $|+\rangle$, $|+\rangle$ will be split into two dressed states $|+G_2\pm\rangle$ by $\boldsymbol{E}_2$ ($\boldsymbol{E}_2'$) (Figure 4.13(c)). At the point far away from the resonance energy level $|+G_2+\rangle$ when scanning Δ_2, the enhancement condition $\Delta_{F1}+\Delta_2'+G_2+|m|k_1v\sin^2(\theta/2)=0$ is satisfied, where Δ_2' is introduced to emphasize the interaction between the two dressing fields, and is the frequency detuning of $\boldsymbol{E}_2$ ($\boldsymbol{E}_2'$) fields relative to the first dressed states $|\pm\rangle$ created by $\boldsymbol{E}_1$ ($\boldsymbol{E}_2'$) satisfying $\boldsymbol{E}_1'$. Therefore, $\boldsymbol{E}_{F1}$ is obviously purely enhanced as shown in Figure 4.13(a1) and (b1). The corresponding dressed energy level diagram is shown in Figure 4.13(c1). Then, with $\boldsymbol{E}_1$ increasing, the transition generating the dressed DFWM signal can first satisfy the enhancement condition $\boldsymbol{E}_1'$ where the DFWM resonates with $|+G_2+\rangle$ and next satisfies the suppression condition $\Delta_{F1}+\Delta_2'+|m|k_1v\sin^2(\theta/2)=0$, where the two-photon resonance of DFWM is matched, leading to the first enhancement and next suppression shown as Figure 4.13(a2), (b2), and (c2). When the DFWM signal resonates with $|+\rangle$ at which any real energy state no longer exists and $|+G_2\pm\rangle$ could not be reached; only the suppressed condition $\Delta_{F1}+\Delta_2'+|m|k_1v\sin^2(\theta/2)=0$ can be satisfied in the processing of scanning Δ_2. So, $\boldsymbol{E}_{F1}$ appears purely suppressed as shown in Figure 4.13(a3), (b3), and (c3). Then, as Δ_1 further increases, first suppression and next enhancement is obtained by scanning Δ_2 because the DFWM transition first satisfies the suppression condition $\Delta_{F1}+\Delta_2'+|m|k_1v\sin^2(\theta/2)=0$, that is, the two-photon resonance condition, and next satisfies the enhancement condition $\Delta_{F1}+\Delta_2'-G_2+|m|k_1v\sin^2(\theta/2)=0$ where the DFWM resonates with $\boldsymbol{E}_1$ as shown in Figure 4.13(a4), (b4), and (c4). At the point $\boldsymbol{E}_1'$, the DFWM and the dressing fields all reach the resonance energy level $|1\rangle$, where the strong resonant absorption leads to the weakened suppression; so, here, we obtain the dual enhancement experimentally in Figure 4.13(a5) rather than the theoretically calculated pure suppression in Figure 4.13(b5). When Δ_1 changes in its positive region, the DFWM signal shows symmetric behavior of the evolution in the negative region. So, the results in this case are similar to that around $|+\rangle$, as shown in Figure 4.13(a6) to (a9) and (b6) to (b9). The experimental results have good agreement with the theoretical results.

While the powers of $\boldsymbol{E}_1$ ($\boldsymbol{E}_1'$), $\boldsymbol{E}_3$, and $\boldsymbol{E}_2$ ($\boldsymbol{E}_2'$) are set to be 10, 20, and 25 μW, the nested-cascade DFWM is obtained. The two global peaks remain as obtained by scanning Δ_1 in the experiment as shown in Figure 4.14(a) (experimental results) and Figure 4.14(b) (theoretical simulation). If Δ_1 is arranged at different discrete values in order from the negative to the positive while Δ_2 is scanned, the evolution of the intensity of the DFWM is obtained as shown in Figure 4.14(a1) to (a9). Here, there are two points of difference between Figure 4.13(a) and Figure 4.14(a) worth being noted. One is that the pure enhancement in Figure 4.13(a) is not obtained at large Δ_1 in Figure 4.14(a). The other is that the dual enhancement at $\Delta_1 = 0$ in Figure 4.13(a) turns into pure suppression in Figure 4.14(a). Because other evolving processes of enhancement or suppression in such cases are similar to those in the sequential doubly-dressed case in Figure 4.13(a), we concentrate here only on the differences between these two dressing schemes. As in the nested-cascade DFWM, $\boldsymbol{E}_3$ is not only the probe field but also the dressing field; therefore, it could not touch the states $|G_3 + G_2 +\rangle$ and $|G_3 - G_2 -\rangle$, which leads to the first difference described. The other difference that the dual enhancement in Figure 4.13(a) changes into suppression in Figure 4.14(a5) and (b5) at $\Delta_1 = 0$ is due to the effect that the inner dressing field $\boldsymbol{E}_3$ can largely strengthen the suppression effect on the DFWM signal in the experiment. The experimental and theoretical results fit essentially.

In addition, the spatial splitting in y-direction of the FWM signal beams due to XPM or EIG is observed simultaneously, as shown in Figure 4.14(c). The number of the splitting spots changes when the FWM intensity is suppressed. When $\boldsymbol{E}_1$ and $\boldsymbol{E}_1'$ are set in the middle of the oven, $\boldsymbol{E}_{\mathrm{F1}}$ exactly overlaps with $\boldsymbol{E}_1'$ in the y-direction because of the phase-matching condition. As described in our previous investigation [29], for XPM, the spatial splitting of the FWM beam can be controlled by the intensities of the involved laser beams, the cross-Kerr nonlinear coefficients, and the atomic density according to the nonlinear phase shift $\phi = 2k_{\mathrm{F1}} n_2 z I_1' \mathrm{e}^{-r^2/2}/(n_0 I_{\mathrm{F1}})$ where n_2 is the nonlinear cross-Kerr refractive index, I_{F1} is the intensity of DFWM, and I_1' is the intensity of $\boldsymbol{E}_1'$. Furthermore, the EIG, which is formed by a strong standing wave interacting with the atomic medium, can diffract a weak field into a high-order direction. On the basis of the theory, a vertical EIG is induced in the y-direction with interference fringes spacing $\Lambda = \lambda/\theta_{1y}$, where $\theta_{1y}(\approx 0.05^\circ)$ is the angle between $\boldsymbol{E}_1$ and $\boldsymbol{E}_1'$ in the y-direction. Because θ_{1y} is far less than the angle of $\boldsymbol{E}_1$ and $\boldsymbol{E}_1'$ in the x-direction, Λ is big enough for observing the splitting in y-direction due to EIG. When n_2 is relatively small at the large detuning shown as in Figure 4.14(c1) and (c9), the y-direction splitting is obviously observed because the overlap of $\boldsymbol{E}_1'$ and $\boldsymbol{E}_{\mathrm{F1}}$ in y-direction and the nonlinear cross-Kerr effect of $\boldsymbol{E}_1'$ are dominant. When Δ_2 is scanned in the experiment, the nonlinear phase shift ϕ is primarily determined by I_{F1}. When the suppression condition is satisfied, and I_{F1} reaches its minimum, the spatial splitting will become stronger as shown in Figure 4.14(c1) and (c9). While in the enhancement condition with the larger I_{F1}, ϕ is decreased, and therefore the splitting is weakened correspondingly. From Figure 4.14(c2) to (c8), the splitting in the y-direction due to the EIG is obtained because n_2 is big enough for forming the grating in the y-direction. Correspondingly, a larger spot of the $\boldsymbol{E}_{\mathrm{F1}}$ beam is split into more parts passing through the grating in the y-direction.

Comparing the doubly-dressed (Figure 4.13 and Figure 4.14) and single-dressed (Figure 4.12) DFWM processes, we find that three symmetric centers of enhancements and suppressions appear in the former when scanning the dressing field (Figure 4.13(a3), (a5), and (a7) and Figure 4.14(a3), (a5), and (a7)), whereas only one symmetric center appears in the latter (Figure 4.12(b3)). For the single-dressed DFWM, the symmetric center at $\Delta_1 = 0$ is caused by the state $|1\rangle$ and the evolution of enhancement and suppression is induced by the dressing effect. For the doubly-dressed DFWM, the profile and three symmetric centers are due to the primary dressing effect of one dressing field $\boldsymbol{E}_1$ ($\boldsymbol{E}_1'$) for Figure 4.13 and $\boldsymbol{E}_3$ for Figure 4.14, while the evolution of enhancement and suppression is induced by the interaction between two dressing fields.

From this analyses, the two kinds of doubly-dressed (sequential-cascade and nested-cascade) schemes have similarities and dissimilarities. First, the two dressing fields in them have similar dressing effects. Second, for the nested-cascade scheme, when the inner dressing field is zero, there is no outer dressing influence, whereas for the sequential-cascade scheme, one dressing field always has effects no matter whether another dressing field is zero or not. Third, two dressing fields in the sequential-cascade case affect the corresponding FWM signal independently, whereas the two in the nested-cascade case affect the corresponding FWM signal dependently. So, we can conclude that the interaction between two dressing fields is stronger in the nested-cascade scheme and weaker in the sequential-cascade scheme.

4.3.2.3 **Triply-Dressed DFWM**

In order to investigate the triply-dressed scheme, the powers of $\boldsymbol{E}_1$ and $\boldsymbol{E}_3$ are increased respectively. (We can investigate different triply-dressed schemes in the case that the powers of $\boldsymbol{E}_1$, $\boldsymbol{E}_3$ are large, respectively, and the case that the powers of $\boldsymbol{E}_1$ and $\boldsymbol{E}_3$ are all large.) First, when $\boldsymbol{E}_1$ and $\boldsymbol{E}_3$ are set to be strong and weak relative to each other, and Δ_2 is scanned at all the selected discrete Δ_1 values (the pure enhancement is dominant as shown in Figure 4.15(a)), while the dual enhancement is obtained only at $\Delta_1 = 0$. In order to analyze these results, the DFWM triply dressed by $\boldsymbol{E}_1$ and $\boldsymbol{E}_2$ ($\boldsymbol{E}_2'$) is studied. Here, the bichromatic field dressing effects are ignored because the intensities of $\boldsymbol{E}_1'$ and $\boldsymbol{E}_3$ are too weak. The modified third-order nonlinear density matrix element is shown as $\rho_{10}^{(3)} = g/\{\Gamma_{00}[d_1 + |G_2|^2/d_2 + |G_1|^2/(\Gamma_{00} + |G_1|^2/d_6)]^2\}$, where $d_6 = \Gamma_{01} - i\Delta_1$. When the power of $\boldsymbol{E}_1$ is very large, the field can be nested by itself in a nested-cascade scheme. After $\boldsymbol{E}_2$ ($\boldsymbol{E}_2'$) splitting $|1\rangle$ into $|G_2 \pm\rangle$, the self-dressing effect split the state $|1\rangle$ into $|G_2 \pm G_1 \pm\rangle$ to control FWM directly, while the self-dressing effect of $\boldsymbol{E}_1$ ($\boldsymbol{E}_1'$) can split $|0\rangle$ to control FWM indirectly, as shown in Figure 4.15(b1) and (b2). As $\boldsymbol{E}_1$ ($\boldsymbol{E}_1'$) play roles in the latter two splitting steps in this three-steps scheme, the distances between the pairs of dressed states split by $\boldsymbol{E}_1$ successively are equal, and the two split states by $\boldsymbol{E}_1$ in the first dressed step are approximately symmetrical with respect to the point the $\boldsymbol{E}_1$ reaches. Therefore, the DFWM can resonate with one of the two dressed states split by $\boldsymbol{E}_1$ in the second dressed step, as shown in Figure 4.15(b2). As a result, the DFWM signal will be enhanced all the time. The enhancement condition in this case can be expressed as $\Delta_{F1} + \Delta_2 + G_2 + 2G_1 = 0$.

Next, when the powers of $\boldsymbol{E}_1$ ($\boldsymbol{E}_1'$) and $\boldsymbol{E}_3$ are set to be small and large, respectively, the pure suppressions are observed at all the discrete Δ_1 values by scanning Δ_2 except at significantly large Δ_1 where the pure enhancements are observed as shown in Figure 4.15(c). To explain these phenomena, the DFWM triply-dressed by $\boldsymbol{E}_3$ and $\boldsymbol{E}_2$ ($\boldsymbol{E}_2'$) is studied. Considering the relatively weak

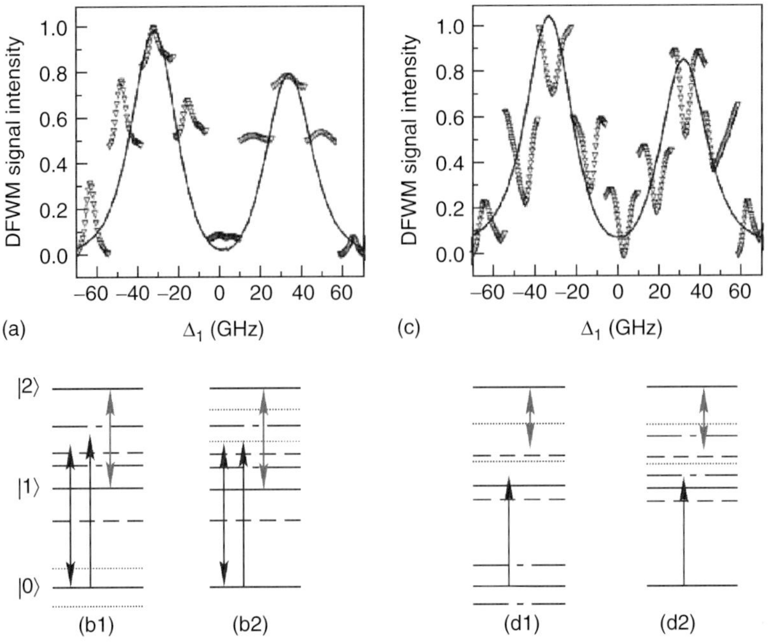

Figure 4.15 (a) The experimentally measured intensities of the DFWM multi-dressed by the pump fields and dressing fields, versus Δ_2 at different Δ_1 values of $\Delta_1 = -63.4$, -50.5, -31.7, -16.5, 0, 15.8, 35.5, 50.2, and 65.8 GHz, from the left to right side, respectively. The involved powers are set at $P_1 = 22$ μW, $P_2 = 25$ μW, and $P_3 = 8$ μW. The background profile is the double-peak DFWM signal $\boldsymbol{E}_{F1}$ versus Δ_1. (b1) and (b2) The dressed-state energy level diagrams of the DFWM signal multi-dressed by the pump fields and dressing fields. The first-order dressed states are dashed lines; the second-order dressed states are dot-dashed lines, and the third-order dressed states are short-dashed lines respectively. (c) The experimentally measured intensities of the DFWM multi-dressed by the probe field and dressing fields, versus Δ_2 at different Δ_1 values of $\Delta_1 = -65.4$, -50.5, -31.7, -16.5, 0, 15.8, 35.5, 50.2, and 64.8 GHz, from the left to right side, respectively. The involved powers are set at $P_1 = 8$ μW, $P_2 = 25$ μW, and $P_3 = 30$ μW. The background profile is the double-peak DFWM signal $\boldsymbol{E}_{F1}$ versus Δ_1. (d1) and (d2) The dressed-state energy level diagrams of the DFWM signal multi-dressed by the probe field and dressing fields. The first-order dressed states are dashed lines; the second-order dressed states are dot-dashed lines; and the third-order dressed states are short-dashed lines respectively.

intensity of $\boldsymbol{E}_1$ ($\boldsymbol{E}_1'$), we do not take the bichromatic field dressing effects into account. The corresponding density matrix element for such a DFWM process is $\rho_{10}^{(3)} = g/\{\Gamma_{00}[d_1 + |G_2|^2/d_2 + |G_3|^2/(\Gamma_{00} + |G_3|^2/d_6)]^2\}$. As $\boldsymbol{E}_3$ has a large power, it is nested with itself to form a nested-cascade scheme, in which $|1\rangle$ is split twice by $\boldsymbol{E}_3$ field to be the dressed states $|G_3 \pm G_3 \pm\rangle$ before $\boldsymbol{E}_2$ ($\boldsymbol{E}_2'$) split it as shown in Figure 4.15(d1) and (d2), just as what $\boldsymbol{E}_1$ has done in Figure 4.15(b1) and (b2). As a result, the DFWM will resonate to the dressed state split by $\boldsymbol{E}_3$ in the second dressing step. Therefore, the DFWM signal will be suppressed at almost all values of Δ_1 except the extreme values far away from the resonant point. The suppression condition is $\Delta_{F1} + \Delta_2'' + G_2 + 2G_3 = 0$, where Δ_2'' is the frequency detuning of $\boldsymbol{E}_2$ ($\boldsymbol{E}_2'$) field relative to $|G_3 \pm G_3 \pm\rangle$. As an inertial property of the system, resonating absorption results in the two global peaks in the DFWM signal in experiment as shown in Figure 4.15(a) and (c).

Finally, the triply-dressed scheme by $\boldsymbol{E}_1$ ($\boldsymbol{E}_1'$), $\boldsymbol{E}_3$, and $\boldsymbol{E}_2$ ($\boldsymbol{E}_2'$) is investigated. When the powers of $\boldsymbol{E}_1$ ($\boldsymbol{E}_1'$), $\boldsymbol{E}_3$, and $\boldsymbol{E}_2$ ($\boldsymbol{E}_2'$) are all large, three global peaks are obtained by scanning Δ_1. The evolution of the intensity of $\boldsymbol{E}_{F1}$ is obtained by scanning Δ_2 at different Δ_1 values, as shown in Figure 4.16(a). In order to explain the results, the triply-dressed DFWM is analyzed theoretically. Here, the bichromatic field dressing effects on the weak DFWM signals induced by the strong fields $\boldsymbol{E}_1$ and $\boldsymbol{E}_3$ with different Doppler shifts should be considered, and the bichromatic field dressing effect induced by $\boldsymbol{E}_1$ and $\boldsymbol{E}_1'$ also should be considered. First, we obtain two dressed states $|\pm\rangle$ because of the bichromatic field ($\boldsymbol{E}_1$ and $\boldsymbol{E}_3$) dressing effects, then the bichromatic fields ($\boldsymbol{E}_1$ and $\boldsymbol{E}_1'$) can dress the states $|\pm\rangle$ to $|+\pm\rangle$ or $|-\pm\rangle$, and at last $|\pm\pm\rangle$ can be split into dressed states $|\pm\pm G_2 \pm\rangle$ by $\boldsymbol{E}_2$ ($\boldsymbol{E}_2'$), as shown as in Figure 4.16(b). As Δ_1 is scanned with $G_2 = 0$, the DFWM signal with three global peaks can be obtained when the DFWM resonates to three dressed states, such as $|++\rangle$, $|+-\rangle$, and $|-\rangle$, where the bichromatic fields ($\boldsymbol{E}_1$ and $\boldsymbol{E}_1'$) dress the state $|+\rangle$ dressed by bichromatic fields ($\boldsymbol{E}_1$ and $\boldsymbol{E}_3$) into $|++\rangle$ and $|+-\rangle$. The results by scanning Δ_2 at different discrete Δ_1 values are like the series connection of three single-dressing

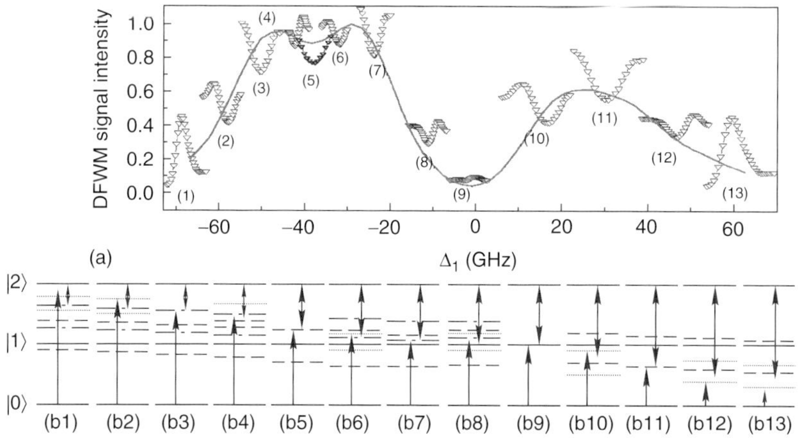

Figure 4.16 (a) The experimentally measured intensities of the DFWM multi-dressed by the probe field, pump fields, and dressing fields, versus Δ_2 at different Δ_1 with the powers $P_1 = 22$ μW, $P_2 = 25$ μW, and $P_3 = 25$ μW. (1) $\Delta_1 = -68.8$ GHz, (2) $\Delta_1 = -60.2$ GHz, (3) $\Delta_1 = -50.7$ GHz, (4) $\Delta_1 = -42.8$ GHz, (5) $\Delta_1 = -36.5$ GHz, (6) $\Delta_1 = -31.7$ GHz, (7) $\Delta_1 = -20.3$ GHz, (8) $\Delta_1 = -16.5$ GHz, (9) $\Delta_1 = 0$ GHz, (10) $\Delta_1 = 15.8$ GHz, (11) $\Delta_1 = 35.5$ GHz, (12) $\Delta_1 = 50.2$ GHz, and (13) $\Delta_1 = 65.8$ GHz, respectively. The background profile is three-peak DFWM signal $\boldsymbol{E}_{F1}$ versus Δ_1. (b1)–(b13) The dressed-state energy level diagrams of the DFWM signal. The first-, second-, and third-order dressed state are the dashed, dot-dashed, and short-dashed lines, respectively.

states, as shown in Figure 4.16(a1) to (a13). The dressed states are shown as in Figure 4.16(b). Correspondingly, owing to the resonating of DFWM with the dressed states split by the dressing fields, the enhancement conditions in this multi-dressed system are $\Delta_{F1} + \Delta_2''' + |m_1|k_1v \pm |m_2|k_1v\sin^2(\theta/2) \pm G_2 = 0$ for $\Delta_1 < 0$ and $\Delta_{F1} + \Delta_2' - |m_1|k_1v \pm G_2 = 0$ for $\Delta_1 > 0$, where Δ_2''' is the frequency detuning of $\boldsymbol{E}_2$ ($\boldsymbol{E}_2'$) fields relative to the second dressed states $|+\pm\rangle$. Under the two-photon resonance, the suppression conditions are $\Delta_{F1} + \Delta_2''' + |m_1|k_1v \pm |m_2|k_1v\sin^2(\theta/2) = 0$ for $\Delta_1 < 0$ and $\Delta_{F1} + \Delta_2' - |m_1|k_1v = 0$ for $\Delta_1 > 0$. Here, m_1 and m_2 express the two bichromatic fields, respectively.

Comparing the single-dressed (Figure 4.12) and doubly-dressed (Figure 4.13 and Figure 4.14) FWM processes, we find that five symmetric centers of enhancement and suppression signals appear in the triply-dressing type when scanning the dressing field (Figure 4.16(a3), (a5), (a7), (a9), and (a11)). The two symmetric centers shown as Figure 4.16(a9) and (a11) result from the primary dressing effect of one dressing field, while the three symmetric centers shown as Figure 4.16(a3), (a5), and (a7) are due to the interaction between the primary and secondary dressing effect of the two dressing fields. The evolution of enhancement and suppression is induced by the interaction between three dressing fields.

4.3.2.4 Power Switching of Enhancement and Suppression

As stated in the preceding text, the enhancement and suppression of DFWM can be switched flexibly by changing the powers of probe, pump, and dressing fields. Also, the suppression of DFWM will become more significant while the power of $\boldsymbol{E}_3$ changes from little to large compared with $\boldsymbol{E}_1$ ($\boldsymbol{E}_1'$), which can strengthen the enhancement extent of the DFWM signal. However, in Figure 4.17(a), with the decreasing of $\boldsymbol{E}_3$ power, it is obvious that the pure suppression changes into pure enhancement. Here, we set $P_1 = 10$ μW, $P_2 = 25$ μW. When the $\boldsymbol{E}_3$ power is very large, the multi-dressed suppression condition $\Delta_{F1} + \Delta_2' + 2G_3 = 0$ is satisfied, and therefore the pure suppression will appear at first as shown in Figure 4.17(a8) and (a9). While the $\boldsymbol{E}_3$ power becomes lower, the doubly-dressed suppression condition

Figure 4.17 (a) Enhancement and suppression of the DFWM signals versus Δ_2 at the point $\Delta_1 = -58.8$ GHz when the probe power P_3 is set at (a1) 5.99 μW, (a2) 13.51 μW, (a3) 17.27 μW, (a4) 21.03 μW, (a5) 24.79 μW, (a6) 28.55 μW, (a7) 32.31 μW, (a8) 36.07 μW, and (a9) 43 μW. The other parameters are $P_1 = 10$ μW and $P_2 = 25$ μW. (b) Enhancement and suppression of the DFWM signals versus Δ_2 at $\Delta_1 = -40.5$ GHz when the dressing fields power P_2 is (b1) 3.03 μW, (b2) 5.78 μW, (b3) 8.53 μW, (b4) 11.28 μW, (b5) 14.03 μW, (b6) 16.78 μW, (b7) 19.53 μW, (b8) 22.28 μW, and (b9) 25.00 μW. The other parameters are $P_1 = 20$ μW and $P_3 = 10$ μW. (c1)–(c3) The dressed-state energy level diagrams of the DFWM signal. The first and second-order dressed states switches controlled by the power of the dressing fields $\boldsymbol{E}_2$ ($\boldsymbol{E}_2'$) are expressed by the dashed and dot-dashed lines, respectively. (d) Enhancement and suppression of the DFWM signals versus Δ_2 at the point $\Delta_1 = -60.3$ GHz when the pump fields power P_1 is (d1) 3.48 μW, (d2) 7.78 μW, (d3) 9.93 μW, (d4) 12.09 μW, (d5) 14.24 μW, (d6) 16.39 μW, (d7) 18.54 μW, (d8) 22.85 μW, and (d9) 25.00 μW. The other parameters are $P_3 = 10$ μW and $P_2 = 25$ μW. (e) The DFWM signal images when $\boldsymbol{E}_1$ and $\boldsymbol{E}_1'$ is set in the middle of the heat-pipe oven. The condition is same as that in (b).

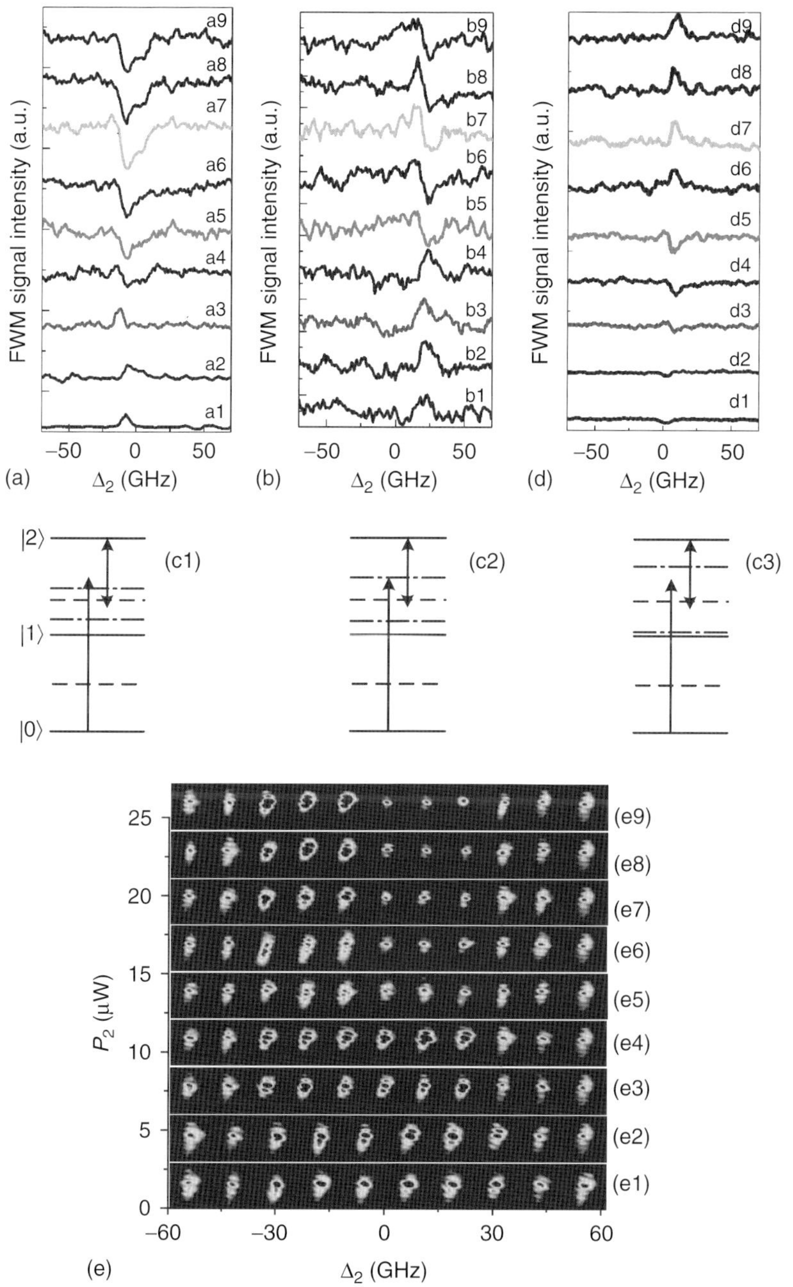

FWM signal intensity (a.u.)
a9
a8
a7
a6
a5
a4
a3
a2
a1
−50 0 50
(a)
Δ2 (GHz)
b9
b8
b7
b6
b5
b4
b3
b2
b1
(b)
d9
d8
d7
d6
d5
d4
d3
d2
d1
(d)
|2⟩
|1⟩
|0⟩
(c1)
(c2)
(c3)
(e9)
(e8)
(e7)
(e6)
(e5)
(e4)
(e3)
(e2)
(e1)
P2 (μW)
0 5 10 15 20 25
−60 −30 0 30 60
(e)
Δ2 (GHz)

$\Delta_{F1} + \Delta_2' + G_3 = 0$ and enhancement condition $\Delta_{F1} + \Delta_2' + G_3 + G_2 = 0$ can be satisfied; therefore, partial enhancement/suppression will be obtained as shown in Figure 4.17(a4) to (a7). As the $\boldsymbol{E}_3$ power decreases to the least value, the single-dressed enhancement condition $\Delta_{F1} + \Delta_2 + G_2 = 0$ is satisfied, and therefore the pure enhancement is obtained finally as shown in Figure 4.17(a1) to (a3). In this process, the power switching between the enhancement and suppression of DFWM is achieved by changing the power of $\boldsymbol{E}_3$. Similar phenomena can also be observed when changing the power of $\boldsymbol{E}_1$ ($\boldsymbol{E}_1'$). The suppression will be switched into the enhancement when the power of $\boldsymbol{E}_1$ ($\boldsymbol{E}_1'$) increases as shown in Figure 4.17(d). Although $\boldsymbol{E}_2$ ($\boldsymbol{E}_2'$) is always a dressing field whether its power is small or large, the change in its power can affect the distance between the dressed states, further leading to the switching of suppression and enhancement. In other words, if the power of $\boldsymbol{E}_2$ ($\boldsymbol{E}_2'$) changes, the distance can be broadened or narrowed and therefore the enhancement or suppression condition also changes correspondingly. When $\boldsymbol{E}_2$ ($\boldsymbol{E}_2'$) power is low and the power of $\boldsymbol{E}_1$ ($\boldsymbol{E}_1'$) is $P_1 = 20$ μW, the intensity of DFWM signal is enhanced while the pure enhancement condition $\Delta_{F1} + \Delta_2' + G_2 + |m|k_1 v \sin^2(\theta/2) = 0$ is satisfied, as shown in Figure 4.17(b1) to (b4) and (c1). On the other hand, considering the expressions of the modified third-order matrix element under this dressed scheme, we can find that G_2 could also influence the enhancement and suppression of the DFWM signal. So, while $\boldsymbol{E}_2$ ($\boldsymbol{E}_2'$) power becomes higher, the dressed states move too far away from the position resonating with the DFWM signal, which destroys the pure enhancement condition, but results in that the enhancement condition $\Delta_{F1} + \Delta_2' + G_2 + |m|k_1 v \sin^2(\theta/2) = 0$ and the suppression condition $\Delta_{F1} + \Delta_2' + |m|k_1 v \sin^2(\theta/2) = 0$ are satisfied, as shown in Figure 4.17(b5) to (b9) and (c2). With these analyses, we can conclude that the pure suppression condition $\Delta_{F1} + \Delta_2' + |m|k_1 v \sin^2(\theta/2) = 0$ will be satisfied, with the distance of the dressed states significantly broadening under larger powers of $\boldsymbol{E}_2$ ($\boldsymbol{E}_2'$). As a result, we will obtain pure suppression finally. The corresponding dressed states diagram is as shown in Figure 4.17(c3). The y-direction spatial splitting images of the DFWM signal are obtained simultaneously as shown in Figure 4.17(e).

4.4
Enhancement and Suppression of Two Coexisting SWM Processes

In this section, we show the enhancement and suppression of two coexisting multi-dressed SWM signals by the external-dressing effect and the AT splitting by the self-dressing effect in the EIT window. We explore a new technique to directly observe the dressing effects in the experiment, which is to change the probe field frequency detuning as the dressing field frequency detuning is scanned. This technique can also clearly distinguish the self-dressing effect and the external-dressing effect. So, we can use this technique as a direct measurement of dressing effects.

4.4.1 Theoretical Model and Experimental Scheme

The experiment is performed in a ^{85}Rb vapor cell involving five energy levels consisting of $5S_{1/2}(F=3)$, $5S_{1/2}(F=2)$, $5P_{3/2}$, $5D_{3/2}$, and $5D_{5/2}$ (Figure 4.18(b)). The atomic vapor cell temperature is set at 60 °C, corresponding to the typical density of 2×10^{11} cm^{-3}. The laser beams are aligned spatially as shown in Figure 4.18(a). First, a weak probe beam E_1 with a wavelength of 780.245 nm (frequency ω_1, wave vector $\mathbf{k}_1$, Rabi frequency G_1, and frequency detuning Δ_1, where $\Delta_i = \Omega_i - \omega_i$, Ω_i is the resonance frequency) propagates through the atomic medium, which probes the lower transition $5S_{1/2}(F=3)$ ($|0\rangle$) to $5P_{3/2}$ ($|1\rangle$). The probe beam E_1 is from an ECDL (ECDL-Toptica DL100L), which is horizontally polarized. Two weak coupling beams E_3 (ω_3, $\mathbf{k}_3$, G_3, and Δ_3) and E_3' (ω_3, $\mathbf{K}_3'$, G_3', and Δ_3) with a wavelength of 780.235 nm propagate in the opposite direction with a small angle (~0.3°) and drive the transition $5P_{3/2}$ ($|1\rangle$) to $5S_{1/2}(F=2)$ ($|3\rangle$). The laser beams E_3 and E_3' are split from a tapered-amplifier diode laser (Thorlabs TCLDM9) with equal power and vertical polarization. The diameters of the probe beam E_1 and the coupling beams E_3 and E_3' are about 0.3 and 0.5 mm at the cell center, respectively. Thus, one FWM signal beam E_F with phase-matching condition $\mathbf{k}_F = \mathbf{k}_1 + \mathbf{k}_3 - \mathbf{k}_3'$ is generated by E_1, E_3, and E_3'. Next, we add two strong dressing fields. One strong dressing laser beam E_2 (ω_2, $\mathbf{k}_2$, G_2, and Δ_2) with a wavelength of 775.978 nm propagates in the same direction as the beam E_3, and drives the transition $5P_{3/2}$

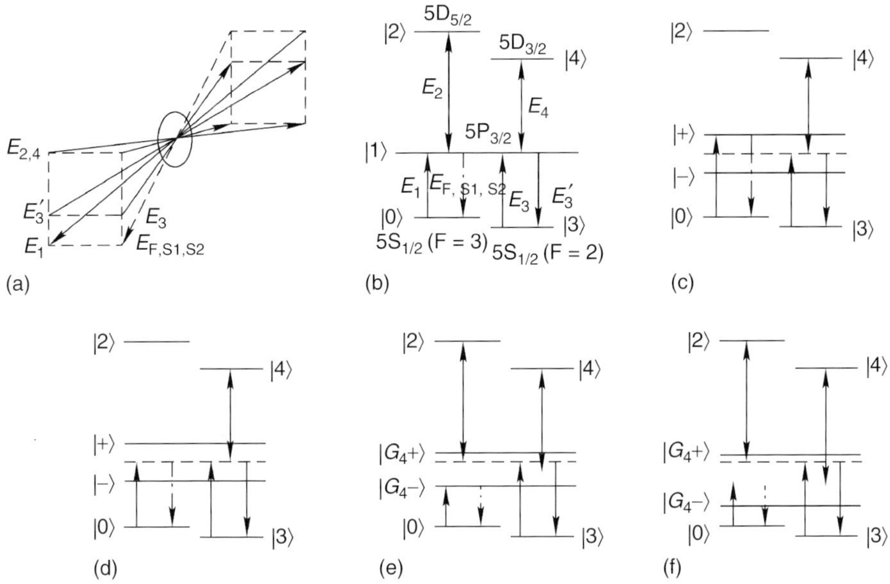

Figure 4.18 (a) Spatial beam geometry used in the experiment. (b) The diagram of relevant ^{85}Rb energy levels. (c) and (d) The dressed-state pictures for the dressing field E_2. (e) and (f) The dressed-state pictures for the dressing field E_4.

($|1\rangle$) to $5D_{5/2}$ ($|2\rangle$). The other strong dressing laser beam E_4 (ω_4, $\mathbf{k}_4$, G_4, and Δ_4) with a wavelength of 776.157 nm propagates in the same direction as the beam E_3, and drives the transition $5P_{3/2}$ ($|1\rangle$) to $5D_{3/2}$ ($|4\rangle$). The dressing beams E_2 and E_4 are from two ECDLs (Hawkeye Optoquantum and UQEL100) with vertical polarization. The diameter of the dressing beams E_2 and E_4 are about 0.5 mm at the cell center. Thus, there exist two ladder-type EIT windows, and two SWM signals form simultaneously and interplay each other in the five-level atomic system (Figure 4.18(b)). Specifically, one SWM signal, S1, is generated by E_1, E_3, E_3', and E_4 in the $|0\rangle - |1\rangle - |4\rangle$ ladder-type EIT window, which satisfies the phase-matching condition $\mathbf{k}_{S1} = \mathbf{k}_1 + \mathbf{k}_3 - \mathbf{k}_3' + \mathbf{k}_4 - \mathbf{k}_4$, and the other SWM signal, S2, is generated by E_1, E_3, E_3', and E_2 in the $|0\rangle - |1\rangle - |2\rangle$ ladder-type EIT window, which satisfies the phase-matching condition $\mathbf{k}_{S2} = \mathbf{k}_1 + \mathbf{k}_3 - \mathbf{k}_3' + \mathbf{k}_2 - \mathbf{k}_2$. Finally, the generated FWM, SWM signals, with the same horizontal polarization and the transmitted probe beam, are detected by an avalanche photodiode detector and a photodiode, respectively. Note that, the FWM signal without the EIT window can be neglected.

In general, the expression of the density matrix element $\rho_{10}^{(5)}$ related to the SWM process can be obtained by solving the coupled density matrix equations. When the field E_2 is blocked, via the Liouville pathway: $\rho_{00}^{(0)} \xrightarrow{\omega_1} \rho_{G_4\pm0}^{(1)} \xrightarrow{-\omega_3} \rho_{30}^{(2)} \xrightarrow{\omega_3} \rho_{G_4\pm0}^{(3)} \xrightarrow{\omega_4} \rho_{40}^{(4)} \xrightarrow{-\omega_4} \rho_{10}^{(5)}$, we can obtain $\rho_{10a}^{(5)} = G_A e^{i\mathbf{k}_{S1}\times\mathbf{r}}/[d_1 d_3 d_4 (d_1 + |G_4|^2/d_4)^2]$ for the SWM process S1 (the intensity $I_{S1} \propto |\rho_{10a}^{(5)}|^2$), where $G_A = iG_1 G_3 G_3'^* G_4 G_4^*$, $d_1 = i\Delta_1 + \Gamma_{10}$, $d_3 = i(\Delta_1 - \Delta_3) + \Gamma_{30}$, $d_4 = i(\Delta_1 + \Delta_4) + \Gamma_{40}$, and Γ_{ij} is the transverse relaxation rate between states $|i\rangle$ and $|j\rangle$. In the SWM process S1, E_4 is both the dressing field of the S1 signal and the field to generate it, so we refer to the dressing of E_4 as the self-dressing. Similarly, when the field E_4 is blocked, via the Liouville pathway: $\rho_{00}^{(0)} \xrightarrow{\omega_1} \rho_{G_2\pm0}^{(1)} \xrightarrow{-\omega_3} \rho_{30}^{(2)} \xrightarrow{\omega_3} \rho_{G_2\pm0}^{(3)} \xrightarrow{\omega_2} \rho_{20}^{(4)} \xrightarrow{-\omega_2} \rho_{10}^{(5)}$, we can obtain $\rho_{10b}^{(5)} = G_B e^{i\mathbf{k}_{S2}\times\mathbf{r}}/[d_1 d_2 d_3 (d_1 + |G_2|^2/d_2)^2]$ with self-dressing of the field E_2 for the SWM process S2 (the intensity $I_{S2} \propto |\rho_{10b}^{(5)}|^2$), where $G_B = iG_1 G_2 G_2^* G_3 G_3'^*$ and $d_2 = i(\Delta_1 + \Delta_2) + \Gamma_{20}$. When all laser beams are turned on, two multi-dressed SWM signals form simultaneously and there exist an interplay between them. For the SWM process S1 dressed by E_2 (defined as D1), via the dressed Liouville pathway: $\rho_{00}^{(0)} \xrightarrow{\omega_1} \rho_{(G_2\pm G_4\pm)0}^{(1)} \xrightarrow{-\omega_3} \rho_{30}^{(2)} \xrightarrow{\omega_3} \rho_{(G_2\pm G_4\pm)0}^{(3)} \xrightarrow{\omega_4} \rho_{40}^{(4)} \xrightarrow{-\omega_4} \rho_{(G_2\pm)0}^{(5)}$, the solved expression of D1 is $\rho_{10Da}^{(5)} = G_A e^{i\mathbf{k}_{S1}\cdot\mathbf{r}}/[d_3 d_4 (d_1 + |G_2|^2/d_2))(d_1 + |G_2|^2/d_2 + |G_4|^2/d_4)^2]$ (the intensity $I_{D1} \propto \rho_{10Da}^{5)}$). In the D1 process, E_2 is only the dressing field, so we refer to the dressing of E_2 as the external dressing. Similarly, the SWM process S2 is externally dressed by E_4 (defined as D2), and via the dressed Liouville pathway: $\rho_{00}^{(0)} \xrightarrow{\omega_1} \rho_{(G_2\pm G_4\pm)0}^{(1)} \xrightarrow{-\omega_3} \rho_{30}^{(2)} \xrightarrow{\omega_3} \rho_{(G_2\pm G_4\pm)0}^{(3)} \xrightarrow{\omega_2} \rho_{20}^{(4)} \xrightarrow{-\omega_2} \rho_{(G_4\pm)0}^{(5)}$, the solved expression of D2 is M $\rho_{10Db}^{(5)} = G_B e^{i\mathbf{k}_{S2}\cdot\mathbf{r}}/[d_2 d_3 (d_1 + |G_4|^2/d_4))(d_1 + |G_2|^2/d_2 + |G_4|^2/d_4)^2]$ (the intensity $I_{D2} \propto \rho_{10Db}^{(5)}$). From the expressions of the D1 and D2 signals, one can see that the two SWM processes are closely connected by mutual dressing.

4.4.2 Experimental Results

First, we study the transmission property of the probe field E_1 by scanning the dressing field frequency detuning Δ_4 at different designated probe frequency detuning (Δ_1) values (note: do not scan Δ_1) as shown in Figure 4.19(a). Note that the values in the abscissa axis in Figure 4.19(a1) to (a3) and (b1) to (b3), Figure 4.20, and Figure 4.21(a), (c), and (e) stand for the discrete points of Δ_1. The inset in Figure 4.19(a) is from the group of experimental data in the rectangle, which is obtained by scanning Δ_4 at $\Delta_1 = 16$ MHz. When the laser beam E_2 is blocked and E_4 is turned on, each EIT peak (see Figure 4.19(a1)) denotes the transparent degree (without Doppler absorption background), and is induced by the dressing field E_4 in $|0\rangle - |1\rangle - |4\rangle$ ladder subsystem satisfying the condition $\omega_1 + \omega_4 = \Omega_1 + \Omega_4$ (i.e.,

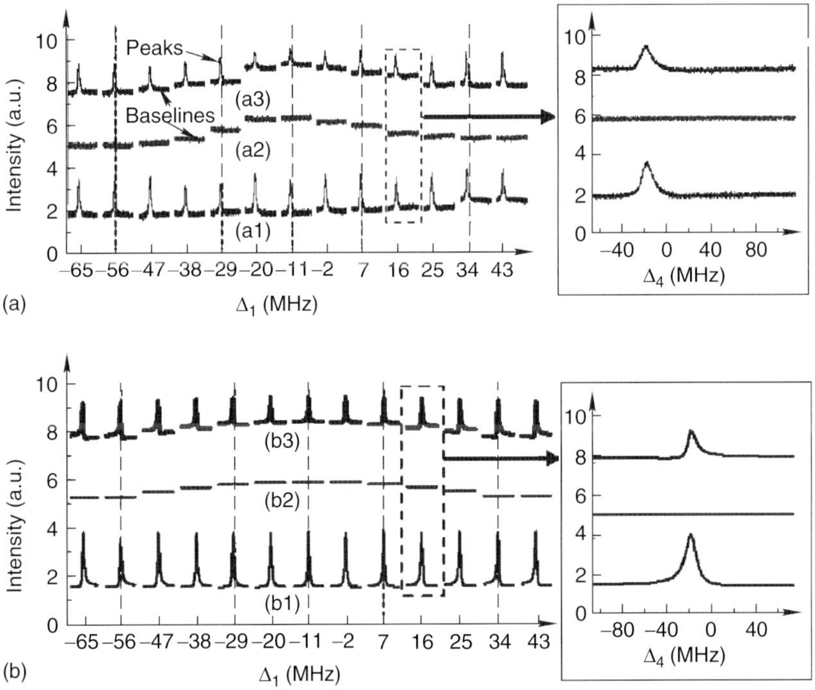

Figure 4.19 (a) Measured probe transmission signals and (b) the calculated results obtained by scanning Δ_4 for different discrete points of Δ_1. (a1) and (b1) Signals obtained with the laser beam E_2 blocked and the others turned on. (a2) and (b2) Signals obtained with the laser beam E_4 blocked and the others turned on. (a3) and (b3) Signals obtained with all laser beams on. The inset in (a) is from the group of experimental data (the data of (a1), (a2), and (a3) in the rectangle) versus Δ_4 at $\Delta_1 = 16$ MHz. The parameters used in the calculated results are $G_1 = 11.8$ MHz, $G_2 = 45.6$ MHz, $G_4 = 56.7$ MHz, $\Gamma_{10} = 18.5$ MHz, $\Gamma_{20} = \Gamma_{40} = 3$ MHz, $\Delta_2 = 1.8$ MHz, and $\Delta_1 = -65$, -56, -47, -38, -29, -20, -11, -2, 7, 16, 25, 34, and 43 MHz. Source: Adapted from Ref. [45].

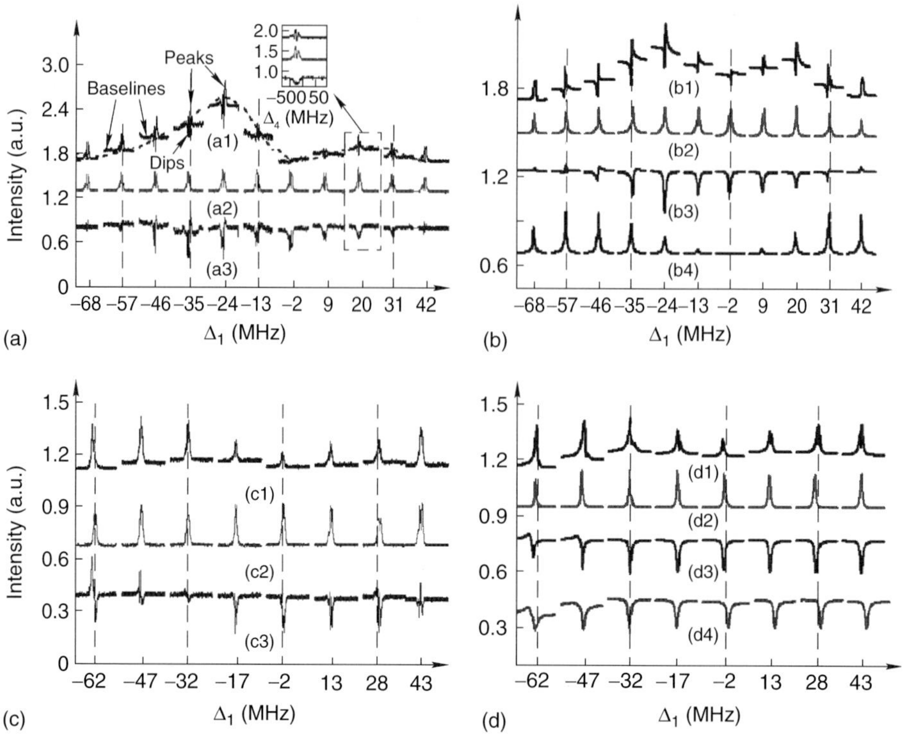

Figure 4.20 (a) and (c) Measured SWM signals and (b) and (d) the calculated results obtained by scanning Δ_4 for different discrete points of Δ_1. (a1)–(d1) Signals obtained with all laser beams on. The dashed line represents the profile of all baselines of solid curves in (a1). (a2)–(d2) Signals obtained with the laser beam E_2 blocked and the others turned on. (a3)–(d3) The sum of $I_{D1} - I_{S1}$ and $I_{D2} - I_{S2}$. (b4) The calculated results of I_{D1}. (d4) The calculated results of I_{D2}. The inset in (a) is from the group of experimental data (the data of (a1)–(a3) in the rectangle) versus Δ_4 at $\Delta_1 = 20$ MHz. Powers of all laser beams are 3.7 mW (E_1), 50 mW (E_2), 44 mW ($E_3 + E'_3$), 6 mW (E_4) for (a) and 3.7 mW (E_1), 5.6 mW (E_2), 44 mW ($E_3 + E'_3$), 75 mW (E_4) for (c). The parameters used in (b) are $G_2 = 43.5$ MHz, $G_4 = 15$ MHz, and $\Delta_1 = -68, -57, -46, -35, -24, -13, -2, 9, 20, 31$, and 42 MHz. The parameters used in (d) are $G_2 = 14.5$ MHz, $G_4 = 53.3$ MHz, and $\Delta_1 = -62, -47, -32, -17, -2, 13, 28$, and 43 MHz. The other parameters are $G_1 = 11.8$ MHz, $G_3 = G'_3 = 28.9$ MHz, $\Gamma_{10} = 18.5$ MHz, $\Gamma_{20} = \Gamma_{40} = 3$ MHz, $\Gamma_{30} = 0.6$ MHz, $\Delta_2 = 1.8$ MHz, and $\Delta_3 = 300$ MHz.

$\Delta_1 + \Delta_4 = 0$) with the expression $\rho_{10}^{(1)} = iG_1 \exp(i\mathbf{k}_1 \cdot \mathbf{r})/(d_1 + |G_4|^2/d_4)$. When E_2 is turned on (set $\Delta_2 = 1.8$ MHz) and E_4 is blocked, the measured probe transmission signal is shown in Figure 4.19(a2). The height of each baseline is smaller when $|\Delta_1|$ is large, while it becomes larger when $|\Delta_1|$ tends to 0. One can easily find that the profile of all the baselines in Figure 4.19(a2) reveals the EIT window induced by E_2 in $|0\rangle - |1\rangle - |2\rangle$ ladder subsystem satisfying $\omega_1 + \omega_2 = \Omega_1 + \Omega_2$ (i.e., $\Delta_1 + \Delta_2 = 0$) with the expression $\rho_{10}^{(1)} = iG_1 \exp(i\mathbf{k}_1 \cdot \mathbf{r})/(d_1 + |G_2|^2/d_2)$. When all laser beams are turned on, the measured probe transmission signal is shown in Figure 4.19(a3). We can see that the curve (a3) is the sum of curves (a1) and (a2). Each baseline represents the transparent degree of E_1 induced by E_2 and the peak over each baseline denotes the transparent degree of E_1 induced by E_4. The total EIT peak reaches the highest point at $\Delta_1 = -\Delta_4 = -\Delta_2 \approx 0$. More interestingly, compared to Figure 4.19(a1), the EIT peaks induced by E_4 near $|\Delta_1| = 0$ are suppressed in Figure 4.19(a3) because of the strong cascade dressing interaction between E_2 and E_4 with the expression $\rho_{10}^{(1)} = iG_1 \exp(i\mathbf{k}_1 \cdot \mathbf{r})/(d_1 + |G_2|^2/d_2 + |G_4|^2/d_4)$. Figure 4.19(b) shows the theoretical curves corresponding to Figure 4.19(a), which are in good agreement with the experimental results.

Next, when all five laser beams are turned on, two SWM signals (S1 and S2) in two separate EIT windows form simultaneously. When we tune the two SWM signals to overlap, there exist interplays between them. Three groups of experiments ($G_2 \gg G_4$, $G_2 \ll G_4$, and $G_2 \approx G_4$) have been considered in order to investigate the self- and external-dressing in two SWM signals.

Figure 4.20 shows the measured SWM signals versus the dressing field detuning Δ_4 at different probe detuning (Δ_1) values. The experimental results for $G_2 \gg G_4$ are shown in Figure 4.20(a), while Figure 4.20(b1) to (b3) shows the theoretical results corresponding to Figure 4.20(a1) to (a3). The inset in Figure 4.20(a) is from the group of experimental data in the rectangle, which is obtained by scanning Δ_4 at $\Delta_1 = 20$ MHz. Each solid curve in Figure 4.20(a1) shows the total signals with all laser beams on by scanning the dressing field detuning Δ_4 at different designated probe detuning (Δ_1) values. It includes two components: I_{D1} and I_{D2}. The height of the baseline of each solid curve represents the intensity of the SWM signal S2 at one designated probe detuning value, as a DC background. The profile (the dashed line of Figure 4.20(a1)) of these baselines shows the AT splitting due to the self-dressing effect of the beam E_2. The left and right peaks (at $\Delta_1 = \pm 20$ MHz in the dashed line) of the AT splitting correspond to the dressed states $|+\rangle$ and $|-\rangle$, respectively (Figure 4.18(c)). The strong asymmetry between the two peaks of AT splitting is induced by the value of frequency detuning Δ_2. On the other hand, in each solid curve of Figure 4.20(a1), peaks and dips on each baseline (note that the baseline is regarded as the reference point) include the two components: I_{D1} and $I_{D2} - I_{S2}$. Here, $I_{D2} - I_{S2}$ represents the enhancement or suppression of the S2 signal caused by the dressing field E_4. When we block the beam E_2, the measured SWM signal I_{S1} is shown in Figure 4.20(a2). When the laser beam E_2 is turned on and strong, because we cannot identify the dressed SWM signal D1 from the total signal in the experiment, we show the theoretical D1 signal in Figure 4.20(b4). In the curve (b4), when $\Delta_1 = -\Delta_2$, as the field E_1 touches the virtual energy level created by the external-dressing field E_2 as shown in Figure 4.18(d), the suppression of the SWM signal S1 is strongest. While $\Delta_1 = -\Delta_2 \pm G_2$, the field E_1 touches the dressed-state energy level $|\pm\rangle$ (Figure 4.18(c)) and then the SWM signal S1 is enhanced. By subtracting the height of each baseline in the solid curve of Figure 4.20(a1) and the corresponding SWM signal S1 in Figure 4.20(a2) from the total signals in the solid curve of Figure 4.20(a1), one can obtain the sum of interplays $I_{D1} - I_{S1}$ and $I_{D2} - I_{S2}$, as shown in Figure 4.20(a3). Here, $I_{D1} - I_{S1}$ expresses the enhancement

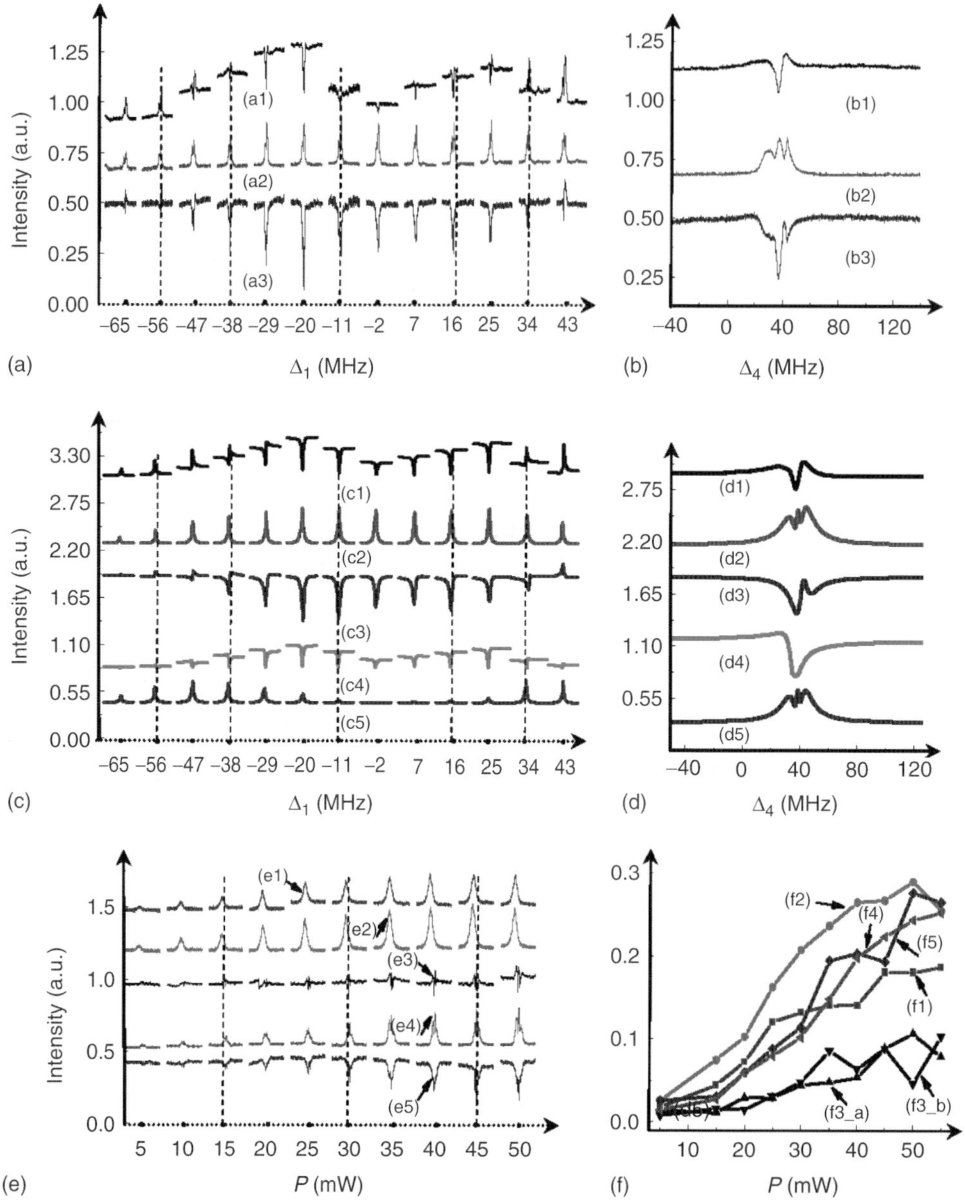
Intensity (a.u.)
1.25
1.00
0.75
0.50
0.25
0.00
(a1)
(a2)
(a3)
−65 −56 −47 −38 −29 −20 −11 −2 7 16 25 34 43
(a)
Δ_1 (MHz)
1.25
1.00
0.75
0.50
0.25
(b1)
(b2)
(b3)
−40 0 40 80 120
(b)
Δ_4 (MHz)
Intensity (a.u.)
3.30
2.75
2.20
1.65
1.10
0.55
0.00
(c1)
(c2)
(c3)
(c4)
(c5)
−65 −56 −47 −38 −29 −20 −11 −2 7 16 25 34 43
(c)
Δ_1 (MHz)
2.75
2.20
1.65
1.10
0.55
(d1)
(d2)
(d3)
(d4)
(d5)
−40 0 40 80 120
(d)
Δ_4 (MHz)
Intensity (a.u.)
1.5
1.0
0.5
0.0
(e1)
(e2)
(e3)
(e4)
(e5)
5 10 15 20 25 30 35 40 45 50
(e)
P (mW)
0.3
0.2
0.1
0.0
(f2)
(f4)
(f5)
(f1)
(f3_a)
(f3_b)
10 20 30 40 50
(f)
P (mW)

or suppression of the SWM signal S1 caused by the field E_2. Considering the power set of $G_2 \gg G_4$ in Figure 4.20(a), because the external-dressing effect by the beam E_4 is weak, $I_{D2} - I_{S2}$ is quite small. Therefore, the signals in the curve (a3) mainly represent $I_{D1} - I_{S1}$ caused by the external-dressing field E_2. One can see from the curve (a3) that when Δ_1 is near $-\Delta_2$, the depth of the dip is approximately maximum. It means that the suppression of the SWM signal S1 is largest. On the other hand, when Δ_1 is near $-\Delta_2 \pm G_2$, the SWM signal S1 is enhanced as shown by the small peak. The variation rules in Figure 4.20(a3) and (b4) show that our theoretical analysis is in accordance with the experimental observation.

The experimental results for $G_2 \ll G_4$ are shown in Figure 4.20(c). Figure 4.20(c) is obtained in the same way as Figure 4.20(a) while Figure 4.20(d1) to (d3) are the theoretical results corresponding to Figure 4.20(c1) to (c3). In Figure 4.20(c), the curve (c1) representing the total signal with all laser beams on consists of peaks without any dip. The reason is that the external-dressing effect of the beam E_2 is weak and the suppression of the SWM signal S1 is not obvious. On the other hand, from the profile of baselines in Figure 4.20(c1), one can see that the AT splitting is not obvious as the self-dressing effect of the beam E_2 is weak and can be neglected. The curve (c2) represents I_{S1} with E_2 blocked, where each peak shows the AT splitting caused by the beams E_3 and E_4. Owing to the weak dressing effect of E_2, the curve (c3) mainly represents the component $I_{D2} - I_{S2}$ induced by the external-dressing effect of the beam E_4. We also show the theoretical D2 signal in Figure 4.20(d4) as we cannot identify the dressed SWM signal D2 from the total signal with all laser beams on in the experiment. The height of each baseline in Figure 4.20(d4) represents the intensity of the SWM signal S2. Obviously, the result obtained by subtracting the height of the corresponding baseline from each curve in Figure 4.20(d4) represents $I_{D2} - I_{S2}$, which is basically consistent with the result in Figure 4.20(d3) or (c3). One can see from Figure 4.20(d3) or (c3) that the

Figure 4.21 (a) Measured SWM signals obtained by scanning Δ_4 for different discrete points of Δ_1. (a1) Signals obtained with all laser beams on. (a2) Signals obtained with the laser beam E_2 blocked. (a3) The sum of $I_{D1} - I_{S1}$ and $I_{D2} - I_{S2}$. (b) SWM signals taken at $\Delta_1 = -38$ MHz in (a) versus Δ_4. (c1)–(c3) Calculated results of (a1)–(a3). (c4) Calculated results of I_{D2}. (c5) Calculated results of I_{D1}. (d) Signals taken at $\Delta_1 = -38$ MHz in (c) versus Δ_4. Powers of all laser beams are 3.7 mW (E_1), 55 mW (E_2), 44 mW ($E_3 + E_3'$), 85 mW (E_4) for (a). The parameters used in (c) are $G_1 = 11.8$ MHz, $G_2 = 45.6$ MHz, $G_3 = G_3' = 28.9$ MHz, $G_4 = 56.7$ MHz, $\Gamma_{10} = 18.5$ MHz, $\Gamma_{20} = \Gamma_{40} = 3$ MHz, $\Gamma_{30} = 0.6$ MHz, $\Delta_2 = 1.8$ MHz, $\Delta_3 = 300$ MHz, and $\Delta_1 = -65, -56, -47, -38, -29, -20, -11, -2, 7, 16, 25, 34$, and 43 MHz. (e) Measured probe transmission signals and SWM signals versus different field (E_4) intensities, from 5 to 50 mW with the step of 5 mW. (e1) EIT signals obtained with all laser beams on. (e2) EIT signals obtained with the laser beam E_2 blocked. (e3) SWM signals obtained with all laser beams on. (e4) SWM signals obtained with the laser beam E_2 blocked. (e5) The sum of $I_{D1} - I_{S1}$ and $I_{D2} - I_{S2}$. (f) Experimental curves of power dependence of signals in (e). (f1) and (f2) Power dependences of heights of peaks in the curves (e1) and (e2), respectively. (f3_a) and (f3_b) Power dependences of heights of peaks and depths of dips in the curve (e3), respectively. (f4) Power dependence of heights of peaks in the curves (e4). (f5) Power dependence of depths of dips in the curves (e5). Source: Adapted from Ref. [45].

signal $I_{D2} - I_{S2}$ changes from all-suppressed to half-suppressed and half-enhanced with $|\Delta_1|$ increasing. On the basis of the expression $\rho^{(5)}_{10Db}$, the enhancement and suppression conditions can be obtained as $\Delta_4 = (|G_4|^2 - \Delta_1{}^2)/\Delta_1$ and $\Delta_1 + \Delta_4 = 0$, respectively. In the dressed-state picture (Figure 4.18(e) and (f)), the dressed-state energy level $|G_4 \pm\rangle$ is created by the beam E_4. When E_1 meets the dressed state energy level $|G_4 -\rangle$, one-photon resonance arises as shown in Figure 4.18(e). Thus, the enhancement condition can be satisfied and the SWM signal S2 is enhanced. When the beam E_4 is tuned to $-\Delta_1$, two-photon resonance arises as shown in Figure 4.18(f). So, the suppression condition is satisfied and the SWM signal S2 is suppressed.

The experimental results for $G_2 \approx G_4$ are shown in Figure 4.21(a) where we can observe the self- and external-dressing effects of both E_2 and E_4. The SWM signals in the curves (a1) and (a2) of Figure 4.21(a) are obtained under the EIT windows as shown in the curves (a3) and (a1) of Figure 4.19(a), respectively. Figure 4.21(c1) to (c3) are the theoretical results corresponding to Figure 4.21(a1) to (a3). We can see the self-dressing effect of E_2 (as shown by the AT splitting vs Δ_1 in the profile of baselines of Figure 4.21(a1)) and the external-dressing effect of E_2 (as shown by the enhancement and suppression in Figure 4.21(c5)). Simultaneously, we can also observe the self-dressing effect of E_4 (as shown by the AT splitting vs Δ_4 in Figure 4.21(b2)) and the external-dressing effect of E_4 (as shown by the enhancement and suppression in Figure 4.21(c4)). In the following, we focus on observing the external-dressing effect of E_4 from the experimental results in Figure 4.21(a). Comparing Figure 4.20(a) with Figure 4.21(a), the component $I_{D2} - I_{S2}$ becomes obvious because of the strong external-dressing effect of E_4 in addition to the component $I_{D1} - I_{S1}$ induced by the external-dressing effect of E_2 in Figure 4.21(a). For example, first, the depths of the dips in Figure 4.21(a1) (or Figure 4.21(a3)) are larger than those of the dips in Figure 4.20(a1) (or Figure 4.20(a3)). Second, the total SWM signal at $\Delta_1 = -2$ MHz in Figure 4.21(a1) shows the dip, which disappears at the same position in Figure 4.20(a1). In addition, when scanning the dressing field detuning, we can generally obtain the enhancement and suppression of the dressed SWM signals shown in Figure 4.21(c4). We find that the dressed SWM signals show left-enhancement and right-suppression shapes in the half suppression and half enhancement region when $\Delta_1 < 0$, but left-suppression and right-enhancement shapes when $\Delta_1 > 0$. However, Figure 4.21(a1) shows left-suppression and right-enhancement shapes in both $\Delta_1 < 0$ and $\Delta_1 > 0$ regions. We believe that it results from the interaction between two SWM signals. When all laser beams are turned on, both I_{D1} and I_{D2} coexist. The superposition of I_{D1} and I_{D2} causes the left-suppression and right-enhancement shapes in the experiment, as shown in Figure 4.21(a1). Specifically, Figure 4.21(b) and (d) shows the detailed structure of SWM signals versus Δ_4 at $\Delta_1 = -38$ MHz. Figure 4.21(b1) and (b2) shows the SWM signals obtained with all laser beams on and with E_2 blocked, respectively, where the line width of the observed peaks is about 30 MHz. Figure 4.21(b3) is the sum of $I_{D1} - I_{S1}$ and $I_{D2} - I_{S2}$. Figure 4.21(d1) to (d3) shows the theoretical results corresponding to Figure 4.21(b1) to (b3). In Figure 4.21(d), the curve (d1) can also be obtained by the sum of the curve (d4) representing the theoretical I_{D2} and the curve (d5) representing the theoretical I_{D1}. In detail, the superposition of the dip of the curve (d4) and the one of the curve (d5) form the dip of the curve (d1) as the dips of those curves are in the same position (satisfying $\Delta_4 = -\Delta_1$). Similarly, the peaks of the curve (d1) are caused by the peaks of the AT splitting in the curve (d5) as the positions of their peaks are all at $\Delta_4 = -\Delta_1 \pm G_4$ based on the expression $\rho_{10Da}^{(5)}$. It can be seen from Figure 4.21(b1) and (d1) that the calculated results are well in accordance with the experimental observations.

Last, we measure the power dependences of probe transmission signals and SWM signals by scanning Δ_4 near $\Delta_1 = 0$ (Figure 4.21(e) and (f)). The transparent degree of the probe transmission signals increases as the power of the field E_4 gets large as shown in both Figure 4.21(e1) with all laser beams on and Figure 4.21(e2) with E_2 blocked. It can also be clearly seen from Figure 4.21(f1) and (f2) which shows power dependencies of heights of peaks in the curves (e1) and (e2), respectively. The curve (e3) is the total signal with all laser beams on. In the curve (e3), on the one hand, the peaks representing I_{D1} tend to zero when the power of the field E_4 is small and then become strong as the power of the field E_4 increases. Also, Figure 4.21(f3_a) shows the power dependence of heights of peaks in the curve (e3). On the other hand, the dips in Figure 4.21(e3) representing the pure suppression on the S2 signal caused by the external-dressing effect of E_4 will become deeper as the power of the field E_4 increases. It can also be seen from Figure 4.21(f3_b), which shows the power dependence of depths of dips in the curve (e3). The curve (e4) represents I_{S1} at a different power of E_4. The heights of peaks of I_{S1} become large as the power of E_4 increases as shown in Figure 4.21(f4). Subtracting the curve (e4) from the curve (e3), we obtain the curve (e5), which consists of the components $I_{D1} - I_{S1}$ and $I_{D2} - I_{S2}$. As both $I_{D1} - I_{S1}$ and $I_{D2} - I_{S2}$ get strong with E_4 increasing, we can see that the depths of dips in the curve (e5) becomes large, which can also be seen from the power dependence (Figure 4.21(f5)) of depths of dips in the curve (e5).

4.4.3 Conclusion

In summary, we discuss the enhancement and suppression of two mutual dressing SWM processes via the atomic coherence. By scanning the dressing field frequency detuning at different probe field frequency detuning, the enhancement and suppression of two SWM signals caused by the external-dressing effect arise simultaneously. We also observe the AT splitting due to the self-dressing effect. In addition, we have measured the power dependences of the enhancement and suppression of SWM signals. Our calculated results are in good agreement with the experimental data. Such studies about the enhancement and suppression of SWM processes have applications in optical switch, optical communication, and quantum information processing [46–49].

Problems

4.1 Explain why the enhancement and suppression MWM signal can be switched and optimized.

4.2 Try to use dark state theory to give the corresponding relationship between enhancement/suppression and AT Splitting of the MWM signal, as well as that between EIT/EIA of probe transmission and enhancement/suppression of MWM signal.

References

1. Harris, S.E. (1997) Electromagnetically induced transparency. *Phys. Today*, **50**, 36.
2. Gea-Banacloche, J., Li, Y., Jin, S., and Xiao, M. (1995) Electromagnetically induced transparency in ladder-type inhomogeneously broadened media: theory and experiment. *Phys. Rev. A*, **51**, 576–584.
3. Braje, D.A., Balić, V., Goda, S., Yin, G., and Harris, S. (2004) Frequency mixing using electromagnetically induced transparency in cold atoms. *Phys. Rev. Lett.*, **93**, 183601.
4. Zibrov, A.S., Matsko, A.B., Kocharovskaya, O., Rostovtsev, Y.V., Welch, G.R., and Scully, M.O. (2002) Transporting and time reversing light via atomic coherence. *Phys. Rev. Lett.*, **88**, 103601.
5. Li, Y. and Xiao, M. (1996) Enhancement of nondegenerate four-wave mixing based on electromagnetically induced transparency in rubidium atoms. *Opt. Lett.*, **21**, 1064–1066.
6. Hemmer, P., Katz, D., Donoghue, J., Cronin-Golomb, M., Shahriar, M., and Kumar, P. (1995) Efficient low-intensity optical phase conjugation based on coherent population trapping in sodium. *Opt. Lett.*, **20**, 982–984.
7. Kang, H., Hernandez, G., and Zhu, Y. (2004) Slow-light six-wave mixing at low light intensities. *Phys. Rev. Lett.*, **93**, 073601.
8. Zuo, Z., Sun, J., Liu, X., Jiang, Q., Fu, G., Wu, L.-A., and Fu, P. (2006) Generalized n-photon resonant 2n-wave mixing in an (n+1)-level system with phase-conjugate geometry. *Phys. Rev. Lett.*, **97**, 193904.
9. Zhang, Y., Brown, A.W., and Xiao, M. (2007) Opening four-wave mixing and six-wave mixing channels via dual electromagnetically induced transparency windows. *Phys. Rev. Lett.*, **99**, 123603.
10. Zhang, Y., Anderson, B., Brown, A.W., and Xiao, M. (2007) Competition between two four-wave mixing channels via atomic coherence. *Appl. Phys. Lett.*, **91**, 061113.
11. Li, C., Zheng, H., Zhang, Y., Nie, Z., Song, J., and Xiao, M. (2009) Observation of enhancement and suppression in four-wave mixing processes. *Appl. Phys. Lett.*, **95**, 041103.
12. Nie, Z., Zheng, H., Li, P., Yang, Y., Zhang, Y., and Xiao, M. (2008) Interacting multiwave mixing in a five-level atomic system. *Phys. Rev. A*, **77**, 063829.
13. Zhang, Y., Li, P., Zheng, H., Wang, Z., Chen, H., Li, C., Zhang, R., and Xiao, M. (2011) Observation of Autler-Townes splitting in six-wave mixing. *Opt. Express*, **19**, 7769–7777.
14. Zhang, Y., Nie, Z., Wang, Z., Li, C., Wen, F., and Xiao, M. (2010) Evidence of Autler–Townes splitting in high-order nonlinear processes. *Opt. Lett.*, **35**, 3420–3422.
15. Wang, B., Han, Y., Xiao, J., Yang, X., Xie, C., Wang, H., and Xiao, M. (2006) Multi-dark-state resonances in cold multi-Zeeman-sublevel atoms. *Opt. Lett.*, **31**, 3647–3649.
16. Zhu, C.J., Senin, A.A., Lu, Z.H., Gao, J., Xiao, Y., and Eden, J.G. (2005) Polarization of signal wave radiation generated by parametric four-wave mixing in rubidium vapor: ultrafast (~150-fs) and nanosecond time scale excitation. *Phys. Rev. A*, **72**, 023811.

17. Magno, W.C., Prandini, R.B., Nussenzveig, P., and Vianna, S.S. (2001) Four-wave mixing with Rydberg levels in rubidium vapor: observation of interference fringes. *Phys. Rev. A*, **63**, 063406.
18. Lipsich, A., Barreiro, S., Akulshin, A.M., and Lezama, A. (2000) Absorption spectra of driven degenerate two-level atomic systems. *Phys. Rev. A*, **61**, 053803.
19. Ling, H.Y., Li, Y.-Q., and Xiao, M. (1996) Coherent population trapping and electromagnetically induced transparency in multi-Zeeman-sublevel atoms. *Phys. Rev. A*, **53**, 1014–1026.
20. Li, C., Zhang, Y., Nie, Z., Du, Y., Wang, R., Song, J., and Xiao, M. (2010) Controlling enhancement and suppression of four-wave mixing via polarized light. *Phys. Rev. A*, **81**, 033801.
21. Li, Y.-q., Jin, S.-z., and Xiao, M. (1995) Observation of an electromagnetically induced change of absorption in multilevel rubidium atoms. *Phys. Rev. A*, **51**, R1754–R1757.
22. Nie, Z., Zhang, Y., Zhao, Y., Yuan, C., Li, C., Tao, R., Si, J., and Gan, C. (2011) Enhancing and suppressing four-wave mixing in electromagnetically induced transparency window. *J. Raman Spectrosc.*, **42**, 1–4.
23. Li, H., Sautenkov, V.A., Rostovtsev, Y.V., Welch, G.R., Hemmer, P.R., and Scully, M.O. (2009) Electromagnetically induced transparency controlled by a microwave field. *Phys. Rev. A*, **80**, 023820.
24. Yan, M., Rickey, E.G., and Zhu, Y. (2001) Observation of absorptive photon switching by quantum interference. *Phys. Rev. A*, **64**, 041801.
25. Lukin, M.D., Matsko, A.B., Fleischhauer, M., and Scully, M.O. (1999) Quantum noise and correlations in resonantly enhanced wave mixing based on atomic coherence. *Phys. Rev. Lett.*, **82**, 1847–1850.
26. Agrawal, G.P. (1990) Induced focusing of optical beams in self-defocusing nonlinear media. *Phys. Rev. Lett.*, **64**, 2487–2490.
27. Bennink, R.S., Wong, V., Marino, A.M., Aronstein, D.L., Boyd, R.W., Stroud, C. Jr., Lukishova, S., and Gauthier, D.J. (2002) Honeycomb pattern formation by laser-beam filamentation in atomic sodium vapor. *Phys. Rev. Lett.*, **88**, 113901.
28. Stentz, A.J., Kauranen, M., Maki, J.J., Agrawal, G.P., and Boyd, R.W. (1992) Induced focusing and spatial wave breaking from cross-phase modulation in a self-defocusing medium. *Opt. Lett.*, **17**, 19–21.
29. Yanpeng, Z., Zhiguo, W., Huaibin, Z., Chenzhi, Y., Changbiao, L., Keqing, L., and Min, X. (2010) Four-wave-mixing gap solitons. *Phys. Rev. A*, **82**, 053837.
30. Wang, H., Goorskey, D., and Xiao, M. (2001) Enhanced Kerr nonlinearity via atomic coherence in a three-level atomic system. *Phys. Rev. Lett.*, **87**, 73601.
31. Korsunsky, E.A. and Kosachiov, D.V. (1999) Phase-dependent nonlinear optics with double-Λ atoms. *Phys. Rev. A*, **60**, 4996–5009.
32. Vudyasetu, P.K., Camacho, R.M., and Howell, J.C. (2008) Storage and retrieval of multimode transverse images in hot atomic rubidium vapor. *Phys. Rev. Lett.*, **100**, 123903.
33. Boyer, V., Marino, A.M., Pooser, R.C., and Lett, P.D. (2008) Entangled images from four-wave mixing. *Science*, **321**, 544–547.
34. Krolikowski, W., Ostrovskaya, E.A., Weilnau, C., Geisser, M., McCarthy, G., Kivshar, Y.S., Denz, C., and Luther-Davies, B. (2000) Observation of dipole-mode vector solitons. *Phys. Rev. Lett.*, **85**, 1424–1427.
35. Zhang, Y., Khadka, U., Anderson, B., and Xiao, M. (2009) Temporal and spatial interference between four-wave mixing and six-wave mixing channels. *Phys. Rev. Lett.*, **102**, 13601.
36. Li, C., Zhang, Y., Zheng, H., Wang, Z., Chen, H., Sang, S., Zhang, R., Wu, Z., Li, L., and Li, P. (2011) Controlling cascade dressing interaction of four-wave mixing image. *Opt. Express*, **19**, 13675–13685.
37. Moseley, R.R., Shepherd, S., Fulton, D.J., Sinclair, B.D., and Dunn, M.H. (1996) Electromagnetically-induced focusing. *Phys. Rev. A*, **53**, 408.

38. Harris, S. and Yamamoto, Y. (1998) Photon switching by quantum interference. *Phys. Rev. Lett.*, **81**, 3611–3614.
39. Lukin, M., Yelin, S., Fleischhauer, M., and Scully, M. (1999) Quantum interference effects induced by interacting dark resonances. *Phys. Rev. A*, **60**, 3225–3228.
40. Ling, H.Y., Li, Y.Q., and Xiao, M. (1998) Electromagnetically induced grating: homogeneously broadened medium. *Phys. Rev. A*, **57**, 1338.
41. Zhang, J., Hernandez, G., and Zhu, Y. (2008) Optical switching mediated by quantum interference of Raman transitions. *Opt. Express*, **16**, 19112–19117.
42. Li, C., Zhang, Y., Nie, Z., Zheng, H., Zuo, C., Du, Y., Song, J., Lu, K., and Gan, C. (2010) Controlled multi-wave mixing via interacting dark states in a five-level system. *Opt. Commun.*, **283**, 2918–2928.
43. Zhang, Y., Anderson, B., and Xiao, M. (2008) Coexistence of four-wave, six-wave and eight-wave mixing processes in multi-dressed atomic systems. *J. Phys. B: At. Mol. Opt. Phys.*, **41**, 045502.
44. Wang, J., Zhu, Y., Jiang, K., and Zhan, M. (2003) Bichromatic electromagnetically induced transparency in cold rubidium atoms. *Phys. Rev. A*, **68**, 063810.
45. Wang, Z., Zhang, Y., Chen, H., Wu, Z., Fu, Y., and Zheng, H. (2011) Enhancement and suppression of two coexisting six-wave-mixing processes. *Phys. Rev. A*, **84**, 013804.
46. Phillips, D., Fleischhauer, A., Mair, A., Walsworth, R., and Lukin, M. (2001) Storage of light in atomic vapor. *Phys. Rev. Lett.*, **86**, 783–786.
47. Imamoglu, A. and Lukin, M.D. (2001) Controlling photons using electromagnetically induced transparency. *Nature*, **413**, 273–276.
48. Liu, C., Dutton, Z., Behroozi, C.H., and Hau, L.V. (2001) Observation of coherent optical information storage in an atomic medium using halted light pulses. *Nature*, **409**, 490–493.
49. Duan, L.M., Lukin, M., Cirac, I., and Zoller, P. (2001) Long-distance quantum communication with atomic ensembles and linear optics. *Nature*, **414**, 413 Arxiv preprint quant-ph/0105105.

5
Controllable Polarization of MWM Process via Dark State

Highlights

Intensities and polarization of MWM can be controlled by changing the polarization states of the coupled laser beams, which are essential for high-precision measurements, coherent quantum control, and quantum information processing.

As the real atomic systems usually have multiple Zeeman sublevels, one can use the polarization states of the coupled laser beams to modulate the strength of the MWM processes. By manipulating the dark state or EIT windows with the polarization states of the laser beams, the MWM processes can be modified and controlled. Such studies of controllable intermixing between different-order nonlinear optical processes with polarization states of the laser beams can be very important in high-precision measurements, coherent quantum control, and quantum information processing. In this chapter, some examples of polarization-controlled MWM processes in multi-level atomic system are described. First, we discuss the enhancement/suppression and the efficiencies of FWM processes, which can be controlled by changing the polarization of the pumping beams. Then, periodic spatial splitting of FWM, which is due to periodic change of the pumping beam's polarization states, are introduced. After that, we focus on the intensities and polarization states of eight coexisting FWM signals when the polarization configurations and frequency detuning of the incident fields are changed. And at last, the suppression/enhancement as well as the AT splitting of two coexisting SWM processes in multi-Zeeman level atomic system are introduced, under the different polarization of the probe beam. Such studies of intermixing between different-order nonlinear optical processes with controllable phase delay can have important applications in high-precision measurements, coherence quantum control, and quantum information processing.

5.1
Enhancement and Suppression of FWM via Polarized Light

The FWM is a powerful technique for generating coherent radiations and studying a variety of coherent optical phenomena. In recent years, FWM has been widely

Quantum Control of Multi-Wave Mixing, First Edition. Yanpeng Zhang, Feng Wen, and Min Xiao.

used to observe atomic coherence [1, 2], generate entangled photon pairs [3, 4], and to coherently control the field-matter interactions [5]. In these processes, the intensities of the FWM signals are related to the polarizations of the incident lasers. That is because that the variation of the incidence polarization leads to different transition pathways among degenerate Zeeman sublevels. Different transitions generally have different coupling strength values which are indicated by CG coefficients, and different FWM transition pathways are dressed by different dressing fields. So, we can coherently control the nonlinear signal by suitably designing the polarizations of the incident laser beams.

The polarization properties of two-photon resonant FWM have been well investigated previously [6–9]. Several previous experimental and theoretical studies have shown that EIT [10–12] and FWM processes can be effectively controlled by changing the polarization states and frequency detuning of the involved laser beams [1, 6, 10, 13–16]. Also, when the FWM processes are modulated by different polarizations of the strong coupling fields, selective transitions among polarization dark states can occur [10, 13, 16–19]. In our previous experiments, we have shown the enhancement and suppression of FWM by controlling the dressing laser beams in the multi-level atomic systems [10, 16, 18–20]. Recently, we studied the polarization dependencies of FWM and dressing effects in two-level and cascade three-level atomic systems [21], as well as the MWM processes in reversed Y-type system with EIT windows at different polarization configurations [15].

In this section, we show that the DFWM caused by two strong pumping beams and a weak probe beam in a two-level Zeeman-degenerate atomic system can be modified by the polarization states of the two pumping beams, and by an additional dressing beam interacting with an adjacent atomic transition, as shown in Figure 5.1(a). The DFWM process is enhanced or suppressed owing to the combined polarization and dressing effects. The polarizations of the pumping beams select the transitions among different Zeeman levels, which usually have different transition strengths [15], and the dressing beam determines the effective frequency detuning of the probe beam from the multi-Zeeman levels. The experimental observations clearly show the evolution of the DFWM enhancement and suppression versus pump field polarizations.

5.1.1
Theoretical Model and Analysis

Three energy levels in sodium atoms (in a heat-pipe oven) are employed in the experiment (Figure 5.1(a)). The pulse laser beams are spatially aligned as shown in Figure 5.1(b). The pumping laser beams E_1 (ω_1, $\mathbf{k}_1$, and Rabi frequency $G_{g,M}$) and E_1' (ω_1, k_1', and $G_{g,M}'$) (having a small angle of 0.3°) are tuned to the transition $|0\rangle$ ($3S_{1/2}$) to $|1\rangle$ ($3P_{3/2}$), and E_1 propagates in the opposite direction of the weak probe field E_3 (ω_1, $\mathbf{k}_3$, and $G_{p,M}$), where M denotes the magnetic quantum number of the lower state in transition. These three laser beams are from the same near-transform-limited dye laser (10 Hz repetition rate, 5 ns pulse width, and 0.04 cm^{-1} line width) with the frequency detuning $\Delta_1 = \omega_{10} - \omega_1$, where ω_{10} is the atomic

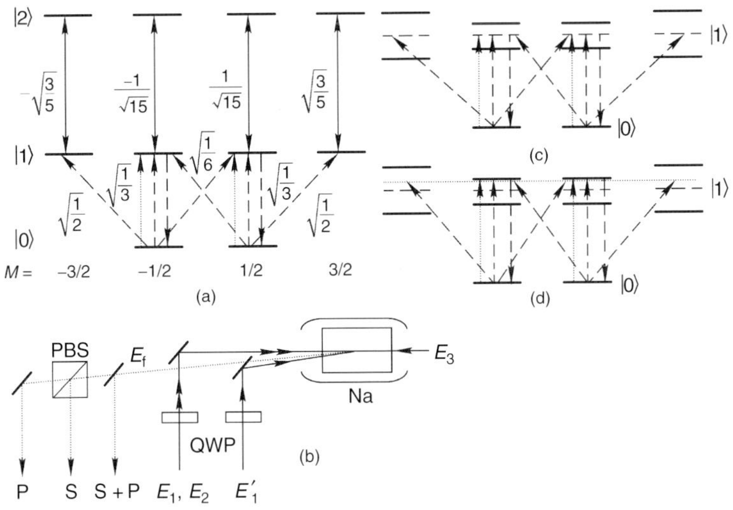

Figure 5.1 (a) Zeeman structure of the three-level ladder-type atomic system in the experiment and various transition pathways in it. Solid line: dressing field, G_d; short-dashed lines: when the pumping fields are linearly polarized, G_g, G'_g; long-dashed lines: when the pumping fields are circularly polarized; dotted line: the probe field, G_p. (b) The schematic diagram of the experiment. (c) and (d) Schematic diagrams for suppression and enhancement of the DFWM in the dressed-state picture.

transition frequency between $|0\rangle$ and $|1\rangle$. E_1 and E'_1 (both with frequency ω_1) in beams 1 and 2 induce a population grating between states $|0\rangle$ and $|1\rangle$, which is probed by beam 3 (E_3) with the same frequency (ω_1). This interaction generates a DFWM signal E_f (Figure 5.1(a)) satisfying the phase-matching condition [22], $\mathbf{k}_f = \mathbf{k}_3 + \mathbf{k}_1 - \mathbf{k}'_1$. Then, an additional dressing field E_2 (ω_2, $\mathbf{k}_2$, and $G_{d,M}$) is applied to the transition between $|1\rangle$ and the third level $|2\rangle$ ($4D_{3/2,5/2}$) with a frequency detuning Δ_2 ($= \omega_{21} - \omega_2$). E_2 is from another similar dye laser. Two QWPs are used for changing the polarizations of the pumping fields $\mathbf{k}_1$, $\mathbf{k}'_1$. The generated DFWM signal is split into two equal components by a 50% beam splitter (BS) before detection, one is detected directly (denoted as I_T) and the other is further decomposed into P- and S-polarized components by a polarized beam splitter (PBS), which are denoted as I_P and I_S, respectively.

Figure 5.1(c) depicts the dressed-state picture with split $3P_{3/2}$ Zeeman sublevels, which corresponds to the DFWM suppression case when fields $\mathbf{k}_3$, $\mathbf{k}_1$, and $\mathbf{k}'_1$ are on resonance with transition $|0\rangle \to |1\rangle$. Figure 5.1(d) shows the enhancement case when these fields are tuned to near the dressed energy level. For most cases in this work, only one QWP is used to modify the polarization state of $\mathbf{k}_1$; so, it can be decomposed into linearly and circularly polarized components while all other fields are kept as linearly polarized (Figure 5.1(a)). In fact, we assume P-polarization direction as the quantization axis and the component perpendicular to it (S-polarization) is decomposed into balanced left- and right-circularly polarized parts, while the component parallel to it (P-polarization) keeps linear polarization.

Then, the generated FWM signals will also contain linearly and circularly polarized components denoted as I_L and I_C, and they associate the detected intensities in the P- and S-polarizations with the equations, namely, the detected intensities of I_P, I_S, and total intensity I_T in the real experiment can be written as, $I_P = I_L \cos^2\alpha + I_C/2$, $I_S = I_L \sin^2\alpha + I_C/2$, and $I_T = I_S + I_P = I_L + I_C$, where α is the angle between the P-polarization and the direction of the linearly polarized signal. As the CG coefficients may be different for different transitions between Zeeman sublevels, the Rabi frequencies are different even with the same laser field [15]. For example, considering CG coefficient values [23], we can obtain $|G^{\pm}_{g,\pm 3/2}|^2/|G^{\pm}_{g,\pm 1/2}|^2 = 3$, which indicates that the circularly polarized DFWM signal is mainly dressed by $G^0_{d,\pm 3/2}$, not by $G^0_{g,\pm 3/2}$. And also from CG coefficients, we can obtain that $|G^0_{d,\pm 3/2}|^2 = 9|G^0_{d,\pm 1/2}|^2$, which indicates that the dressing effects in the circularly polarized subsystems are far greater than in the linearly polarized subsystems.

On the basis of the discussion in the preceding text, we can get the expressions for I_L and I_C. As Figure 5.1(a) shows, there are two linearly polarized subsystems $(|0_M\rangle \xleftrightarrow{G^0_{g,M},(G^0_{g,M})^*,G^0_{p,M}} |1_M\rangle \quad (M = \pm 1/2))$, which can generate linearly polarized DFWM, and is dressed by the linearly polarized dressing transition with $|G^0_{d,\pm 1/2}|^2$. By simply substituting the corresponding dressing terms into Eq. (5.4) of Ref. [24], we can obtain an expression of density matrix element that induces the FWM signal of the linearly polarized component. To simplify the expression, the symmetry of CG coefficients is considered, namely, $|G^0_{p(g,d)_M}| = |G^0_{p(g,d)_{-M}}|$ and $|G^+_{p(c,d)_M}| = |G^-_{p(c,d)_{-M}}|$. Moreover, if $G^{0,\pm}_{p(c,d)_M} \gg \Gamma_{0(1,2),0(1,2)}$, we can have the conditions of $\Gamma_{0(1,2),0(1,2)} \approx \Gamma_{0_M(1_M,2_M),0_M(1_M,2_M)}$. Consequently, the simplified expression is given by $\rho^{(3)}_L = -2i|G^0_{gM}|^2 G^0_{pM}(A_1 + 2A_2)[1/(A_7 + A_3)^2 + 1/(\Delta_1^2 + \Gamma_{10}^2 + |G_{d_M}|^4/A_4 + 2A_5|G_{d_M}|^2/A_6)]$, where $A_1 = 1/\Gamma_{00} + 1/\Gamma_{11}$, $A_2 = \Gamma_{21}|G_{d_M}|^2/(\Delta_2^2 + \Gamma_{21}^2)$, $A_3 = |G_{d_M}|^2/[i(\Delta_1 + \Delta_2) + \Gamma_{21}]$, $A_4 = (\Delta_1 + \Delta_2)^2 + \Gamma_{21}^2$, $A_5 = -\Delta_1\Delta_2 - \Delta_1^2 + \Gamma_{10}\Gamma_{20}$, $A_6 = (\Delta_1 + \Delta_2)^2 + \Gamma_{20}^2$, and $A_7 = i\Delta_1 + \Gamma_{10}$.

On the other hand, the circularly polarized subsystems are more complicated [15]. In addition, besides being dressed by $|G^0_{d,\pm 1/2}|^2$, they are also dressed by $|G^0_{d,\pm 3/2}|^2$. Also, by inserting the dressing terms into Eq. (5.6) of Ref. [24] and under the same simplified conditions, we can obtain the expression of the density matrix element which induces the FWM signal of the circularly polarized component as $\rho^{(3)}_C = -2B_1/[\Gamma_{00}(A_7 + B_2)^2] - \sum_{M=\pm 1/2} 2B_3/[\Gamma_{00}(A_7 + |G^0_{d_M}|^2/A_8)(A_7 + |G^0_{d_{M+1}}|^2/A_8)]$, where $A_8 = i(\Delta_1 + \Delta_2) + \Gamma_{20}$, $B_1 = iG^0_{p_{-1/2}} G^+_{g_{-1/2}} (G^0_{g_{-1/2}})^*$, $B_2 = |G^0_{d_{-1/2}}|^2/A_8$, and $B_3 = iG^0_{pM}(G^0_{gM})^* G^+_{gM}$. Therefore, the intensities of the FWM in the P- and S-polarization directions are $I_L \propto |\rho^{(3)}_L|^2$ and $I_C \propto |\rho^{(3)}_C|^2$, respectively.

5.1.2
Experimental Results

In the ladder-type three-level system (with Zeeman sublevels), as shown in Figure 5.1(a), the pumping fields E_1 and E'_1 (with diameter of 0.8 mm and power of 3 μW) and the probe field E_3 (with a diameter of 0.8 mm and power of 5 μW) are tuned to the line center (589.0 nm) of the lower $|0\rangle$ to $|1\rangle$ transition, which generate the DFWM signal E_f at frequency ω_1 by using one photon each from fields E_1, E_1', and E_3. The dressing field E_2 (with a diameter of 1.1 mm and power of 100 μW) scans from 568.5 to 569.1 nm (crossing the upper $|1\rangle$ to $|2\rangle$ transition) to dress the DFWM process.

The suppression and enhancement of the DFWM processes happen as the probe field is set at different frequency detuning conditions. For example, when $\Delta_1 = 0$ (Figure 5.1(c)), the DFWM signal is suppressed by the dressing field. To clearly understand the influences of the incident beams to suppression and enhancement of FWM processes, we investigate the signals in P- and S-polarizations separately, while the total intensity is the sum of intensities in these two polarizations components, as shown in Figure 5.2(a1) to (a3). The background represents the signal strength of the pure DFWM with no dressing field while the dips represent that the signal was suppressed at different polarizations of the pumping beam. When Δ_1 gets large enough (as in Figure 5.1(d)), the DFWM signal is enhanced by the dressing field, as shown in Figure 5.3(a1) to (a3) and (d1) to (d3). When Δ_1 is set at a proper position that is not too far from the resonant position, both suppression (dips lower than background) and enhancement (peaks higher than background) can occur at the same time, as shown in Figure 5.4(a1) to (a3). The line widths of the measured suppressed dips and enhanced peaks of FWM spectra are about 20 GHz.

Let us first consider the experimental results of the DFWM suppression. Figure 5.2(a1) to (a3) presents the DFWM spectra (with scanned dressing field Δ_2)

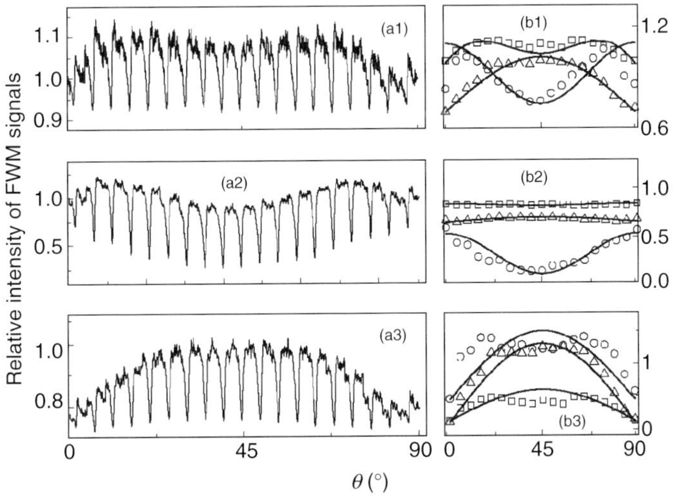

Figure 5.2 Polarization dependence of the suppressed DFWM signals. (a1)–(a3) variations of I_T, I_P, and I_S (by scanning Δ_2) versus rotation angle θ (0–90° per 5°), respectively. (b1)–(b3) Dependence curves of the background, minimums of the dips, and suppression depths for I_T (squares), I_P (circles), and I_S (triangles), respectively. The solid curves in (b1)–(b3) are the corresponding theory results $\Delta_1 = 0$. Source: Adapted from Ref. [25].

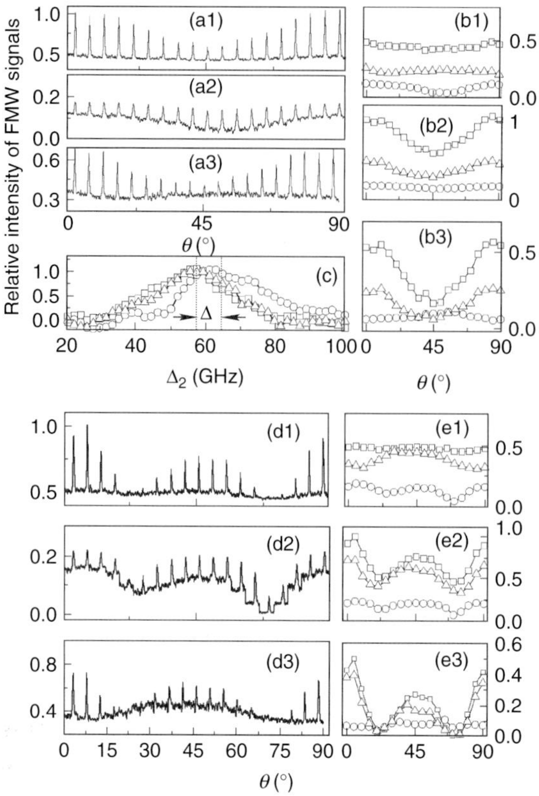

Figure 5.3 Polarization dependence of DFWM enhancement versus θ. (a1)–(a3) and (b1)–(b3) DFWM enhancement with conditions parallel to Figure 5.2 except at $\Delta_1 = -67$ GHz. (c) I_P with scanning Δ_2 for $\theta = 0°$ (squares), 45° (circles), and 90° (triangles), when $\Delta_1 = 67$ GHz. (d1)–(d3) and (e1)–(e3) are for I_T, I_P, and I_S polarization dependencies, respectively, when both polarizations of the $\mathbf{k}_1$ and k'_1 beams are rotated simultaneously, with $\Delta_1 = -67$ GHz. Source: Adapted from Ref. [25].

from $\theta = 0°$ to $\theta = 90°$ per 5°, which is the polarization angle of the pumping field E_1. The dips below the background represent the suppressed DFWM by the dressing field. Figure 5.2(b1) to (b3) presents the θ-dependence curves of the background, the minimum of the suppressed dips, and the depth of the suppressed dips (background minus minimum) in Figure 5.2(a1) to (a3), respectively. The dressing effect is clearly revealed by Figure 5.2(b3), which shows that the suppression depths in P- and S-polarizations are both ascending as the QWP is rotated from 0° to 45°. This can be explained by changing the DFWM subsystems from linearly polarized ones to circularly polarized ones, and then calculating the intensities I_P, I_S, and I_T. In fact, as Figure 5.2(a1) to (a3) shows, DFWM signals are mainly generated in the linearly polarized subsystems, which are dressed by $G^0_{d,\pm 1/2}$ when $\mathbf{k}_1$ is linearly polarized ($\theta = 0$). As QWP is rotated, the linearly polarized transitions gradually transform into circularly polarized ones, which then involve the dressing

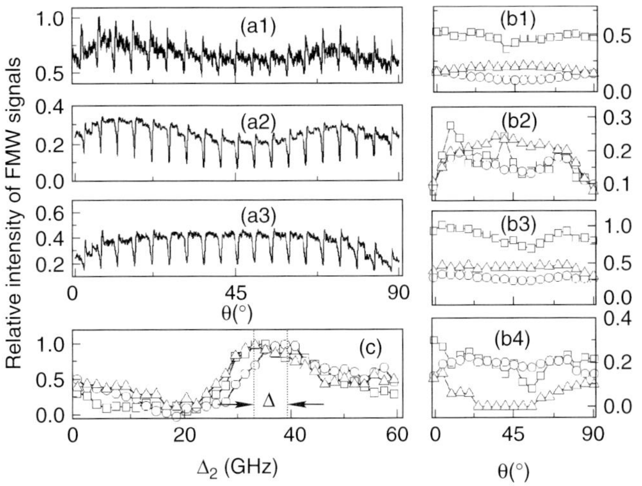

Figure 5.4 Polarization dependence of DFWM versus the rotation angle θ. (a1)–(a3) Half-enhancement and half-suppression with the condition parallel to Figure 5.2(a1) to (a3) except $\Delta_1 = -30$ GHz. (b1) and (b2) Dependencies of the minimum and maximum of each part on θ, (b3) depths of the suppressed dips, and (b4) heights of the enhanced peak for I_T (squares), I_P (circles), and I_S (triangles), respectively. (c) I_P as scanning Δ_2 for 0° (squares), 45° (circles), and 90° (triangles).

transitions $G^0_{d,\pm 3/2}$ partly instead of $G^0_{d,\pm 1/2}$. Consequently, the dressing effect gets larger and the suppression dips become deeper as QWP is rotated from 0° to 45°. Furthermore, the suppression condition ($\Delta_1 + \Delta_2 = 0$) for DFWM in all the subsystems is uniform because it contains no term relating to the Zeeman structure, which results in the similar dependence curves for S- and P-polarizations, as well as the total intensity, as shown in Figure 5.2(a1) to (a3).

Figure 5.2(b1) presents the polarization dependence of the background as well as the pure DFWM. The shapes of the curves for the P- and S-polarizations and the total intensity basically follow the well-expected classical polarization spectroscopy [1, 6]. Figure 5.2(b2) shows the polarization dependence of the dressed DFWM signal peak values, which include the pure DFWM and the suppression dips.

For DFWM enhancement when $\mathbf{k}_3$, $\mathbf{k}_1$, and $\mathbf{k}'_1$ are far detuned (Figure 5.1(d)), as shown in Figure 5.3, the polarization dependence of the enhanced peak heights (maximum minus background) for the S-polarization is different (Figure 5.3(b3) triangle points); it descends as QWP is rotated from 0° to 45°. Comparing expressions of I_L and I_C, we can see that at far detuning condition for $\mathbf{k}_3$, $\mathbf{k}_1$, and $\mathbf{k}'_1$, the polarization variation of $\mathbf{k}_1$ enlarges α. It means that S-polarized components projecting from linearly polarized FWM are increasing while P-polarization components are decreasing gradually as rotating QWP. Consequently, the dressing efficiency of the S-polarization is relatively reduced as compared with the condition when $\mathbf{k}_1$ is linearly polarized. On the other side, the P-polarization component is relatively enhanced.

As discussed in the preceding text, the dressing field Rabi frequencies for different Zeeman sublevels may be different (e.g., $|G^0_{d,\pm 3/2}|^2 = 9|G^0_{d,\pm 1/2}|^2$), which will induce different splitting distances for different sublevels. The exact expression of the split sublevel positions is $\delta_M = (\Delta_2 \pm \sqrt{\Delta_2^2 + 4|G_{dM}|^2})/2$. The enhanced peaks appear when the splitting sublevels are on resonance with the generating fields G_g, G'_g, and the probe field G_p. This then satisfies the enhancement condition $\Delta_1 + \delta_M = 0$ [26]. Combining it with $\delta_M = (\Delta_2 \pm \sqrt{\Delta_2^2 + 4|G_{dM}|^2})/2$, we can obtain the positions of the enhanced peaks in the plotted figure, $O_M = (\Delta_1^2 - |G^0_{d,M}|^2)/\Delta_1$. There should be two distinct enhanced peaks $O_{\pm 3/2}$ and $O_{\pm 1/2}$, which are covered in the wide power-broadened profile. However, when $\mathbf{k}_1$ is linearly ($\theta = 0^\circ$) and circularly ($\theta = 45^\circ$) polarized, the enhanced peaks are primarily created by $M = \pm 1/2$ and $M = \pm 3/2$, which are at $O_{\pm 1/2}$ and $O_{\pm 3/2}$, respectively, as shown in Figure 5.3(c). By using O_M expression and the CG coefficients, we can calculate the shift distance between the enhanced peaks as $\Delta = O_{3/2} - O_{1/2} = (|G^0_{d,3/2}|^2 - |G^0_{d,1/2}|^2)/\Delta_1 \approx 8.8$ GHz. The measured shift distance between the enhanced peaks in Figure 5.3(c) is about 7.5 GHz.

When two QWPs are used to change the polarizations of the $\mathbf{k}_1$ and $\mathbf{k}'_1$ beams simultaneously, as shown in Figure 5.3(d1) and (d2), the variation period is reduced to half of the case with changing $\mathbf{k}_1$ only. Also, the enhancement peak gets close to 0 at about $\theta = 22.5^\circ$.

Finally, we set the frequency detuning of $\mathbf{k}_3$, $\mathbf{k}_1$, and $\mathbf{k}'_1$ at an intermediate position (about 30 GHz, smaller than the value in the enhancement case), half-enhancement and half-suppression appear when the frequency of the dressing field is scanned [22], which is also modified by the polarization variation of $\mathbf{k}_1$, as shown in Figure 5.4. The variation rules also follow the ones discussed in the preceding text, the background obeys traditional laws [1, 6], the dependencies of the suppression and enhancement curves on the polarization are similar to the results in suppression (Figure 5.2) and enhancement (Figure 5.3) parts, respectively.

5.1.3
Conclusion

We have discussed evolutions of dressed DFWM effects when the polarizations of the pumping fields are changed. In the suppressed DFWM case, the generated DFWM signals in P- and S-polarizations are both ascending as the QWP changes from 0° to 45°, which is caused by different dressing strengths for the linearly polarized and circularly polarized DFWM signals. In the enhanced DFWM case, the dependence curve for the S-polarized DFWM signal descends while the P-polarization component ascends as the QWP is rotated. The experimentally measured data are in good agreement with the results from dressed-state analysis involving all relevant Zeeman sublevels. In addition, the dressing effects strongly depend on the dipole moments of the transitions, which can provide an easy and qualitative way to determine the orders of magnitude of the effective dipole moments for different transitions by measuring the shifted distances between two enhanced peaks when the pump field's polarization is changed. Such studies provide detailed physical mechanisms to control and optimize the efficiencies of the MWM processes in multi-level atomic systems.

5.2 Polarization-Controlled Spatial Splitting of FWM

In this section, we discuss the modulated beam intensities and spatial splitting of the FWM signal beams induced by changing the polarization states of the pumping laser beams in the ladder-type three-level atomic system. Different dressing conditions can control the spatial splitting in the transverse (x and y) directions. Also, periodic spatial splitting of both S- and P-polarized components of the FWM beam has been observed.

5.2.1 Theoretical Model and Experimental Scheme

In the ladder-type three-level system (with Zeeman sublevels), as shown in Figure 5.5(a) and (b), five laser beams with same diameter of about 0.2 mm are applied to the atomic system with the spatial configuration given in Figure 5.5(c)

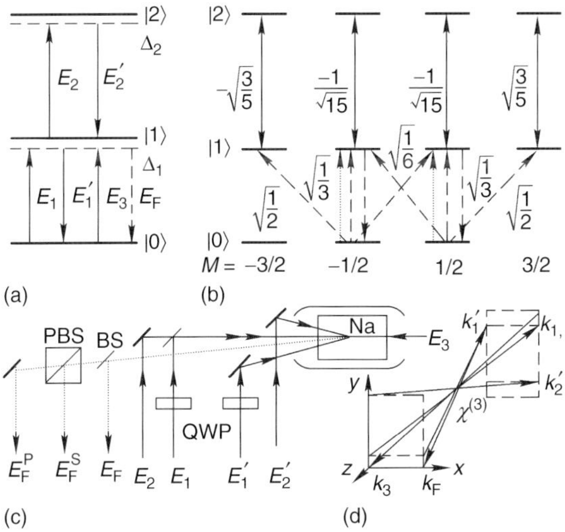

Figure 5.5 (a) Energy level diagram to generate FWM signal $\boldsymbol{E}_F$ in a ladder-type three-level atomic system. (b) The relevant Zeeman levels in the experiment and various transition pathways. Solid line: dressing fields G_2 and G_2'; short-dashed lines: the linearly polarized pumping fields G_1 and G_1'; long-dashed lines: the circularly polarized pumping fields $G_1^{\pm}$ and $G_1'^{\pm}$; dotted line: the P-polarized probe field G_3. (c) The schematic diagram of the experimental configuration. (d) Spatial geometry for the laser beams used in the experiment.

and (d). The pumping laser beams $\boldsymbol{E}_1$ (ω_1, $\mathbf{k}_1$, and Rabi frequency G_1) and $\boldsymbol{E}_1'$ (ω_1, $\mathbf{k}_1'$, and G_1') (with a small angle of 0.3° between them) are tuned to drive the transition $|0\rangle$ ($3S_{1/2}$) to $|1\rangle$ ($3P_{3/2}$). $\boldsymbol{E}_1$ propagates in the opposite direction of the weak probe field $\boldsymbol{E}_3$ (ω_1, $\mathbf{k}_3$, and G_3), as shown in Figure 5.5(d). These three laser beams are from the same near-transform-limited dye laser with the same frequency detuning $\Delta_1 = \omega_{10} - \omega_1$, where ω_{10} is the transition frequency between $|0\rangle$ and $|1\rangle$, and generate an efficient DFWM signal $\boldsymbol{E}_F$ ($\mathbf{k}_F = \mathbf{k}_1 - \mathbf{k'}_1 + \mathbf{k}_3$) in the direction shown at the lower right corner of Figure 5.5(d). Another pair of beams, $\boldsymbol{E}_2$ (ω_2, $\mathbf{k}_2$, and G_2) and $\mathbf{E}_2'$ (ω_2, $\mathbf{k}_2'$, and G_2'), are the dressing beams with $\boldsymbol{E}_2$ propagating in the same direction as $\boldsymbol{E}_1$ and $\mathbf{E}_2'$ having a small angle (0.3°) from $\boldsymbol{E}_2$. $\boldsymbol{E}_2$ and $\boldsymbol{E}_2'$ have the same frequency detuning $\Delta_2 = \omega_{21} - \omega_2$ (from the same laser), which are tuned to the transition from $|1\rangle$ ($3P_{3/2}$) to $|2\rangle$ ($4D_{3/2,5/2}$). Two QWPs are used to control the polarizations of the pumping fields $\boldsymbol{E}_1$ and $\boldsymbol{E}_1'$. The generated FWM signal is split into two parts by a 50% BS. One is detected directly (denoted as $\boldsymbol{E}_F$), and the other decomposed into P- and S-polarized components by a polarization beam splitter (PBS), which are denoted as $\boldsymbol{E}_F^P$ and $\boldsymbol{E}_F^S$, respectively (Figure 5.5(c)).

The experiment was done in Na vapor in a heat pipe at the temperature of 250 °C. The atomic density is about $2.45 \times 10^{13}\,\text{cm}^{-3}$. The ground state ($|0\rangle$) is the $3S_{1/2}$ energy level and the first excited state ($|1\rangle$) is the $3P_{3/2}$ level. The upper excited state ($|2\rangle$) is the $3D_{3/2}$ energy level. Both lasers (for frequencies ω_1 and ω_2) are near-transform-limited dye lasers with a repetition rate of 10 Hz and pulse width of 3.5 ns. One laser beam is split to produce beams $\boldsymbol{E}_1$, $\boldsymbol{E}_1'$, and $\boldsymbol{E}_3$ with frequency ω_1, which are tuned to the line center (589.0 nm) of the $|0\rangle$ to $|1\rangle$ transition to generate the FWM signal $\boldsymbol{E}_F$. Another laser is used for beams $\boldsymbol{E}_2$ and $\boldsymbol{E}_2'$, which are tuned to the line center (568.8 nm) of the $|1\rangle$ to $|2\rangle$ transition to dress the FWM signal $\boldsymbol{E}_F$. These five laser beams are carefully aligned in the spatial configuration as shown in Figure 5.5(d). In order to optimize the beam shift and splitting effects, the field $\mathbf{E}_1'$ (energy 11 μJ) is set to be the strongest, approximately six times larger than $\boldsymbol{E}_1$ and $\boldsymbol{E}_2$ (energy 2 μJ), 14 times larger than the beam $\boldsymbol{E}_2'$ (energy 0.8 μJ), and 55 times larger than the weak probe beam $\boldsymbol{E}_3$ (energy 0.2 μJ), which is as weak as the generated FWM signal beam. These weak beams are recorded by a CCD.

To understand the observed changes in beam intensities and splitting of the FWM beams, we need to consider the polarization states of the beams and various SPM and XPM processes. The spatial beam breaking is mainly due to the overlap between the weak FWM beam and the strong dressing or pumping beams [27–30]. The propagation equations for the S- and P-polarizations of the generated FWM beam are

$$\frac{\partial E_F^P}{\partial z} - \frac{i\nabla_\perp^2 E_F^P}{2k_F} = \frac{ik_F}{n_0}\left[n_2^{P1}\left|E_F^P\right|^2 + 2n_2^{P2}|E_1'|^2 + 2n_2^{P3}|E_2'|^2 + 2n_2^{P4}|E_1|^2 + 2n_2^{P5}|E_2|^2\right]E_F^P \tag{5.1}$$

$$\frac{\partial E_F^S}{\partial z} - \frac{i\nabla_\perp^2 E_F^S}{2k_F} = \frac{ik_F}{n_0}\left[n_2^{S1}\left|E_F^S\right|^2 + 2n_2^{S2}|E_1'|^2 + 2n_2^{S3}|E_2'|^2 + 2n_2^{S4}|E_1|^2 + 2n_2^{S5}|E_2|^2\right]E_F^S \tag{5.2}$$

where z is the propagation distance, $k_F = \omega_1 n_0/c$ is the wave vector of the FWM beam, n_0 is the linear refractive index at ω_1, and $n_2^{P1,S1}$ are the self-Kerr coefficients for the S and P components of $\boldsymbol{E}_F$ and $n_2^{P2-P5,S2-S5}$ are the P- and S-polarized cross-Kerr coefficients of $\boldsymbol{E}_F$ induced by $\boldsymbol{E}'_{1,2}$ and $\boldsymbol{E}_{1,2}$, respectively. The Kerr coefficients can be defined as $n_2 = C\text{Re}\chi^{(3)}$, where $C = (\varepsilon_0 c n_0)^{-1}$ and the Kerr nonlinear susceptibility is expressed as $\chi^{(3)} = D\rho_{10}^{(3)}$, where $D = N\mu_{10}^4/(\hbar^3\varepsilon_0 G_1|G_1|^2)$ for $n_2^{P1,S1}$ and $D = N\mu_{10}^4/(\hbar^3\varepsilon_0 G_1|G_3|^2)$ for $n_2^{P2-P5,S2-S5}$ [31]. N is the atomic density of the medium (determined by the cell temperature) and μ_{10} (μ_{20}) is the dipole matrix element between energy levels $|0\rangle$ and $|1\rangle$ ($|2\rangle$).

Different pumping field polarizations can generate FWM signals with different polarizations. The S-polarized component of $\boldsymbol{E}_1$ or $\boldsymbol{E}'_1$ can be decomposed into balanced left- and right-circularly polarized parts, while the P component keeps linear polarization. Such polarization configuration results in S- and P-polarized FWM beams generated from different transition pathways among various Zeeman sublevels, as shown in Figure 5.5(b). With different dipole moments for the transitions between $|0\rangle-|1\rangle$ and $|1\rangle-|2\rangle$, the S- and P-polarized FWM components have different appearances in both intensity modulations and spatial patterns due to nonlinear Kerr effects.

One can obtain the expression for the P-polarized FWM ($\boldsymbol{E}_F^P$) intensity when changing the polarization of $\boldsymbol{E}_1$, $I^P \propto I(\sin^4\theta + \cos^4\theta)$. When changing the polarizations of both the $\boldsymbol{E}_1$ and $\boldsymbol{E}'_1$ beams, the P-polarized FWM intensity is given by $I^P \propto I(\sin^4\theta + \cos^4\theta)[\sin^4(\theta+\theta_0) + \cos^4(\theta+\theta_0)]$ [31]. θ is the polarization angle of $\boldsymbol{E}_1$ or E'_1 controlled by QWP. θ_0 is the polarization angle difference between $\boldsymbol{E}_1$ and $\boldsymbol{E}'_1$ fields. Also, one can solve the coupled density matrix equations to obtain $\rho_{10}^{(3)}$ for n_2^P induced by the $\boldsymbol{E}_1$ and $\boldsymbol{E}'_1$ fields, $n_2^P \propto n_2^a(\sin^4\theta + \cos^4\theta)$ [31], where $n_2^a \propto \text{Re}(-iG_F^P F_a^P)$. Similarly, the expression of the S-polarized FWM ($\boldsymbol{E}_F^S$) intensity, when changing the $\boldsymbol{E}_1$ polarization, is $I^S \propto I\sin^2\theta\cos^2\theta$. When changing polarizations of both $\boldsymbol{E}_1$ and $\boldsymbol{E}'_1$ beams, it becomes $I^S \propto I(\sin^4\theta + \cos^4\theta)\sin^2(\theta+\theta_0)\cos^2(\theta+\theta_0)$, where $I \propto (\rho_{10}^{(3)})^2 = (-iG_3F_a^{S,P})^2$. n_2^S, induced by the $\boldsymbol{E}_1$ and E'_1 fields, is given by $n_2^S \propto n_2^b \sin^2\theta\cos^2\theta$, where $n_2^b \propto \text{Re}(-iG_F^S F_a^S)$. The coefficients are $F_a^S = (G_1^{\mp})^2/F_1^2 F_2$, $F_a^S = (G_1^{\mp} + G_1'^{\mp})^2/F_1^2F_2$ ($F_a^P = G_1^2/F_1^2F_2$, $F_a^P = (G_1 + G'_1)^2/F_1^2F_2$) for $\boldsymbol{E}_F^S$ ($\boldsymbol{E}_F^P$) beam due to the $\boldsymbol{E}_1$ and $\boldsymbol{E}_1$ and $\boldsymbol{E}'_1$ dressings, respectively. Here, $F_i(i = 1, 2)$ is the function of the detuning and relaxation rates for different perturbation chains.

If we neglect the diffraction terms, the SPM and the small XPM contributions, and assume that all the beams involved are initially Gaussian with different centers, amplitudes, and half-widths, Eq. (5.1) and Eq. (5.2) can be readily solved to obtain the XPM-induced phase shift $\phi = 2k_F n_2 z I_1 e^{-r^2/2}/(n_0 I_0)$ imposed on the FWM beams by the pump fields [29]. The additional transverse propagation wave vector is $\delta\mathbf{k}_r = (\partial\phi/\partial r)\boldsymbol{r}$ and its direction is always toward the beam center of the strong pump field with positive n_2. Therefore, the weak $\boldsymbol{E}_F$ field is shifted to the pump field center and split globally. The locally focusing and defocusing due to the spatially varied phase-front curvature $\partial^2\phi/\partial r^2$ in the $\boldsymbol{E}_F$ beam further leads to its local splitting. The expression of the nonlinear phase shift shows that the strong spatial splitting can occur with increased I_1, n_2, and decreased I_0, where I_1 is the dressing field intensity and I_0 is the intensity of the FWM beam. In those expressions,

$r = x/w_0$ and y/w_0 are the transverse coordinates, respectively, and $\boldsymbol{r}$ is the unit vector along the transverse axes. w_0 is the spot size of the FWM beam.

5.2.2
Spatial Splitting of FWM Beam

In the first experiment, we only turn on $\boldsymbol{E}_1$, $\boldsymbol{E}_1'$, and $\boldsymbol{E}_3$ beams. The FWM signal $\boldsymbol{E}_F$ can be obtained without $\boldsymbol{E}_2$ and $\boldsymbol{E}_2'$ dressing. A QWP is used with a rotation angle θ to change the polarization state of the $\boldsymbol{E}_1$ field; we can obtain the splitting of the $\boldsymbol{E}_F^S$ beam for the S-polarized FWM beam (Figure 5.6). Such beam splitting formed in the atomic medium is obtained with flexible and easy-to-control parameters, such as atomic density, intensities of the dressing and FWM beams, and nonlinear dispersion [31]. Figure 5.6 presents the experimentally recorded splitting spots and the y cross-section intensity curves of the $\boldsymbol{E}_F^S$ beam at different polarization states of $\boldsymbol{E}_1$ (with $\theta = 0°$, 22.5°, 45°, 67.5°, and 90°, respectively) in the x-direction and different frequency detuning (with $\Delta_1 = -60.2$, -55.1, -50.7, and -46.4 GHz, respectively) in the y-direction. The inverted triangle dots show that the periodic intensity change of the $\boldsymbol{E}_F$ beam is decided by the polarization of $\boldsymbol{E}_1$ (at $\Delta_1 = -46.4$ GHz). Under this condition, $\boldsymbol{E}_1$ is strongest at 0° and 90° while weakest at 45°. At 45°, the nonlinear phase shift ϕ reaches its maximum value with the largest I_1/I_0 (having the constant $I_1 \propto (\boldsymbol{E}_1')^2$ and smallest $I_0 \propto (\boldsymbol{E}_F^S)^2$), leading to the strongest splitting in the y-direction caused by $\boldsymbol{E}_1'$, which we use the splitting number to measure. Correspondingly, the weakest y-direction splitting appears at the 0° and 90° polarization states, as shown by the y-direction splitting (circle dots) of $\boldsymbol{E}_F^S$ in Figure 5.6. With different frequency detuning, this phenomenon changes

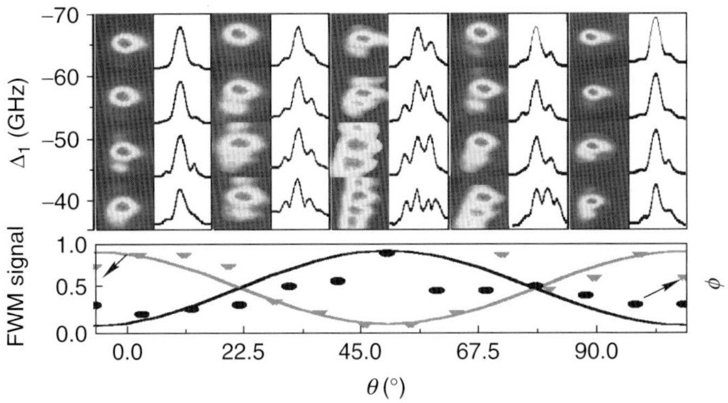

Figure 5.6 The experimentally measured spots and the y cross-section intensity curves of $\boldsymbol{E}_F^S$ beam versus different polarization states of $\boldsymbol{E}_1$ with different frequency detuning from up to down ($G_1 = 15$ GHz and $G_1' = 20$ GHz). The curves below are the y-direction splitting number of $\boldsymbol{E}_F^S$ (circle dots) and the fitted ϕ solid curve; also the normalized intensity values (inverted triangle) and the corresponding theoretical solid curve versus different polarization states, with $\Delta_1 = -46.4$ GHz. Source: Adapted from Ref. [32].

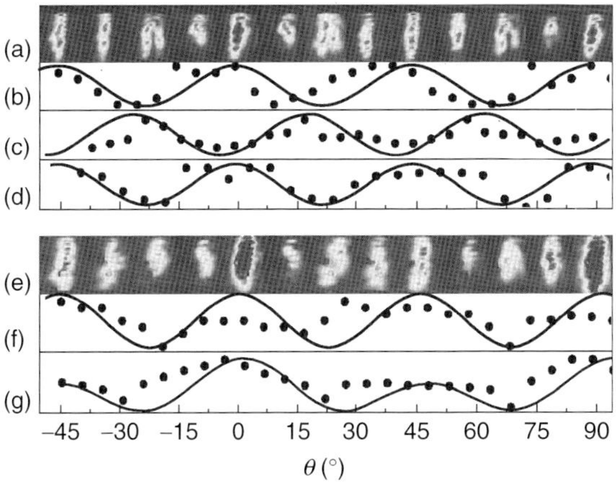

Figure 5.7 The FWM beam spots (a, e), the splitting numbers in the beam's *x* (c) and *y* (b, f) directions, the intensities (d, g) of the E_F^S (a–d) and E_F^P (e–g) components versus different polarization states of both E_1 and E_1' ($G_1 = 22$ *GHz* and $G_1' = 20$ *GHz*). The solid dots are the measured results and the solid curves are the fitted theoretical ϕ values.

gradually. In the self-focusing medium and near the resonance frequency, ϕ reaches its maximum, so E_F^S splits into more parts with corresponding polarization states of E_1.

Next, in the same beam configuration (without E_2 and E_2' dressing beams), we change the polarization states of both E_1 and E_1' beams (E_1' is set at 45° polarization angle before E_1). Figure 5.7 shows the experimentally measured spots, which split strongly in the *y*-direction at −45°, 0°, 45°, 90°, and weakly at −22.5°, 22.5°, and 67.5° for both the S-polarized and P-polarized FWM beams (Figure 5.7(a) and (e)). The splitting in the *y*-direction changes with a 45° period (Figure 5.7(b) and (f)). However, there only exists splitting in the *x*-direction for the S-polarized FWM beam (Figure 5.7(a) and (c)). Moreover, the intensities of the E_F^S and E_F^P beams are shown in Figure 5.7(d) and (g), respectively, which change with a 45° period as the polarization states of the E_1' and E_1 beams changing. These results match reasonably well with the theoretical calculations (using split-step Fourier method) based on the coupled propagation Eq. (5.1) and Eq. (5.2) for the S- and P-polarized FWM components.

Let us first consider the behavior of E_F^S, as shown in Figure 5.7(a) to (d). It is interesting to note that when $\theta = \perp 45°$, 0°, and 90°, the beam splitting in the *y*-direction is strong, while the splitting becomes stronger in the *x*-direction and weaker in the *y*-direction when $\theta = \pm 22.5°$ or 67.5° in Figure 5.7(a), although the splitting in both directions have the same 45° period. This phenomenon can be explained by the relative positions between the weak FWM beams and the strong dressing beams shown in Figure 5.5(b). When the polarizations of both E_1 and E_1' beams (E_1' is set at 45° before E_1) are changed, the intensities of E_1 and E_1' alternatively reach their maximum values in the period of 90° (the E_1' beam is strongest at −45° and 45°, while E_1' is strongest at 0° and 90°). First, at the 0°

(or 90°) polarization state of $\boldsymbol{E}_1$, with little dressing by the weakest $\boldsymbol{E}_1$, the $\boldsymbol{E}_\mathrm{F}$ beam overlaps on the strongest $\boldsymbol{E}_1'$ beam in the y-direction at a special alignment, where ϕ reaches its maximum caused by $I_1 \propto (\boldsymbol{E}_1')^2$, leading to the strongest y-direction splitting (Figure 5.7(a)). From $\theta = 0°$ to $\pm 45°$, as $\boldsymbol{E}_1'$ decreasing and $\boldsymbol{E}_1$ increasing, the $\boldsymbol{E}_\mathrm{F}$ beam is shifted to the left direction and gets closer to $\boldsymbol{E}_1$ due to the attraction ($n_2 > 0$) of the strong $\boldsymbol{E}_1$ beam and the probe beam $\boldsymbol{E}_3$ (Figure 5.5(d)). Second, at $\theta = \pm 22.5°$, $\boldsymbol{E}_\mathrm{F}$ partially overlaps with $\boldsymbol{E}_1$ and $\boldsymbol{E}_1'$, causing beam splitting in the x-direction (Figure 5.7(a)). Third, when $\boldsymbol{E}_1$ reaches its maximum, the y-direction overlaps with the strongest $\boldsymbol{E}_1$ at $\theta = \pm 45°$ (here $I_1 \propto (\boldsymbol{E}_1)^2$), with little dressing from the weakest $\boldsymbol{E}_1'$, producing strong y-direction splitting. Thus, as the polarization state periodically changes, the $\boldsymbol{E}_\mathrm{F}$ beam shifts back and forth between $\boldsymbol{E}_1$ and $\boldsymbol{E}_1'$ beams, so a similar phenomenon periodically appears.

Then, Figure 5.7(e) to (g) presents the experimentally measured spots, the periodic splitting in the y-direction, and the beam intensity of $\boldsymbol{E}_\mathrm{F}^\mathrm{P}$. Figure 5.7(f) and (g) is similar to Figure 5.7(b) and (d), respectively. However, the differences between $\boldsymbol{E}_\mathrm{F}^\mathrm{S}$ and $\boldsymbol{E}_\mathrm{F}^\mathrm{P}$ can also be observed, which mainly appear on the x-direction splitting ($\theta = \pm 22.5°$ or $67.5°$) and the beam intensity ($0°$ or $90°$). Here, the dipole moments of the transitions between $|0\rangle$ and $|1\rangle$ of the S-polarized laser beams are different from that of the P-polarized beams for choosing different transition pathways among various Zeeman sublevels as shown in Figure 5.5(b), where $F_1 = (\Gamma_{10} + i\Delta_1) + (G_1^\pm + G_1'^\pm)^2/\Gamma_{00} + (G_1^\pm + G_1'^\pm)^2/\Gamma_{11}$ and $F_2 = \Gamma_{00} + (G_1^\pm + G_1'^\pm)^2/(\Gamma_{10} + i\Delta_1) + (G_1^\pm + G_1'^\pm)^2/(\Gamma_{01} - i\Delta_1)$ for the S-polarized beam pathways, and $F_1 = (\Gamma_{10} + i\Delta_1) + (G_1 + G_1')^2/\Gamma_{00} + (G_1 + G_1')^2/\Gamma_{11}$ and $F_2 = \Gamma_{00} + (G_1 + G_1')^2/(\Gamma_{10} + i\Delta_1) + (G_1 + G_1')^2/(\Gamma_{01} - i\Delta_1)$ for the P-polarized beam pathways, resulting in $n_2^\mathrm{S}(1.5 \times 10^{-8}\,\mathrm{cm^2\,W^{-1}}) > n_2^\mathrm{P}(1.2 \times 10^{-8}\,\mathrm{cm^2\,W^{-1}})$ and $I_0^\mathrm{S} < I_0^\mathrm{P}$ for $|\rho_{10}^\mathrm{S}|^2 < |\rho_{10}^\mathrm{P}|^2$. So, one can obtain $\phi^\mathrm{S} > \phi^\mathrm{P}$, where ϕ^S and ϕ^P are the nonlinear phase shifts of the $\boldsymbol{E}_\mathrm{F}^\mathrm{S}$ and $\boldsymbol{E}_\mathrm{F}^\mathrm{P}$ beams, respectively. From Figure 5.7(e), the intensity of $\boldsymbol{E}_\mathrm{F}^\mathrm{P}$ is much stronger than that of $\boldsymbol{E}_\mathrm{F}^\mathrm{S}$ at $\theta = 0°$ and $90°$ due to $I_0^\mathrm{S} < I_0^\mathrm{P}$. Then, the y-direction splitting of the $\boldsymbol{E}_\mathrm{F}^\mathrm{P}$ beam is weaker than $\boldsymbol{E}_\mathrm{F}^\mathrm{S}$ due to $\phi^\mathrm{S} > \phi^\mathrm{P}$. Compared with $\boldsymbol{E}_\mathrm{F}^\mathrm{S}$ at $\theta = \pm 22.5°$ or $67.5°$, the x-direction splitting of the $\boldsymbol{E}_\mathrm{F}^\mathrm{P}$ beam is too weak to be observed in Figure 5.7(e).

When the $\boldsymbol{E}_2$ and E_2' beams are turned on, Figure 5.8 shows different dressing effects for the $\boldsymbol{E}_\mathrm{F}^\mathrm{S}$ and $\boldsymbol{E}_\mathrm{F}^\mathrm{P}$ components with changing $\boldsymbol{E}_1$ and $\boldsymbol{E}_1'$ polarization states. Specifically, the splitting of both $\boldsymbol{E}_\mathrm{F}^\mathrm{S}$ and $\boldsymbol{E}_\mathrm{F}^\mathrm{P}$ beams have the same 45° period, as shown in Figure 5.7. Comparing (i) with (iv) in Figure 5.8(b) and (c), we can see that both $\boldsymbol{E}_\mathrm{F}^\mathrm{S}$ and $\boldsymbol{E}_\mathrm{F}^\mathrm{P}$ intensities in (i) are weaker owing to the suppressions of $\boldsymbol{E}_2$ and $\boldsymbol{E}_2'$ ($\Delta_1 + \Delta_2 = 0$ and $\Delta_1 = -10$ GHz), which causes a larger splitting in the y-direction ((ii) in Figure 5.8(b) and (c)). For the $\boldsymbol{E}_\mathrm{F}^\mathrm{S}$ beam, at $\theta = \pm 22.5°$ or $67.5°$ without $\boldsymbol{E}_2$ and $\boldsymbol{E}_2'$, it has the x-direction splitting ((iv) and (vi) in Figure 5.8(b)). However, with five beams all on, $\boldsymbol{E}_\mathrm{F}^\mathrm{S}$ is shifted to the upper direction by $\boldsymbol{E}_1'$, and hardly overlaps with both $\boldsymbol{E}_1$ and $\boldsymbol{E}_1'$ beams, causing the x-direction splitting to be weak ((i) and (iii) in Figure 5.8(b)).

Moreover, in Figure 5.8(b) and (c), the contrast ratio of the y-direction splitting $\eta = N_1/N_2$ turns to be $\eta = 0.29$ and $\eta = 0.65$ for $\boldsymbol{E}_\mathrm{F}^\mathrm{S}$ ($\boldsymbol{E}_\mathrm{F}^\mathrm{P}$) with or without $\boldsymbol{E}_2$ and G_2' dressing, respectively. Here, N_1 and N_2 are the numbers of the y-direction splitting

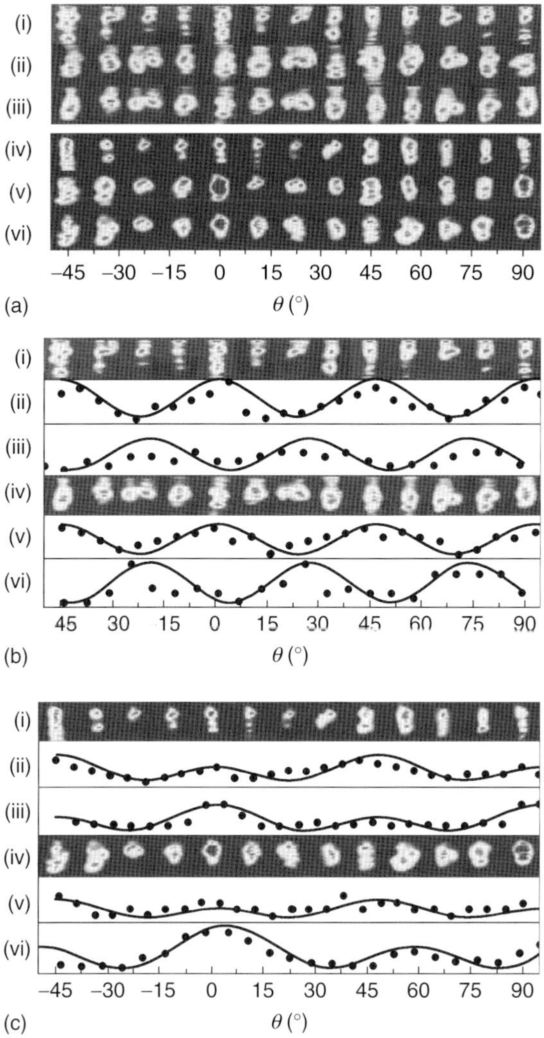

Figure 5.8 (a) The measured beam spots of the $\boldsymbol{E}_F^S$ (i–iii) and $\boldsymbol{E}_F^P$ (iv–vi) fields with all five beams on (i, iv), without $\boldsymbol{E}_2$ (ii, v), and without $\boldsymbol{E}_2$ and $\boldsymbol{E}_2'$ (iii, vi) versus different polarization states of $\boldsymbol{E}_1$ and $\boldsymbol{E}_1'$. (b) The measured beam spots and the beam splitting number of both the x (iii, vi), and y (ii, v) directions (solid dots) of the $\boldsymbol{E}_F^S$ beam with all five beams (i–iii), and without $\boldsymbol{E}_2$ and $\boldsymbol{E}_2'$ (iv–vi). (c) The measured beam spots, the beam splitting number of the y (ii, v) direction (solid dots) and the intensity (iii, vi) of the $\boldsymbol{E}_F^P$ beam with all five beams (i–iii), and without $\boldsymbol{E}_2$ and $\boldsymbol{E}_2'$ (iv–vi). The solid curves are the fitted theoretical ϕ values ($G_1 = 22$ GHz, $G_1' = 20$ GHz, $G_2 = 2.0$ GHz, and $G_2' = 1.8$ GHz). Source: Adapted from Ref. [32].

spots at valley and peak points of the curves (ii) in Figure 5.8(b) and (c), respectively. We can observe that the contrast of periodic splitting in the y-direction becomes better with dressing fields $\boldsymbol{E}_2$ and $\boldsymbol{E}_2'$ ((i) and (iv) in Figure 5.8(b) and (c)). This behavior occurs because different polarization states of the pumping beams can select different transitions among Zeeman sublevels and have different dressing strengths [17].

5.3
Coexisting Polarized FWM

In this section, the polarization properties of several coexisting FWM signals in the two-level system are investigated. In the presence of additional coupling laser fields, more FWM processes can coexist in the same system. In this case, several interesting physical phenomena can occur, such as the quantum interference, competition, and mutual dressing among these FWM signals. We obtain the intensities and polarizations of these coexisting FWM signals under different polarization configurations and different frequency detuning of the incident fields. Moreover, the polarization dependence of mutual-dressing effect and the interaction among coexisting FWM signals are investigated. These results verify that the coexisting FWM processes can be modulated via the polarization configurations and frequency detuning of the incident fields. Such controlled FWM signals are important for optical communication and quantum information processes.

5.3.1
Experiment Setup

The experiments shown in Figure 5.9(a) are carried out in a Na atom vapor oven (the sodium atomic density is about $1.5 \times 10^{13}\text{cm}^{-3}$ ($T = 235\,^{\circ}\text{C}$)). As shown in Figure 5.9(b), energy levels $|a\rangle(3S_{1/2})$ and $|b\rangle(3P_{3/2})$ form a two-level atomic system, and the resonant frequency of which is ω_0. Six laser beams are all driving the

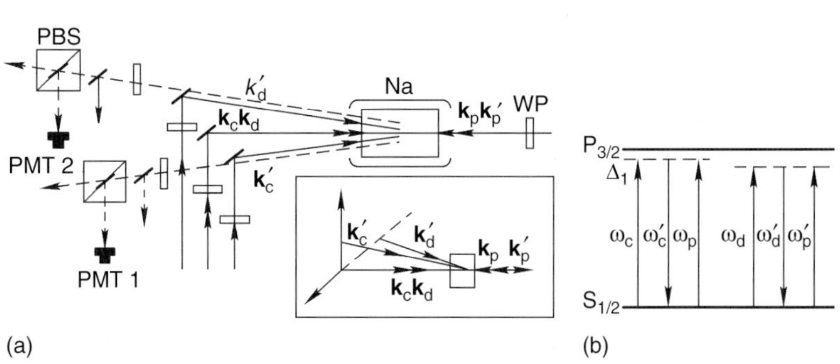

Figure 5.9 (a) and (b) Schematic diagrams of the experimental setup and the relevant energy levels in Na atom.

Table 5.1 Wave vectors (first column) and frequencies (second column) of the generated FWM signals detected by PMT1 and PMT2.

PMT1	$\mathbf{k}_{s1}=\mathbf{k}_p+\mathbf{k}_c-\mathbf{k}'_c$	$\omega_{s1}=\omega_c$
	$\mathbf{k}_{s2}=\mathbf{k}_p+\mathbf{k}_d-\mathbf{k}'_c$	$\omega_{s2}=\omega_d$
	$\mathbf{k}_{s3}=\mathbf{k}'_p+\mathbf{k}_c-\mathbf{k}'_c$	$\omega_{s3}=\omega_d$
	$\mathbf{k}_{s4}=\mathbf{k}'_p+\mathbf{k}_d-\mathbf{k}'_c$	$\omega_{s4}=2\omega_d-\omega_c$
PMT2	$\mathbf{k}_{s5}=\mathbf{k}'_p+\mathbf{k}_d-\mathbf{k}'_d$	$\omega_{s5}=\omega_d$
	$\mathbf{k}_{s6}=\mathbf{k}'_p+\mathbf{k}_c-\mathbf{k}'_d$	$\omega_{s6}=\omega_c$
	$\mathbf{k}_{s7}=\mathbf{k}_p+\mathbf{k}_d-\mathbf{k}'_d$	$\omega_{s7}=\omega_c$
	$\mathbf{k}_{s8}=\mathbf{k}_p+\mathbf{k}_c-\mathbf{k}'_d$	$\omega_{s8}=2\omega_c-\omega_d$

transition between $|a\rangle$ and $|b\rangle$. Two laser beams E_c (ω_c, $\mathbf{k}_c$, Rabi frequency G_c, and intensity $I=4.4$ W cm^{-2}) and E'_c (ω_c, k'_c, G'_c, and $I=4.4$ W cm^{-2}) propagate in the direction opposite to the weak probe beam E_p (ω_p, $\mathbf{k}_p$, G_p, and 0.3 W cm^{-2}). These three laser beams come from the same dye laser DL1 (10 Hz repetition rate, 5 ns pulse width, and 0.04 cm^{-1} line width) with a frequency detuning $\Delta_1=\omega_0-\omega_c$, pumped by the second-harmonic beam of a Nd:YAG laser. The other three laser beams E_d (ω_d, $\mathbf{k}_d$, G_d, and 3.2 W cm^{-2}), E'_d (ω_d, $\mathbf{k}'_d$, G'_d, and 3.2 W cm^{-2}), and E'_p (ω'_p, $\mathbf{k}'_p$, G'_p, and 0.2 W cm^{-2}) are from another dye laser DL2 (which has the same characteristics as the DL1) with a frequency detuning $\Delta_2=\omega_0-\omega_d$. In this case, there are eight FWM signals coexisting in one atomic system. The phase-matching conditions and frequencies of generated FWM signals are tabulated in Table 5.1. These FWM signals propagate in two directions (FWM signals $\mathbf{k}_{s1}$, $\mathbf{k}_{s2}$, $\mathbf{k}_{s3}$, and $\mathbf{k}_{s4}$ propagate in the direction opposite to $\mathbf{k}'_c$. FWM signals $\mathbf{k}_{s5}$, $\mathbf{k}_{s6}$, $\mathbf{k}_{s7}$, and $\mathbf{k}_{s8}$ propagate in the opposite direction of $\mathbf{k}'_d$). All the FWM signals are first split into two equal components by a splitter, in which one is detected directly (denoted as I_T), and the other is decomposed into P- and S-polarized components by a polarization beam splitter (PBS). Two photomultiplier tube (PMT) detectors are used to receive the P or S component of these FWM signals in the direction opposite to $\mathbf{k}'_c$ (PMT1) and $\mathbf{k}'_d$ (PMT2), respectively. A half-wave plate (HWP) and a QWP are selectively used (in different experiments, respectively) to control the polarization states of the incident fields.

5.3.2
Theoretical Model

When the six laser beams are all turned on, there are eight FWM signals coexisting in one atomic system. The quantum constructive or destructive interference between different pathways can result in the mutual-dressing effect between these coexisting FWM signals. For the reason of the application of several wave plates to modify the polarization states of the incident fields, Zeeman sublevels of each involved energy level will play an important role in the interaction between atoms and polarized fields. So, we theoretically investigate the generated FWM signals by

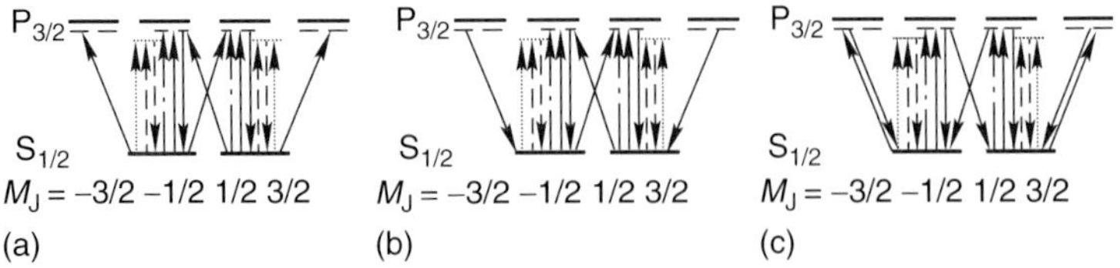

Figure 5.10 Schematic of two-level system configuration consisting of Zeeman sublevels. (a) QWP changes the field $\mathbf{k}_c$, (b) QWP changes the field k'_c, and (c) QWP changes both the fields $\mathbf{k}_c$ and k'_c. Solid lines: the coupling fields $\mathbf{k}_c$ and k'_c; dashed lines: the coupling fields $\mathbf{k}_d$ and k'_d; dash-dot lines: the probe field $\mathbf{k}_p$; dotted lines: the probe field k'_p. Source: Adapted from Ref. [24].

considering its generation process among various Zeeman sublevels in the semi-classical framework. The transition pathways of the generated FWM are presented in Figure 5.10. It is based on the fact that different polarization schemes can excite different transition pathways in the Zeeman-degenerate atomic systems. As a sample, Table 5.2 lists all the perturbation chains of the FWM signals when the fields $\mathbf{k}_c$ and $\mathbf{k}'_i$ are changed by QWP. The total FWM signal can be considered as the summed contribution of each perturbation chain. According to the experimental setup, the x-axis is the original polarization direction of all the incident fields, and it is also the quantization axis. We then decompose an arbitrary field into two components, parallel to and perpendicular to the x-axis, respectively. When this field interacts with a two-level atom, the perpendicular component can be decomposed into equally left-circularly and right-circularly polarized components. The generated FWM signals contain linearly polarized component I_L and circularly polarized component I_C. We have $I_P = I_L \sin^2\alpha + I_C/2$ (where α is the angle between the P-polarization and the polarization of the linearly polarized signal), $I_S = I_L \cos^2\alpha + I_C/2$, and $I_T = I_S + I_P = I_L + I_C$ [25].

Using the method of perturbation chain [33–35], we can obtain the expressions of various density matrix elements corresponding to the third-order nonlinear susceptibilities under different polarization schemes. When the polarizations of $\mathbf{k}_c$ and $\mathbf{k}'_c$ are changed by QWP, the corresponding density matrix elements of undressed FWM signals in P- and S-polarization are

$$\begin{aligned}\rho^{k_1,k'_1}_{\mathrm{P(PMT1)}} = &-i\sum_{M=\pm1/2}\left[\frac{\left|G^0_{\mathrm{c}_M}\right|^2}{\Gamma_{a_Ma_M}d_1}\left(\frac{G^0_{\mathrm{p}_M}}{d_1}+\frac{G^{0\prime}_{\mathrm{p}M}}{d_2}\right)+\frac{(G^0_{\mathrm{c}_M})^*G^0_{\mathrm{d}_M}}{d_3d_2}\left(\frac{G^0_{\mathrm{p}_M}}{d_2}+\frac{G^{0\prime}_{\mathrm{p}M}}{d_4}\right)\right]\\ &-i\sum_{M=\pm1/2}\frac{1}{\Gamma_{a_Ma_M}}\left(\frac{(G^-_{\mathrm{c}_M})^*G^-_{\mathrm{c}_M}}{d_{11}}+\frac{(G^+_{\mathrm{c}_M})^*G^+_{\mathrm{c}_M}}{d_{12}}\right)\left(\frac{G^0_{\mathrm{p}_M}}{d_1}+\frac{G^{0\prime}_{\mathrm{p}M}}{d_2}\right)\end{aligned} \tag{5.3}$$

$$\rho^{k_1,k'_1}_{\mathrm{s(PMT1)}} = -i\sum_{M=\pm1/2}\left[\frac{G^0_{\mathrm{c}_M}(G'^{\pm}_{\mathrm{c}_M})^*}{\Gamma_{a_Ma_{-M}}d_1}\left(\frac{G'^0_{\mathrm{p}M}}{d_{14}}+\frac{G^0_{\mathrm{p}_M}}{d_{13}}\right)+\frac{G^0_{\mathrm{d}_M}(G'^{\pm}_{\mathrm{c}M})^*}{d_7d_2}\left(\frac{G^0_{\mathrm{p}_M}}{d_{14}}+\frac{G'^0_{\mathrm{p}M}}{d_8}\right)\right] \tag{5.4}$$

Table 5.2 Perturbation chains of horizontally polarized components of FWM signals when fields $\mathbf{k}_c$ and $\mathbf{k}'_c$ are changed by QWP.

The subsystem generating P-polarization signal	$\lvert a_{-1/2}\rangle \xrightarrow{G_{c1}^{-}} \lvert b_{-3/2}\rangle \xrightarrow{(G'^{-}_{c1})^{*}} \lvert a_{-1/2}\rangle \xrightarrow{G_{p1}^{0}} \lvert b_{-1/2}\rangle \xrightarrow{(G_{F1}^{0})^{*}} \lvert a_{-1/2}\rangle : \mathbf{k}_c - \mathbf{k}'_c + \mathbf{k}_p$
	$\lvert a_{-1/2}\rangle \xrightarrow{G_{c1}^{-}} \lvert b_{-3/2}\rangle \xrightarrow{(G'^{-}_{c1})^{*}} \lvert a_{-1/2}\rangle \xrightarrow{G'^{0}_{p1}} \lvert b_{-1/2}\rangle \xrightarrow{(G_{F1}^{0})^{*}} \lvert a_{-1/2}\rangle : \mathbf{k}_c - \mathbf{k}'_c + \mathbf{k}'_p$
	$\lvert a_{-1/2}\rangle \xrightarrow{G_{c1}^{0}} \lvert b_{-1/2}\rangle \xrightarrow{G'^{0}_{c1}} \lvert a_{-1/2}\rangle \xrightarrow{G_{p1}^{0}} \lvert b_{-1/2}\rangle \xrightarrow{(G_{F1}^{0})^{*}} \lvert a_{-1/2}\rangle : \mathbf{k}_c - \mathbf{k}'_c + \mathbf{k}_p$
	$\lvert a_{-1/2}\rangle \xrightarrow{G_{c1}^{0}} \lvert b_{-1/2}\rangle \xrightarrow{G'^{0}_{c1}} \lvert a_{-1/2}\rangle \xrightarrow{G'^{0}_{p1}} \lvert b_{-1/2}\rangle \xrightarrow{(G_{F1}^{0})^{*}} \lvert a_{-1/2}\rangle : \mathbf{k}_c - \mathbf{k}'_c + \mathbf{k}'_p$
	$\lvert a_{-1/2}\rangle \xrightarrow{G_{c1}^{0}} \lvert b_{-1/2}\rangle \xrightarrow{G'^{0}_{d1}} \lvert a_{-1/2}\rangle \xrightarrow{G_{p1}^{0}} \lvert b_{-1/2}\rangle \xrightarrow{(G_{F1}^{0})^{*}} \lvert a_{-1/2}\rangle : \mathbf{k}_c - \mathbf{k}'_d + \mathbf{k}_p$
	$\lvert a_{-1/2}\rangle \xrightarrow{G_{c1}^{0}} \lvert b_{-1/2}\rangle \xrightarrow{G'^{0}_{d1}} \lvert a_{-1/2}\rangle \xrightarrow{G'^{0}_{p1}} \lvert b_{-1/2}\rangle \xrightarrow{(G_{F1}^{0})^{*}} \lvert a_{-1/2}\rangle : \mathbf{k}_c - \mathbf{k}'_d + \mathbf{k}'_p$
	$\lvert a_{-1/2}\rangle \xrightarrow{G_{c1}^{+}} \lvert b_{1/2}\rangle \xrightarrow{(G'^{+}_{c1})^{*}} \lvert a_{-1/2}\rangle \xrightarrow{G_{p2}^{0}} \lvert b_{-1/2}\rangle \xrightarrow{(G_{F2}^{0})^{*}} \lvert a_{-1/2}\rangle : \mathbf{k}_c - \mathbf{k}'_c + \mathbf{k}_p$
	$\lvert a_{-1/2}\rangle \xrightarrow{G_{c1}^{+}} \lvert b_{1/2}\rangle \xrightarrow{(G'^{+}_{c1})^{*}} \lvert a_{-1/2}\rangle \xrightarrow{G'^{0}_{p2}} \lvert b_{-1/2}\rangle \xrightarrow{(G_{F2}^{0})^{*}} \lvert a_{-1/2}\rangle : \mathbf{k}_c - \mathbf{k}'{}_c + \mathbf{k}'_p$
	$\lvert a_{-1/2}\rangle \xrightarrow{G_{d1}^{0}} \lvert b_{-1/2}\rangle \xrightarrow{G'^{0}_{c1}} \lvert a_{-1/2}\rangle \xrightarrow{G_{p1}^{0}} \lvert b_{-1/2}\rangle \xrightarrow{(G_{F1}^{0})^{*}} \lvert a_{-1/2}\rangle : \mathbf{k}_d - \mathbf{k}'_c + \mathbf{k}_p$
	$\lvert a_{-1/2}\rangle \xrightarrow{G_{d1}^{0}} \lvert b_{-1/2}\rangle \xrightarrow{G'^{0}_{c1}} \lvert a_{-1/2}\rangle \xrightarrow{G'^{0}_{p1}} \lvert b_{-1/2}\rangle \xrightarrow{(G_{F1}^{0})^{*}} \lvert a_{-1/2}\rangle : \mathbf{k}_d - \mathbf{k}'_c + \mathbf{k}'_p$
	$\lvert a_{-1/2}\rangle \xrightarrow{G_{d1}^{0}} \lvert b_{-1/2}\rangle \xrightarrow{G'^{0}_{d1}} \lvert a_{-1/2}\rangle \xrightarrow{G_{p1}^{0}} \lvert b_{-1/2}\rangle \xrightarrow{(G_{F1}^{0})^{*}} \lvert a_{-1/2}\rangle : \mathbf{k}_d - \mathbf{k}'_d + \mathbf{k}_p$
	$\lvert a_{-1/2}\rangle \xrightarrow{G_{d1}^{0}} \lvert b_{-1/2}\rangle \xrightarrow{G'^{0}_{d1}} \lvert a_{-1/2}\rangle \xrightarrow{G'^{0}_{p1}} \lvert b_{-1/2}\rangle \xrightarrow{(G_{F1}^{0})^{*}} \lvert a_{-1/2}\rangle : \mathbf{k}_d - \mathbf{k}'_d + \mathbf{k}'_p$
	$\lvert a_{1/2}\rangle \xrightarrow{G_{c2}^{-}} \lvert b_{-1/2}\rangle \xrightarrow{(G'^{-}_{c2})^{*}} \lvert a_{1/2}\rangle \xrightarrow{G_{p2}^{0}} \lvert b_{1/2}\rangle \xrightarrow{(G_{F2}^{0})^{*}} \lvert a_{1/2}\rangle : \mathbf{k}_c - \mathbf{k}'_c + \mathbf{k}_p$
	$\lvert a_{1/2}\rangle \xrightarrow{G_{c2}^{-}} \lvert b_{-1/2}\rangle \xrightarrow{(G'^{-}_{c2})^{*}} \lvert a_{1/2}\rangle \xrightarrow{G'^{0}_{p2}} \lvert b_{1/2}\rangle \xrightarrow{(G_{F2}^{0})^{*}} \lvert a_{1/2}\rangle : \mathbf{k}_c - \mathbf{k}'_c + \mathbf{k}'_p$
	$\lvert a_{1/2}\rangle \xrightarrow{G_{c2}^{0}} \lvert b_{1/2}\rangle \xrightarrow{G'^{0}_{c2}} \lvert a_{1/2}\rangle \xrightarrow{G_{p2}^{0}} \lvert b_{1/2}\rangle \xrightarrow{(G_{F2}^{0})^{*}} \lvert a_{1/2}\rangle : \mathbf{k}_c - \mathbf{k}'_c + \mathbf{k}_p$
	$\lvert a_{1/2}\rangle \xrightarrow{G_{c2}^{0}} \lvert b_{1/2}\rangle \xrightarrow{G'^{0}_{c2}} \lvert a_{1/2}\rangle \xrightarrow{G'^{0}_{p2}} \lvert b_{1/2}\rangle \xrightarrow{(G_{F2}^{0})^{*}} \lvert a_{1/2}\rangle : \mathbf{k}_c - \mathbf{k}'_c + \mathbf{k}'_p$
	$\lvert a_{1/2}\rangle \xrightarrow{G_{c2}^{0}} \lvert b_{1/2}\rangle \xrightarrow{G'^{0}_{d2}} \lvert a_{1/2}\rangle \xrightarrow{G_{p2}^{0}} \lvert b_{1/2}\rangle \xrightarrow{(G_{F2}^{0})^{*}} \lvert a_{1/2}\rangle : \mathbf{k}_c - \mathbf{k}'_d + \mathbf{k}_p$
	$\lvert a_{1/2}\rangle \xrightarrow{G_{c2}^{0}} \lvert b_{1/2}\rangle \xrightarrow{G'^{0}_{d2}} \lvert a_{1/2}\rangle \xrightarrow{G'^{0}_{p2}} \lvert b_{1/2}\rangle \xrightarrow{(G_{F2}^{0})^{*}} \lvert a_{1/2}\rangle : \mathbf{k}_c - \mathbf{k}'_d + \mathbf{k}'_p$
	$\lvert a_{1/2}\rangle \xrightarrow{G_{c2}^{+}} \lvert b_{3/2}\rangle \xrightarrow{(G'^{+}_{c2})^{*}} \lvert a_{1/2}\rangle \xrightarrow{G_{p2}^{0}} \lvert b_{1/2}\rangle \xrightarrow{(G_{F2}^{0})^{*}} \lvert a_{1/2}\rangle : \mathbf{k}_c - \mathbf{k}'_c + \mathbf{k}_p$
	$\lvert a_{1/2}\rangle \xrightarrow{G_{c2}^{+}} \lvert b_{3/2}\rangle \xrightarrow{(G'^{+}_{c2})^{*}} \lvert a_{1/2}\rangle \xrightarrow{G'^{0}_{p2}} \lvert b_{1/2}\rangle \xrightarrow{(G_{F2}^{0})^{*}} \lvert a_{1/2}\rangle : \mathbf{k}_c - \mathbf{k}'_c + \mathbf{k}'_p$
	$\lvert a_{1/2}\rangle \xrightarrow{G_{d2}^{0}} \lvert b_{1/2}\rangle \xrightarrow{G'^{0}_{c2}} \lvert a_{1/2}\rangle \xrightarrow{G_{p2}^{0}} \lvert b_{1/2}\rangle \xrightarrow{(G_{F2}^{0})^{*}} \lvert a_{1/2}\rangle : \mathbf{k}_d - \mathbf{k}'{}_c + \mathbf{k}_p$
	$\lvert a_{1/2}\rangle \xrightarrow{G_{d2}^{0}} \lvert b_{1/2}\rangle \xrightarrow{G'^{0}_{c2}} \lvert a_{1/2}\rangle \xrightarrow{G'^{0}_{p2}} \lvert b_{1/2}\rangle \xrightarrow{(G_{F2}^{0})^{*}} \lvert a_{1/2}\rangle : \mathbf{k}_d - \mathbf{k}'_c + \mathbf{k}'{}_p$
	$\lvert a_{1/2}\rangle \xrightarrow{G_{d2}^{0}} \lvert b_{1/2}\rangle \xrightarrow{G'^{0}_{d2}} \lvert a_{1/2}\rangle \xrightarrow{G_{p2}^{0}} \lvert b_{1/2}\rangle \xrightarrow{(G_{F2}^{0})^{*}} \lvert a_{1/2}\rangle : \mathbf{k}_d - \mathbf{k}'_d + \mathbf{k}_p$
	$\lvert a_{1/2}\rangle \xrightarrow{G_{d2}^{0}} \lvert b_{1/2}\rangle \xrightarrow{G'^{0}_{d2}} \lvert a_{1/2}\rangle \xrightarrow{G'^{0}_{p2}} \lvert b_{1/2}\rangle \xrightarrow{(G_{F2}^{0})^{*}} \lvert a_{1/2}\rangle : \mathbf{k}_d - \mathbf{k}'_d + \mathbf{k}'_p$

(continued overleaf)

Table 5.2 (*Continued*)

The subsystem generating S-polarization signal	$\lvert a_{-1/2}\rangle \xrightarrow{G^0_{c1}} \lvert b_{-1/2}\rangle \xrightarrow{G'^-_{c1}} \lvert a_{1/2}\rangle \xrightarrow{G^0_{p1}} \lvert b_{1/2}\rangle \xrightarrow{(G^+_{F2})^*} \lvert a_{-1/2}\rangle : \mathbf{k}_c\text{-}\mathbf{k}'_c + \mathbf{k}_p$
	$\lvert a_{-1/2}\rangle \xrightarrow{G^0_{c1}} \lvert b_{-1/2}\rangle \xrightarrow{G'^-_{c1}} \lvert a_{1/2}\rangle \xrightarrow{G'^0_{p1}} \lvert b_{1/2}\rangle \xrightarrow{(G^+_{F2})^*} \lvert a_{-1/2}\rangle : \mathbf{k}_c\text{-}\mathbf{k}'_c + \mathbf{k}'_p$
	$\lvert a_{-1/2}\rangle \xrightarrow{G^+_{c1}} \lvert b_{1/2}\rangle \xrightarrow{G'^0_{c2}} \lvert a_{1/2}\rangle \xrightarrow{G^0_{p2}} \lvert b_{1/2}\rangle \xrightarrow{(G^+_{F2})^*} \lvert a_{-1/2}\rangle : \mathbf{k}_c\text{-}\mathbf{k}'_c + \mathbf{k}_p$
	$\lvert a_{-1/2}\rangle \xrightarrow{G^+_{c1}} \lvert b_{1/2}\rangle \xrightarrow{G'^0_{c2}} \lvert a_{1/2}\rangle \xrightarrow{G'^0_{p2}} \lvert b_{1/2}\rangle \xrightarrow{(G^+_{F2})^*} \lvert a_{-1/2}\rangle : \mathbf{k}_c\text{-}\mathbf{k}'_c + \mathbf{k}'_p$
	$\lvert a_{-1/2}\rangle \xrightarrow{G^+_{c1}} \lvert b_{1/2}\rangle \xrightarrow{G'^0_{d2}} \lvert a_{1/2}\rangle \xrightarrow{G^0_{p2}} \lvert b_{1/2}\rangle \xrightarrow{(G^+_{F2})^*} \lvert a_{-1/2}\rangle : \mathbf{k}_c\text{-}\mathbf{k}'_d + \mathbf{k}_p$
	$\lvert a_{-1/2}\rangle \xrightarrow{G^+_{c1}} \lvert b_{1/2}\rangle \xrightarrow{G'^0_{d2}} \lvert a_{1/2}\rangle \xrightarrow{G'^0_{p2}} \lvert b_{1/2}\rangle \xrightarrow{(G^+_{F2})^*} \lvert a_{-1/2}\rangle : \mathbf{k}_c\text{-}\mathbf{k}'_d + \mathbf{k}'_p$
	$\lvert a_{-1/2}\rangle \xrightarrow{G^0_{d1}} \lvert b_{-1/2}\rangle \xrightarrow{G'^-_{c1}} \lvert a_{1/2}\rangle \xrightarrow{G^0_{p2}} \lvert b_{1/2}\rangle \xrightarrow{(G^+_{F2})^*} \lvert a_{-1/2}\rangle : \mathbf{k}_d\text{-}\mathbf{k}'_c + \mathbf{k}_p$
	$\lvert a_{-1/2}\rangle \xrightarrow{G^0_{d1}} \lvert b_{-1/2}\rangle \xrightarrow{G'^-_{c1}} \lvert a_{1/2}\rangle \xrightarrow{G'^0_{p2}} \lvert b_{1/2}\rangle \xrightarrow{(G^+_{F2})^*} \lvert a_{-1/2}\rangle : \mathbf{k}_d\text{-}\mathbf{k}'_c + \mathbf{k}'_p$
	$\lvert a_{1/2}\rangle \xrightarrow{G^-_{c2}} \lvert b_{-1/2}\rangle \xrightarrow{G'^0_{c1}} \lvert a_{-1/2}\rangle \xrightarrow{G^0_{p1}} \lvert b_{-1/2}\rangle \xrightarrow{(G^+_{F2})^*} \lvert a_{1/2}\rangle : \mathbf{k}_c\text{-}\mathbf{k}'_c + \mathbf{k}_p$
	$\lvert a_{1/2}\rangle \xrightarrow{G^-_{c2}} \lvert b_{-1/2}\rangle \xrightarrow{G'^0_{c1}} \lvert a_{-1/2}\rangle \xrightarrow{G'^0_{p1}} \lvert b_{-1/2}\rangle \xrightarrow{(G^+_{F2})^*} \lvert a_{1/2}\rangle : \mathbf{k}_c\text{-}\mathbf{k}'_c + \mathbf{k}'_p$
	$\lvert a_{1/2}\rangle \xrightarrow{G^-_{c2}} \lvert b_{-1/2}\rangle \xrightarrow{G'^0_{d1}} \lvert a_{-1/2}\rangle \xrightarrow{G^0_{p1}} \lvert b_{-1/2}\rangle \xrightarrow{(G^+_{F2})^*} \lvert a_{1/2}\rangle : \mathbf{k}_c\text{-}\mathbf{k}'_d + \mathbf{k}_p$
	$\lvert a_{1/2}\rangle \xrightarrow{G^-_{c2}} \lvert b_{-1/2}\rangle \xrightarrow{G'^0_{d1}} \lvert a_{-1/2}\rangle \xrightarrow{G'^0_{p1}} \lvert b_{-1/2}\rangle \xrightarrow{(G^+_{F2})^*} \lvert a_{1/2}\rangle : \mathbf{k}_c\text{-}\mathbf{k}'_d + \mathbf{k}'_p$
	$\lvert a_{1/2}\rangle \xrightarrow{G^0_{c2}} \lvert b_{1/2}\rangle \xrightarrow{G'^+_{c1}} \lvert a_{-1/2}\rangle \xrightarrow{G^0_{p1}} \lvert b_{-1/2}\rangle \xrightarrow{(G^+_{F2})^*} \lvert a_{1/2}\rangle : \mathbf{k}_c\text{-}\mathbf{k}'_c + \mathbf{k}_p$
	$\lvert a_{1/2}\rangle \xrightarrow{G^0_{c2}} \lvert b_{1/2}\rangle \xrightarrow{G'^+_{c1}} \lvert a_{-1/2}\rangle \xrightarrow{G'^0_{p1}} \lvert b_{-1/2}\rangle \xrightarrow{(G^+_{F2})^*} \lvert a_{1/2}\rangle : \mathbf{k}_c\text{-}\mathbf{k}'_c + \mathbf{k}'_p$
	$\lvert a_{1/2}\rangle \xrightarrow{G^0_{d2}} \lvert b_{1/2}\rangle \xrightarrow{G'^+_{c1}} \lvert a_{-1/2}\rangle \xrightarrow{G^0_{p1}} \lvert b_{-1/2}\rangle \xrightarrow{(G^+_{F2})^*} \lvert a_{1/2}\rangle : \mathbf{k}_d\text{-}\mathbf{k}'_c + \mathbf{k}_p$
	$\lvert a_{1/2}\rangle \xrightarrow{G^0_{d2}} \lvert b_{1/2}\rangle \xrightarrow{G'^+_{c1}} \lvert a_{-1/2}\rangle \xrightarrow{G'^0_{p1}} \lvert b_{-1/2}\rangle \xrightarrow{(G^+_{F2})^*} \lvert a_{1/2}\rangle : \mathbf{k}_d\text{-}\mathbf{k}'_c + \mathbf{k}'_p$

$$\rho^{k_1,k_1'}_{\mathrm{P(PMT2)}} = -i\sum_{M=\pm 1/2}\left[\frac{\left|G^0_{\mathrm{d}M}\right|^2}{\Gamma_{a_M a_M} d_2}\left(\frac{G^0_{\mathrm{P}M}}{d_1}+\frac{G'^0_{\mathrm{p}M}}{d_2}\right)+\frac{G^0_{\mathrm{c}M}\left(G^0_{\mathrm{d}M}\right)^*}{d_5 d_1}\left(\frac{G^0_{\mathrm{P}M}}{d_6}+\frac{G'^0_{\mathrm{p}M}}{d_1}\right)\right] \tag{5.5}$$

$$\rho^{k_1,k_1'}_{\mathrm{s(PMT2)}} = -i\sum_{M=\pm 1/2}\left[\frac{G^{\mp}_{\mathrm{c}M}\left(G'^0_{\mathrm{c}M}\right)^*}{\Gamma_{a_M a_{-M}} d_{13}}\left(\frac{G^0_{\mathrm{P}M}}{d_{13}}+\frac{G'^0_{\mathrm{p}M}}{d_{14}}\right)+\frac{G^{\mp}_{\mathrm{c}M}\left(G'^0_{\mathrm{d}M}\right)^*}{d_9 d_{13}}\left(\frac{G^0_{\mathrm{P}M}}{d_{10}}+\frac{G'^0_{\mathrm{p}M}}{d_{13}}\right)\right] \tag{5.6}$$

where $G_i = -\mu_i E_i/\hbar$ $(i = c,\ d,\ p)$ is the Rabi frequency, $d_1 = i\Delta_1 + \Gamma_{b_M a_M}$, $d_2 = i\Delta_2 + \Gamma_{b_M a_M}$, $d_3 = i(\Delta_2 - \Delta_1) + \Gamma_{a_M a_M}$, $d_4 = i(2\Delta_2 - \Delta_1) + \Gamma_{b_M a_M}$, $d_5 = i(\Delta_1 - \Delta_2) + \Gamma_{a_M a_M}$, $d_6 = i(2\Delta_1 - \Delta_2) + \Gamma_{b_M a_M}$, $d_7 = i(\Delta_2 - \Delta_1) + \Gamma_{a_M a_{-M}}$, $d_8 = i(2\Delta_2 - \Delta_1) + \Gamma_{b_{-M} a_M}$, $d_9 = \Gamma_{a_M a_{-M}} + i(2\Delta_2 - \Delta_1)$, $d_{10} = i(2\Delta_1 - \Delta_2) + \Gamma_{b_{-M} a_M}$, $d_{11} =$

$i\Delta_1 + \Gamma_{b_{M-1}a_M}$, $d_{12} = i\Delta_1 + \Gamma_{b_{M+1}a_M}$, $d_{13} = i\Delta_1 + \Gamma_{b_{-M}a_M}$, $d_{14} = i\Delta_2 + \Gamma_{b_{-M}a_M}$, Γ_{ab} and Γ_{ba} are the transverse relaxation rates, and Γ_{aa} is the longitudinal one.

Now we only consider the mutual-dressing effect of coexisting FWM signals. As Figure 5.10 shows, different channels have different dressing strengths. When the rotation angle of the QWP is at 0°, only the transition pathways $|a_{-1/2}\rangle - - |b_{-1/2}\rangle$ and $|a_{1/2}\rangle - - |b_{1/2}\rangle$ are allowed. In this case, the density matrix elements of dressing FWM signals in P-polarization are

$$\rho_{\mathrm{P(PMT1)}}^{k_1,k_1'} = -i\sum_{M=\pm1/2}\left[\frac{\left|G_{\mathrm{c}M}^0\right|^2}{\Gamma_{a_Ma_M}\left(d_1 + \frac{\left|G_{\mathrm{d}M}^0\right|^2}{d_3}\right)}\left(\frac{G_{\mathrm{p}M}^0}{d_1} + \frac{G_{\mathrm{p}M}'^0}{d_2}\right)\right.$$
$$\left. + \frac{\left(G_{\mathrm{c}M}^0\right)^* G_{\mathrm{d}M}^0}{d_3\left(d_2 + \frac{\left|\left(G_{\mathrm{d}M}^0\right)^*\right|^2}{\Gamma_{a_Ma_M}} + \frac{|G_{\mathrm{c}M}^0|^2}{d_5}\right)}\left(\frac{G_{\mathrm{p}M}^0}{d_2} + \frac{G_{\mathrm{p}M}'^0}{d_4}\right)\right] \tag{5.7}$$

$$\rho_{\mathrm{P(PMT2)}}^{k_1,k_1'} = -i\sum_{M=\pm1/2}\left[\frac{\left|G_{\mathrm{d}M}^0\right|^2}{\Gamma'_{a_Ma_M}\left(d_2 + \frac{\left|G_{\mathrm{c}M}^0\right|^2}{d_5}\right)}\left(\frac{G_{\mathrm{p}M}^0}{d_1} + \frac{G_{\mathrm{p}M}'^0}{d_2}\right)\right.$$
$$\left. + \frac{G_{\mathrm{c}M}^0\left(G_{\mathrm{d}M}^0\right)^*}{d_5\left(d_1 + \frac{\left|\left(G_{\mathrm{c}M}^0\right)^*\right|^2}{\Gamma_{a_Ma_M}} + \frac{|G_{\mathrm{d}M}^0|^2}{d_5}\right)}\left(\frac{G_{\mathrm{p}M}^0}{d_6} + \frac{G_{\mathrm{p}M}'^0}{d_1}\right)\right] \tag{5.8}$$

When the rotation angle of the QWP is at 45°, the incident fields can be decomposed into three components with linear, left-circular and right-circular polarizations. As shown in Figure 5.10, there are six considerable transition pathways in this system, $|a_{-1/2}\rangle - - |b_{-3/2}\rangle$, $|a_{-1/2}\rangle - - |b_{-1/2}\rangle$, $|a_{-1/2}\rangle - - |b_{1/2}\rangle$, $|a_{1/2}\rangle - - |b_{-1/2}\rangle$, $|a_{1/2}\rangle - - |b_{1/2}\rangle$, and $|a_{1/2}\rangle - - |b_{3/2}\rangle$. Each transition pathway corresponds to a different dressing field, and the density matrix elements of dressing FWM signals in P-polarization can be expressed as

$$\rho_{\mathrm{P(PMT1)}}^{k_1,k_1'} = -i\sum_{M=\pm1/2}\left[\frac{\left|G_{\mathrm{c}M}^0\right|^2}{\Gamma_{a_Ma_M}(d_1 + |G_{\mathrm{d}M}^0|^2/d_3)}\left(\frac{G_{\mathrm{p}M}^0}{d_1} + \frac{G_{\mathrm{p}M}'^0}{d_2}\right)\right.$$
$$\left. + \frac{\left(G_{\mathrm{c}M}^0\right)^* G_{\mathrm{d}M}^0}{d_3(d_2 + |(G_{\mathrm{d}M}^0)^*|^2/\Gamma_{a_Ma_M} + (|G_{\mathrm{c}M}^-|^2 + |G_{\mathrm{c}M}^+|^2)/d_3)}\left(\frac{G_{\mathrm{p}M}^0}{d_2} + \frac{G_{\mathrm{p}M}'^0}{d_4}\right)\right]$$
$$-i\sum_{M=\pm1/2}\frac{1}{\Gamma_{a_Ma_M}}\left(\frac{\left(G_{\mathrm{c}M}^-\right)^* G_{\mathrm{c}M}^-}{d_{13} + |G_{\mathrm{d}M}^0|^2/d_3} + \frac{(G_{\mathrm{c}M}^+)^* G_{\mathrm{c}M}^+}{d_{12} + |G_{\mathrm{d}M}^0|^2/d_3}\right)\left(\frac{G_{\mathrm{p}M}^0}{d_1} + \frac{G_{\mathrm{p}M}'^0}{d_2}\right) \tag{5.9}$$

$$\rho_{P(PMT2)}^{k_1,k_1'} = -i \sum_{M=\pm1/2} \frac{|G_{dM}^0|^2}{\Gamma_{a_M a_M}(d_2 + (|G_{cM}^-|^2 + |G_{cM}^+|^2)/d_5)} \left(\frac{G_{pM}^0}{d_1} + \frac{G_{pM}'^0}{d_2} \right)$$
$$-i \sum_{M=\pm1/2} \frac{G_{cM}^0 (G_{dM}^0)^*}{d_5(d_1 + (|(G_{cM}^-)^*|^2 + |(G_{cM}^+)^*|^2)/\Gamma_{a_M a_M} + |G_{dM}^0|^2/d_5)} \left(\frac{G_{pM}^0}{d_6} + \frac{G_{pM}'^0}{d_1} \right) \quad (5.10)$$

According to Eq. (5.7), Eq. (5.8), Eq. (5.9), and Eq. (5.10), the dressing FWM signals can be modulated via the polarizations of the incident laser beams. Simultaneously, the dressing effects also depend on the frequency detuning Δ_1 and Δ_2. Suppression and enhancement derived from the dressing effect can be obtained by adjusting the detuning difference Δ ($\Delta = \Delta_1 - \Delta_2$) of input laser beams [20].

5.3.3 Results and Discussions

In order to observe the intensity and the dressing effect of each FWM signal, we set Δ_2 at $-0.3\,\mathrm{cm}^{-1}$ and scan the detuning Δ_1. Figure 5.11(a) and (b) gives the undressed FWM and k_d' dressed FWM signals, which are detected by PMT1. With the changing of Δ_1, an emission peak can be observed in each FWM curve. We can see that the DFWM signal $\mathbf{k}_{s1}$ (normalized intensity $I_1 = 1$) is much stronger than the three NDFWM signals (relative intensities are $I_2 = 0.15 \pm 0.05$, $I_3 = 0.11 \pm 0.04$, and $I_4 = 0.02 \pm 0.006$, respectively), and shows a dip at the resonance position $\Delta_1 = 0$. This is attributed to the resonance absorption of the FWM signal. When the dressing field k_d' is opened, the left peak of signal $\mathbf{k}_{s1}$ is suppressed and the right one is enhanced. On the same condition, other FWM signals are all suppressed. Figure 5.11(c) and (d) gives the coexisting FWM signals $\mathbf{k}_{s1} + \mathbf{k}_{s2}$ and $\mathbf{k}_{s3} + \mathbf{k}_{s4}$, respectively. Compared with that of the single FWM signal $\mathbf{k}_{s1}$, the intensity of the coexisting signal $\mathbf{k}_{s1} + \mathbf{k}_{s2}$ is decreased owing to the mutual-dressing effect. When another dressing field k_d' is turned on, the coexisting signal is further suppressed.

Next, we investigate the interactions among these coexisting FWM signals by modifying the polarizations of the incident fields. On the basis of the results of Figure 5.11, we set Δ_2 at $-0.3\,\mathrm{cm}^{-1}$, Δ_1 at $-0.4\,\mathrm{cm}^{-1}$ and detect the P-polarized FWM components (Figure 5.12 and Figure 5.13). In this case, four FWM signals can be observed simultaneously and the suppressed condition $\Delta_1/m - \Delta_2 = 0$ is satisfied, where m is the modified factor.

Figure 5.12 shows the dependence of the coexisting FWM signal intensity on the rotation angle θ of the HWP, which is set on the path of the laser beam $\mathbf{k}_c'$, while other beams keep horizontal polarization. Figure 5.12(a) and (b) gives the FWM signals detected by PMT1. There are four coexisting FWM signals, namely, $\mathbf{k}_{s1}$, $\mathbf{k}_{s2}$, $\mathbf{k}_{s3}$, and $\mathbf{k}_{s4}$. Beam $\mathbf{k}_c'$ acts as the coupling field for these FWM signals. The dependence of these FWM intensities on θ follows $(\cos\,2\theta)^2$ [24]. In order to explore the interactions among these FWM signals, a different laser beam is blocked in each case. We can see in Figure 5.12(a) that the total signal intensity decreases when the field $\mathbf{k}_c$ is turned off, but increases when the field $\mathbf{k}_d$ is off. In fact, the coupling

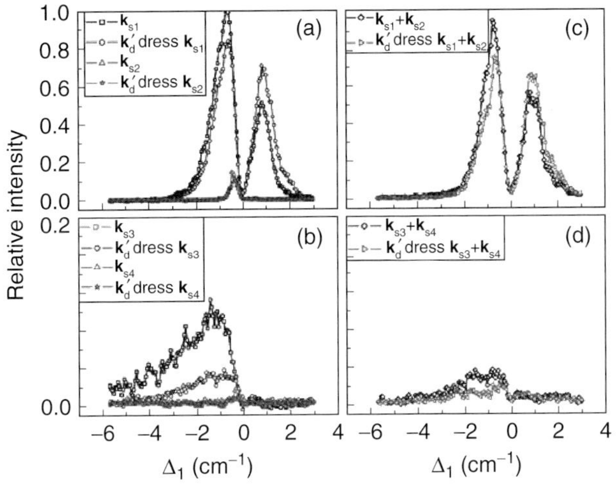

Figure 5.11 The reative intensities of four FWM signals ($\mathbf{k}_{s1}$, $\mathbf{k}_{s2}$, $\mathbf{k}_{s3}$, and $\mathbf{k}_{s4}$) versus Δ_1 with $\Delta_2 = -0.3$ cm^{-1}. (a) Undressed FWM signals $\mathbf{k}_{s1}$, $\mathbf{k}_{s2}$, and $\mathbf{k}'_d$-dressed $\mathbf{k}_{s1}$, $\mathbf{k}_{s2}$. (b) Undressed FWM signals $\mathbf{k}_{s3}$, $\mathbf{k}_{s4}$, and $\mathbf{k}'_d$-dressed $\mathbf{k}_{s3}$, $\mathbf{k}_{s4}$. (c) Coexisting FWM signals $\mathbf{k}_{s1}+\mathbf{k}_{s2}$ and $\mathbf{k}'_d$-dressed $\mathbf{k}_{s1}+\mathbf{k}_{s2}$. (d) Coexisting FWM signals $\mathbf{k}_{s3}+\mathbf{k}_{s4}$ and $\mathbf{k}'_d$-dressed $\mathbf{k}_{s3}+\mathbf{k}_{s4}$. Source: Adapted from Ref. [24].

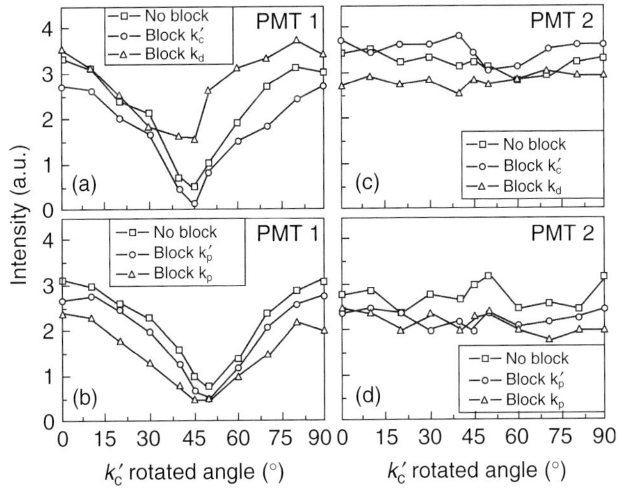

Figure 5.12 Dependence of the FWM signal intensity on the rotation angle of the HWP put on the path of the field $\mathbf{k}'_c$. (a)–(c) The FWM signals when the coupling field $\mathbf{k}_c$ or $\mathbf{k}_d$ is blocked. Squares: six laser beams are all turned on; circles: $\mathbf{k}_c$ is blocked; triangles: $\mathbf{k}_d$ is blocked. (b)–(d) The FWM signals when the probe field $\mathbf{k}_p$ or $\mathbf{k}'_p$ is blocked. Squares: six laser beams are all turned on; circles: $\mathbf{k}'_p$ is blocked; triangles: $\mathbf{k}_p$ is blocked.

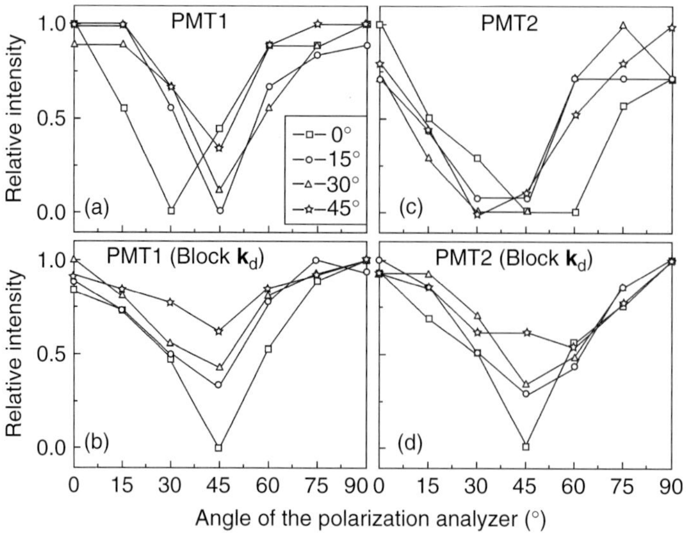

Figure 5.13 Dependence of the relative FWM signal intensity on the rotation angle of the polarization analyzer for four values of the ellipticity of field $\mathbf{k}_c$. (a) and (c) The FWM signals detected by PMT1 and PMT2, respectively. (b) and (d) The FWM signals detected by PMT1 and PMT2 when $\mathbf{k}_d$ is blocked. Squares: $\theta = 0^\circ$ (θ is the polarization angle of $\mathbf{k}_c$); circles: $\theta = 15^\circ$; triangles: $\theta = 30^\circ$; and asterisks: $\theta = 45^\circ$.

field $\mathbf{k}_c$ of signals $\mathbf{k}_{s1}$ and $\mathbf{k}_{s3}$ acts as the dressing field for signals $\mathbf{k}_{s2}$ and $\mathbf{k}_{s4}$. When field $\mathbf{k}_c$ is blocked, FWM signals $\mathbf{k}_{s1}$ and $\mathbf{k}_{s3}$ disappear, but FWM signals $\mathbf{k}_{s2}$ and $\mathbf{k}_{s4}$ become stronger because of the absence of suppression effect of dressing field. However, as mentioned in the preceding text, DFWM signal $\mathbf{k}_{s1}$ is much stronger than NDFWM signals, so the total intensity decreases. On the contrary, when the field $\mathbf{k}_d$ is blocked, the total signal intensity increases. Figure 5.12(b) shows the cases of blocking probe fields $\mathbf{k}_p$ and $\mathbf{k}'_p$, respectively. In these two cases, the total signal intensities are all decreased. This is because the two signals $\mathbf{k}_{s1}$, $\mathbf{k}_{s2}$ (or $\mathbf{k}_{s3}$, $\mathbf{k}_{s4}$) disappear when $\mathbf{k}_p$ (or $\mathbf{k}'_p$) is blocked, but the dressing field does not change. The different phenomenon in each case clearly shows the mutual dressings among coexisting FWM signals. For the signals detected by PMT2, field $\mathbf{k}'_c$ acts as the dressing field. The signal intensities show little change with the rotation angle θ in Figure 5.12(c) and (d). This means the polarization direction of the FWM signals mainly depends on the coupling field. The total signal intensity detected by PMT2 increases when the field $\mathbf{k}_c$ is turned off, but decreases when the field $\mathbf{k}_d$ is off. The result can also be explained by the mutual-dressing effect of coexisting FWM signals.

Then, we measure the ellipticity of the coexisting FWM signals. A QWP was used to modulate the ellipticity of the incident field $\mathbf{k}_c$. In order to detect the polarization states of the FWM signals, a special combination HWP + PBS is used as a polarization analyzer put on the path of the FWM signals. Figure 5.13 illustrates the dependence of the relative FWM signal intensity on the rotation angle of the polarization analyzer. We can see in Figure 5.13(a) and (b) that the oscillation amplitudes of the signals in PMT1 change with the ellipticity of $\mathbf{k}_c$, and this change becomes more obvious when $\mathbf{k}_d$ is blocked (only $\mathbf{k}_{s1}$ and $\mathbf{k}_{s3}$ exist). As mentioned in the preceding text, $\mathbf{k}_c$ is the coupling field for the signals $\mathbf{k}_{s1}$ and $\mathbf{k}_{s3}$. When $\mathbf{k}_c$ is rotated from $\theta = 0^\circ$ to $\theta = 45^\circ$, the polarizations of $\mathbf{k}_{s1}$ and $\mathbf{k}_{s3}$ change from linear polarization to elliptic polarization [24]; thus, the oscillation amplitudes of $\mathbf{k}_{s1}$ and $\mathbf{k}_{s3}$ decrease clearly. However, the polarizations and the oscillation amplitudes of signals $\mathbf{k}_{s2}$ and $\mathbf{k}_{s4}$ do not change under the polarization rotation of the dressing field $\mathbf{k}_c$. Therefore, there is no remarkable decrease in the oscillation amplitude at $\theta = 45^\circ$ when the four FWM signals coexist. For the signals detected by PMT2, field $\mathbf{k}_c$ mainly acts as a dressing field. The curves exhibit little sensitivity to the ellipticity of $\mathbf{k}_c$ (as shown in Figure 5.13(c)). However, when $\mathbf{k}_d$ is blocked, the oscillation amplitude changes with the elasticity of $\mathbf{k}_c$. These results mean the ellipticity of the FWM signals is determined mainly by the coupling field, while the polarization states of the dressing field show little influence on the ellipticity of the FWM signals.

Now we investigate the polarization dependence of the dressing strength of the FWM signals. The polarization of the dressing field keeps linear polarization and the ellipticity of the coupling field (and the FWM signals) is changed by QWP. In order to show different dressing effects, we set Δ_1 at different values and scan Δ_2.

First, when Δ_1 is set at a small value (the suppressed condition $\Delta_1/m - \Delta_2 = 0$ is satisfied), FWM signals are suppressed by the dressing field. Figure 5.14 shows the polarization dependence of the suppressed FWM signals in PMT1 when probe field $\mathbf{k}'_p$ is blocked and coupling field $\mathbf{k}_c$ is modulated by QWP.

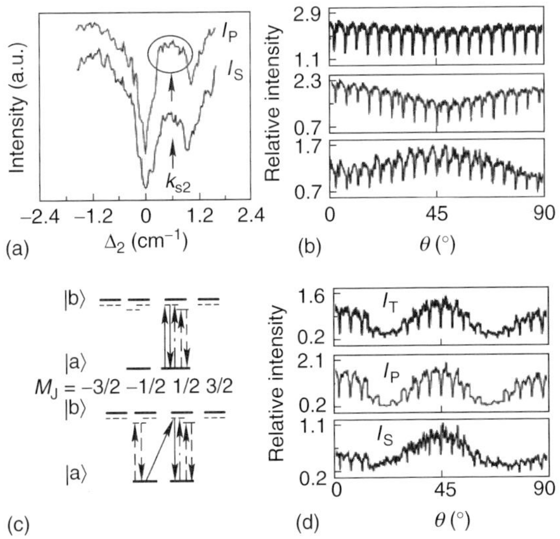

Figure 5.14 Polarization dependence of the suppression of FWM signals versus the rotation angle of QWP. (a) FWM signals when $\mathbf{k}_c$ at 45°. (b) The field $\mathbf{k}_c$ is modulated by QWP. (c) Zeeman sublevel schemes. (d) Fields $\mathbf{k}_c$ and $\mathbf{k}'_c$ are modulated by QWP, simultaneously. Source: Adapted from Ref. [24].

Figure 5.14(a) gives one curve detected at 45°. In such cases, there are two FWM signals $\mathbf{k}_{s1}$ and $\mathbf{k}_{s2}$ dressed by $\mathbf{k}_d$ and $\mathbf{k}'_d$. With the detuning Δ_2 scanned, NDFWM signal $\mathbf{k}_{s2}$ presents an emission peak at $\Delta_2 = \Delta_1$, and DFWM signal $\mathbf{k}_{s1}$ presents a suppression dip at $\Delta_2 = \Delta_1/m$. Figure 5.14(b) shows the curves detected at different polarization angles (from 0° to 90° per 5°). The background represents the signal intensity of the FWM without the dressing field while the dips represent that the signal was suppressed by the dressing field. We can see that the suppression dips become deeper when the polarization angle rotated from 0° to 45°. Such a phenomenon indicates that the dressing field has different dressing strengths for FWM signals with different ellipticity. The result can be explained by the mutual-dressing effect. As discussed in the preceding text, the DFWM signal $\mathbf{k}_{s1}$ can be generated through two balanced transition subsystems $|a_{1/2}\rangle \xrightarrow{G^0_{c2}} |b_{1/2}\rangle \xrightarrow{G'^0_{c2}} |a_{1/2}\rangle \xrightarrow{G^0_{p2}} |b_{1/2}\rangle \xrightarrow{(G^0_{F2})^*} |a_{1/2}\rangle$ and $|a_{-1/2}\rangle \xrightarrow{G^0_{c1}} |b_{-1/2}\rangle \xrightarrow{G'^0_{c1}} |a_{-1/2}\rangle \xrightarrow{G'^0_{p1}} |b_{-1/2}\rangle \xrightarrow{(G^0_{F1})^*} |a_{-1/2}\rangle$ when $\theta = 0°$. Figure 5.14(c) shows the former transition pathway. If only mutual-dressing effect was being considered, the corresponding density matrix element of the signal $\mathbf{k}_{s1}$ can be expressed as

$$\rho_1^{k_1} = \frac{-iG^0_{p2}G^0_{c2}(G^0_{c2})^*}{\Gamma_{a_{1/2}a_{1/2}}(i\Delta_1 + \Gamma_{b_{1/2}a_{1/2}} + |G^0_{d2}|^2/(i(\Delta_1 - \Delta_2) + \Gamma_{a_{1/2}a_{1/2}}))(i\Delta_1 + \Gamma_{b_{1/2}a_{1/2}})} \tag{5.11}$$

When $\theta = 45°$, the two transition pathways generating the signal $\mathbf{k}_{s1}$ are $|a_{-1/2}\rangle \xrightarrow{G^+_{c1}} |b_{1/2}\rangle \xrightarrow{G'^0_{c2}} |a_{1/2}\rangle \xrightarrow{G^0_{p2}} |b_{1/2}\rangle \xrightarrow{(G^+_{F2})^*} |a_{1/2}\rangle$ and $|a_{1/2}\rangle \xrightarrow{G^-_{c2}} |b_{-1/2}\rangle \xrightarrow{G'^0_{c1}} |a_{-1/2}\rangle \xrightarrow{G^0_{p1}} |b_{-1/2}\rangle \xrightarrow{(G^-_{F1})^*} |a_{-1/2}\rangle$. Figure 5.14(c) also shows the first transition pathway. The corresponding density matrix element can be written as

$$\rho_5^{k_1} = \frac{-iG^0_{p2}G^+_{c1}(G^0_{c2})^*}{\Gamma_{a_{1/2}a_{1/2}}(i\Delta_1 + \Gamma_{b_{1/2}a_{-1/2}} + |G^0_{d2}|^2/(i(\Delta_1 - \Delta_2) + \Gamma_{a_{-1/2}a_{-1/2}}))^2} \tag{5.12}$$

Comparing the denominator of Eq. (5.11) with that of Eq. (5.12), the dressing term $|G^0_{d2}|^2/[i(\Delta_1 - \Delta_2) + \Gamma_{a_{\pm1/2}a_{\pm1/2}}]$ exists once in Eq. (5.11), but it is quadratic in Eq. (5.12). We can conclude that the FWM signal is dressed once by G_d at 0°, while it is dressed two times by G_d at 45°. So the dressing efficiency at 45° is higher than that at 0°.

Figure 5.14(d) presents the experimental results when the polarizations of fields $\mathbf{k}_c$ and $\mathbf{k}'_c$ are changed by QWP simultaneously. In this case, the background curve obeys the formula $I_p \propto I(\sin^4\theta + \cos^4\theta)[\sin^4(\theta + \theta_0) + \cos^4(\theta + \theta_0)]$, where θ_0 is the polarization angle difference between $\mathbf{k}_c$ and $\mathbf{k}'_c$. The period of the curve is $\pi/4$ [31]. The field $\mathbf{k}'_c$ is set at 45° polarization angle before field $\mathbf{k}_c$ ($\theta_0 = 45°$). When the rotation angle of QWP is at $\theta = 0°$, the polarization angle of field $\mathbf{k}_c$ is at 0° and that of $\mathbf{k}'_c$ is at 45°. According to their transition pathways (Table 5.2) and Eq. (5.12), the FWM signal is dressed two times by G_d. When QWP rotated to $\theta = 45°$, field $\mathbf{k}_c$ is at 45°45°-polarization, and field $\mathbf{k}'_c$ is at 90°-polarization. The FWM signal is

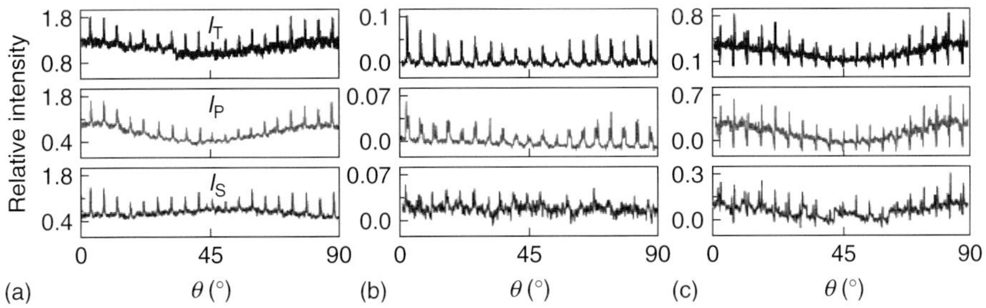

Figure 5.15 Polarization dependence of the enhancement of FWM signals versus the rotation angle of QWP. (a)The field $\mathbf{k}_c$ is modulated by QWP. (b) The field $\mathbf{k}_c$ is blocked, and $\mathbf{k}_d$ is modulated by QWP. (c) The field $\mathbf{k}'_c$ is modulated by QWP.

also dressed two times by G_d. Therefore, there are largest suppression dips both at $\theta = 0^\circ$ and $\theta = 45^\circ$.

Second, when the frequency detuning Δ_1 gets larger, the enhancement condition ($\Delta_1/m - \Delta_2 + G_d = 0$) is satisfied. Figure 5.15(a) gives the variations of the enhancement peaks versus the polarization angle of the field $\mathbf{k}_c$. The enhancement peak is strongly heightened at $\theta = 0^\circ$, but lower clearly at $\theta = 45^\circ$. As discussed in the preceding text, when field $\mathbf{k}_c$ is rotated from 0° to 45°, the polarization of the DFWM signal $\mathbf{k}_{s1}$ changes from linear to elliptic polarization, but the NDFWM signal $\mathbf{k}_{s2}$ keeps linear polarization. In order to show the polarization dependence of the NDFWM signal $\mathbf{k}_{s2}$, the field $\mathbf{k}_c$ is blocked and $\mathbf{k}_d$ is modulated by QWP (as shown in Figure 5.15(b)). The signal $\mathbf{k}_{s2}$ shows an emission peak and its intensity changes with the polarization of the coupling field $\mathbf{k}_d$. However, because the DFWM signal $\mathbf{k}_{s1}$ is much stronger than the NDFWM signal $\mathbf{k}_{s2}$, Figure 5.15(a) mainly presents the variation of the enhancement peak height of $\mathbf{k}_{s1}$. Such a variation can be explained by the self-dressing effect. On the condition of far detuning, both mutual-dressing effect and self-dressing effect should be considered. So, Eq. (5.11) and Eq. (5.12) are corrected into

$$\rho_1^{k_1} = \frac{-iG_{p2}^0 G_{c2}^0 \left(G_{c2}^0\right)^*}{\Gamma_{a_{1/2}a_{1/2}}\left(i\Delta_1 + \Gamma_{b_{1/2}a_{1/2}} + \frac{|G_{d2}^0|^2}{i(\Delta_1-\Delta_2)+\Gamma_{a_{1/2}a_{1/2}}} + \frac{|G_{c2}^0|^2}{\Gamma_{a_{1/2}a_{1/2}}}\right)\left(i\Delta_1 + \Gamma_{b_{1/2}a_{1/2}}\right)} \tag{5.13}$$

$$\rho_5^{k_1} = \frac{-iG_{p2}^0 G_{c1}^+ \left(G_{c2}^0\right)^*}{\Gamma_{a_{1/2}a_{1/2}}\left(i\Delta_1 + \Gamma_{b_{1/2}a_{-1/2}} + \frac{|G_{c1}^0|^2}{\Gamma_{a_{-1/2}a_{-1/2}}} + \frac{|(G_{c1}^0)^+|^2}{\Gamma_{a_{-1/2}a_{-1/2}} + \frac{|G_{d2}^0|^2}{i(-\Delta_2)+\Gamma_{a_{-1/2}b_{-1/2}}}} + \frac{|G_{d2}^0|^2}{i(\Delta_1-\Delta_2)+\Gamma_{a_{-1/2}a_{-1/2}}}\right)^2} \tag{5.14}$$

From Eq. (5.13) and Eq. (5) we can see, when Δ_1 becomes large enough, the value of $|G_{d2}^0|^2/[i(\Delta_1 - \Delta_2) + \Gamma_{a_{\pm 1/2}a_{\pm 1/2}}]$ decreases; thus, the mutual-dressing efficiency of G_d decreases, and the self-dressing field G_c plays a dominant role. As the CG coefficients can be different for different transitions between Zeeman sublevels, if considering multiplied CG coefficients of each transition pathway [25], we can obtain that the Rabi frequency of the self-dressing field G_c at 45° is smaller than that at 0°, and so the self-dressing efficiency at 45° is less than that at 0°.

Finally, when the frequency detuning Δ_1 is adjusted at an intermediate value, the FWM signals show half-enhancement and half-suppression (as shown in Figure 5.15(c)). The dependencies of the suppression dips and enhancement peaks on the polarization are similar to the results obtained under the pure enhancement condition. It indicates that the self-dressing effect also plays an important role in this case.

5.4 Polarized Suppression and Enhancement of SWM

In this section, we discuss theoretical and experimental results about the polarization dependence of the SWM enhancement and suppression as well as the AT splitting in the five-level atomic system of ^{85}Rb. First, we observe the suppression and enhancement as well as the AT splitting in the two coexisting SWM processes. Next, the polarization properties of the enhancement and suppression as well as the AT splitting of two coexisting SWM processes are studied by changing the polarization of the probe beam at a different probe frequency detuning in the multi-Zeeman level atomic system. Different polarization states of the probe beam can select different transitions among multi-Zeeman levels and different dressing strengths. We find that the AT splitting separations get large as the probe beam changes from the linearly polarized state to the circularly polarized one. We also compare the difference of polarization properties of the SWM enhancement and suppression with or without the dressing effect of the probe beam. Owing to the contributing dressing effect of the probe beam, the polarization dependence of the SWM enhancement and suppression shows the inverse variation. At last, we have shown the influence on the SWM enhancement and suppression when the dressing effect of one dressing field is considered or not.

5.4.1 Theoretical Model and Experimental Scheme

The laser beams are aligned spatially as shown in Figure 5.16(a). A weak probe beam E_1 (frequency ω_1, wave vector $\mathbf{k}_1$, frequency detuning Δ_1, and Rabi frequency G_1) is modulated by a QWP and propagates through the atomic medium. Here, we define the frequency detuning $\Delta_i = \Omega_i - \omega_i$, Ω_i is the resonance frequency. Two pump beams E_3 (ω_3, $\mathbf{k}_3$, Δ_3, and G_3), and E_3' (ω_3, k_3', Δ_3, and G_3') propagate in the opposite direction of the beam E_1 with a small angle (~0.3°) between them. Thus, one FWM signal beam E_F with phase-matching condition $\mathbf{k}_F = \mathbf{k}_1 + \mathbf{k}_3 - \mathbf{k}_3'$ is generated by the beams E_1, E_3, and E_3'. Next, we add two strong dressing laser

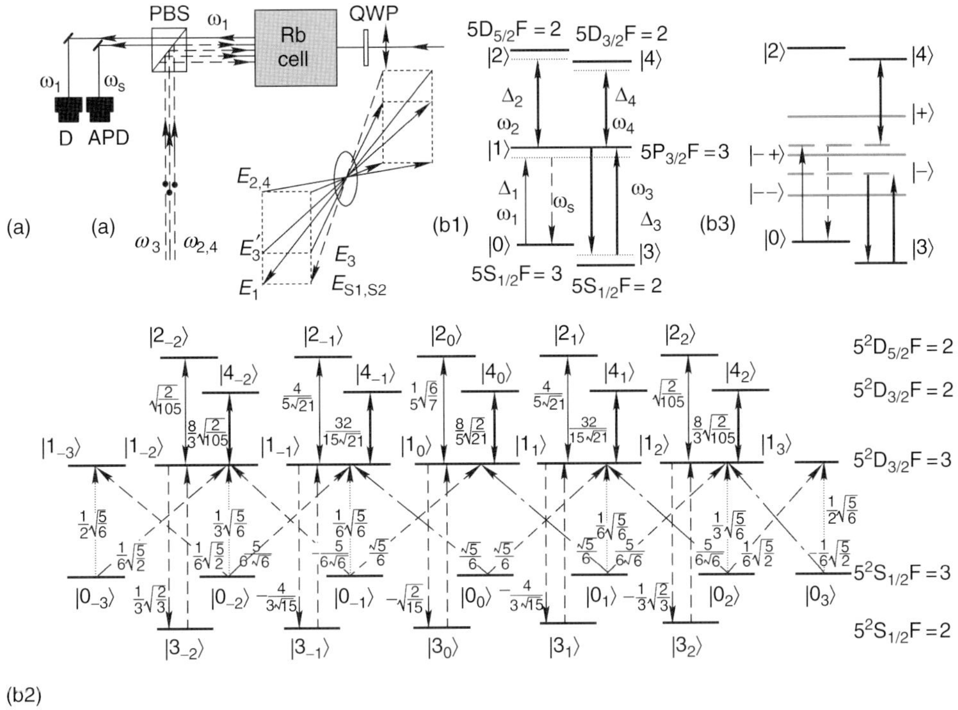

Figure 5.16 (a) The schematic diagram of the experiment. D denotes the photodiode, APD denotes the avalanche diode detector and PBS denotes the polarized beam splitter. Inset: the spatial alignment of the laser beams. (b1) Relevant ^{85}Rb energy levels and (b2) the corresponding Zeeman sublevels with various transition pathways. Double-arrow thin lines: dressing field E_2; double-arrow thick lines: dressing field E_4; linearly (dot lines), left (dot-dash lines) circularly and right (dash lines) circularly polarized probe fields; long-dash lines: the coupling fields E_3 and E_3'. (b3) Dressed-state picture of the experimental atomic system.

beams E_2 (ω_2, $\mathbf{k}_2$, Δ_2, and G_2) and E_4 (ω_4, $\mathbf{k}_4$, Δ_4, and G_4). They propagate in the same direction as the beam E_3 to generate and influence two SWM signals S1 (E_{S1}) and S2 (E_{S2}) with the phase-matching conditions $\mathbf{k}_{S1} = \mathbf{k}_1 + \mathbf{k}_3 - \mathbf{k}_3' + \mathbf{k}_2 - \mathbf{k}_2$ and $\mathbf{k}_{S2} = \mathbf{k}_1 + \mathbf{k}_3 - \mathbf{k}_3' + \mathbf{k}_4 - \mathbf{k}_4$, respectively. The two SWM signals S1 and S2 fall into the $|0\rangle - |1\rangle - |2\rangle$ and $|0\rangle - |1\rangle - |4\rangle$ ladder-type EIT windows, respectively.

For most cases in this work, only the polarization state of the probe beam E_1 is modified by the rotation angle θ of one QWP; therefore, the probe beam can be decomposed into linearly and circularly polarized components while all other fields are kept as linearly polarized (Figure 5.16(a)). In fact, we assume P-polarization direction as the quantization axis and the component perpendicular to it (S-polarization) is decomposed into linearly polarized, left-hand circularly and right-hand circularly polarized parts, while the component parallel to it (P polarization) keeps linear polarization. Figure 5.16(b2) depicts the SWM channels among $5S_{1/2}$ (F = 2, 3), $5P_{3/2}$ (F = 3), $5D_{3/2}$, and $5D_{5/2}$ (F = 2) sublevels. We can see that the dotted lines represent the linear P-polarized probe fields, the dot-dashed lines represent the left-hand circularly polarized probe fields, the dashed lines represent the right-hand circularly polarized probe fields, the long-dashed lines are the linear P-polarized E_3 fields, the double-arrow thin lines are the linear P-polarized E_2 fields, and the double-arrow thick lines are the linear P-polarized E_4 fields.

There exist many transitions to generate the different polarized SWM signals for the different polarized probe fields. In detail, when E_1 is linear P-polarization, according to Figure 5.16(b2) we have four pathways to generate linearly P-polarized SWM signals. We take one pathway, for example, that is, $|0_{-1}\rangle \xrightarrow{G_1^0} |1_{-1}\rangle_{G_1^0 G_i^{00}\pm} \xrightarrow{(G'^{00}_3)^*} |3_{-1}\rangle \xrightarrow{G_3^{00}} |1_{-1}\rangle_{G_1^0 G_i^{00}\pm} \xrightarrow{G_2^{00}} |2_{-1}\rangle \xrightarrow{(G_2^{00})^*} |1_{-1}\rangle_{G_1^0 G_i^{00}\pm}$ $(i=2, 3, 4)$ for E^0_{S1} and $|0_{-1}\rangle \xrightarrow{G_1^0} |1_{-1}\rangle_{G_1^0 G_i^{00}\pm} \xrightarrow{(G'^{00}_3)^*} |3_{-1}\rangle \xrightarrow{G_3^{00}} |1_{-1}\rangle_{G_1^0 G_i^{00}\pm} \xrightarrow{G_4^{00}} |4_{-1}\rangle \xrightarrow{(G_4^{00})^*} |1_{-1}\rangle_{G_1^0 G_i^{00}\pm}$ $(i=2, 3, 4)$ for E^0_{S2}. Here $|0_{-1}\rangle$, $|1_{-1}\rangle$, $|2_{-1}\rangle$, $|3_{-1}\rangle$, and $|4_{-1}\rangle$ represent the Zeeman sublevels $|0\rangle$, $|1\rangle$, $|2\rangle$, $|3\rangle$, and $|4\rangle$ with $M=-1$ (M denotes the magnetic quantum number). $|1_{-1}\rangle_{G_1^0 G_i^{00}\pm}$ $(i=2, 3, 4)$ represents the linear P-polarized dressing fields E_1 (between $|0_{-1}\rangle$ and $|1_{-1}\rangle$), E_2 (between $|1_{-1}\rangle$ and $|2_{-1}\rangle$), E_3 (between $|1_{-1}\rangle$ and $|3_{-1}\rangle$), and E_4 (between $|1_{-1}\rangle$ and $|4_{-1}\rangle$) dress the Zeeman sublevel $|1_{-1}\rangle$ to produce dressed states $|+_{-1}\rangle$ and $|-_{-1}\rangle$. G_1^0, G_2^{00}, G_3^{00}, and G_4^{00} denote Rabi frequencies of linear polarized fields E_1, E_2, E_3, and E_4 in the case of linear polarized probe field. According to those mentioned in the preceding text, we can obtain the corresponding density matrix elements of the linear P-polarized SWM signals as

$$\rho^{(5)0}_{S1_M} = \frac{G_M}{(d_{2_M} d_{3_M} A^3_M)} \quad (M = -2, -1, 1, 2) \tag{5.15}$$

for $E^0_{S1_M}$ and

$$\rho^{(5)0}_{S2_M} = \frac{G'_M}{(d_{3_M} d_{4_M} A^3_M)} \quad (M = -2, -1, 1, 2) \tag{5.16}$$

for $E^0_{S2_M}$, where $G_M = iG^0_{1_M} G^{00}_{3_M} (G'^{00}_{3_M})^* (G^{00}_{2_M})^2$, $G'_M = iG^0_{1_M} G^{00}_{3_M} (G'^{00}_{3_M})^* (G^{00}_{4_M})^2$, $A_M = d_{1_M} + (G^0_{1_M})^2/\Gamma_{0_M 0_M} + (G^{00}_{2_M})^2/d_{2_M} + (G^{00}_{3_M})^2/d_{3_M} + (G^{00}_{4_M})^2/d_{4_M}$, $d_{1_M} = \Gamma_{1_M 0_M} + i\Delta_1$, $d_{2_M} = \Gamma_{2_M 0_M} + i(\Delta_1 + \Delta_2)$, $d_{3_M} = \Gamma_{3_M 0_M} + i(\Delta_1 - \Delta_3)$, and $d_{4_M} = \Gamma_{4_M 0_M} + i(\Delta_1 + \Delta_4)$. The signal intensity I is proportional to $|N\mu_1\rho|^2$. So, the intensities of the linear P-polarized SWM signals are

$$I_{L1} \propto \left| \sum_{M=2,1,1,2} \mu_{1_M 0_M} \rho^{(5)0}_{S1_M} \right|^2 \tag{5.17}$$

for E^0_{S1} and

$$I_{L2} \propto \left| \sum_{M=2,1,1,2} \mu_{1_M 0_M} \rho^{(5)0}_{S2_M} \right|^2 \tag{5.18}$$

for E^0_{S2}. $\mu_{i_M j_M}$ is the transition dipole moments between various Zeeman sublevels. In the SWM process E^0_{S1}, the beams E_1, E_2, and E_3 are both the dressing field of the signal E^0_{S1} and the field to generate it, so we refer to the dressing effects of E_1, E_2, and E_3 as the self-dressing effects. However, the beam E_4 is only the dressing field, so we refer to the dressing effect of E_4 as the external-dressing effect. Similarly, in the SWM process E^0_{S2}, the beams E_1, E_3, and E_4 are the self-dressing fields and the beam E_2 is the external-dressing field.

On the other hand, when E_1 is right-hand or left-hand circularly polarized, we have five pathways to generate right-hand or left-hand circularly polarized SWM signals. We take one pathway, for example, that is, $|0_{-2}\rangle \xrightarrow{G_1^+} |1_{-1}\rangle_{G_1 G_i^{0+}\pm} \xrightarrow{(G_3'^{0+})^*} |3_{-1}\rangle \xrightarrow{G_3^{0+}} |1_{-1}\rangle_{G_1 G_i^{0+}\pm} \xrightarrow{G_2^{0+}} |2_{-1}\rangle \xrightarrow{(G_2^{0+})^*} |1_{-1}\rangle_{G_1 G_i^{0+}\pm}$ $(i=2, 3, 4)$ for E^+_{S1}, $|0_{-2}\rangle \xrightarrow{G_1^+} |1_{-1}\rangle_{G_1 G_i^{0+}\pm} \xrightarrow{(G_3'^{0+})^*} |3_{-1}\rangle \xrightarrow{G_3^{0+}} |1_{-1}\rangle_{G_1 G_i^{0+}\pm} \xrightarrow{G_4^{0+}} |4_{-1}\rangle \xrightarrow{(G_4^{0+})^*} |1_{-1}\rangle_{G_1 G_i^{0+}\pm}$ $(i=2, 3, 4)$ for E^+_{S2}, $|0_2\rangle \xrightarrow{G_1^-} |1_1\rangle_{G_1 G_i^{0-}\pm} \xrightarrow{(G_3'^{0-})^*} |3_1\rangle \xrightarrow{G_3^{0-}} |1_1\rangle_{G_1 G_i^{0-}\pm} \xrightarrow{G_2^{0-}} |2_1\rangle \xrightarrow{(G_2^{0-})^*} |1_1\rangle_{G_1 G_i^{0-}\pm}$ $(i=2, 3, 4)$ for E^-_{S1}, and $|0_2\rangle \xrightarrow{G_1^-} |1_1\rangle_{G_1 G_i^{0-}\pm} \xrightarrow{(G_3'^{0-})^*} |3_1\rangle \xrightarrow{G_3^{0-}} |1_1\rangle_{G_1 G_i^{0-}\pm} \xrightarrow{G_4^{0-}} |4_1\rangle \xrightarrow{(G_4^{0-})^*} |1_1\rangle_{G_1 G_i^{0-}\pm}$ $(i=2, 3, 4)$ for E^-_{S2}. G_1^+ (G_1^-), G_2^{0+} (G_2^{0-}), G_3^{0+} (G_3^{0-}), and G_4^{0+} (G_4^{0-}) denote Rabi frequencies of circularly polarized field E_1 and linear polarized fields E_2, E_3, and E_4 in the case of right-hand (or left-hand) circularly polarized probe field. According to these pathways, we can obtain the corresponding density matrix elements of the right-hand or left-hand circularly polarized SWM signals as

$$\rho^{(5)+}_{S1_M} = \frac{G_M^+}{[d^+_{2_M} d^+_{3_M} (A^+_M)^2]} \quad (M = -2, -1, 0, 1, 2) \tag{5.19}$$

for $E^+_{S1_M}$, and

$$\rho^{(5)+}_{S2_M} = \frac{G'^+_M}{[d^+_{3_M} d^+_{4_M} (A^+_M)^3]} \quad (M = -2, -1, 0, 1, 2) \tag{5.20}$$

for $E^+_{S2_M}$, where $G^+_M = iG^+_{1_{M-1}} G^{0+}_{3_M} (G'^{0+}_{3_M})^* (G^{0+}_{2_M})^2$, $G'^+_M = iG^+_{1_{M-1}} G^{0+}_{3_M} (G'^{0+}_{3_M})^* (G^{0+}_{4_M})^2$, $A^+_M = d^+_{1_M} + (G^+_{1_{M-1}})^2/\Gamma_{0_{M-1}0_{M-1}} + (G^-_{1_{M+1}})^2/\Gamma_{0_{M-1}0_{M+1}} + (G^0_{1_M})^2/\Gamma_{0_M 0_{M-1}} + (G^{0+}_{2_M})^2/d^+_{2_M} + (G^{0+}_{3_M})^2/d^+_{3_M} + (G^{0+}_{4_M})^2/d^+_{4_M}$, $d^+_{1_M} = \Gamma_{1_M 0_{M-1}} + i\Delta_1$, $d^+_{2_M} = \Gamma_{2_M 0_{M-1}} + i(\Delta_1 + \Delta_2)$, $d^+_{3_M} = \Gamma_{3_M 0_{M-1}} + i(\Delta_1 - \Delta_3)$, and $d^+_{4_M} = \Gamma_{4_M 0_{M-1}} + i(\Delta_1 + \Delta_4)$, as well as

$$\rho^{(5)-}_{S1_M} = \frac{G^-_M}{[d^-_{2_M} d^-_{3_M} (A^-_M)^3]} \quad (M = -2, -1, 0, 1, 2) \tag{5.21}$$

for $E^-_{S1_M}$, and

$$\rho^{(5)-}_{S2_M} = \frac{G'^-_M}{[d^-_{3_M} d^-_{4_M} (A^-_M)^3]} \quad (M = -2, -1, 0, 1, 2) \tag{5.22}$$

for $E^-_{S2_M}$, where $G^-_M = iG^-_{1_{M+1}} G^{0-}_{3_M} (G'^{0-}_{3_M})^* (G^{0-}_{2_M})^2$, $G'^-_M = iG^-_{1_{M+1}} G^{0-}_{3_M} (G'^{0-}_{3_M})^* (G^{0-}_{4_M})^2$, $A^-_M = d^-_{1_M} + (G^+_{1_{M-1}})^2/\Gamma_{0_{M-1}0_{M+1}} + (G^-_{1_{M+1}})^2/\Gamma_{0_{M+1}0_{M+1}} + (G^0_{1_M})^2/\Gamma_{0_M 0_{M+1}} + (G^{0-}_{2_M})^2/d^-_{2_M} + (G^{0-}_{3_M})^2/d^-_{3_M} + (G^{0-}_{4_M})^2/d^-_{4_M}$, $d^-_{1_M} = \Gamma_{1_M 0_{M+1}} + i\Delta_1$, $d^-_{2_M} = \Gamma_{2_M 0_{M+1}} + i(\Delta_1 +$

$\Delta_2)$, $d_{3_M}^{-} = \Gamma_{3_M 0_{M+1}} + i(\Delta_1 - \Delta_3)$, and $d_{4_M}^{-} = \Gamma_{4_M 0_{M+1}} + i(\Delta_1 + \Delta_4)$. Thus, the intensities of circularly polarized SWM signals are given by

$$I_{C1} \propto \left| \sum_{M=-2}^{2} \left(\mu_{1_M 0_{M-1}} \rho_{S1_M}^{(5)+} + \mu_{1_M 0_{M+1}} \rho_{S1_M}^{(5)-} \right) \right|^2 \tag{5.23}$$

for $E_{S1}^{\pm}$, and

$$I_{C2} \propto \left| \sum_{M=-2}^{2} \left(\mu_{1_M 0_{M-1}} \rho_{S2_M}^{(5)+} + \mu_{1_M 0_{M+1}} \rho_{S2_M}^{(5)-} \right) \right|^2 \tag{5.24}$$

for $E_{S2}^{\pm}$.

The generated SWM signals contain linearly polarized component I_L ($I_L = I_{L1} + I_{L2}$) and circularly polarized component I_C ($I_C = I_{C1} + I_{C2}$). They associate the detected intensities in the P- and S-polarizations with the equations, namely, the detected intensities of I_P, I_S, and the total intensity I_T in the experiment can be written as $I_P = I_L \cos^2\phi + I_C \sin^2\phi$, $I_S = I_L \sin^2\phi + I_C \cos^2\phi$ (where ϕ is the angle between the P-polarization and the direction of the linearly polarized signal), and $I_T = I_S + I_P = I_L + I_C$. According to Eq. (5.15), Eq. (5.16), Eq. (5.19) Eq. (5.20), Eq. (5.21), and Eq. (5.22), the generated SWM signals can be modulated by the polarizations of the incident laser beams. Simultaneously, the dressing effects on the SWM signals also depend on the frequency detuning. The suppression and enhancement by the dressing effect can be obtained by adjusting the frequency detuning of the probe field.

5.4.2
Polarized Suppression and Enhancement

The experiment was done in an atomic vapor cell of ^{85}Rb. The energy levels $5S_{1/2}$ (F = 2), $5S_{1/2}$ (F = 3), $5P_{3/2}$ (F = 3), $5D_{3/2}$ (F = 2), and $5D_{5/2}$ (F = 2) form a five-level system (Figure 5.16(b1)). The probe laser beam E_1 (wavelength of 780.245 nm, connecting the transition $5S_{1/2}$(F = 3) ($|0\rangle$) to $5P_{3/2}$ (F = 3) ($|1\rangle$)) with the horizontal polarization is from an ECDL. The laser beam E_2 (wavelength of 775.978 nm, connecting the transition $5P_{3/2}$ (F = 3) ($|1\rangle$) to $5D_{5/2}$ (F = 2) ($|2\rangle$)) with the vertical polarization is from the second ECDL. The beams E_3 and E_3' (wavelength of 780.235 nm, connecting the transition $5P_{3/2}$ (F = 3) ($|1\rangle$) to $5S_{1/2}$ (F = 2) ($|3\rangle$)), are split from the third ECDL with the vertical polarization. The laser beam E_4 (wavelength of 776.157 nm, connecting transition $5P_{3/2}$ (F = 3) ($|1\rangle$) to $5D_{3/2}$ (F = 2) ($|4\rangle$)) is from the fourth ECDL with the vertical polarization. Great care was taken in aligning the five laser beams with spatial overlaps and wave vector phase-matching conditions with small angles between them, as indicated in Figure 5.16(a). Two diffracted SWM signals (satisfying phase-matching conditions $\mathbf{k}_{S1} = \mathbf{k}_1 + \mathbf{k}_3 - \mathbf{k}_3' + \mathbf{k}_2 - \mathbf{k}_2$ and $\mathbf{k}_{S2} = \mathbf{k}_1 + \mathbf{k}_3 - \mathbf{k}_3' + \mathbf{k}_4 - \mathbf{k}_4$, respectively) appear in the direction of $E_{S1,S2}$ as shown in the square-box-pattern beam geometry of Figure 5.16(a). The P-polarized SWM component I_P is detected by an avalanche

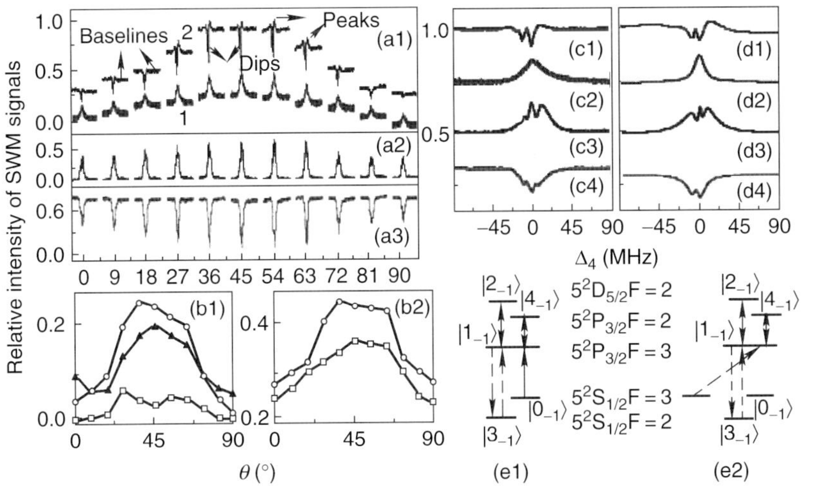

Figure 5.17 (a1) Measured probe transmission signals (the curve 1) and SWM signals (the curve 2) with all laser beams on and (a2) measured SWM signals with the laser beam E_2 blocked by scanning Δ_4 at different designated rotation angles θ (increases from 0° to 90° with the variation step of 9°) when $\Delta_1 = 0$. (a3) Results obtained by subtracting the intensities of baselines of the curve 2 in (a1) and the curve in (a2) from the curve 2 in (a1). (b1) Dependence curves of heights (squares) of peaks and depths (circles) of dips in the curve 2 of (a1), heights (triangles) of enhanced peaks in the curve 1 of (a1). (b2) Dependence curves of heights (squares) of peaks in (a2) and depths (circles) of suppressed dips in (a3). (c1)–(c4) Measured signals versus Δ_4, which are from signals at $\theta = 72°$ in (a1)–(a3). (d1)–(d4) The calculated results corresponding to (c1)–(c4). Two transition pathways of the SWM processes (e1) with the linearly polarized probe field and (e2) with the right-circularly polarized probe field. Source: Adapted from Ref. [36].

photodiode detector. The transmitted probe beam is detected by a photodiode. However, the FWM signal E_F without the EIT window can be neglected.

First, when the dressing effect of the probe beam E_1 is not considered, we study the influence of different probe beam polarization configurations on the probe transmission signals and the SWM signals at three different probe frequency detuning (Figure 5.17 and Figure 5.18). When the probe detuning Δ_1 is set at 0 MHz, the experimental results are shown in Figure 5.17, which presents the probe transmission signals and the SWM signals obtained by scanning the dressing field detuning Δ_4 at different designated rotation angles θ of the QWP. Note that the values in the abscissa axis in Figure 5.17(a1) to (a3) stand for the discrete points of θ. In Figure 5.17(a1), the curve 1 is the probe transmission signals obtained by scanning the dressing field detuning Δ_4 with all beams on at different rotation angles θ. Each peak in the curve 1 shows the EIT phenomenon at $\Delta_4 = -\Delta_1$ induced by the dressing field E_4 according to the first-order density matrix element $\rho_{10}^{(1)} = \sum_{M=-2}^{2} iG_{1M}^{0(\pm)}/[d_{1M}^{(\pm)} + (G_{2M}^{00(\pm)})^2/d_{2M}^{(\pm)} + (G_{4M}^{00(\pm)})^2/d_{4M}^{(\pm)}]$. As the rotation angle θ changes, the heights of the EIT peaks are lowest at $\theta = 0°$, then

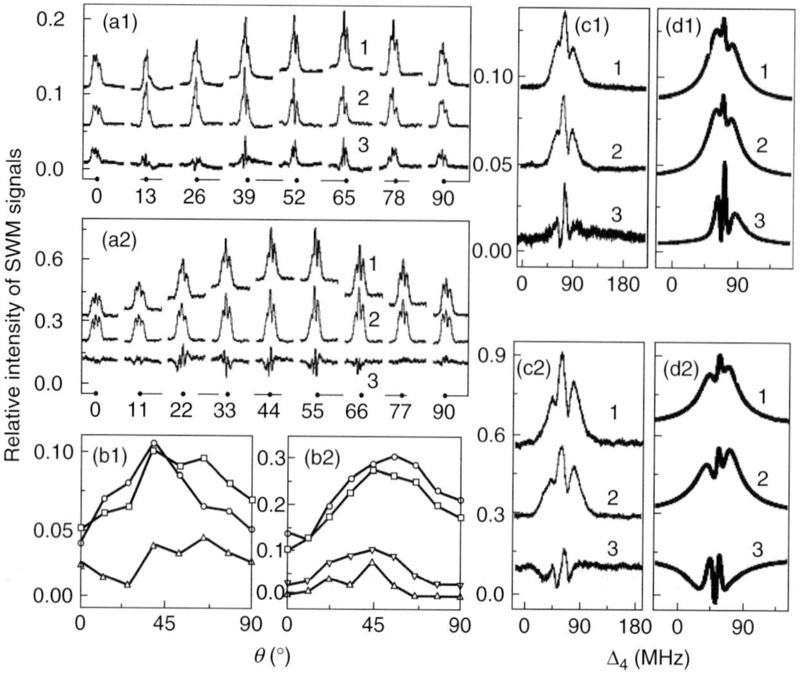

Figure 5.18 (a1) and (a2) Measured SWM signals with all laser beams on (the curve 1) and with the laser beam E_2 blocked (the curve 2) by scanning Δ_4 at different designated rotation angles θ when $\Delta_1 = -65$ MHz and $\Delta_1 = -48$ MHz, respectively. The curves 3 are obtained by subtracting the intensities of baselines of the curve 1 and the curve 2 from the curve 1. (b1) Dependence curves of heights of peaks in the curve 1 (squares), the curve 2 (circles), and the curve 3 (triangles) in (a1). (b2) Dependence curves of heights of peaks in the curve 1 (squares), the curve 2 (circles), and the curve 3 (triangles) and depths of suppressed dips in the curve 3 (inverse triangles) in (a2). (c1) and (c2) Measured signals versus Δ_4, which are from signals at $\theta = 39^\circ$ in (a1) and at $\theta = 44^\circ$ in (a2), respectively. (d1) and (d2) The calculated results corresponding to (c1) and (c2). Source: Adapted from Ref. [36].

reach the highest point at $\theta = 45^\circ$, and finally decrease to the original height when $\theta = 90^\circ$, as shown by the triangles in Figure 5.17(b1). The variation of the EIT peaks can be interpreted as follows. There are combinations of different transition paths between Zeeman sublevels for different polarization schemes as shown in Figure 5.16(b2). So, it is necessary to consider the CG coefficients associated with the various transitions between Zeeman sublevels in all pathways when analyzing the dressing effect. Just as the CG coefficients may be different for different transitions between Zeeman sublevels, the Rabi frequencies are different even with the same laser field. For example, considering the CG coefficients, we can obtain $(|G_4^{0\pm}|)^2/(|G_4^{00}|)^2 = \sum_{M=-2}^{M=2} \mu_{1_M 4_M}^2 / \sum_{M=-2,-1,1,2} \mu_{1_M 4_M}^2 = 35/26$, where the coefficients $\mu_{1_M 4_M}$ $(M = -2, \ldots, 2)$ are $8\sqrt{2}/3\sqrt{105}$, $32/15\sqrt{21}$, $8\sqrt{2}/5\sqrt{21}$, $32/15\sqrt{21}$, and $8\sqrt{2}/3\sqrt{105}$, respectively. So, one can find that the dressing effects of $G_4^{0\pm}$ are greater than those of G_4^{00}, which make the heights of the EIT peaks highest at $\theta = 45^\circ$ and lowest at $\theta = 0^\circ$. On the other hand, the baseline (horizontal constant background of the EIT peak) at $\theta = 0^\circ$ in the curve 1 represents the transparent degree of E_1 caused by E_2 according to the expression $\rho_{10}^{(1)}$. The variation of all the baselines with the rotation angle θ is determined by the function $\sin^4(\theta + \theta_0) + \cos^4(\theta + \theta_0)$, where $\theta_0 (= \pi/4)$ is the initial phase angle of the P-polarization direction.

The curve 2 in Figure 5.17(a1) shows the total SWM signals consisting of the SWM processes S1 and S2 with all beams on by scanning the dressing field detuning Δ_4 at different rotation angles θ. In detail, the baseline at $\theta = 0^\circ$ in the curve 2 represents the SWM signal E_{PS1} (i.e., E_{S1} with $G_{4_M}^{00(\pm)} = 0$) with the self-dressing effect of E_2 and without the external-dressing effect of E_4. The variation in all the baselines with θ also follows the function $\sin^4(\theta + \theta_0) + \cos^4(\theta + \theta_0)$, where $\theta_0 = \pi/4$. Peaks and dips on each baseline in the curve 2 (note that the baseline is regarded as the reference point and its intensity should be subtracted here) are the sum of $E_{S1} - E_{PS1}$ and E_{S2}. On the basis of the expression (5.15), when the condition $\Delta_1 + \Delta_4 = 0$ is satisfied, the SWM signal S1 is suppressed by the external-dressing effect of E_4 and the suppressed SWM signal is denoted as E_{S1}. Therefore, $E_{S1} - E_{PS1}$ represents the pure suppression of the SWM signal S1 caused by the beam E_4. In addition, E_{S2} is the SWM signal S2 with the self-dressing effect of E_4 and the external-dressing effect of E_2.

If we block the beam E_2, only the SWM signal E_{PS2} (i.e., E_{S2} with $G_{2_M}^{00(\pm)} = 0$) is obtained as shown in Figure 5.17(a2). The multiple peaks in the signal E_{PS2} present the AT splitting induced by the self-dressing effects of E_3 and E_4 [37], which can be seen more clearly from Figure 5.17(c3). Figure 5.17(c1) to (c4) shows the detailed structure of measured signals, which are from signals at $\theta = 72^\circ$ in Figure 5.17(a1) to (a3), versus Δ_4. Figure 5.17(d1) to (d4) shows the theoretical results corresponding to Figure 5.17(c1) to (c4). From Figure 5.17(c3), one can see that there are three peaks in the AT splitting. The three peaks correspond, from left to right, to the primarily dressed state $|+\rangle$, the secondarily dressed states $|-+\rangle$ and $|--\rangle$, respectively, as shown in Figure 5.16(b3). The two primarily dressed states $|\pm\rangle$ are caused by the beam E_4. They can be written as $|\pm\rangle = \sin\alpha_{1\pm}|1\rangle + \cos\alpha_{1\pm}|4\rangle$. The secondarily dressed states are caused by the beams E_1 and E_3. But only the beam E_3 is considered as the beam E_1 is weak. So, the secondarily dressed states can be given by $|-\pm\rangle = \sin\alpha_{2\pm}|-\rangle + \cos\alpha_{2\pm}|3\rangle$, where $\tan\alpha_{1\pm} = -b_{1\pm}/G_4^{00(\pm)}$, $\tan\alpha_{2\pm} = -b_{2\pm}/G_3^{00(\pm)}$, $b_{1\pm} = \Delta_4 - \lambda_\pm^{(3)}$, and $b_{2\pm} = \Delta_3 - \lambda_-^{(3)} - \lambda_{-\pm}^{(3)}$. One can obtain the eigenvalues $\lambda_\pm^{(3)} = (\Delta_4 \pm \sqrt{\Delta_4^2 + 4|G_4^{00(\pm)}|^2})/2$ (measured from level $|1\rangle$) of $|\pm\rangle$, and $\lambda_{-\pm}^{(3)} = (\Delta_3' \pm \sqrt{\Delta_3'^2 + 4|G_3^{00(\pm)}|^2})/2$ (measured from level $|-\rangle$) of $|-\pm\rangle$, where $\Delta_3' = -\Delta_3 - \lambda_-^{(3)}$. The AT splitting separation Δ_a (satisfying $\Delta_a = \lambda_{-+}^{(3)} - \lambda_{--}^{(3)} \approx 2|G_3^{00(\pm)}|$) between the right and middle peaks mainly results from the beam E_3. The AT splitting separation Δ_b (satisfying $\Delta_b = \lambda_+^{(3)} - \lambda_{-+}^{(3)} \approx 2|G_4^{00(\pm)}|$) between the middle and left peaks mainly results from the beam E_4. Now, let us consider the polarization dependence of the AT splitting separation on the rotation angle θ. As shown in Figure 5.17(a2), when θ changes from 0° to 45°, the AT splitting separations Δ_a and Δ_b get large. This is determined by the self-dressing efficiencies of E_3 and E_4. When the dressing effect of the beam E_1 is not considered, Eq. (5.16), Eq. (5.20), and Eq. (5.22) can be written in the uniform expression as $\rho^{(5)0(\pm)}_{S2_M} = G'^{(\pm)}_M/\{d^{(\pm)}_{3_M} d^{(\pm)}_{4_M}[d^{(\pm)}_{1_M} + (G^{00(\pm)}_{3_M})^2/d^{(\pm)}_{3_M} + (G^{00(\pm)}_{4_M})^2/d^{(\pm)}_{4_M}]^3\}$ as the beam E_2 is blocked. As discussed in the preceding text, the dressing efficiencies of E_3 and E_4 at $\theta = 45^\circ$ are stronger than those at $\theta = 0^\circ$ by comparing $|G_4^{0\pm}|$ (or $|G_3^{0\pm}|$) with $|G_4^{00}|$ (or $|G_3^{00}|$). Therefore, the AT splitting separations ($\Delta_a \approx 2|G_3^{0\pm}|$ and $\Delta_b \approx 2|G_4^{0\pm}|$) at $\theta = 45^\circ$ are larger than those ($\Delta_a \approx 2|G_3^{00}|$ and $\Delta_b \approx 2|G_4^{00}|$) at $\theta = 0^\circ$.

Figure 5.17(a3) is obtained by subtracting the intensities of its baselines from the curve 2 in Figure 5.17(a1) and then subtracting the curve in Figure 5.17(a2) from the result in the first subtraction. In Figure 5.17(a3), the depth of each dip represents the interaction between the SWM processes S1 and S2, which includes the sum of $E_{S1} - E_{PS1}$ and $E_{S2} - E_{PS2}$. According to latter, when the condition $\Delta_1 + \Delta_2 = 0$ is satisfied, the SWM signal S2 is suppressed by the external-dressing effect of E_2 and the suppressed SWM signal is denoted as E_{S2}. Therefore, $E_{S2} - E_{PS2}$ is the pure suppression of the SWM signal S2 caused by the external-dressing effect of E_2.

Now, let us consider the polarization dependence of SWM signals on the rotation angle θ as shown in the curve 2 of Figure 5.17(a1). We divide the curve 2 into two parts, peaks representing E_{S2} and dips representing E_{S1}. Let us first consider the variation of peaks as squares shown in Figure 5.17(b1). When $\theta = 0^\circ$, E_{PS2} is small, as shown in Figure 5.17(a2), owing to the external-dressing effect of E_2, E_{S2} is suppressed close to zero (i.e., $E_{S2} \approx 0$) [22]. This can be confirmed from the fact that peaks do not arise at $\theta = 0^\circ$ in the curve 2 of Figure 5.17(a1). When θ changes from 0° to 45°, the heights of the peaks in the curve 2 gradually get larger. This results from the increase of E_{PS2} which basically follows the well-expected classical polarization spectroscopy [6] as shown in Figure 5.17(a2). With respect to the variation of dips, as shown by the circles in Figure 5.17(b1), it is because the polarization variation transforms the energy distribution among the various transition paths, for example, linearly polarized transitions to circularly polarized ones. As different SWM transition paths are dressed by different transition dressing fields, the total dressing strength is generally different. In detail, Figure 5.17(e1) shows one linearly polarized transition pathway $|0_{-1}\rangle \xrightarrow{G_1^0} |1_{-1}\rangle \xrightarrow{(G_3'^{00})^*} |3_{-1}\rangle \xrightarrow{G_3^{00}} |1_{-1}\rangle \xrightarrow{G_2^{00}} |2_{-1}\rangle \xrightarrow{(G_2^{00})^*} |1_{-1}\rangle$ corresponding to the case of $\theta = 0^\circ$ which is the subsystem from Figure 5.16(b2). When the dressing effect of the probe beam E_1 is not considered, the corresponding fifth-order density matrix element can be written as

$$\rho^{(5)0}_{S1} = \frac{iG^0_{1_{-1}} G^{00}_{3_{-1}} (G'^{00}_{3_{-1}})^* (G^{00}_{2_{-1}})^2}{\{[d_{1_{-1}} + (G^{00}_{2_{-1}})^2/d_{2_{-1}} + (G^{00}_{3_{-1}})^2/d_{3_{-1}} + (G^{00}_{4_{-1}})^2/d_{4_{-1}}]^3 d_{2_{-1}} d_{3_{-1}}\}} \quad (5.25)$$

By comparison, Figure 5.17(e2) shows one right-hand circularly polarized transition pathway $|0_{-2}\rangle \xrightarrow{G_1^+} |1_{-1}\rangle \xrightarrow{(G_3'^{0+})*} |3_{-1}\rangle \xrightarrow{G_3^{0+}} |1_{-1}\rangle \xrightarrow{G_2^{0+}} |2_{-1}\rangle \xrightarrow{(G_2^{0+})*} |1_{-1}\rangle$ corresponding to the case of $\theta = 45^\circ$. The corresponding fifth-order density matrix element can be written as

$$\rho_{S1}^{(5)+} = \frac{iG_{1\_2}^{+} G_{3\_1}^{0+} (G_{3-1}^{\prime 0+})^{*} (G_{2\_1}^{0+})^{2}}{\{[d_{1\_1}^{+} + (G_{2\_1}^{0+})^{2}/d_{2\_1}^{+} + (G_{3\_1}^{0+})^{2}/d_{3\_1}^{+} + (G_{4\_1}^{0+})^{2}/d_{4\_1}^{+}]^{3} d_{2\_1}^{+} d_{3\_1}^{+}\}} \tag{5.26}$$

As discussed in the preceding text, the suppression on the signal S1, which induces the dips in the curve 2 of Figure 5.17(a1), is mainly caused by the external-dressing effect of E_4. Comparing Eq. (5.25) with Eq. (5.26), one can find that the dressing efficiency of E_4 at $\theta = 45^{\circ}$ is stronger than that at $\theta = 0^{\circ}$ due to $(|G_4^{0+}|)^2/(|G_4^{00}|)^2 = 35/26$. Therefore, the depths of the dips get large when θ changes from 0° to 45° as shown by the circles in Figure 5.17(b1). In Figure 5.17(b2), the circles depict the variation of dips in Figure 5.17(a3), which give the total amount of the mutual suppression between the SWM signals S1 and S2 versus the rotation angle θ. Its variation rule is the same as that of the dips in the curve 2 of Figure 5.17(a1).

When Δ_1 is set at -65 and -48 MHz, the other two groups of experimental results are shown in Figure 5.18(a1) and (a2), respectively. The curves 1–3 in Figure 5.18(a1) and (a2) are obtained in the same way as the curve 2 in Figure 5.17(a1) and the curves in Figure 5.17(a2) and (a3).

The curve 1 in Figure 5.18(a1) and (a2) shows total SWM signals with all beams on at different the rotation angle θ. First, we observe the polarization property of the AT splitting of the SWM signal. When the rotation angle θ changes from 0° to 45°, the AT splitting separations Δ_a and Δ_b get large as shown by the curve 2 in Figure 5.18(a1) or (a2). This also results from the different dressing efficiencies of E_3 and E_4 at difference the rotation angle θ as discussed in Figure 5.17(a2). Next, let us consider the enhancement and suppression of SWM signals. Here, as Δ_1 is far from the resonant position, $E_{S2} - E_{PS2}$ caused by the external-dressing effect of E_2 is small, that is, E_{S2} and is almost equal to E_{PS2}. Therefore, the peaks of the curve 3 in Figure 5.18(a1) mainly represent $E_{S1} - E_{PS1,}$ which shows the pure enhancement. The enhancement condition can be obtained as $\Delta_4 = (|G_{4M}^{00(\pm)}|^2 - \Delta_1^2)/\Delta_1$ according to Eq. (5.25) and Eq. (5.26). Similarly, the peaks and dips of the curve 3 in Figure 5.18(a2) also mainly represent $E_{S1} - E_{PS1}$, but here it shows the case of half-suppression and half-enhancement due to the different Δ_1. The variation rules of the triangles in Figure 5.18(b1) and (b2) are similar to those of the circles in Figure 5.17(b2). It also indicates that the dressing efficiency of E_4 at $\theta = 45^{\circ}$ is stronger than that at $\theta = 0^{\circ}$ owing to different SWM transition paths dressed by different transition dressing fields in these two cases. Figure 5.18(c1) and (c2) shows the detailed structure of measured SWM signals versus Δ_4, which are from signals at $\theta = 39^{\circ}$ in Figure 5.18(a1) and at $\theta = 44^{\circ}$ in Figure 5.18(a2), respectively. Figure 5.18(d1) and (d2) shows the corresponding theoretical results, which are in good agreement with Figure 5.18(c1) and (c2).

We can see that the variations in the enhanced or suppressed signals (the curve 2 of Figure 5.17(a1), the curve 3 of Figure 5.18(a1) and (a2)) with θ are more obvious than those of the AT splitting separations (the curve of Figure 5.17(a2), the curve 2 of Figure 5.18(a1) and (a2)). It can be well understood for the following reason. The enhanced or suppressed signals obtained by scanning the external-dressing field frequency detuning directly represent the dressing effect, which shows quantum signals. Its variations with θ are dramatic. And the AT splitting signals obtained by scanning the probe field frequency detuning contain both the classical SWM signal and the dressing effect (quantum component). In the AT splitting signals, the variations in the AT splitting separations (quantum component) with θ are not sensitive owing to the existence of the classical SWM signal.

Next, when the intensity of the beam E_1 becomes large and its dressing effect is also considered, we perform the two groups of experiments similar to Figure 5.17 and Figure 5.18 at $\Delta_1 = -20$ MHz. The first group of experimental results is shown in Figure 5.19(a), where the dressing effect of the beam E_2 is not considered as its intensity is decreased here. The curve 1 in Figure 5.19(a) shows total SWM signals with all beams on at different the rotation angle θ. In the curve 2 of Figure 5.19(a), we show the SWM signal E_{PS2} and the AT splitting about it. When θ changes from 0° to 45°, the AT splitting separations Δ_a and Δ_b get large owing to the

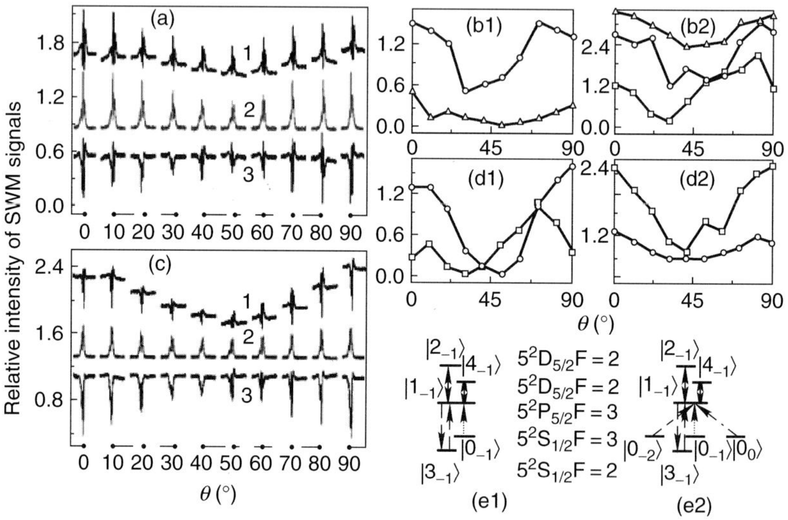

Figure 5.19 (a) and (c) Measured SWM signals with all laser beams on (the curve 1) and with the laser beam E_2 blocked (the curve 2) by scanning Δ_4 at different designated rotation angles θ when $\Delta_1 = -20$ MHz for different E_2 intensities, (a) 5 mW and (c) 55 mW. Intensities of other laser beams are 40 mW (E_1), 44 mW ($E_3 + E_3'$), and 70 mW (E_4). The curves 3 are obtained by subtracting the intensities of baselines of the curve 1 and the curve 2 from the curve 1. (b1) Dependence curves of heights of peaks (circles) and depths of dips (triangles) in the curve 1 of (a). (b2) Dependence curves of heights of peaks in the curve 2 (triangles) of (a), Heights of enhanced peaks (squares) and depths of suppressed dips (circles) in the curve 3 of (a). (d1) Dependence curves of heights of enhanced peaks (squares) and depths of suppressed dips (circles) in the curve 1 of (c). (d2) Dependence curves of heights of peaks (circles) in the curve 2 of (c) and depths of suppressed dips (squares) in the curve 3 of (c). Two transition pathways of the SWM processes (e1) with the linearly polarized probe field and (e2) with the left- and right-circularly polarized probe fields.

different dressing efficiencies of E_1, E_3, and E_4 at difference the rotation angle θ. Here, because the dressing effect of E_1 is also considered, the AT splitting separation Δ_a satisfies $\Delta_a = \lambda_{-+}^{(3)} - \lambda_{--}^{(3)} \approx 2\sqrt{|G_1^{00(\pm)}|^2 + |G_3^{00(\pm)}|^2}$, which results from the beams E_1 and E_3. However, the AT splitting separation Δ_b still satisfies $\Delta_b = \lambda_{+}^{(3)} - \lambda_{-+}^{(3)} \approx 2|G_4^{00(\pm)}|$. On the other hand, the enhancement and suppression of SWM signals is also observed. The peaks and dips in the curve 3 of Figure 5.19(a) mainly represent the signal $E_{S1} - E_{PS1}$, which shows the case of half-suppression and half-enhancement. The variation rule of the signal intensities along with the rotation angle θ in the curve 3 of Figure 5.19(a) is opposite to that in the curve 3 of Figure 5.18(a2). Specifically, when θ changes from 0° to 45°, the signal intensities in the curve 3 of Figure 5.19(a) get small, but those in the curve 3 of Figure 5.18(a2) gradually get large. The squares and circles in Figure 5.19(b2) give dependencies of the heights of enhancement peaks and the depths of suppression dips on θ, respectively, which denote that the dressing efficiency of the field E_4 decreases as E_1 changes from the linearly polarized state to the circularly polarized one. These opposite behaviors can be accounted for by considering the dressing effect of the beam E_1. According to the transition pathways in Figure 5.19(e1) and (e2), Eq. (5.25) and Eq. (5.26) can be replaced by

$$\rho_{S1}^{(5)0} = \frac{G_{-1}}{[(A_{-1})^3 d_{2_{-1}} d_{3_{-1}}]} \tag{5.27}$$

$$\rho_{S1}^{(5)+} = \frac{G_{-1}^{+}}{[(A_{-1}^{+})^3 d_{2_{-1}}^{+} d_{3_{-1}}^{+}]} \tag{5.28}$$

where $G_{-1} = iG_{1_{-1}}^{0} G_{3_{-1}}^{00} (G_{3_{-1}}^{\prime 00})^{*} (G_{2_{-1}}^{00})^2$, $A_{-1} = d_{1_1} + (G_{1_{-1}}^{0})^2/\Gamma_{0_{-1}0_{-1}} + (G_{3_{-1}}^{00})^2/d_{3_{-1}} + (G_{4_{-1}}^{00})^2/d_{4_{-1}}$, $G_{-1}^{+} = iG_{1_{-2}}^{+} G_{3_{-1}}^{0+} (G_{3_{-1}}^{\prime 0+})^{*} (G_{2_{-1}}^{0+})^2$, and $A_{-1}^{+} = d_{1_{-1}}^{+} + (G_{1_{-2}}^{+})^2/\Gamma_{0_{-2}0_{-2}} + (G_{1_0}^{-})^2/\Gamma_{0_{-2}0_0} + (G_{1_{-1}}^{0})^2/\Gamma_{0_{-1}0_{-2}} + (G_{3_{-1}}^{0+})^2/d_{3_{-1}}^{+} + (G_{4_{-1}}^{0+})^2/d_{4_{-1}}^{+}$.

According to Eq. (5.27) and Eq. (5.28), the enhancement and suppression of the SWM E_{S1} is mainly caused by the beam E_4. But the beam E_1 restrains the dressing effect of E_4 at certain frequencies owing to the resonance dressing effect of E_1. From Eq. (5.27) and Eq. (5.28), we can obtain the enhancement condition as $\Delta_1 + \Delta_4' \pm G_{1_M}^{0(\pm)} \pm G_{4_M}^{00(\pm)} = 0$ and the suppression condition as $\Delta_1 + \Delta_4' \pm G_{1_M}^{0(\pm)} = 0$. Here, Δ_4' is the frequency detuning of E_4 relative to the dressed states created by E_1. As discussed in the preceding text, considering the CG coefficients, we can obtain $(|G_4^{0\pm}|)^2/(|G_4^{00}|)^2 = 35/26$. Similarly, as for the probe field E_1, we can also obtain the proportional relation $(|G_1^{\pm}|)^2/(|G_1^{0}|)^2 = 5$ from Figure 5.16(b2). Compared with E_4, Rabi frequency of E_1 increases more remarkably with the variation of θ from 0° to 45°. This results in the suppression on the dressing effect of $G_4^{0\pm}$ caused by $G_1^{\pm}$, which is stronger than that on the dressing effect of G_4^{00} caused by G_1^{0}. Thus, the effective dressing efficiency of the beam E_4 is lower at 45° than at 0°. In addition, in the curve 1 of Figure 5.19(a), the variation in the baselines (horizontal constant background) with the rotation angle θ is determined by the function $\sin^4(\theta + \theta_0) + \cos^4(\theta + \theta_0)$, where $\theta_0 = 0$.

Comparing with Figure 5.19(a), we only increase the intensity of the beam E_2 in order to observe the influence of the dressing effect of E_2 on measured SWM signals in Figure 5.19(c). Compared with the curve 1 in Figure 5.19(a), it can be seen that the peak of the signal at $\theta = 0^\circ$ get smaller in the curve 1 of Figure 5.19(c). This is because the signal E_{S2} in the curve 1 of Figure 5.19(c) is weaker than the one in the curve 1 of Figure 5.19(a) owing to the external-dressing effect of E_2. This experimental result well confirms the dressed-state analysis on measured SWM signals. Figure 5.19(d1) shows dependence curves of heights of enhanced peaks (squares) and depths of suppressed dips (circles) in the curve 1 of (c). In addition, as shown by the squares in Figure 5.19(d2). Figure 5.19(b1) shows dependence curves of heights of peaks (circles) and depths of dips (triangles) in the curve 1 of (a). The variation rule of the signal intensities along with the rotation angle θ is the same as that in Figure 5.19(b2). It also indicates that the effective dressing efficiency of E_4 is lower at 45° than at 0° after considering the dressing effect of the beam E_1.

5.4.3
Conclusion

We have discussed the polarization properties of the suppression and enhancement as well as the AT splitting of two coexisting SWM processes when the polarization of the probe beam is changed. First, we observe the suppression and enhancement of two coexisting SWM signals as well as the AT splitting of the SWM signals in detail. Next, we study the variations in the suppression and enhancement as well as the AT splitting of two coexisting SWM processes along with the polarization of the probe beam. Without the dressing effect of the probe beam E_1, the amplitudes of the suppression and enhancement of SWM signals increase as the rotation angle θ of QWP changes from 0° to 45° at different probe frequency detuning. It is caused by different dressing strengths for the linear polarized and circularly polarized SWM signals. In the case of the dressing effect of E_1, the dependence curves of the amplitudes of the suppression and enhancement decrease with the variation of θ from 0° to 45°, which is because the suppression caused by E_1 on the dressing effect of E_4 gradually strengthens. The experimental results are in good agreement with those from the dressed-state analysis involving all relevant Zeeman sublevels. Such studies provide detailed physical mechanisms to control and optimize the efficiencies of the MWM processes in multi-level atomic systems.

Problems

5.1 Answer the question why the enhancement/suppression of MWM processes can be controlled by the polarization dressing states among Zeeman sublevels.

5.2 Explain why the polarization dressing states of Zeeman sublevels can be used to control the intensities and spatial splitting of MWM signal beams.

References

1. Magno, W.C., Prandini, R.B., Nussenzveig, P., and Vianna, S.S. (2001) Four-wave mixing with Rydberg levels in rubidium vapor: observation of interference fringes. *Phys. Rev. A*, **63**, 063406.

2. Chapple, P.B., Baldwin, K.G.H., and Bachor, H.A. (1989) Interference between competing quantum-mechanical pathways for four-wave mixing. *J. Opt. Soc. Am. B*, **6**, 180–183.
3. Du, S., Wen, J., Rubin, M.H., and Yin, G.Y. (2007) Four-wave mixing and biphoton generation in a two-level system. *Phys. Rev. Lett.*, **98**, 053601.
4. Du, S., Oh, E., Wen, J., and Rubin, M.H. (2007) Four-wave mixing in three-level systems: interference and entanglement. *Phys. Rev. A*, **76**, 013803.
5. McCormack, E.F. and Sarajlic, E. (2001) Polarization effects in quantum coherences probed by two-color, resonant four-wave mixing in the time domain. *Phys. Rev. A*, **63**, 023406.
6. Zhu, C.J., Senin, A., Lu, Z.H., Gao, J., Xiao, Y., and Eden, J. (2005) Polarization of signal wave radiation generated by parametric four-wave mixing in rubidium vapor: ultrafast (~150-fs) and nanosecond time scale excitation. *Phys. Rev. A*, **72**, 023811.
7. Museur, L., Olivero, C., Riedel, D., and Castex, M.C. (2000) Polarization properties of coherent VUV light at 125 nm generated by sum-frequency four-wave mixing in mercury. *Appl. Phys. B*, **70**, 499–503.
8. Tsukiyama, K. (1996) Parametric four-wave mixing in Kr. *J. Phys. B: At. Mol. Opt. Phys.*, **29**, L345.
9. Ishii, J., Ogi, Y., Tanaka, Y., and Tsukiyama, K. (1996) Observation of the two-photon resonant parametric four-wave mixing in the NO $C^2\Pi$ ($v = 0$) state. *Opt. Commun.*, **132**, 316–320.
10. Wang, B., Han, Y., Xiao, J., Yang, X., Xie, C., Wang, H., and Xiao, M. (2006) Multi-dark-state resonances in cold multi-Zeeman-sublevel atoms. *Opt. Lett.*, **31**, 3647–3649.
11. Harris, S.E. (1997) Electromagnetically induced transparency. *Phys. Today*, **50**, 36.
12. Gea-Banacloche, J., Li, Y., Jin, S., and Xiao, M. (1995) Electromagnetically induced transparency in ladder-type inhomogeneously broadened media: theory and experiment. *Phys. Rev. A*, **51**, 576–584.
13. Ling, H.Y., Li, Y.-Q., and Xiao, M. (1996) Coherent population trapping and electromagnetically induced transparency in multi-Zeeman-sublevel atoms. *Phys. Rev. A*, **53**, 1014–1026.
14. Yuratich, M. and Hanna, D. (1976) Nonlinear atomic susceptibilities. *J. Phys. B: At. Mol. Phys.*, **9**, 729.
15. Zheng, H., Zhang, Y., Khadka, U., Wang, R., Li, C., Nie, Z., and Xiao, M. (2009) Modulating the multi-wave mixing processes via the polarizable dark states. *Opt. Express*, **17**, 15468–15480.
16. Li, S., Wang, B., Yang, X., Han, Y., Wang, H., Xiao, M., and Peng, K.C. (2006) Controlled polarization rotation of an optical field in multi-Zeeman-sublevel atoms. *Phys. Rev. A*, **74**, 033821.
17. Akulshin, A., Barreiro, S., and Lezama, A. (1998) Electromagnetically induced absorption and transparency due to resonant two-field excitation of quasidegenerate levels in Rb vapor. *Phys. Rev. A*, **57**, 2996.
18. Wang, Z., Zhang, Y., Chen, H., Wu, Z., Fu, Y., and Zheng, H. (2011) Enhancement and suppression of two coexisting six-wave-mixing processes. *Phys. Rev. A*, **84**, 013804.
19. Lipsich, A., Barreiro, S., Akulshin, A.M., and Lezama, A. (2000) Absorption spectra of driven degenerate two-level atomic systems. *Phys. Rev. A*, **61**, 053803.
20. Li, C., Zheng, H., Zhang, Y., Nie, Z., Song, J., and Xiao, M. (2009) Observation of enhancement and suppression in four-wave mixing processes. *Appl. Phys. Lett.*, **95**, 041103.
21. Wang, R., Du, Y., Zhang, Y., Zheng, H., Nie, Z., Li, C., Li, Y., Song, J., Xiao, M., and Balakin, A.V. (2009) Polarization spectroscopy of dressed four-wave mixing in a three-level atomic system. *J. Opt. Soc. Am. B*, **26**, 1710–1719.
22. Zhang, Y., Anderson, B., Brown, A.W., and Xiao, M. (2007) Competition between two four-wave mixing channels via atomic coherence. *Appl. Phys. Lett.*, **91**, 061113.
23. Steck, D.A. (2001) Rubidium 87 D Line Data.
24. Wang, R., Wu, Z., Sang, S., Song, J., Zheng, H., Wang, Z., Li, C., and Zhang, Y. (2011) Coexisting polarized

four-wave mixing processes in a two-level atomic system. *J. Opt. Soc. Am. B*, **28**, 2940–2946.

25. Li, C., Zhang, Y., Nie, Z., Du, Y., Wang, R., Song, J., and Xiao, M. (2010) Controlling enhancement and suppression of four-wave mixing via polarized light. *Phys. Rev. A*, **81**, 033801.
26. Zuo, Z., Sun, J., Liu, X., Jiang, Q., Fu, G., Wu, L.-A., and Fu, P. (2006) Generalized n-photon resonant 2n-wave mixing in an (n + 1)-level system with phase-conjugate geometry. *Phys. Rev. Lett.*, **97**, 193904.
27. Bennink, R.S., Wong, V., Marino, A.M., Aronstein, D.L., Boyd, R.W., Stroud, C. Jr.,, Lukishova, S., and Gauthier, D.J. (2002) Honeycomb pattern formation by laser-beam filamentation in atomic sodium vapor. *Phys. Rev. Lett.*, **88**, 113901.
28. Stentz, A.J., Kauranen, M., Maki, J.J., Agrawal, G.P., and Boyd, R.W. (1992) Induced focusing and spatial wave breaking from cross-phase modulation in a self-defocusing medium. *Opt. Lett.*, **17**, 19–21.
29. Agrawal, G.P. (1990) Induced focusing of optical beams in self-defocusing nonlinear media. *Phys. Rev. Lett.*, **64**, 2487–2490.
30. Hickmann, J.M., Gomes, A., and de Araújo, C.B. (1992) Observation of spatial cross-phase modulation effects in a self-defocusing nonlinear medium. *Phys. Rev. Lett.*, **68**, 3547–3550.
31. Zhang, Y., Zuo, C., Zheng, H., Li, C., Nie, Z., Song, J., Chang, H., and Xiao, M. (2009) Controlled spatial beam splitter using four-wave-mixing images. *Phys. Rev. A*, **80**, 055804.
32. Wang, Z., Zhang, Y., Li, P., Sang, S., Yuan, C., Zheng, H., Li, C., and Xiao, M. (2011) Observation of polarization-controlled spatial splitting of four-wave mixing in a three-level atomic system. *Appl. Phys. B*, **104**, 633–638.
33. Li, C., Zhang, Y., Nie, Z., Zheng, H., Zuo, C., Du, Y., Song, J., Lu, K., and Gan, C. (2010) Controlled multi-wave mixing via interacting dark states in a five-level system. *Opt. Commun.*, **283**, 2918–2928.
34. Nie, Z., Zheng, H., Li, P., Yang, Y., Zhang, Y., and Xiao, M. (2008) Interacting multiwave mixing in a five-level atomic system. *Phys. Rev. A*, **77**, 063829.
35. Zhang, Y. and Xiao, M. (2007) Generalized dressed and doubly-dressed multi-wave mixing. *Opt. Express*, **15**, 7182–7189.
36. Wang, Z., Zheng, H., Chen, H., Li, P., Sang, S., Lan, H., Li, C., and Zhang, Y. (2012) Polarized suppression and enhancement of six-wave mixing in electromagnetically induced transparency window. *IEEE J. Quantum Electron.*, **48**, 669–677.
37. Zhang, Y., Li, P., Zheng, H., Wang, Z., Chen, H., Li, C., Zhang, R., and Xiao, M. (2011) Observation of Autler-Townes splitting in six-wave mixing. *Opt. Express*, **19**, 7769–7777.

6
Exploring Nonclassical Properties of MWM Process

Highlights
Ultra-narrow two-photon fluorescence signal, strong three-photon correlation as well as vacuum induced Rabi splitting and optical bistability in a coupled atom-cavity system are experimentally observed, which are essential for various quantum information processing protocols.

Many nonclassical properties can be explored in multi-dressed multi-level atomic systems, which will bring significant influence on the MWM signals. In this chapter, the formation of two-photon fluorescence process, three-photon correlation and dark-state-assisted atom-cavity system of MWM and their various effects are discussed. First, we discuss the probe transmission, FWM and fluorescence signals with dressing effects, in which an ultranarrow two-photon fluorescence signal is obtained with line width much narrower than the Doppler-free EIT window. Then, the phase-controlled switching from pure dark state to pure bright state in the FWM and fluorescence channels is introduced. Afterwards, we focus on the dynamics of the images of the probe, the generated FWM and fluorescence signals when the angles between pumping fields and dressing fields are changed. This is followed by three-photon correlation between two coexisting FWM signals and the probe signal in a coherently prepared rubidium vapor. And at last, VRS and OB of the MWM process in a collective atomic-cavity coupling system is discussed in detail. Understanding the high efficiencies in generating ultranarrow fluorescence, strongly correlated or anticorrelated, can be useful in controlling them for various applications, such as coherent quantum control, nonlinear optical spectroscopy, precise measurements, and quantum information processing.

6.1
Opening Fluorescence and FWM via Dual EIT Windows

Enhanced FWM process has been experimentally studied in multi-level atomic systems [1–3]. Therein, AT splitting has been investigated [4]. The keys in such enhanced nonlinear optical processes include greatly reduced linear absorption of

Quantum Control of Multi-Wave Mixing, First Edition. Yanpeng Zhang, Feng Wen, and Min Xiao.

the generated optical fields due to EIT. A great deal of attention has been paid to observe and understand EIT and related effects in multi-level atomic systems interacting with electromagnetic fields [5–9]. Under EIT conditions, not only can the FWM [10, 11] signals be allowed to transmit through the atomic medium but also the fluorescence induced by spontaneous emission can be generated [12–14]. In the fluorescence spectrum, the AT splitting has also been studied in lithium molecules [13].

In this section, we compare the probe transmission, FWM, and fluorescence signals under dressing effects. The ultranarrow two-photon fluorescence signal, which is sheared twice by the EIT window, is obtained in the ladder or Y-type atomic system. Such fluorescence with very high coherence and monochromaticity can be potentially applied in metrology, long-distance quantum communication, and quantum correlation. Also, we investigate the interaction effect between two ladder subsystems on the measured signals. Moreover, the amplitude of the signals can be effectively controlled by the incident beam intensity and frequency detuning.

6.1.1
Theory and Experimental Scheme

The experiment is carried out in atomic vapor of ^{85}Rb. The energy levels $5S_{1/2}(F=3)$ ($|0\rangle$), $5P_{3/2}$ ($|1\rangle$), $5D_{5/2}$ ($|2\rangle$), and $5D_{3/2}$ ($|3\rangle$) form the four-level Y-type system as shown in Figure 6.1(a). The vapor cell is set at 60 °C. A weak laser field E_1 (ω_1, $\mathbf{k}_1$ Rabi frequency G_1, wavelength 780 nm) from an ECDL with horizontal polarization probes the lower transition $|0\rangle$ to $|1\rangle$. Two strong coupling fields, E_2 (ω_2, $\mathbf{k}_2$, G_2) and E_2' (ω_2, $\mathbf{k}_2'$, and G_2'), each with a vertical polarization and wavelength 775.98 nm, are split from a continuous wave (cw) Ti:sapphire laser, driving the upper transition $|1\rangle$ to $|2\rangle$. Another two strong coupling fields, E_3 (ω_3, $\mathbf{k}_3$, and G_3) and E_3' (ω_3, $\mathbf{k}_3'$, and G_3'), each with a vertical polarization and wavelength 776.16 nm, are split from another ECDL laser, driving the upper transition $|1\rangle$ to $|3\rangle$. The laser beams are aligned spatially in the square pattern (Figure 6.1(b)). Four coupling beams propagate through Rb vapor in the same direction with small angles (~0.3°) between one

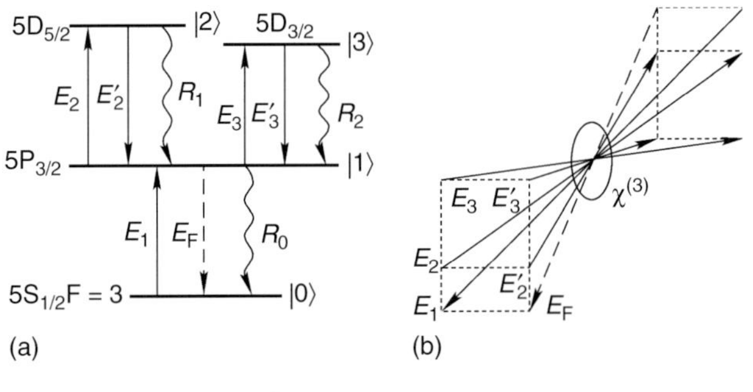

Figure 6.1 (a) Relevant ^{85}Rb four-level Y-type atomic system. (b) Spatial beam geometry used in the experiment.

another, and the probe field E_1 propagates in the direction opposite from them. Thus, two ladder-type subsystems ($|0\rangle - |1\rangle - |2\rangle$ and $|0\rangle - |1\rangle - |3\rangle$) form and two EIT windows appear in the probe transmission spectrum [4], which is measured by a silicon photodiode. Also, two FWM signals, F1 (generated by E_1, E_2, and E_2') and F2 (generated by E_1, E_3, and E_3'), can occur simultaneously within the two EIT windows satisfying the phase-matching condition $\mathbf{k}_{F1} = \mathbf{k}_1 + \mathbf{k}_2 - \mathbf{k}_2'$ or $\mathbf{k}_{F2} = \mathbf{k}_1 - \mathbf{k}_3' + \mathbf{k}_3$, which are detected by an avalanche photo diode (APD). Besides, three types of fluorescence due to spontaneous emission are detected in our experiment: the decay of photons from $|1\rangle$ to $|0\rangle$ will generate single-photon fluorescence signal R_0 (wavelength 780 nm), and the decay of photons from $|2\rangle$ and $|3\rangle$ to $|1\rangle$ will generate two-photon fluorescence signals R_1 and R_2 (wavelength 776 nm), as shown in Figure 6.1(a). Compared with FWM, the fluorescence signals are nondirectional and detected by another photodiode. The two-photon fluorescence signals can also fall into the EIT windows and form the Doppler-free sharp peaks.

In general, the expressions of the density matrix element $\rho_{10}^{(3)}$ related to the FWM processes and density matrix elements $\rho_{11}^{(2)}$, $\rho_{22}^{(4)}$, and $\rho_{33}^{(4)}$ related to the fluorescence processes can be obtained by solving the coupled density matrix equations. First, with E_1, E_2, and E_2' on (E_3 and E_3' blocked), the simple FWM process F1 via the Liouville pathway $\rho_{00}^{(0)} \xrightarrow{E_1} \rho_{10}^{(1)} \xrightarrow{E_2} \rho_{20}^{(2)} \xrightarrow{(E_2')^*} \rho_{10}^{(3)}$ gives $\rho_{F1}^{(3)} = -iG_a e^{i\mathbf{k}_{F1}\cdot\mathbf{r}}/(d_1^2 d_2)$, where $G_a = G_1 G_2 (G_2')^*$, $d_1 = \Gamma_{10} + i\Delta_1$, $d_2 = \Gamma_{20} + i(\Delta_1 + \Delta_2)$ with frequency detuning $\Delta_i = \Omega_i - \omega_i$ (Ω_i is the atomic resonance frequency for the corresponding transition), and Γ_{ij} is the transverse relaxation rate between states $|i\rangle$ and $|j\rangle$. Considering the self-dressing effect of E_2 and E_2', the energy level $|1\rangle$ could be split into two dressed states $|+\rangle$ and $|-\rangle$, which can be described via the Liouville pathway $\rho_{00}^{(0)} \xrightarrow{E_1} \rho_{\pm 0}^{(1)} \xrightarrow{E_2} \rho_{20}^{(2)} \xrightarrow{(E_2')^*} \rho_{\pm 0}^{(3)}$, and the expression of $\rho_{F1}^{(3)}$ can be modified as $\rho_{F1SD}^{(3)} = -iG_a e^{i\mathbf{k}_{F1}\cdot\mathbf{r}}/[(d_1 + |G_2|^2/d_2)^2 d_2]$. Next, when the coupling field E_3 is turned on, its dressing effect should also be considered, resulting in the doubly-dressed FWM process as $\rho_{F1DD}^{(3)} = -iG_a e^{i\mathbf{k}_{F1}\cdot\mathbf{r}}/[(d_1 + |G_2|^2/d_2 + |G_3|^2/d_3)^2 d_2]$, where $d_3 = \Gamma_{30} + i(\Delta_1 + \Delta_3)$. Similarly, with E_1, E_3, and E_3' on, the other FWM process F2 via the Liouville pathway $\rho_{00}^{(0)} \xrightarrow{E_1} \rho_{10}^{(1)} \xrightarrow{E_3} \rho_{30}^{(2)} \xrightarrow{(E_3')^*} \rho_{10}^{(3)}$ gives $\rho_{F2}^{(3)} = -iG_b e^{i\mathbf{k}_{F2}\cdot\mathbf{r}}/(d_1^2 d_3)$ with $G_b = G_1 G_3 (G_3')^*$, which can also be singly or doubly dressed.

For the fluorescence signals, the single-photon fluorescence R_0 is described by $\rho_{00}^{(0)} \xrightarrow{E_1} \rho_{10}^{(1)} \xrightarrow{(E_1)^*} \rho_{11}^{(2)}$. By solving the coupled density matrix equations, the expression of the density matrix element $\rho_{11}^{(2)}$ can be obtained as $\rho_{11}^{(2)} = -|G_1|^2/(d_1\Gamma_{11})$, the amplitude squared of which is proportional to the intensity of R_0. When the beams E_2 and E_3 are turned on, the fluorescence process R_0 can be doubly dressed, and the expression of $\rho_{11}^{(2)}$ should be modified as $\rho_{11DD}^{(2)} = -|G_1|^2/[\Gamma_{11}(d_1 + |G_2|^2/d_2 + |G_3|^2/d_3)]$. For two-photon fluorescence R_1, via the Liouville pathway $\rho_{00}^{(0)} \xrightarrow{E_1} \rho_{10}^{(1)} \xrightarrow{E_2} \rho_{20}^{(2)} \xrightarrow{(E_1)^*} \rho_{21}^{(3)} \xrightarrow{(E_2)^*} \rho_{22}^{(4)}$, we can obtain the density matrix element $\rho_{22}^{(4)}$ as $\rho_{22}^{(4)} = |G_1|^2|G_2|^2/(\Gamma_{22}d_1 d_2 d_4)$, where $d_4 = \Gamma_{21} + i\Delta_2$, the amplitude squared of which is proportional to the intensity of R_1. Considering the self-dressing effect of E_2 and E_2', the expression of $\rho_{22}^{(4)}$ should be modified as $\rho_{22SD}^{(4)} = |G_1|^2 G_2{}^2/[\Gamma_{22}d_1 d_4(d_2 + |G_2|^2/d_1)]$. When the beam E_3 is turned on, its dressing effect should be additionally considered and the doubly-dressed fluorescence process R_1 is given as $\rho_{22DD}^{(4)} = |G_1|^2|G_2|^2/[\Gamma_{22}d_1d_4(d_2 + |G_2|^2/(d_1 + |G_3|^2/d_3))]$. Correspondingly, the fluorescence signal R_2 is related with the density matrix element $\rho_{33}^{(4)} = |G_1|^2|G_3|^2/(\Gamma_{33}d_1d_3d_5)$ where $d_5 = \Gamma_{31} + i\Delta_3$. Similarly with R_1, the singly-dressed R_2 process is given as $\rho_{33SD}^{(4)} = |G_1|^2|G_3|^2/[\Gamma_{33}d_1d_5(d_3 + |G_3|^2/d_1)]$ and doubly-dressed R_2 process is given as $\rho_{33DD}^{(4)} = |G_1|^2|G_3|^2/[\Gamma_{33}d_1d_5(d_3 + |G_3|^2/(d_1 + |G_2|^2/d_2))]$.

6.1.2
Fluorescence and FWM via EIT Windows

Figure 6.2 presents the measured FWM and fluorescence signals opened via the EIT windows versus the probe detuning Δ_1. First, with all five beams on and $P_1 = 1$ mW, the measured curves under different Δ_2 are depicted in Figure 6.2(a1) to (a4). In the probe transmission, two EIT windows arise at $\Delta_1 = -\Delta_2$ and $\Delta_1 = -\Delta_3$ (labeled as P_1 and P_2) within the Doppler absorption background. The FWM signals F1 and F2 fall into the two EIT windows, respectively. As Δ_2 changes, the EIT window P_1 and FWM signal F1 shift from left to right, and overlap with the fixed EIT window P_2 and FWM signal F2 at $\Delta_2 = \Delta_3 = 0$ (Figure 6.2(a3)). For the fluorescence signals, the big background curve represents the single-photon fluorescence R_0 ($\rho_{11}^{(2)}$). The other two small sharp peaks on it are the two-photon fluorescence R_1 ($\rho_{22}^{(4)}$) and R_2 ($\rho_{33}^{(4)}$) falling into the EIT windows. The intensity of fluorescence R_1 reaches its maximum at the resonant point ($\Delta_2 = 0$ MHz, Figure 6.2(a3)) and decreases gradually as Δ_2 is set farther from resonance (from Figure 6.2(a3) to (a1)), due to the effect of single-photon term d_1 and d_4 in $\rho_{22}^{(4)}$.

Next, when P_1 increases to 6 mW, the dressed signals are shown in Figure 6.2(b1) to (b4), where E_3' is blocked. Here, the FWM signal F2 disappears owing to the absence of E_3' and the fixed FWM signal F1 shows AT splitting from the self-dressing effect of $E_2(E_2')$, denoted by the dressing term $|G_2|^2/d_2$ in $\rho_{F1DD}^{(3)}$. When the EIT windows P_1 and P_2 overlap in Figure 6.2(b3), F1 is suppressed obviously due to the dressing effect of E_3, denoted by the dressing term $|G_3|^2/d_3$. In the fluorescence signals, two suppression dips that are lower than the background curve containing sharp peaks R_1 and R_2 can be observed. These suppression dips are induced by the dressing effects of $E_2(E_2')$ and E_3 individually, described by the terms $|G_2|^2/d_2$ and $|G_3|^2/d_3$ in $\rho_{11DD}^{(2)}$. Such dressing effects can be modulated by E_1 according to $\rho_{11DD}^{(2)}$; therefore, when P_1 is small (Figure 6.2(a)), the dips are invisible. When Δ_3 moves far from resonance, the dip at $\Delta_1 = -\Delta_3$ gradually becomes shallower, corresponding with the weakened EIT. On the other hand, the R_2 peak gets slightly higher with Δ_3 increasing (from Figure 6.2(b2) to (b1)), which is entirely different from the case in Figure 6.2(a) where R_1 peak weakens with Δ_2 increasing. This is due to the fact that as the power of E_1 increases, its dressing effect on the two-photon term d_3 in $\rho_{33}^{(4)}$ should be considered, expressed as $d_3 + |G_1|^2/d_5$. Hence, the fluorescence peaks R_1 and R_2 are suppressed around the resonant point. More importantly, the fluorescence peak R_2 within the dip can

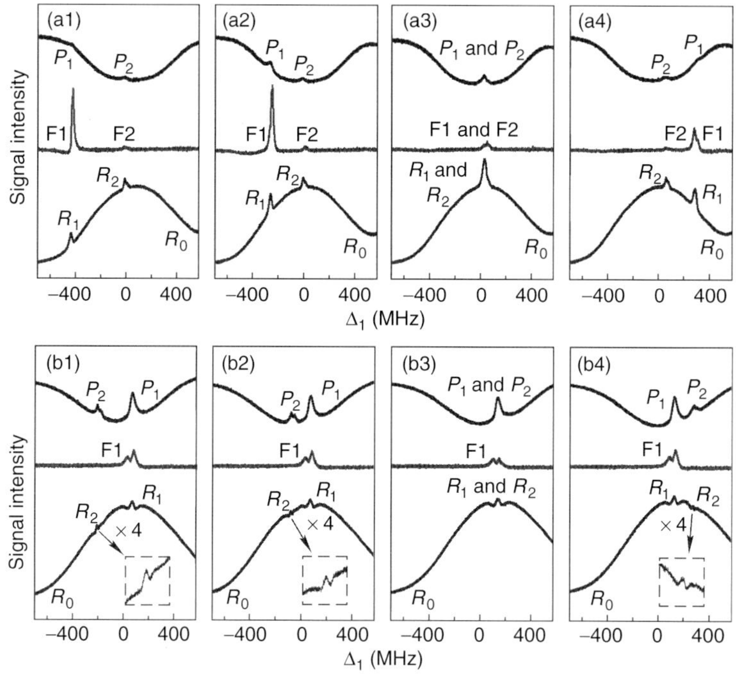

Figure 6.2 Measured probe transmission (the upper curves), FWM (the middle curves), and fluorescence (the bottom curves) versus the probe detuning Δ_1. (a) $P_1 = 1$ mW and Δ_2 set as (a1) 400 MHz, (a2) 250 MHz, (a3) 0 MHz, and (a4) −250 MHz, respectively, with $\Delta_3 = 0$ MHz. (b) $P_1 = 6$ mW and Δ_3 set as (b1) 250 MHz, (b2) 100 MHz, (b3) −100 MHz, and (b4) −250 MHz, respectively, with $\Delta_2 = -100$ MHz and E_3' blocked. The other experimental parameters are $P_2 = 12$ mW, $P_2' = 8$ mW, $P_3 = 22$ mW, $P_3' = 16$ mW. Source: Adapted from Ref. [15].

be seen with ultranarrow line width (about 10 MHz), which is much narrower than the EIT windows (about 50 MHz). Such high-resolution fluorescence is generated because it has been sheared twice by the EIT window P_2. First, owing to the two-photon dressing term $|G_3|^2/d_3$, the single-photon term d_1 of $\rho_{11}^{(2)}$ is clipped out, resulting in the suppression dip on R_0, which is of the same width as the EIT window P_2. Further, such a clipped single-photon term, as a factor of the two-photon term d_2, participates in the process of two-photon fluorescence R_2 ($\rho_{33}^{(4)}$), which also stays in the EIT window P_2. Therefore, the fluorescence is sheared for the second time to an ultranarrow peak. Similarly, in the $|0\rangle - |1\rangle - |2\rangle$ subsystem, the fluorescence R_1 can also be sheared twice by the EIT window P_1 and, therefore, with ultranarrow line width.

In the following, we observe the dressing effects and the interplay between two ladder subsystems by scanning coupling detuning Δ_2. When E_1, E_2, and E_2' are turned on, we first study the singly-dressed signals in the $|0\rangle - |1\rangle - |2\rangle$ subsystem, as depicted in Figure 6.3(a1) to (a3) with different Δ_1 values. In the probe transmission signals, the heights of the baselines (the horizontal background)

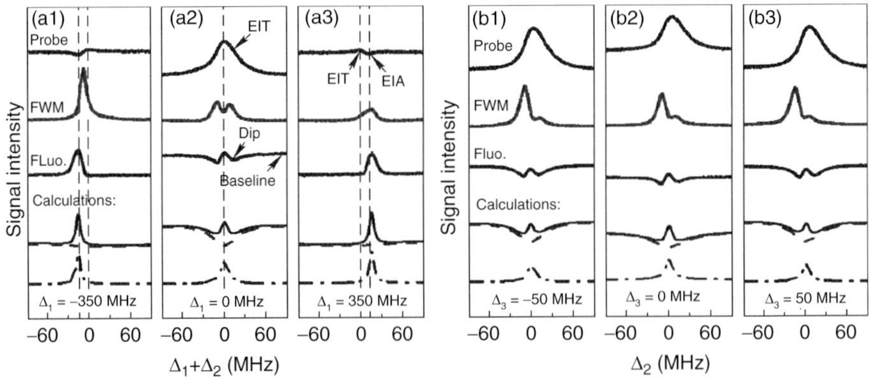

Figure 6.3 Measured probe transmission, FWM and fluorescence versus Δ_2 with (a) E_3 and E_3' blocked and Δ_1 set as (a1) −350 MHz, (a2) 0 MHz, and (a3) 350 MHz. (b) E_3' blocked and Δ_3 set as (b1) −50 MHz, (b2) 0 MHz, and (b3) 50 MHz with fixed $\Delta_1 = 0$. The other parameters are $P_1 = 4$ mW, $P_2 = 14$ mW, $P_2' = 7$ mW, and $P_3 = 25$ mW. The calculations of fluorescence are presented below the measured curves. Especially, the dash lines represent the calculated R_0, while the dash-dot lines represent the calculated R_1.

represent the Doppler absorption background at corresponding Δ_1. EIT peaks higher than the baselines appear at $\Delta_1 + \Delta_2 = 0$ and EIA dips lower than baselines satisfying $\Delta_1 + \Delta_2 = |G_2|^2/\Delta_1$ can be observed at large detuning (Figure 6.3(a1) and (a3)). The FWM signal F1 with a double-peak structure can be observed in Figure 6.3(a2) because of the synthesis of the self-dressing effect and the two-photon emission feature. Such a double-peak structure becomes unobvious in Figure 6.3(a1) and (a3), as the dressing effect weakens at the large frequency detuning. For the detected fluorescence signal, the baselines with suppression dips represent fluorescence R_0, and the peaks within the dips are fluorescence R_1. With Δ_1 set far from resonance, the dip gradually becomes shallower, and eventually almost invisible at large detuning (Figure 6.3(a1) and (a3)), as corresponds to the weakening process of EIT. On the contrary, the peak gets stronger with Δ_1 increasing, for R_1 is suppressed around the resonant point according to the dressed term $d_2 + |G_2|^2/d_1$ in $\rho_{22SD}^{(4)}$. Moreover, the suppression dips just fall into the EIT windows, satisfying $\Delta_1 + \Delta_2 = 0$, and fluorescence peaks at large detuning are in alignment with EIA satisfying $\Delta_1 + \Delta_2 = |G_2|^2/\Delta_1$. In order to demonstrate the phenomena more clearly, we present the corresponding calculations of fluorescence signals below the experimental curves. Especially, the calculated R_0 (the dashed lines) and R_1 (the dashed-dot lines) are displayed separately. Such theoretical calculations confirm our experimental analysis stated in the preceding text.

When E_3 is also turned on, the $|0\rangle - |1\rangle - |2\rangle$ and $|0\rangle - |1\rangle - |3\rangle$ subsystems will interplay with each other, resulting in some interesting phenomena as shown in Figure 6.3(b1)–(b3). In the probe transmission, the profile of the baselines reveals the EIT induced by E_3 at $\Delta_3 = -\Delta_1 = 0$ and the peaks over each baseline are EIT induced by $E_2(E_2')$. It is obvious that the EIT induced by $E_2(E_2')$ is smaller at $\Delta_3 = 0$ (Figure 6.3(b2)) than Δ_3 is detuned (Figure 6.3(b1) and (b3)). This is the result of the strong cascade-dressing interaction between $E_2(E_2')$ and E_3 near $\Delta_1 = 0$, according to the doubly-dressed element $\rho^{(1)}_{10DD} = iG_1/(d_1 + |G_2|^2/d_2 + |G_3|^2/d_3)$. The FWM signal F1 shows a double-peak structure induced by $E_2(E_2')$, and is additionally suppressed by external-dressing field E_3 when $\Delta_3 = -\Delta_1$ (Figure 6.3(b2)). The fluorescence R_0 is also suppressed by E_3 as depicted by the lower fluorescence baseline in Figure 6.3(b2), in addition to suppression effect of $E_2(E_2')$. Corresponding to the EIT window, the suppression dip induced by $E_2(E_2')$ is shallower at $\Delta_3 = 0$ (Figure 6.3(b2)) than Δ_3 is detuned (Figure 6.3(b1) and (b3)). On the other hand, the fluorescence peak R_1 in Figure 6.3(b2) is slightly stronger than the ones in Figure 6.3(b1) and (b3), resulting from the enhancement effect of E_3 around $\Delta_3 = -\Delta_1$ by considering the nest-dressing term $|G_2|^2/(d_1 + |G_3|^2/d_3)$ in $\rho^{(4)}_{22DD}$. The corresponding calculations of fluorescence are also presented in Figure 6.3(b), which are in agreement with the experimental results.

Finally, we concentrate on the signal intensity depending on the power of the laser beams by scanning Δ_2 with $\Delta_1 = \Delta_3 = 0$. First, when the power of the beam E_2 changes from small to large, we arrange the experimental curves from bottom to top in Figure 6.4(a1) to (a3). In this case, beam E_2' is blocked and other beams are turned on, so that the signals in Figure 6.4(a2) represent the suppression degree induced by E_2 on the FWM signal F2. In accordance with expectation, both the height of the EIT peak (Figure 6.4(a1)) and the suppression degree of F2 (Figure 6.4(a2)) get larger with P_2 increasing, owing to the fact that the function of the two-photon dressing term $|G_2|^2/d_2$ in $\rho^{(1)}_{10DD}$ and $\rho^{(3)}_{F2DD}$ becomes stronger as P_2 increases. For the fluorescence signal (Figure 6.4(a3)), on the one hand, the suppression dip deepens as P_2 increases, also due to the increasing dressing effect of E_2 on $\rho^{(2)}_{11}$. On

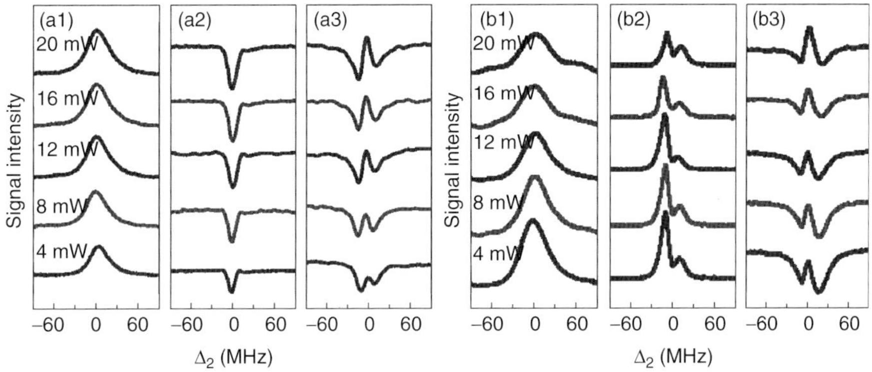

Figure 6.4 Measured probe transmission ((a1) and (b1)), FWM ((a2) and (b2)), and fluorescence signals ((a3) and (b3)) versus Δ_2 with $\Delta_1 = \Delta_3 = 0$, for (a1)–(a3) blocking E_2' and increasing $P_2 = 4$, 8, 12, 16, and 20 mW when $P_1 = 6$ mW and $P_3 = P_3' = 20$ mW, and (b1)–(b3) blocking E_3' and increasing $P_3 = 4$, 8, 12, 16, and 20 mW from bottom to top when $P_1 = 6$ mW and $P_2 = P_2' = 20$ mW. Source: Adapted from Ref. [15].

the other hand, the fluorescence peak R_1, which is mainly dependent on the beam E_2 intensity according to $\rho_{22}^{(4)}$, gets greatly larger with P_2 increasing.

Next, when the power of the other coupling beam E_3 is changed with E_3' blocked, the results are strikingly different (Figure 6.4(b1) to (b3)). As P_3 increases, we find the EIT peak induced by $E_2(E_2')$ decreases, as shown from bottom to top in Figure 6.4(b1), which is contrary to the case of changing the power of the self-dressing field P_2, where the EIT peak increases from bottom to top in Figure 6.4(a1). This is due to the fact that the interaction between the dressing effect of $E_2(E_2')$ and E_3 on $\rho_{10}^{(1)}$, which has been explicated in Figure 6.3(b), becomes stronger with P_3 increasing. The FWM signal F1 in Figure 6.4(b2) also weakens as P_3 increases because of the stronger suppression effect of E_3 denoted by the dressing term $|G_3|^2/d_3$ in $\rho_{\mathrm{F1DD}}^{(3)}$. When we turn to the measured fluorescence signals (Figure 6.4(b3)), we find that the suppression dip related to $\rho_{11}^{(2)}$ become shallower as P_3 increases, also due to the stronger interaction between the dressing term $|G_3|^2/d_3$ and $|G_2|^2/d_2$ in $\rho_{\mathrm{11DD}}^{(2)}$. In contrast, the peaks within the dips become slightly larger from bottom to top in Figure 6.4(b3), as the enhancement effect of E_3 on R_1 gets stronger according to the nest-dressing term $|G_2|^2/(d_1 + |G_3|^2/d_3)$ in $\rho_{\mathrm{22DD}}^{(4)}$.

6.2
Phase Control of Bright and Dark States in FWM and Fluorescence Channels

The switch between EIT (dark state) and EIA (bright state) [16] had been observed by changing the phase difference between the two circularly polarized components of a single coherent field. Moreover, the switch between suppression (dark state) and enhancement (bright state) of FWM had been demonstrated by changing the probe detuning [17, 18] and dressing field power [19]. Meanwhile, the fluorescence signal induced by the spontaneous emission within EIT window [12, 13] had been observed.

In this section, we discuss the switch between bright and dark states by controlling the relative phase in a four-level ^{85}Rb atomic vapor. In the relative phase control, the enhancement and suppression conditions of the measured signals are significantly modified, and, therefore, we can effectively modulate the transmitted probe, FWM, and fluorescence signals.

6.2.1
Theory and Experimental Scheme

Our experiment was implemented in a cell with rubidium atomic vapor, in which the relevant ^{85}Rb energy levels are $5S_{1/2}(F=3, |0\rangle)$, $5P_{3/2}(|1\rangle)$, $5D_{5/2}(|2\rangle)$, and $5D_{3/2}(|3\rangle)$, as shown in Figure 6.5(a). A weak probe laser beam E_1 (frequency ω_1, wave vector $\mathbf{k}_1$) probes the lower transition $|0\rangle$ to $|1\rangle$. A pair of coupling laser beams E_2 (ω_2, $\mathbf{k}_2$) and E_2' (ω_2, $\mathbf{k}_2'$) connect the upper transition $|1\rangle$ to $|2\rangle$, and another pair of coupling laser beams E_3 (ω_3, $\mathbf{k}_3$) and E_3' (ω_3, $\mathbf{k}_3'$) drive another upper transition $|1\rangle$ to $|3\rangle$. In normal spatial configuration, the laser beams are

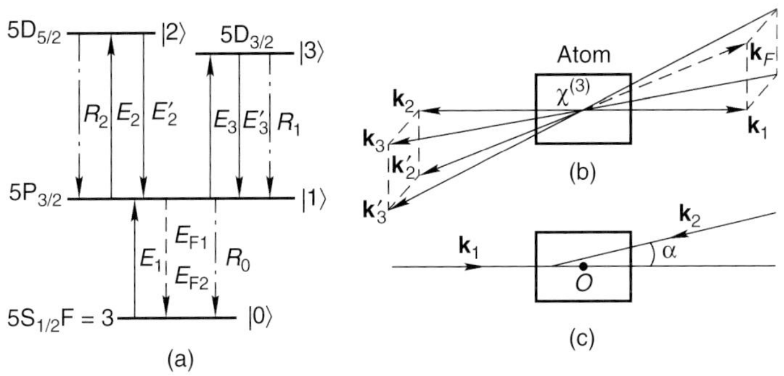

Figure 6.5 (a) Four-level atomic system used in experiment. (b) Normal-phase-matching configuration with $\mathbf{k}_2$ propagating in the opposite direction of $\mathbf{k}_1$. (c) Abnormal configuration where $\mathbf{k}_1$ and $\mathbf{k}_2$ intersect with a small angle α.

aligned in the square-box pattern shown in Figure 6.5(b), where all coupling laser beams propagate through Rb vapor in the same direction with small angles ($\sim 0.3^\circ$) between them, whereas the probe field E_1 propagates in the direction opposite to E_2. However, in this section, the normal configuration is modified by introducing an extra small angle α between E_2 and the direction opposite to E_1 (abnormal configuration), as shown in Figure 6.5(c). In both configurations, two FWM signals E_{F1} (generated by E_1, E_3, and E_3') and E_{F2} (generated by E_1, E_2, and E_2') are generated in ladder-type $|0\rangle - |1\rangle - |3\rangle$ (satisfying $\mathbf{k}_{F1} = \mathbf{k}_1 + \mathbf{k}_3 - \mathbf{k}_3'$) and $|0\rangle - |1\rangle - |2\rangle$ (satisfying $\mathbf{k}_{F2} = \mathbf{k}_1 + \mathbf{k}_2 - \mathbf{k}_2'$) subsystems. Meanwhile, the single-photon fluorescence signal R_0, two-photon fluorescence signals R_1 and R_2 due to spontaneous emissions from $|1\rangle$ to $|0\rangle$, $|3\rangle$ to $|1\rangle$, and $|2\rangle$ to $|1\rangle$ respectively, are also generated (dashed-dot lines in Figure 6.5(a)).

Generally, the density matrix elements $\rho_{10}^{(1)}$ (related to the transmitted probe signal), $\rho_{10}^{(3)}$ (FWM signal), and $\rho_{11}^{(2)}$, $\rho_{22}^{(4)}$, and $\rho_{33}^{(4)}$ (fluorescence signals) can be obtained by solving the density matrix equations. The undressed perturbation chain $\rho_{00}^{(0)} \xrightarrow{\omega_1} \rho_{10}^{(1)}$ gives $\rho_{10}^{(1)} = iG_1/d_1$, where $G_i = \mu_i E_i/\hbar$ is the Rabi frequency with transition dipole moment μ_i, $d_1 = \Gamma_{10} + i\Delta_1$ with frequency detuning $\Delta_i = \Omega_i - \omega_i$ (Ω_i is the resonance frequency of the transition driven by E_i) and transverse relaxation rate Γ_{ij} between $|i\rangle$ and $|j\rangle$. This element can be singly dressed as $\rho_{10SD}^{(1)} = iG_1/(d_1 + |G_2|^2 e^{i\Delta\Phi}/d_2)$ (with $d_2 = \Gamma_{20} + i(\Delta_1 + \Delta_2)$) or doubly dressed as $\rho_{10DD}^{(1)} = iG_1/(d_1 + |G_3|^2/d_3 + |G_2|^2 e^{i\Delta\Phi}/d_2)$ (with $d_3 = \Gamma_{30} + i(\Delta_1 + \Delta_3)$) by the coupling fields. Here, an additional phase factor $e^{i\Delta\Phi}$ is introduced into the dressing term $|G_2|^2/d_2$ by means of altering the angle α, by which the relative phase $\Delta\Phi$ related to the orientations of induced dipole moments μ_1 and μ_2 [20], could be manipulated. For FWM signals, the doubly dressed E_{F2} can be described as $\rho_{10F2DD}^{(3)} = G_{F2}/[(d_1 + |G_2|^2 e^{i\Delta\Phi}/d_2 + |G_3|^2/d_3)^2 d_2]$ with $G_{F2} = -iG_1 G_2 G_2'^*$. FWM signal E_{F1} is similar to E_{F2}. The single-photon fluorescence signal R_0 can be obtained as $\rho_{11}^{(2)} = -|G_1|^2/(\Gamma_{11} d_1)$ via $\rho_{00}^{(0)} \xrightarrow{\omega_1} \rho_{10}^{(1)} \xrightarrow{-\omega_1} \rho_{11}^{(2)}$, which can also be doubly dressed as $\rho_{11DD}^{(2)} = -|G_1|^2/[\Gamma_{11}(d_1 + |G_3|^2/d_3 + |G_2|^2 e^{i\Delta\Phi}/d_2)]$. The two-photon fluorescence signal R_2 can be obtained as $\rho_{22}^{(4)} = |G_1|^2|G_2|^2/(\Gamma_{22} d_1 d_2 d_4)$

(with $d_4 = \Gamma_{21} + i\Delta_2$) via $\rho_{00}^{(0)} \xrightarrow{\omega_1} \rho_{10}^{(1)} \xrightarrow{\omega_2} \rho_{20}^{(2)} \xrightarrow{-\omega_1} \rho_{21}^{(3)} \xrightarrow{-\omega_2} \rho_{22}^{(4)}$, which can be modified as $\rho_{22DD}^{(4)} = |G_1|^2|G_2|^2/[\Gamma_{22}d_1d_4(d_2 + |G_2|^2 e^{i\Delta\Phi}/(d_1 + |G_3|^2/d_3))]$ with self- and external-dressing effects. The other two-photon fluorescence signal R_1 is similar to R_2. According to these expressions, it is obvious that the dressing effect can be modulated by manipulating the relative phase $\Delta\Phi$, and so the switch between bright and dark states could be achieved.

6.2.2
Theory and Experimental Results

When blocking E_3 and E_3', we show the evolutions of the transmitted probe, FWM, and fluorescence signals under three typical relative phase $\Delta\Phi$, as shown in Figure 6.6(a1) to (a3). Under the normal configuration where $\alpha = 0$, the relative phase $\Delta\Phi = 0$ corresponds to the factor $e^{i\Delta\Phi} = 1$. Therefore, the dressing terms will degenerate to the normal ones that have been investigated in Ref. [18].

The signal under such normal conditions is depicted in Figure 6.6(a1), from which we can see an EIT window appearing at $\Delta_1 = -\Delta_2$ in the transmitted probe signal, a relatively lower FWM signal E_{F2} falling into this EIT window, and the two-photon fluorescence signal R_2, revealing as a sharp peak on the big background of the single-photon fluorescence signal R_0, also falling within the EIT window. With $\Delta\Phi$ changed to $-\pi/2$ (Figure 6.6(a2)), the EIT window in the transmitted probe signal almost disappears and a small EIA dip arises. When $\Delta\Phi$ further changes to $-\pi$ (Figure 6.6(a3)), a strong EIA dip appears. Such a switch between dark (EIT) and bright (EIA) states is due to the fact that the dressing effect gets modulated as $\Delta\Phi$ is altered, denoted by the dressing term with a phase

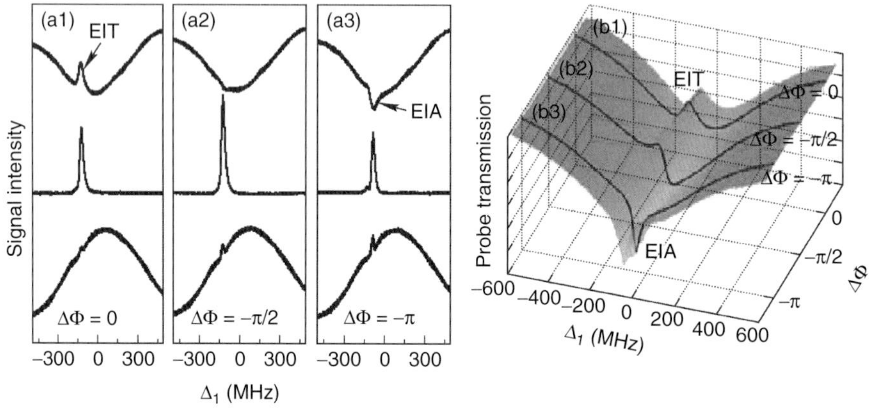

Figure 6.6 (a) Transmitted probe (top curves), FWM (middle curves), and fluorescence (bottom curves) signals versus Δ_1 with E_3 and E_3' blocked and $\Delta_2 = 100$ MHz for different $\Delta\Phi$: (a1) 0, (a2) $-\pi/2$, and (a3) $-\pi$. Powers of participated laser beams are $P_1 = 7$ mW, $P_2 = 17$ mW, and $P_2' = 8$ mW. (b1)–(b3) Calculations of the transmitted probe signals corresponding to the results in (a1)–(a3). Source: Adapted from Ref. [21].

factor $|G_2|^2e^{i\Delta\Phi}/d_2$ in $\rho_{10SD}^{(1)}$. To illustrate such a phase-controlled switch more clearly, we present the corresponding calculations of the transmitted probe signal in Figure 6.6(b). For $\Delta\Phi = 0$, the dressing term $|G_2|^2e^{i\Delta\Phi}/d_2$ behaves positive (satisfying the suppression condition) and the normal dark (EIT) state appears (Figure 6.6(b1)); for $\Delta\Phi = -\pi$, the dressing term behaves negative (satisfying the enhancement condition) and so the dark (EIT) state switches to bright (EIA) state (Figure 6.6(b3)); for $\Delta\Phi = -\pi/2$, the transitional partial EIT/EIA could be seen (Figure 6.6(b2)). This dark/bright state switch could result in the modulation of FWM and fluorescence signals. For the FWM signal E_{F2}, it gets increased when $\Delta\Phi = -\pi/2$ (Figure 6.6(a2)) compared with that when $\Delta\Phi = 0$ (Figure 6.6(a1)), as the arising enhancement dressing effect of $E_2(E_2')$; and gets decreased again when $\Delta\Phi = -\pi$, due to strong absorption. For the two-photon fluorescence signal R_2 (in corresponding to EIA), its intensity could be observed increasing straight from Figure 6.6(a1) to (a3).

Further, we study the phase-controlled switch under double-dressing condition with E_3 also turned on, as shown in Figure 6.7(a) and (b), which separately

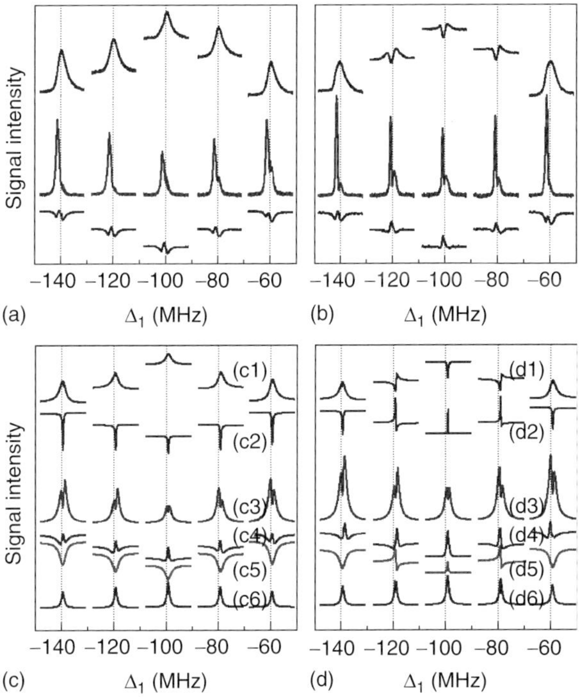

Figure 6.7 (a) and (b) The transmitted probe (top curves), FWM (middle curves), and fluorescence (bottom curves) signals versus Δ_2 at discrete Δ_1 with E_3' blocked and $\Delta_3 = 100$ MHz at $\Delta\Phi = 0$ (a) and $-2\pi/3$ (b). Powers of the participated laser beams are $P_1 = 7$ mW, $P_2 = 17$ mW, $P_2' = 8$ mW, and $P_3 = 60$ mW. (c,d) Theoretical calculations corresponding to (a) and (b).

presents the measured signals at two specified relative phases $\Delta\Phi = 0$ and $-2\pi/3$ by scanning Δ_2. In the transmitted probe signal shown as the top curves in Figure 6.7(a) and (b), the heights of the baselines represent the transparent degree of E_1 induced by E_3, and the peaks (dips) higher (lower) than baseline are EIT (EIA) induced by $E_2(E_2')$. When $\Delta\Phi = 0$, the curves at all Δ_1 points generally behave as EIT peak. When $\Delta\Phi$ is changed to $-2\pi/3$, except EIT, there also appears EIA at $\Delta_1 = -120$, -100, and -80 MHz due to the modulation of the phase factor. For the FWM signals, the global profile of their heights is induced by the dressing effect of E_3, which causes E_{F2} to reach the minimal intensity at $\Delta_1 = -100$ MHz. With $\Delta\Phi$ changed from 0 to $-2\pi/3$, the dressing effect of $E_2(E_2')$ on FWM signal E_{F2} can be switched from suppression to enhancement, in company with the EIT/EIA switch. Therefore E_{F2} is relatively suppressed at $\Delta\Phi = 0$ (Figure 6.7(a)) and enhanced at $\Delta\Phi = -2\pi/3$ (Figure 6.7(b)). Then, for fluorescence signals, no matter at $\Delta\Phi = 0$ or $\Delta\Phi = -2\pi/3$, the profile of their baselines in which the curve is suppressed to its minimum at $\Delta_1 = -100$ MHz, expresses the suppression effect of E_3 (dressing term $|G_3|^2/d_3$ in $\rho^{(2)}_{11DD}$) on R_0. The dip lower than the corresponding baseline represents further suppression effect on R_0 induced by $E_2(E_2')$ ($|G_2|^2 e^{i\Delta\Phi}/d_2$ in $\rho^{(2)}_{11DD}$). The peak within the dip is the two-photon fluorescence signal R_2 ($\rho^{(4)}_{22DD}$). With $\Delta\Phi$ changed from 0 to $-2\pi/3$, we can see the dips get shallower owing to the weakened EIT, and even the arising EIA and the peaks get higher because of the original suppression effect of $E_2(E_2')$ on R_2 switching to enhancement effect at $\Delta\Phi = -2\pi/3$. The theoretical calculations of the signals corresponding to Figure 6.7(a) and (b) are presented in Figure 6.7(c) and (d), in which (c1) and (d1), (c3) and (d3), and (c4) and (d4) are the calculated transmitted probe, FWM, and total fluorescence signal, respectively. The dressed part (bright and dark states) is extracted in theory from the total FWM signal, as shown in (c2) and (d2). And the single- and two-photon fluorescence signals are separately presented in (c5) and (d5) and (c6) and (d6). The bright/dark switch could be clearly observed from the calculated curves, which are in accordance with the experimental observation.

Finally, we concentrate on the variations of the measured signals by continuously changing relative phase $\Delta\Phi$ with different laser beams blocked. Under the conditions of blocking E_3 and E_3' (Figure 6.8(a)), EIT in the transmitted probe signal (Figure 6.8(a1)) can be switched to EIA gradually along with the changes of $\Delta\Phi$. During this process, the strongest EIT and EIA separately appear at $\Delta\Phi = 0$ and $\Delta\Phi = -\pi$. And depending on whether $\Delta\Phi$ is greater than or less than $-\pi/2$, the transmitted probe signal behaves mainly EIT or EIA. The corresponding FWM signal E_{F2} (Figure 6.8(a2)) and fluorescence signal R_2 (Figure 6.8(a3)) are relatively weak in dark (EIT) state, and get their largest intensities in the bright (EIA) state, due to the dressing effect of $E_2(E_2')$ that is switched from destructive to constructive. Next, Figure 6.8(b1) to (b3) shows the doubly-dressed ($E_3(E_3')$ and E_2) case, where the suppression and enhancement of the FWM signal E_{F1} can be observed directly (Figure 6.8(b2)) by scanning the external-dressing field E_2. During the process of changing $\Delta\Phi$, the variation in the transmitted probe signal (Figure 6.8(b1)) is similar to that in Figure 6.8(a1). And the dressing effect of E_2 on the FWM signal E_{F1} turns from suppression (in company with EIT) to enhancement (in company

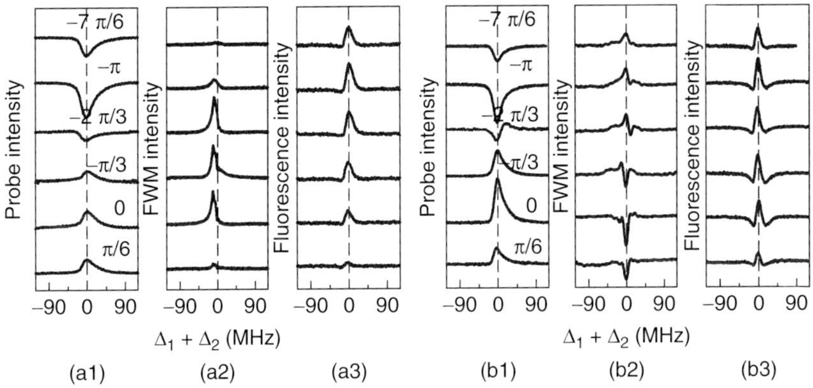

Figure 6.8 The transmitted probe ((a1) and (b1)), FWM ((a2) and (b2)), and fluorescence ((a3) and (b3)) signals versus Δ_2 at different relative phase $\Delta\Phi = \pi/6$, 0, $-\pi/3$, $-2\pi/3$, $-\pi$, and $-7\pi/6$ from the bottom to top. (a) Signals obtained with E_3 and E_3' blocked, $\Delta_1 = -200$ MHz, $P_1 = 7$ mW, $P_2 = 17$ mW, and $P_2' = 8$ mW. (b) Signals obtained with E_2' blocked, $\Delta_3 = 100$ MHz, $\Delta_1 = -120$ MHz, $P_1 = 7$ mW, $P_3 = 50$ mW, $P_3' = 30$ mW, and $P_2 = 20$ mW. Source: Adapted from Ref. [21].

with EIA). For the fluorescence signals (Figure 6.8(b3)), the dip of the single-photon fluorescence signal R_0 induced by E_2 gets deepest at $\Delta\Phi = 0$ due to the strongest EIT, while the two-photon fluorescence peak R_2 gets strongest at $\Delta\Phi = -\pi$ due to the largest EIA.

6.3 Observation of Angle Switching of Dressed FWM Image

6.3.1 Introduction

EIT in multi-level atomic vapors was widely researched in the past two decades [6, 10]. It has been demonstrated that the self- and cross-Kerr nonlinearities can be significantly enhanced and modified because of laser-induced atomic coherence [6], which is crucial for large refractive index modulation [22]. By the refractive index modulation, laser beam self-focusing [23] and pattern formation [24] have been extensively investigated with two laser beams propagating in atomic vapors. And in recent years, we have observed spatial shift [25], spatial AT splitting, gap solitons, and dipole solitons of the FWM beams with dressing effects [26, 27] generated in multi-level atomic systems [28–31]. In addition, we also have observed the evolution of the intensity enhancement and suppression in the FWM signal spectrum by controlling additional laser fields [17]. On the other hand, the fluorescence accompanying the FWM process that can be induced by spontaneous emission under EIT conditions [12, 13] was studied because of its potential applications in metrology and long-distance quantum communication and quantum correlation. It is worth mentioning that the AT splitting in fluorescence spectrum has also been studied in lithium molecules [13].

In this section, we discuss the influence on the images and spectra intensities of probe transmission, FWM, and fluorescence signals brought by the angles between pumping fields and dressing fields in cascade three-level and two-level atomic systems. A remarkable difference in the spatial properties of signals is obtained between the cases, with the dressing field and the pumping field changed. And then, we analyze the experimental results through a qualitative theory. Studies on such controlling spectra intensity and images have potential applications in all optical signal processing.

6.3.2
Theoretical Model and Experimental Scheme

A cascade three-level system ($|0\rangle$ ($3S_{1/2}$), $|1\rangle$ ($3P_{3/2}$), $|2\rangle$ ($4D_{3/2}$)) and a two-level system ($|0\rangle$ ($3S_{1/2}$), $|1\rangle$ ($3P_{3/2}$)) from Na atom (in an 18 cm long heat-pipe oven) are shown in Figure 6.9(a) and (b), respectively. Five laser beams are applied to the atomic system with the spatial configuration given in Figure 6.9(c) and (d). The fields $\boldsymbol{E}_1$ (frequency $\boldsymbol{E}_1'$, wave vector $\mathbf{k}_1$, the Rabi frequency G_1, and frequency detuning $\Delta_1=\Omega_1-\omega_1$, where Ω_1 is the resonant frequency of transition from $|0\rangle$ to $|1\rangle$) and $\boldsymbol{E}_1'$ (ω_1, $\mathbf{K}_1'$, G_1', and Δ_1) are propagating pump fields with a small angle between them in the x–z-plane ($\theta_{1x0}\approx 0.3^\circ$ when they are set at the point B in Figure 6.9(d)) and in the y–z-plane ($\theta_{1y}\approx 0.05^\circ$). The probe field $\boldsymbol{E}_3$ (ω_1, $\mathbf{k}_3$, G_3, Δ_1) propagates in the direction opposite to $\boldsymbol{E}_1$. These three fields ($\boldsymbol{E}_1$, $\boldsymbol{E}_1'$, and $\boldsymbol{E}_3$) with a diameter of about 0.8 mm are from the same near-transform-limited dye laser (10 Hz repetition rate, 5 ns pulse width, and 0.04 cm^{-1} line width) connecting the transition $|0\rangle \leftrightarrow |1\rangle$. The *Rabi frequencies of the beams* are defined as $G_i=\mu_{mn}E_i/\hbar$ ($i=1, 2, 3$) or $G_i'=\mu_{mn}E_i'/\hbar$ ($i=1, 2$) in which μ_{mn} is the dipole moments of the transition $|m\rangle \leftrightarrow |n\rangle$ the field $\boldsymbol{E}_1(\boldsymbol{E}_1')$ drives and $\boldsymbol{E}_1(\boldsymbol{E}_1')$ is the electric field intensity of $\boldsymbol{E}_1(\boldsymbol{E}_1')$. Another pair of dressing fields $\boldsymbol{E}_2$ (ω_2, $\mathbf{k}_2$, G_2, Δ_2) and $\boldsymbol{E}_2'$ (ω_2, $\mathbf{k}_2'$, G_2', Δ_2) propagate in the same direction of $\boldsymbol{E}_1$ having a small angle between them in the y–z-plane ($\theta_{2y0}\approx 0.3^\circ$) and x–z-plane ($\theta_{2x}\approx 0.05^\circ$). Driving the transition $|1\rangle \leftrightarrow |2\rangle$ in the cascade three-level and $|0\rangle \leftrightarrow |1\rangle$ in two-level systems, $\boldsymbol{E}_2$ and $\boldsymbol{E}_2'$ with a diameter of 1 mm are generated from another dye laser with same pulse parameters as the former one. A DFWM signal $\boldsymbol{E}_{F1}$ is produced, satisfying the phase-matching condition $\mathbf{k}_{F1}=\mathbf{k}_1-\mathbf{k}_1'+\mathbf{k}_3$, as shown in Figure 6.9(c) and (d). In order to optimize the beam spatial effects, the intensities of the fields $\boldsymbol{E}_1$ and $\boldsymbol{E}_1'$ are about 0.005 W cm^{-2} in the cascade three-level system and 0.013 W cm^{-2} in the two-level system. $\boldsymbol{E}_2$ and $\boldsymbol{E}_2'$ are approximately 0.005 W cm^{-2}. The spectra intensity of the weak probe beam $\boldsymbol{E}_3$ is less than 0.003 W cm^{-2}. Moreover, all the beams used in the experiment are P-polarized components after a PBS.

In this experiment, we define the reduced angle between $\boldsymbol{E}_1$ and $\boldsymbol{E}_1'$ as $\Delta\theta_1=\theta_{1x}-\theta_{1x0}$, where θ_{1x} is the angle between $\boldsymbol{E}_1$ and $\boldsymbol{E}_1'$ in the x–z-plane. Similarly, we can define the reduced angle between $\boldsymbol{E}_2$ and $\boldsymbol{E}_2'$ as $\Delta\theta_2=\theta_{2y}-\theta_{2y0}$. In Figure 6.9(d), when the intersecting points between the two pumping fields are

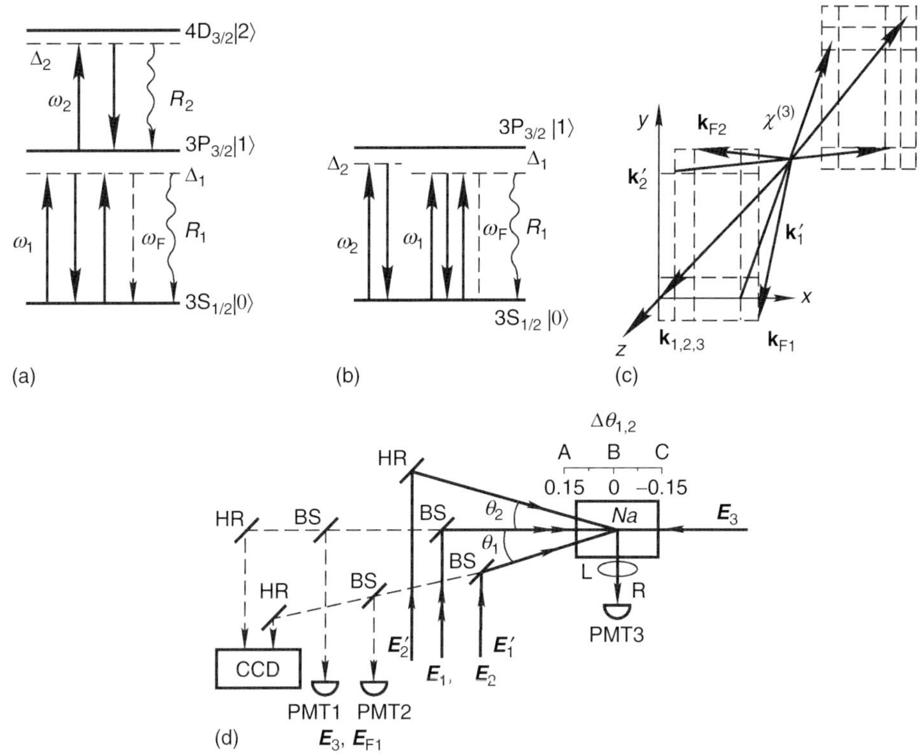

Figure 6.9 (a) and (b) The cascade three-level and two-level diagrams in Na atom, respectively. (c) and (d) Spatial geometry for the laser beams in the experiment and experimental setup, respectively. The scale gives the angle value of the corresponding position in heat-pipe oven as shown in (d). Every optical element can be signified as HR, high reflective mirror; BS, beam splitter (50%); L, convex lens; PMT, photomultiplier tube.

set to move from point C (the back of the heat-pipe oven) to the position A (the front of the heat-pipe oven), the value of $\Delta\theta_{1,2}$ changes from −0.15 to 0.15. The probe transmission and $\boldsymbol{E}_{F1}$ beams are split into two equal components by a 50% beam splitter before being detected. One component is captured by a CCD camera; the other detected by photomultiplier tubes (PMT1 and PMT2) and the fluorescence signal is detected by PMT3 with a fast gated integrator (gate width of 50 ns).

In order to interpret the following experimental results, we perform the theoretical calculation on the probe transmission, $\boldsymbol{E}_{F1}$, and the fluorescence process in the cascade three-level atomic system. For the DFWM signal $\boldsymbol{E}_{F1}$, its generation process can be viewed as a series of transitions steps. The first step is a rising transition from $|0\rangle$ to $|1\rangle$ with absorption of a pump photon $\boldsymbol{E}_1$, the second step is a falling transition from $|1\rangle$ to $|0\rangle$, and the third step is another transition from $|0\rangle$ to $|1\rangle$ with the absorption of a probe photon $\boldsymbol{E}_3$. Then, the last transition is from $|1\rangle$ to $|0\rangle$, which emits an FWM photon at frequency ω_1. According to the Liouville pathway of the pure DFWM $\rho_{00}^{(0)} \xrightarrow{G_1} \rho_{10}^{(1)} \xrightarrow{(G_1')^*} \rho_{00}^{(2)} \xrightarrow{G_3} \rho_{10}^{(3)}$, we can obtain the third-order nonlinear density matrix element $\rho_{10}^{(3)} = g/(\Gamma_{00}d_1^2)$, the magnitude of which determines the spectra intensity of the DFWM signal. Here, $g = -iG_3G_1G_1'^*, d_1 = i\Delta_1' + \Gamma_{10}$, and Γ_{ij} is the transverse relaxation rate between the states $|i\rangle$ and $|j\rangle$. When $\boldsymbol{E}_{F1}$ is also dressed by the fields $\boldsymbol{E}_1(\boldsymbol{E}_1')$, $\boldsymbol{E}_3$ and $\boldsymbol{E}_2(\boldsymbol{E}_2')$, the multi-dressed Liouville pathway in such a case is $\rho_{00}^{(0)} \xrightarrow{(G_1)^*} \rho_{G_1\pm G_2\pm G_30}^{(1)} \xrightarrow{(G_1')^*} \rho_{00}^{(2)} \xrightarrow{G_3} \rho_{G_1\pm G_2\pm G_30}^{(3)}$ (self-dressed by $\boldsymbol{E}_1(\boldsymbol{E}_1')$, $\boldsymbol{E}_3$ and external dressed by $\boldsymbol{E}_2(\boldsymbol{E}_2')$), the expression of the density matrix element $\rho_{10}^{(3)}$ related to such a multi-dressed $\boldsymbol{E}_{F1}$ can be obtained as

$$\rho_{10}^{(3)} = \frac{g}{\begin{array}{c}[d_1 + |G_1|^2/\Gamma_{00} + |G_1|^2/(\Gamma_{00} + A_1) + |G_2|^2/d_2 \\ +|G_3|^2/\Gamma_{00} + |G_3|^2/(\Gamma_{11} + A_1)]^2(\Gamma_{00} + |G_1|^2 + |G_3|^2)/(d_1 + A_2)\end{array}} \tag{6.1}$$

where $d_2 = \Gamma_{20} + i(\Delta_1' + \Delta_2')$, $A_1 = |G_1|^2/(\Gamma_{10} + i\Delta_1')$, and $A_2 = |G_2|^2/[\Gamma_{20} + i(\Delta_1' + \Delta_2')]$. On the other hand, there exist three types of fluorescence signals due to the spontaneous emission of photons from the upper levels. First, the decay of atoms pumped by the beams $\boldsymbol{E}_1(\boldsymbol{E}_1')$ from $|0\rangle$ to $|1\rangle$ will generate a type of fluorescence signal R_1, which can be described by the Liouville pathway $\rho_{00}^{(0)} \xrightarrow{G_1} \rho_{10}^{(1)} \xrightarrow{(G_1)^*} \rho_{11}^{(2)}$. By solving the coupled density matrix equations, the expression of the density matrix element $\boldsymbol{E}_{F1}$ related to the fluorescence process R_1 can be obtained as Δ_1. When the beams $\boldsymbol{E}_2(\boldsymbol{E}_2')$ are turned on, the fluorescence process R_1 is dressed, which can be described by the Liouville pathway $\rho_{00}^{(0)} \xrightarrow{G_1} \rho_{G_1\pm G_2\pm 0}^{(1)} \xrightarrow{(G_1)^*} \rho_{G_1\pm G_2\pm G_1\pm G_2\pm}^{(2)}$ (self-dressed by $\boldsymbol{E}_1(\boldsymbol{E}_1')$ and external dressed by $\boldsymbol{E}_2(\boldsymbol{E}_2')$). The density matrix element $\rho_{11}^{(2)}$ can be modified as

$$\rho_{11}^{(2)} = \frac{-|G_1|^2}{((d_1 + |G_1|^2/\Gamma_{00} + |G_2|^2/d_2)(\Gamma_{11} + |G_1|^2/d_1 + |G_2|^2/d_3 + |G_2|^2/d_4)} \tag{6.2}$$

where $d_3 = \Gamma_{21} + i\Delta_2'$, $d_4 = \Gamma_{12} - i\Delta_2'$. Second, the decay of atoms pumped by the beam $\boldsymbol{E}_1(\boldsymbol{E}_1')$ and $\boldsymbol{E}_2(E_2')$ from $|0\rangle$ to $|2\rangle$ will generate another type of fluorescence signal R_2, which can be described by the Liouville pathway $\rho_{00}^{(0)} \xrightarrow{G_1} \rho_{10}^{(1)} \xrightarrow{G_2} \rho_{20}^{(2)} \xrightarrow{(G_1)^*} \rho_{21}^{(3)} \xrightarrow{(G_2)^*} \rho_{22}^{(4)}$ (or $\rho_{00}^{(0)} \xrightarrow{G_1} \rho_{10}^{(1)} \xrightarrow{G_2'} \rho_{20}^{(2)} \xrightarrow{(G_1)^*} \rho_{21}^{(3)} \xrightarrow{(G_2')^*} \rho_{22}^{(4)}$). We can also obtain the density matrix element $\rho_{22}^{(4)}$ related to the fluorescence process R_2 as $\rho_{22}^{(4)} = |G_1|^2|G_2|^2/(\Gamma_{22}d_1d_2d_4)$. If the dressing effect is also considered, we can further obtain density matrix element for the multi-dressed fluorescence process R_2 as

$$\rho_{22}^{(4)} = \frac{|G_1|^2|G_2|^2}{\Gamma_{22}(d_1 + |G_1|^2/\Gamma_{00} + |G_2|^2/d_2)d_2(d_3 + |G_1|^2/d_2 + |G_2|^2/\Gamma_{22})} \tag{6.3}$$

The fluorescence signal is obviously different from the FWM signals. First, from their Liouville pathway, we can see that the FWM process follows the closed-loop path while the fluorescence process does not. This results in the second difference between them, that is, the FWM signal is of a specific direction but the fluorescence signal is not. Third, the FWM signal is caused by the atomic coherence effect while the fluorescence signal is induced by spontaneous decay of photons pumped to the upper levels. For the probe transmission, the Liouville pathway is $\rho_{00}^{(0)} \xrightarrow{G_3} \rho_{G_1 \pm G_2 \pm 0}^{(1)}$ (dressed by $\boldsymbol{E}_1$, $\boldsymbol{E}_1'$ $\boldsymbol{E}_2$, and $\boldsymbol{E}_2'$), and we can obtain the first-order density matrix element as

$$\rho_{10}^{(1)} = iG_3/[d_1 + |G_1|^2/\Gamma_{00} + |G_2|^2/d_2 + |G_1|^2/(\Gamma_{11} + A_3)], \tag{6.4}$$

the imaginary part of which proportionally determines the absorption of the probe beam in propagation. Here, $A_3 = |G_2|^2/(\Gamma_{12} - i\Delta_2')$.

For the FWM processes in the two-level configuration, there exist several transition paths to generate FWM signals. They can be described by the Liouville pathway

$$\text{(F1)}\ \rho_{00}^{(0)} \xrightarrow{G_1} \rho_{10}^{(1)} \xrightarrow{(G_1')^*} \rho_{00}^{(2)} \xrightarrow{G_3} \rho_{10}^{(3)}$$

$$\text{(F2)}\ \rho_{00}^{(0)} \xrightarrow{G_2} \rho_{10}^{(1)} \xrightarrow{(G_2')^*} \rho_{00}^{(2)} \xrightarrow{G_3} \rho_{10}^{(3)}$$

$$\text{(F3)}\ \rho_{00}^{(0)} \xrightarrow{G_2} \rho_{10}^{(1)} \xrightarrow{(G_1')^*} \rho_{00}^{(2)} \xrightarrow{G_3} \rho_{10}^{(3)}$$

$$\text{(F4)}\ \rho_{00}^{(0)} \xrightarrow{G_1} \rho_{10}^{(1)} \xrightarrow{(G_2')^*} \rho_{00}^{(2)} \xrightarrow{G_3} \rho_{10}^{(3)}$$

respectively. Thus, we can obtain their corresponding dressed forms as

$$\text{(DF1)}\ \rho_{00}^{(0)} \xrightarrow{G_1} \rho_{G2\pm0}^{(1)} \xrightarrow{(G_1')^*} \rho_{00}^{(2)} \xrightarrow{G_3} \rho_{G2\pm0}^{(3)}$$

$$\text{(DF2)}\ \rho_{00}^{(0)} \xrightarrow{G_2} \rho_{G1\pm0}^{(1)} \xrightarrow{(G_2')^*} \rho_{00}^{(2)} \xrightarrow{G_3} \rho_{G1\pm0}^{(3)}$$

$$\text{(DF3)}\ \rho_{00}^{(0)} \xrightarrow{G_2} \rho_{G2\pm0}^{(1)} \xrightarrow{(G_1')^*} \rho_{00}^{(2)} \xrightarrow{G_3} \rho_{G2\pm0}^{(3)}$$

$$\text{(DF4)}\ \rho_{00}^{(0)} \xrightarrow{G_1} \rho_{G1\pm0}^{(1)} \xrightarrow{(G_2')^*} \rho_{00}^{(2)} \xrightarrow{G_3} \rho_{G1\pm0}^{(3)}$$

respectively. The expressions of the corresponding density matrix elements related to the four FWM pathways are

$$\rho_{F1}^{(3)} = \frac{g}{(\Gamma_{00} + |G_2|^2/d_2)(d_1 + |G_1|^2/(\Gamma_{00} + |G_2|^2/d_2))^2} \tag{6.5}$$

$$\rho_{F2}^{(3)} = \frac{-iG_3G_2(G_2')^*}{\Gamma_{00}d_6B_2} \tag{6.6}$$

$$\rho_{F3}^{(3)} = \frac{-iG_2G_3(G_1')^*}{d_7B_3^2} \tag{6.7}$$

$$\rho_{F4}^{(3)} = \frac{-iG_1G_3(G_2')^*}{d_8d_9B_1} \tag{6.8}$$

respectively, where $d_5 = \Gamma_{00} + i(\Delta_1' - \Delta_2')$, $d_6 = \Gamma_{10} + i\Delta_2'$, $d_7 = \Gamma_{00} + i(\Delta_2' - \Delta_1')$, $d_8 = \Gamma_{10} + i(2\Delta_1' - \Delta_2')$, $d_9 = [\Gamma_{00} + i(\Delta_1' - \Delta_2')]$, $A_4 = |G_2|^2/d_6$, $A_5 = G_1^2/\Gamma_{00}$, $A_6 = |G_2|^2/\Gamma_{00}$, $B_1 = d_1 + A_2$, $B_2 = d_1 + A_5$, and $B_3 = d_6 + A_6$. In the experiment, only the signals generated in the pathways DF1 and DF3 propagate opposite to E_1', so only they can be detected.

In the two-level type atomic system, the decay of photons pumped by the beam $\boldsymbol{E}_1(\boldsymbol{E}_1')$ and $\boldsymbol{E}_2(\boldsymbol{E}_2')$ from $|0\rangle$ to $|1\rangle$ will generate the eight types of fluorescence signals, which can be described by the Liouville pathway

$$\text{(R1)}\quad \rho_{00}^{(0)} \xrightarrow{G_1} \rho_{10}^{(1)} \xrightarrow{G_1^*} \rho_{11}^{(2)}$$

$$\text{(R2)}\quad \rho_{00}^{(0)} \xrightarrow{G_2} \rho_{10}^{(1)} \xrightarrow{G_1^*} \rho_{11}^{(2)}$$

$$\text{(R3)}\quad \rho_{00}^{(0)} \xrightarrow{G_1} \rho_{10}^{(1)} \xrightarrow{G_2^*} \rho_{11}^{(2)}$$

$$\text{(R4)}\quad \rho_{00}^{(0)} \xrightarrow{G_2} \rho_{10}^{(1)} \xrightarrow{G_2^*} \rho_{11}^{(2)}$$

$$\text{(R5)}\quad \rho_{00}^{(0)} \xrightarrow{G_2} \rho_{10}^{(1)} \xrightarrow{G_2^*} \rho_{00}^{(2)} \xrightarrow{G_1} \rho_{10}^{(3)} \xrightarrow{G_1^*} \rho_{11}^{(4)}$$

$$\text{(R6)}\quad \rho_{00}^{(0)} \xrightarrow{G_2} \rho_{10}^{(1)} \xrightarrow{G_1^*} \rho_{00}^{(2)} \xrightarrow{G_1} \rho_{10}^{(3)} \xrightarrow{G_2^*} \rho_{11}^{(4)}$$

$$\text{(R7)}\quad \rho_{00}^{(0)} \xrightarrow{G_1} \rho_{10}^{(1)} \xrightarrow{G_1^*} \rho_{00}^{(2)} \xrightarrow{G_2} \rho_{10}^{(3)} \xrightarrow{G_2^*} \rho_{11}^{(4)}$$

$$\text{(R8)}\quad \rho_{00}^{(0)} \xrightarrow{G_1} \rho_{10}^{(1)} \xrightarrow{G_2^*} \rho_{00}^{(2)} \xrightarrow{G_2} \rho_{10}^{(3)} \xrightarrow{G_1^*} \rho_{11}^{(4)}$$

By solving the coupled density matrix equations, the expression of the density matrix element $\rho_{11}^{(2)}$ related to the eight fluorescence processes can be obtained as

$$\rho_{11}^{(2)} = \frac{|G_1|^2}{(d_1 + |G_1|^2/\Gamma_{00} + |G_2|^2/d_9)(\Gamma_{11} + |G_2|^2/d_6 + |G_1|^2/d_1)} \tag{6.9}$$

$$\rho_{11}^{(2)} = \frac{|G_1||G_2|}{(d_6 + |G_2|^2/\Gamma_{00} + |G_2|^2/d_9)(d_{10} + |G_2|^2/d_8 + |G_1|^2/d_6)} \tag{6.10}$$

$$\rho_{11}^{(2)} = \frac{|G_1||G_2|}{(d_1 + |G_1|^2/\Gamma_{00} + |G_2|^2/d_9)(d_{10} + |G_2|^2/d_1 + |G_1|^2/d_6)} \tag{6.11}$$

$$\rho_{11}^{(2)} = \frac{|G_2|^2}{(d_6 + |G_2|^2/\Gamma_{00} + |G_2|^2/d_7)(\Gamma_{11} + |G_2|^2/d_8 + |G_1|^2/d_1)} \tag{6.12}$$

$$\rho_{11}^{(4)} = \frac{|G_1|^2|G_2|^2}{\begin{array}{c}(d_6+|G_2|^2/\Gamma_{00}+|G_1|^2/d_7)\Gamma_{00}(d_1+|G_1|^2/\Gamma_{00}+|G_2|^2/d_9)\\(\Gamma_{11}+|G_1|^2/d_1+|G_1|^2/d_{11}+|G_2|^2/d_6+|G_2|^2/d_{12})\end{array}} \tag{6.13}$$

$$\rho_{11}^{(4)} = \frac{|G_1|^2|G_2|^2}{(d_6+|G_2|^2/\Gamma_{00}+|G_1|^2/d_7)^2 d_7 (\Gamma_{11}+|G_1|^2/d_1+|G_1|^2/d_{11}+|G_2|^2/d_6+|G_2|^2/d_{12})} \tag{6.14}$$

$$\rho_{11}^{(4)} = \frac{|G_1|^2|G_2|^2}{(d_1+|G_1|^2/\Gamma_{00}+|G_2|^2/d_7)\Gamma_{00}(d_6+|G_2|^2/\Gamma_{00}+|G_1|^2/d_7)(\Gamma_{11}+|G_1|^2/d_1+|G_1|^2/d_{11}+|G_2|^2/d_6+|G_2|^2/d_{12})} \tag{6.15}$$

$$\rho_{11}^{(4)} = \frac{|G_1|^2|G_2|^2}{(d_1+|G_1|^2/\Gamma_{00}+|G_2|^2/d_9)^2 d_9 (\Gamma_{11}+|G_1|^2/d_1+|G_1|^2/d_{11}+|G_2|^2/d_6+|G_2|^2/d_{12})} \tag{6.16}$$

where $d_{10}=\Gamma_{11}+i(\Delta_2'-\Delta_1')$, $d_{11}=\Gamma_{01}-i\Delta_1{}'$, $d_{12}=\Gamma_{01}-i\Delta_2{}'$.

In addition, under the experimental condition (Figure 6.9(d)), for a hot atom with velocity v along the angular bisector of the pump fields, the frequency of $\boldsymbol{E}_1$ (or $\boldsymbol{E}_1'$) shifts to $\omega_1 - k_1 v\cos(\theta_{1x}/2)$ when the Doppler effect is phenomenally introduced to the weak signal. In the same way, the frequency of $\boldsymbol{E}_2$ (or $\boldsymbol{E}_2'$) shifts to $\omega_2 - k_2 v\cos(\theta_{2y}/2)$. As a result, the detuning of $\boldsymbol{E}_1$ (or $\boldsymbol{E}_1'$) and $\boldsymbol{E}_2$ (or $\boldsymbol{E}_2'$) shift to $\Delta_1'=\Delta_1+k_1 v\cos(\theta_{1x}/2)$ and $\Delta_2'=\Delta_2+k_2 v\cos(\theta_{2y}/2)$, respectively.

In order to understand the interaction between spatial characteristics and dressed spectra intensity of weak fields, the propagation equations are investigated.

To understand the beam splitting and spatial shift of the probe and FWM beams, the propagation equations, which give the mathematical description of the SPM- and XPM-induced spatial interplay for the probe and FWM beams, are introduced as

$$\frac{\partial u_3}{\partial Z} - \frac{i\partial^2 u_3}{2\partial\xi^2} = \frac{\mathrm{i}k_3^2 w_0 I}{n_0}(n_2^{S1}|u_3|^2 + 2n_2^{X1}|u_1|^2 + 2n_2^{X2}|u_1'|^2 + 2n_2^{X3}|u_2|^2 + 2n_2^{X4}|u_2'|^2)u_3 \tag{6.17}$$

$$\frac{\partial u_{F1}}{\partial Z} - \frac{i\partial^2 u_{F1}}{2\partial\xi^2} = \frac{\mathrm{i}k_{F1}^2 w_0^2 I}{n_0}(n_2^{S2}|u_{F1}|^2 + 2n_2^{X5}|u_1|^2 + 2n_2^{X6}|u_1'|^2 + 2n_2^{X7}|u_2|^2 + 2n_2^{X8}|u_2'|^2)u_{F1} \tag{6.18}$$

where $Z=z/L_D$ ($L_D = k_1 w_0^2$ has the physical meaning of the diffraction length and w_0 is the spot size of probe beam); z is the longitudinal coordinate in the beam propagation direction, $k_3=k_{F1}=\omega_1 n_0/c$; n_0 is the linear refractive index at ω_1; n_2^{S1-S2} are the self-Kerr coefficients of $\boldsymbol{E}_{3,F1}$; n_2^{X1-X8} are the cross-Kerr coefficients of $\boldsymbol{E}_{3,F1}$ induced by $\boldsymbol{E}_1$, $\boldsymbol{E}_1'$, $\boldsymbol{E}_2$, and $\boldsymbol{E}_2'$. The notations $\xi=x/w_0$ and y/w_0 are the two dimensionless coordinates in the dimension transverse to the propagation direction, respectively; $u_{3,F1}=A_{3,F1}/I^{1/2}$, $u_{1,2}=A_{1,2}/I^{1/2}$ and $u_{1,2}'=A_{1,2}'/I^{1/2}$ are the normalized amplitudes of the beams $\boldsymbol{E}_{3,F1}$, $\boldsymbol{E}_{1,2}$, and $\boldsymbol{E}_{1,2}'$, and the intensities of strong fields are both I. Here, on the left-hand side of these equations, the first terms describe the beam longitudinal propagation, and the second terms give the diffraction of the beams. On the right-hand side, the first terms are for the nonlinear self-Kerr effects, and the second as well as the third terms describe the nonlinear cross-Kerr effects. Equation (6.17) and Equation (6.18) can be solved numerically by using the split-step Fourier method that is usually employed.

The Kerr nonlinear coefficient, which is given by the general expression $n_2 \approx \mathrm{Re}\widetilde{\rho}_{10}^{(3)}/(\varepsilon_0 c n_0)$, is negative for self-defocusing and positive for self-focusing. The Kerr nonlinear susceptibility is expressed as $\chi^{(3)} = D\widetilde{\rho}_{10}^{(3)}$, where $D = N\mu_{10}^4/(\hbar^3\varepsilon_0 G_{3,F1}|G_{3,F1}|^2)$ for n_2^{S1-S2} and $D = N\mu_{10}^4/(\hbar^3\varepsilon_0 G_{3,F1}|G_i|^2)$ for n_2^{X1-X8}. N is the atomic density of the medium (determined by the cell temperature) and μ_{10} is the dipole matrix element between $|0\rangle$ and $|1\rangle$. We can solve the coupled density matrix equations to obtain $\widetilde{\rho}_{10}^{(3)}$.

If we neglect the diffraction, and self-Kerr terms, Eq. (6.17) and Eq. (6.18) can be solved to get the solution as $u_{3,F1}(z,\xi) = u_{3,F1}(0,\xi)[e^{-\xi^2/2} + e^{-(\xi-\xi_1)^2/2} + e^{-(\xi-\xi_2)^2/2} + e^{-(\xi-\xi_3)^2/2}]\cos\phi_{NL}$, where the XPM-induced phase shift $\phi_{NL} = 2k_{3,F1}n_2 zI[e^{-\xi^2/2} + e^{-(\xi-\xi_1)^2/2} + e^{-(\xi-\xi_2)^2/2} + e^{-(\xi-\xi_3)^2/2}]/(n_0 I_0)$. So, we obtain the intensity of $\boldsymbol{E}_{3,F1}$ as $I_{3,F1}(z,\xi) = |u_{3,F1}(z,\xi)|^2$. In a cascade three-level system, $I_0 \propto |\rho_{10}^{(1)}|^2$ for $\boldsymbol{E}_3$ and $I_0 \propto |\rho_{10}^{(3)}|^2$ for $\boldsymbol{E}_{F1}$. Moreover, the expression of the nonlinear phase shift shows that the strong spatial splitting can occur with increased I, n_2 and decreased I_0, where I and I_0 are the intensities of the strong and weak beams, respectively.

6.3.3
Experimental Results and Theoretical Analyses

First, we study spectra intensities and spatial properties of signals by changing the angle of the dressing fields in the cascade three-level system. The spectra intensities of the probe, $d_5 = \Gamma_{12} - i\Delta_2$ and fluorescence signals when $\Delta_1 > 0$ is scanned from negative to positive with $\Delta_2 = 55$ GHz and $A_2 = |G_3|^2/d_3$ are shown in Figure 6.10(a), (c), and (e). For comparing signals with $\boldsymbol{E}_2$ and $\boldsymbol{E}_2'$ blocked and with no beams blocked, Figure 6.10(a7), (c7), and (e7) are arranged to display the signals with $\boldsymbol{E}_2$ and $\boldsymbol{E}_2'$ blocked. For the probe signal, the single-photon absorption peak arises at $\Delta\theta_1$ depicted by the right dashed line in Figure 6.10(a). Moreover, the EIT window arises around $\Delta_1 = -55$ GHz and the corresponding EIA dip are depicted by the left dashed curve in Figure 6.10(a). For $d_5 = \Gamma_{12} - i\Delta_2$, in Figure 6.10(c), the primary AT splitting at $\Delta\theta_1$ and secondary one at around $\Delta_1 = -55$ GHz are obtained. For the fluorescence signals, two dips appear on the background peak. One dip is at around $\Delta_1 = -55$ GHz and the other is at $\Delta\theta_1$. Obviously, when $\boldsymbol{E}_2$ and $\boldsymbol{E}_2'$ are blocked, the EIT window of the probe, secondary AT splitting of the DFWM and the dip at around $\Delta_1 = -55$ GHz of the fluorescence disappear, as shown in Figure 6.10(a7), (c7), and (e7). Also, when $\Delta\theta_2$ is changed from -0.15 to 0.15, these signals at around $\Delta_1 = -55$ GHz change correspondingly with the strongest value at $\Delta\theta_2 = 0$.

To explain these results, we use the density matrix elements related to three signals. The Doppler effect and the power broadening effect on the weak signals are considered for the hot atom in the simulation. Determining the transmission signal of $\boldsymbol{E}_3$, Eq. (6.4) includes a term $i\Delta_1' + \Gamma_{10}$ acting as the single-photon resonance term only when Δ_1 is scanned, and $|G_2|^2/[\Gamma_{20} + i(\Delta_1' + \Delta_2')]$ acting as

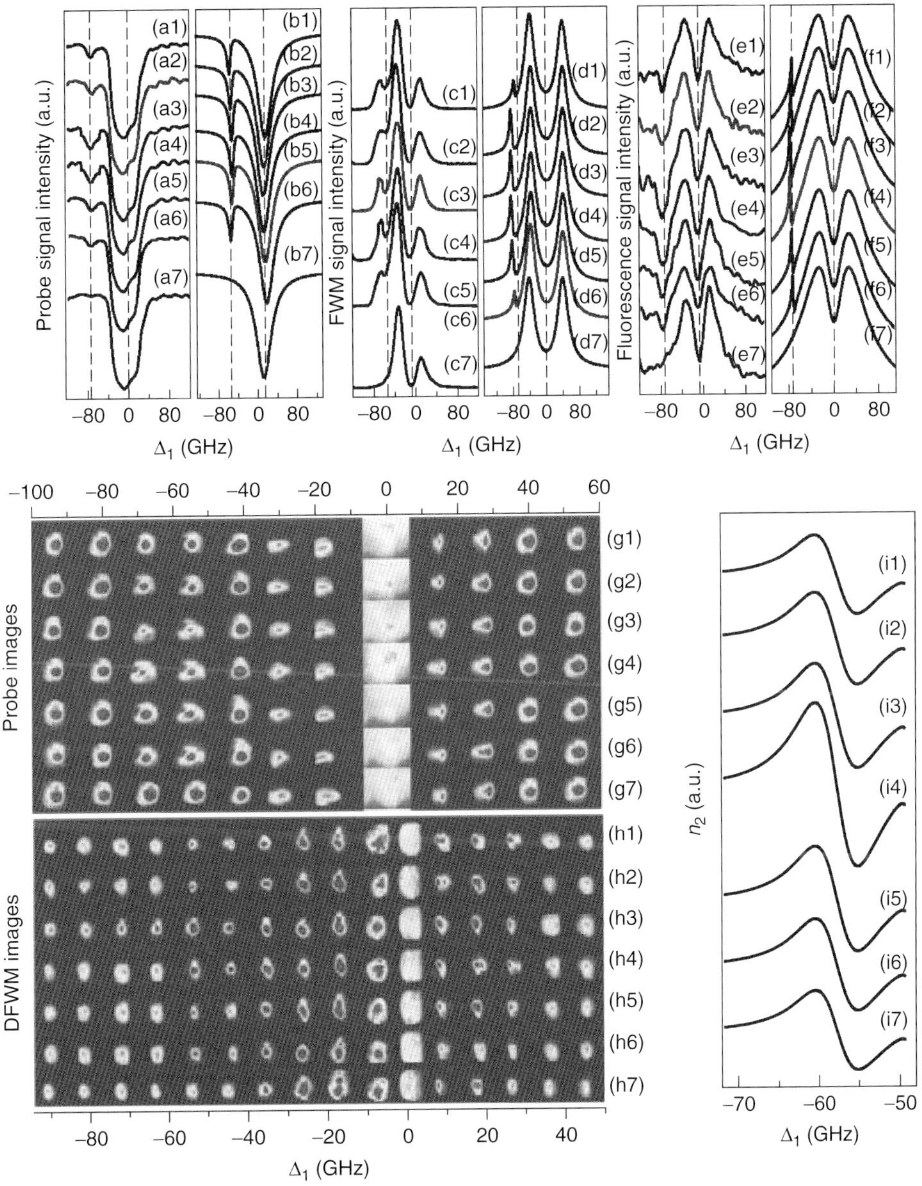

Figure 6.10 (a), (c), and (e) The experimentally measured intensities of the probe, $\boldsymbol{E}_{F1}$ and fluorescence, respectively, versus with $\Delta_2 = 55$ GHz and $\Delta\theta_1 = 0$. From (1) to (7), $\Delta\theta_2$ increases from -0.15 to 0.15. And the involved powers are set at $G_1 = G_1' = 1.2$ GHz, $G_3 = 0.3$ GHz, and $G_2 = G_2' = 1.1$ GHz. In (a7), (c7), and (e7), $\boldsymbol{E}_2$ and $\boldsymbol{E}_2'$ are blocked. (b), (d), and (f) The theoretically calculated intensities corresponding to (a), (c), and (e), respectively. (g) and (h) Images of $\boldsymbol{E}_3$ and $\boldsymbol{E}_{F1}$, the condition is same as that of (a). (i) The theoretical Kerr nonlinear coefficient n_2 of $\boldsymbol{E}_2(\boldsymbol{E}_2')$. Source: Adapted from Ref. [32].

the two-photon dressing term $|G_2|^2/[\Gamma_{20}+i(\Delta_1'+\Delta_2')]$ when either Δ_1 or Δ_2 is scanned. Moreover, the term $|G_1|^2/[\Gamma_{11}+|G_2|^2/(\Gamma_{12}-i\Delta_2')]$ acts as the dressing term only when Δ_2 is scanned, while the term $|G_1|^2/\Gamma_{00}$ is constant to dress the signal. So, when Δ_1 is scanned, the single-photon absorption dip arises at $\Delta\theta_1$ because of $i\Delta_1'+\Gamma_{10}$ in Eq. (6.4) and the EIT window at $\Delta_1=-55$ GHz, where the two-photon resonance condition $\Delta_1'+\Delta_2'=0$ determining the two-photon dressing term $|G_2|^2/[\Gamma_{20}+i(\Delta_1'+\Delta_2')]$ is satisfied, as shown in Figure 6.10(b). Certainly, the appearance of the left dip is due to the satisfaction of the EIA condition $\Delta_1'+\Delta_2'+(-\Delta_2'/2+\sqrt{\Delta_2'^2+4G_2^2/2})=0$. Then, for the $\boldsymbol{E}_{F1}$ signal, the expressions of the corresponding density matrices related to the DFWM processes are given in Eq. (6.1). In the expression, the single-photon emission term $i\Delta_1'+\Gamma_{10}$. The terms Γ_{00}, $|G_1|^2/\Gamma_{00}$, and $|G_3|^2/\Gamma_{00}$ are constant to dress the DFWM signal. Also, DFWM are dressed by the terms $|G_1|^2/(\Gamma_{00}+|G_1|^2/(\Gamma_{10}+i\Delta_1'))$ and $|G_3|^2/(\Gamma_{11}+|G_1|^2/(\Gamma_{10}+i\Delta_1'))$ only when Δ_1 is scanned while dressed by $|G_2|^2/[\Gamma_{20}+i(\Delta_1'+\Delta_2')]$ and $(|G_1|^2+|G_3|^2)/\{i\Delta_1'+\Gamma_{10}+|G_2|^2/[\Gamma_{20}+i(\Delta_1'+\Delta_2')]\}$ when either Δ_1 or Δ_2 is scanned. As a result, as shown in Figure 6.10(d), by scanning Δ_1 the one-photon emission peak forms by $i\Delta_1'+\Gamma_{10}$ in Eq. (6.1). The primary AT splitting at $\Delta_1=0$ is induced by the terms $|G_1|^2/[\Gamma_{00}+|G_1|^2/(\Gamma_{10}+i\Delta_1')]$ and $|G_3|^2/[\Gamma_{11}+|G_1|^2/(\Gamma_{10}+i\Delta_1')]$ in Eq. (6.1). The secondary AT splitting at $\Delta_1=-55$ GHz is caused by the term $|G_2|^2/[\Gamma_{20}+i(\Delta_1'+\Delta_2')]$ in Eq. (6.1) due to the two-photon resonant condition $\Delta_1'+\Delta_2'=0$. Finally, for the fluorescence signal, based on a similar method, we obtained the single-photon emission peak at $\Delta_1=0$ caused by the term $\Gamma_{10}+i\Delta_1'$ Eq. (6.2), while the suppression dip on the peak at $\Delta_1=0$ is caused by $\boldsymbol{E}_1$ and $\boldsymbol{E}_1'$, that is, the term $|G_1|^2/(i\Delta_1'+\Gamma_{10})$ in Eq. (6.2). The other two-photon dip at $\Delta_1=-55$ GHz satisfied $\Delta_1'+\Delta_2'=0$ is mainly induced by $\boldsymbol{E}_2$ and $\boldsymbol{E}_2'$, that is, the term $|G_2|^2/[\Gamma_{20}+i(\Delta_1'+\Delta_2')]$ in Eq. (6.3). When $\Delta\theta_2$ changes from -0.15 to 0.15, the detuning of $\boldsymbol{E}_2$ and $\boldsymbol{E}_2'$ changes with $\Delta\theta_2$ due to $\Delta_2'=\Delta_2+k_2v\cos(\theta_{2y}/2)$; thus, the dressing effect of $\boldsymbol{E}_2$ ($\boldsymbol{E}_2'$) on $\boldsymbol{E}_3$, DFWM, and fluorescence signals changes with $\Delta\theta_2$, too. What is more, the dressing effect is the strongest at $\Delta\theta_2=0$.

In addition, the images of $d_4=\Gamma_{21}+i\Delta_2$ and $\boldsymbol{E}_{F1}$ beams induced by XPM are recorded at different $\Delta\theta_2$ by Δ_1 scanned from negative to positive, as shown in Figure 6.10(g) and (h). The curves in Figure 6.10(i) are the corresponding numerical simulations of nonlinear index n_2 under the dressing effect of $\boldsymbol{E}_2$ and $\boldsymbol{E}_2'$ near $\Delta_1=-55$ GHz with different $\Delta\theta_2$.

Figure 6.10(g4) can be obtained when $\Delta\theta_2=0$ is fixed and Δ_1 is scanned, and the images of $d_4=\Gamma_{21}+i\Delta_2$ show that significant focusing takes up around the EIA windows (Figure 6.10(a)) in the region $\Delta_1<0$, in which n_2 is positive and arrives its maximum, as shown in Figure 6.10(i4). When $|\Delta\theta_2|$ changes from 0 to 0.15, the focusing and defocusing phenomena due to the crossing Kerr nonlinearity from $\boldsymbol{E}_2$ and $\boldsymbol{E}_2'$ become weaker gradually, as shown in Figure 6.10(g1) to (g3) and Figure 6.10(g5) to (g7). There are two reasons for such phenomena. First, the nonlinearity coefficient n_2 due to $\boldsymbol{E}_2$ and $\boldsymbol{E}_2'$, depending on the angle $\Delta\theta_2$, can be expressed as $n_2\propto\widetilde{\rho}_{10}^{(3)}=g/[(d_1+|G_2|^2/d_2)^2d_2]$. Therefore, with angle

$|\Delta\theta_2|$ increasing (from Figure 6.10(i1) to (i7)), we can see $|n_2|$ decreases and reaches its maximum at $\Delta\theta_2 = 0$. So, the focusing and defocusing effect are most significant at $\Delta\theta_2 = 0$. Second, as discussed, the EIA condition $\Delta_1' + \Delta_2' + (-\Delta_2'/2 + \sqrt{\Delta_2'^2 + 4G_2^2/2}) = 0$ is satisfied, which leads to stronger absorption and, therefore, weaker intensity of $d_4 = \Gamma_{21} + i\Delta_2$. The increasing of n_2 and decreasing of the intensity of $d_4 = \Gamma_{21} + i\Delta_2$ can result in a large phase shift to the right by $\boldsymbol{E}_2$ according to ϕ_{NL}.

For $\boldsymbol{E}_{\mathrm{F1}}$, the spatial splitting in the x-direction is observed. In order to explain these results, we consider the relative positions of $\boldsymbol{E}_{\mathrm{F1}}$ and $\boldsymbol{E}_1'$. Because $\boldsymbol{E}_1'$ has a small angle from $\boldsymbol{E}_1$ in the x-direction, as shown in Figure 6.9(d), $\boldsymbol{E}_{\mathrm{F1}}$ overlaps in the x-direction with $\boldsymbol{E}_1'$ due to the phase-matching condition $\mathbf{k}_{\mathrm{F1}} = \mathbf{k}_1 - \mathbf{k}_1' + \mathbf{k}_3$. So, the x-direction splitting is obviously obtained by the nonlinear cross-Kerr effect of $\boldsymbol{E}_1'$.

Next, we investigate the spectra intensity of the probe, DFWM, and fluorescence signals by scanning Δ_2 with $\Delta_1 = -40$ GHz and $\Delta\theta_1 = 0$ at different $\boldsymbol{E}_2'$. The spectra intensity of the probe changes when Δ_2 is scanned from negative to positive, and it becomes weakest at around $\Delta_2 = 40$ GHz, where the absorption is the strongest, that is, an EIA appears because of the dressing term $|G_2|^2/[\Gamma_{20} + i(\Delta_1' + \Delta_2')]$ in Eq. (6.4) as shown in Figure 6.11(a) and the corresponding theoretical simulation shown in Figure 6.11(b). It is obvious that the depth of this EIA window changes with $\boldsymbol{E}_2'$ and reaches its maximum at $\Delta\theta_2 = 0$. Meanwhile, the EIA window can also be obtained when $\boldsymbol{E}_1$ and $\boldsymbol{E}_1'$ are blocked, as shown in Figure 6.11(a1) and (a2). For the FWM signal dressed by $\boldsymbol{E}_2$ ($\boldsymbol{E}_2'$), the switching from pure enhancement (peak higher than the baseline in each curve) to pure suppression (dip lower than the baseline in each curve), and then to pure enhancement with $\boldsymbol{E}_2'$ increased, is observed as shown in Figure 6.11(c) (Fig. 6.11(d) is theoretically simulated according to Fig. 6.11(c)), which is induced by the term $|G_2|^2/[\Gamma_{20} + i(\Delta_1' + \Delta_2')]$ in Eq. (6.1). Obviously, when $\boldsymbol{E}_1$ and $\boldsymbol{E}_1'$ are blocked, the DFWM signal disappears as shown in Figure 6.11(c1) and (c2). For the fluorescence signal as shown in Figure 6.11(e), the main peak is caused by the two-photon emission term $\Gamma_{20} + i(\Delta_1' + \Delta_2')$ and the suppression dip on the main peak is due to the two-photon resonance caused by the terms $|G_2|^2/[\Gamma_{20} + i(\Delta_1' + \Delta_2')]$ and $|G_1|^2/[\Gamma_{20} + i(\Delta_1' + \Delta_2')]$ in Eq. (6.3), which can be verified by the theoretical simulation as shown in Figure 6.11(f). Thus, the intensity of the suppression dip increases with $|\Delta\theta_2|$ decreasing. When $\boldsymbol{E}_1$ and $\boldsymbol{E}_1'$ are blocked, the $\boldsymbol{E}_3$ power is so weak that the two-photon emission and suppression cannot be observed and the fluorescence signal degenerates into a flat line, as shown in Figure 6.11(e1) and (e2). With the $\boldsymbol{E}_1$ and $\boldsymbol{E}_1'$ turned on, the change of fluorescence signal is shown in Figures 6.11(e3)–6.11(e9), which can be explained by the two-photon resonance due to the term $|G_1|^2/[\Gamma_{20} + i(\Delta_1' + \Delta_2')]$.

The images of the probe beam $\boldsymbol{E}_3$ and DFWM after propagation are captured with the detuning Δ_2 scanned in a large region around the two-photon resonant point $\Delta_2' + \Delta_1' = 0$, with different $\boldsymbol{E}_2'$ as shown in Figure 6.11(g) and (h), respectively. The corresponding numerical simulations of n_2 are shown in Figure 6.11(i1) to (i7).

First, we investigate the variation in the backgrounds in Figure 6.11(g), which nearly do not change with changing $\Delta\theta_2$. For the two sides far away from the two-photon resonance ($\Delta_2' + \Delta_1' = 0$), $|n_2|$ induced by $\boldsymbol{E}_2$ ($\boldsymbol{E}_2'$) is very low and can be

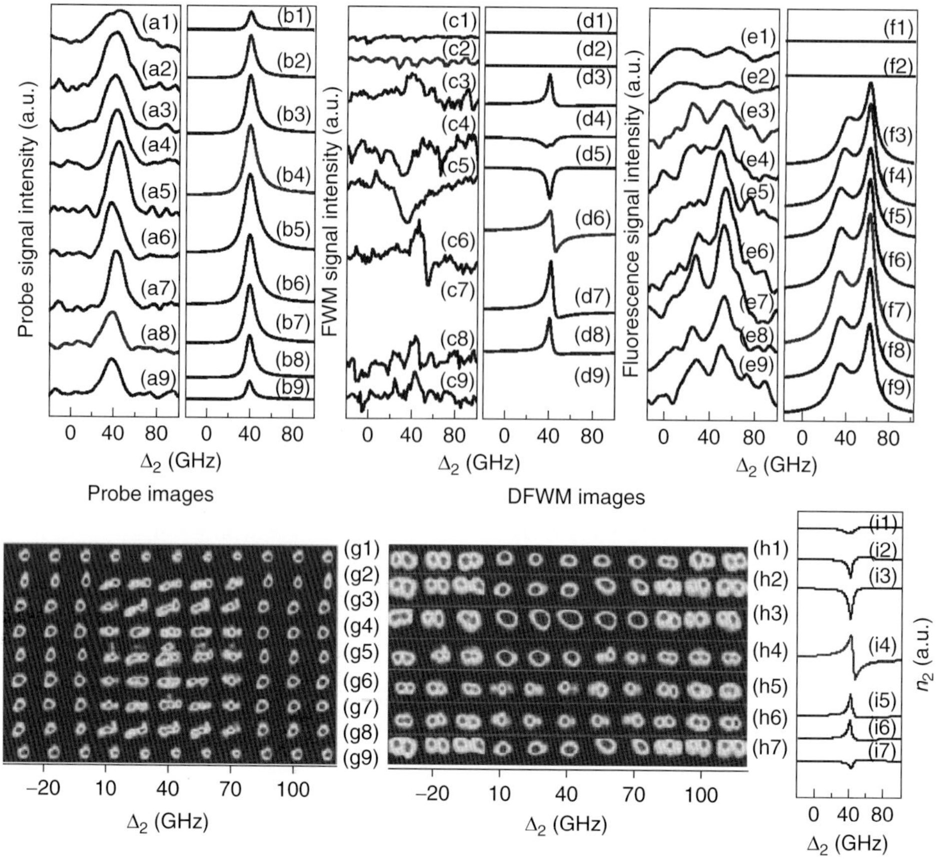

Figure 6.11 (a), (c), and (e) The experimentally measured intensities of the probe transmission, $\boldsymbol{E}_{F1}$ and fluorescence, respectively, versus Δ_2 with $\Delta_1 = -40$ GHz and $\Delta\theta_1 = 0$ in a cascade three-level system. From (1) to (7), $\Delta\theta_2$ increases from −0.15 to 0.15. And the involved powers are set at $G_1 = G_1' = 1.2$ GHz, $G_3 = 0.3$ GHz, and $G_2 = G_2' = 1.1$ GHz. In (a1), (a2), (c1), (c2), (e1), and (e2), $\boldsymbol{E}_1$ and $\boldsymbol{E}_1'$ are blocked. (b), (d), and (f) The theoretically calculated intensities corresponding to (a), (c), and (e), respectively. (g) and (h) Images of $\boldsymbol{E}_3$ and $\boldsymbol{E}_{F1}$, the condition is same as that of (a). (i) The theoretical Kerr nonlinearity of $\boldsymbol{E}_2$ ($\boldsymbol{E}_2'$).

neglected (Figure 6.11(i1) to (i7)). In this case, the cross-Kerr nonlinearity suffered by $\boldsymbol{E}_3$ is mainly due to $\boldsymbol{E}_1$ ($\boldsymbol{E}_1'$), and the refractive index is positive at $\Delta_1 = -40$ GHz, which leads to the focusing of $\boldsymbol{E}_3$. This explanation can be further verified by that fact that the spatial dispersion of $\boldsymbol{E}_3$ conserves in the whole scanned region of Δ_2 with $\boldsymbol{E}_2$ and $\boldsymbol{E}_2'$ blocked.

Second, in Figure 6.11(g), it is obvious that $\boldsymbol{E}_3$ spots suffer from defocusing more significantly in the region around the two-photon resonance point than in other places. This behavior can be explained by the focusing due to the cross-Kerr nonlinearity from $\boldsymbol{E}_2$ ($\boldsymbol{E}_2'$) for different Δ_2. In Figure 6.11(i1) to (i7), the coefficient

n_2 is negative when Δ_2 is scanned from -70 to -20 GHz and reaches its minimum at the resonant point ($\Delta_2 = 40$ GHz), which leads to more significant defocusing. With $|\Delta\theta_2|$ increasing, the beam spots become smaller around the resonant point (Figure 6.11(g1) to (g7)) due to the decrease of $|n_2|$ as shown in Figure 6.11(i1) to (i7). The reason is the same as that in Figure 6.10, which has been discussed. The other reason is that the absorption of $\boldsymbol{E}_3$ is enhanced with the decrease of $|\Delta\theta_2|$, that is, the FWM signal increases, while the transmission of $\boldsymbol{E}_3$ decreases. The two reasons will lead to the decreasing of ϕ_{NL} with the increase of $|\Delta\theta_2|$. Therefore, compared with the cases when $\Delta\theta_2$ set at other value, the splitting of the probe beam is most obvious at $\Delta\theta_2 = 0$.

Finally, in Figure 6.11(h), we concentrate on the spatial characteristics of $\boldsymbol{E}_{F1}$. The spatial images display the switching from focusing to defocusing as $|\Delta\theta_2|$ increases, which is in accordance with the changing of the nonlinear coefficient n_2, as shown in Figure 6.11(i). The phenomenon can also be explained by the reason discussed in the preceding text.

In order to understand the difference between the results with the dressing field and the pumping field changed, the variations in the signals by with the angle of the pumping fields changed is also investigated in the cascade three-level system. We show the spectra intensities of the probe transmission, DFWM, and fluorescence signals with different $\Delta\theta_1$ by scanning Δ_2 from negative to positive as shown in Figure 6.12(a), (c), and (e). Correspondingly, the theoretical simulations in Figure 6.12(b), (d), and (f) have reasonable agreement with the experimental results. In Figure 6.12(a) and (b), the switching from EIA (Figure 6.12(a1) to (a3)) to EIT (Figure 6.12(a4) and (a5)), and then to EIA (Figure 6.12(a6) to (a9)) is obtained with $\Delta\theta_1$ changed. For $\boldsymbol{E}_{F1}$, the switching from pure enhancement (Figure 6.12(c1) and (c2)) to pure suppression (Figure 6.12(c3) to (c5)), to partial enhancement/suppression (Figure 6.12(c6)), and then to pure enhancement (Figure 6.12(c7) to (c9)) with $\Delta\theta_1$ changed is observed. For fluorescence, the two-photon emission peak and two-photon suppression at $\Delta_2 = 30$ GHz are all strongest at $\Delta\theta_1 = 0$. Because the detuning Δ_1 changes with $\Delta\theta_1$ due to the expression $\Delta_1' = \Delta_1 + k_1 v\cos(\theta_{1x}/2)$, the results by scanning E_3 for the probe transmission, DFWM, and fluorescence are easily understood. When $|\Delta\theta_1|$ is not sufficiently large, the enhancement condition $\Delta_1' = -[(\Delta_2'' + \sqrt{\Delta_2''^2 + 4|G_2|^2})/2 + (\Delta_1' + \sqrt{\Delta_1'^2 + 4|G_1|^2})/2]$ is satisfied with Δ_2'' being the frequency detuning of $\boldsymbol{E}_2$ ($\boldsymbol{E}_2'$) relative to the dressed state by $\boldsymbol{E}_1$ ($\boldsymbol{E}_1'$). As a result, the EIA of probe transmission (Figure 6.12(a1) to (a3) and (a6) to (a9)) and pure enhancement of DFWM (Figure 6.12(c1) and (c2) and (c7) to (c9)) are obtained. While $|\Delta\theta_1| \approx 0$, the EIT of the probe transmission (Figure 6.12(a4) and (a5)) and pure suppression of the DFWM (Figure 6.12(c3) to (c6)) are obtained because the suppression condition $\Delta_1' + \Delta_2' = 0$ is satisfied. As a result, the switching from EIA to EIT in probe transmission or from pure enhancement to pure suppression in the DFWM signal is observed.

We also investigate the spatial characteristics of the probe transmission and the DFWM signal when $\Delta\theta_1$ changed and Δ_2 scanned from negative to positive as

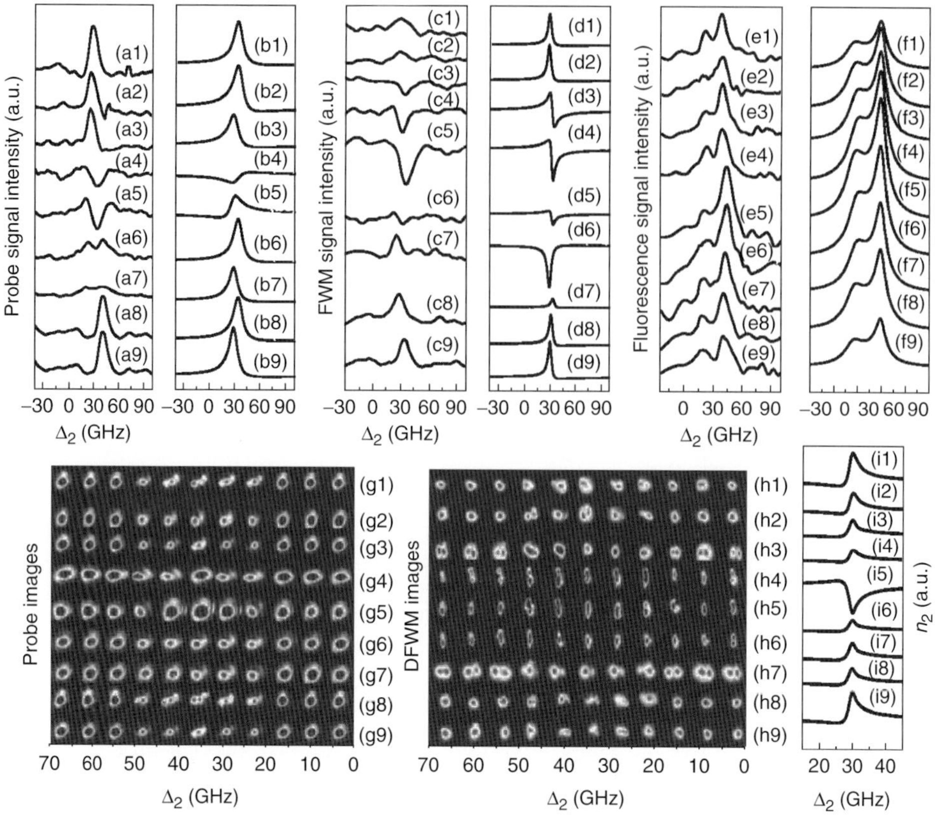

Figure 6.12 (a), (c), and (e) The experimentally measured intensities of the probe transmission, $\boldsymbol{E}_{F1}$ and fluorescence respectively versus Δ_2 with $\Delta_1 = -30$ GHz and $\Delta\theta_2 = 0$ in a cascade three-level system. From (1) to (9), $\Delta\theta_2$ decreases from 0.15 to −0.15. And the involved powers are set at $G_1 = G_1' = 1.2$ GHz, $G_3 = 0.3$ GHz, and $G_2 = G_2' = 1.1$ GHz. (b), (d), and (f) The theoretically calculated intensities corresponding to (a), (c), and (e), respectively. (g) and (h) Images of $\boldsymbol{E}_3$ and $\boldsymbol{E}_{F1}$, the condition is same as that of (a). (i) The theoretical Kerr nonlinear coefficient of $\boldsymbol{E}_1$ ($\boldsymbol{E}_1'$). Adapted from Ref. [32].

shown in Figure 6.12(g) and (h). As discussed in the preceding text, the spots of weak beams $\boldsymbol{E}_3$ and $\boldsymbol{E}_{F1}$ suffer from defocusing more significantly due to the negative n_2 (Figure 6.12(i5)) in the region around the resonance point than those on the two sides. However, n_2 changes with $\Delta\theta_1$ through $n_2 \propto \widetilde{\rho}_{10}^{(3)} = g/[(d_1 + |G_2|^2/d_2)^2 d_2]$, which determines that with $|\Delta\theta_1|$ increasing, the value of $k_1 v \cos(\theta_{1x}/2)$ changes correspondingly, and n_2 varies from positive to negative, as shown in Figure 6.12(i4) and (i6). Therefore, the focusing effect changes to the defocusing effect, and this will lead to the defocusing of spots of $\boldsymbol{E}_3$ and $\boldsymbol{E}_{F1}$ (compared with Figure 6.12(g5) and (h5)) as shown in Figure 6.12(g4), (g6), (h4), and (h6). Moreover, the positive n_2 increases with the increasing of $|\Delta\theta_1|$, as shown in Figure 6.12(i3) to (i1) and (i7) to (i9). It is obvious that E_3 and E_{F1} suffer the most significant focusing effect at $|\Delta\theta_1| = 0.15$ because n_2 reaches its maximum value, as shown in Figure 6.12(g1), (g9), (h1), and (h9), respectively.

For E_{F1}, the transferring between the spatial splitting in the x- and y-directions is observed. It can be explained by the spatial overlapping between the weak signal E_{F1} beam and the strong E_1' beam. Because E_1' has a small angle from E_1 either in the x-direction or the y-direction, as shown in Figure 6.9(c), E_{F1} overlaps in both the x- and y-directions, with E_1' being due to the phase-matching condition $\mathbf{k}_{F1} = \mathbf{k}_1 - \mathbf{k}_1' + \mathbf{k}_3$. With $|\Delta\theta_1| > 0.1$, the overlap of E_{F1} and E_1' in the x-direction is far dominant than that in the y-direction, so E_{F1} only shows the splitting in the x-direction, as shown in Figure 6.12(h1)–(h3) and (h7)–(h9). When $|\Delta\theta_1|$ decreases from 0.15 to 0, the y-direction overlap of E_{F1} and E_1' increases and the x-direction overlap decreases because of $\mathbf{k}_{F1} = \mathbf{k}_1 - \mathbf{k}_1' + \mathbf{k}_3$. So, the x-direction splitting slowly disappears. Finally, with $\Delta\theta_1 = 0$, E_1' overlaps with E_{F1} exactly in the y-direction. Thus, in such a case, the spatial modulation of E_{F1} is only in the y-direction, as shown in Figure 6.12(h4) to (h6).

From these analyses, we can see that the two schemes of angle changing (between the dressing fields and the pumping fields) have similarities and dissimilarities. The two schemes have similar dressing effects on intensity of signals. The angle switching from EIA to EIT for the probe beam or from enhancement to suppression for the FWM is observed in both. However, for spatial properties, it has more a remarkable effect by changing the angle of the pumping fields than changing that of the dressing fields. This is because changing the angle of the pumping fields affects not only the dressing effect but also the relative position of the FWM and strong fields as a result of the phase-matching condition $\mathbf{k}_{F1} = \mathbf{k}_1 - \mathbf{k}_1' + \mathbf{k}_3$ while changing the angle of the dressing fields only affects the dressing effect. So, we can conclude that changing the angle of pumping fields is more effective in controlling the relevant signals than changing the angle of the dressing fields.

In order to research the effect of changing the angles in different systems, the same process is executed in two-level systems. On the basis of the theoretical simulations, for FWM, by scanning Δ_2, the interaction between coexisting FWMs is observed. As shown in Figure 6.13(a) and (b), the left peak is the two-photon emission peak of DF3 satisfying the two-photon resonant condition $\Delta_1' - \Delta_2' = 0$ caused by the term $\Gamma_{00} + i(\Delta_2' - \Delta_1')$ in Eq. (6.7). The right enhancement/suppression of DF1 near $\Delta_2 = 0$ is due to the term $|G_1|^2/[\Gamma_{00} + |G_2|^2/(\Gamma_{20} + i\Delta_2')]$ in Eq. (6.5). The two-photon-like resonant condition is satisfied as $\Delta_1' - m\Delta_2' = 0$ by solving Eq. (6.5), where $m = G_1/G_2$ is determined by the experiment condition. In the experiment, because $m > 1$, the EIT-like condition $\Delta_2 = \Delta_1/m$ is equivalent to moving to the position of EIT to near $\Delta_2 = 0$. Thus, the parameter m means that E_1 (E_1') can modulate the dressing effect by E_2 (E_2'). Finally, in order to investigate the influence of the angle on the FWM and fluorescence in the two-level system, $\Delta\theta_2$ is changed from -0.15 to 0.15 in the experiment. It is observed that the emission peak is highest at $\Delta\theta_2 = 0$. For the DFWM, the transformation from first suppression then enhancement, to pure suppression, and then to first suppression

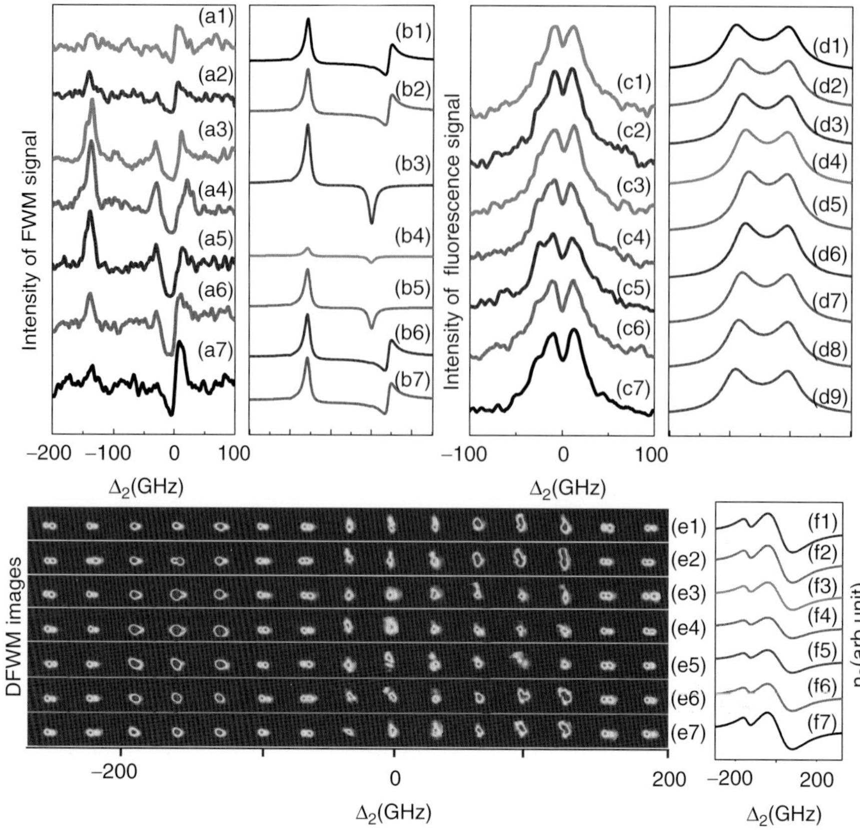

Figure 6.13 (a), (c) Show the experimentally measured intensities of the FWM and fluorescence versus Δ_2, respectively, with $\Delta_1 = -130$ GHz and $\Delta\theta_1 = 0$ in a two-level system. From (1) to (7), $\Delta\theta_2$ increases by 0.15 from −0.15. And the involved powers are $G_1 = G_1' = 1.7$ GHz, $G_3 = 0.3$ GHz, and $G_2 = G_2' = 1.8$ GHz. (b) and (d) The theoretically calculated intensities of the FWM and fluorescence, respectively, versus Δ_2 corresponding to (a), (c). (e) Images of FWM. The condition is same as that of (a). (f) The theoretical Kerr nonlinear coefficient of $\boldsymbol{E}_1$ ($\boldsymbol{E}_1'$).

then enhancement is shown in Figure 6.13(a). For fluorescence, the emission peak and suppression dip are all strongest at $\Delta\theta_2 = 0$. Because Δ_2 changes with $\Delta\theta_2$ due to the expression $\Delta_2' = \Delta_2 + k_2 v\cos(\theta_{2y}/2)$, the results with E_3 scanned for the DFWM and fluorescence are easily understood. When $|\Delta\theta_2|$ is not sufficiently large, the suppression condition is satisfied first and then the enhancement condition. As a result, first suppression and then enhancement of the DFWM are obtained (Figure 6.13(a1), (a2), (a6), and (c7)). While $|\Delta\theta_2| \approx 0$, the pure suppression of DFWM (Figure 6.13(a3) to (a5)) is obtained because the suppression condition is satisfied. As a result, the switching from enhancement/suppression to pure suppression is observed. For fluorescence and emission peak, the dressing effect and emission peak are strongest at $\Delta\theta_1 = 0$.

The images of $\boldsymbol{E}_{F1}$ at $\Delta_1 = -130$ GHz and $\Delta\theta_1 = 0$ with Δ_2 scanned (from top to bottom, $\Delta\theta_2$ changes from -0.15 to 0.15) in the two-level system, are shown in Figure 6.13(e). Correspondingly, the theoretical simulations of n_2 are shown in Figure 6.13(f). When $\Delta\theta_2 = 0$, Figure 6.13(e4) shows that $\boldsymbol{E}_{F1}$ images evolve from focusing to defocusing by increasing Δ_2 gradually, because n_2 varies from negative (in $\Delta_2 < 0$) to positive (in $\Delta_2 > 0$) (Figure 6.13(f4)). As discussed in the preceding text, the decreasing of $|n_2|$ due to $|\Delta\theta_2|$ will result in the transformation between focusing and defocusing, as shown in Figure 6.13(e5) and (e3). However, when $|\Delta\theta_2|$ increases further, n_2 suffered by $\boldsymbol{E}_{F1}$ will be positive, that is, $n_2 > 0$ in $\Delta_2 < 0$, and $n_2 < 0$ in $\Delta_2 > 0$, as shown in Figure 6.13(f2) and (f6), respectively. Therefore, the images of $\boldsymbol{E}_{F1}$ will correspondingly switch from focusing to defocusing when Δ_2 is scanned, as shown in Figure 6.13(e2) and (e6). Moreover, when $|\Delta\theta_2| = 0.15$, the images of $\boldsymbol{E}_{F1}$ will suffer from the most significant focusing in $\Delta_2 < 0$ and defocusing in $\Delta_2 > 0$ (Figure 6.13(e1) and (e7)) as $|n_2|$ reaches its maximum, as shown in Figure 6.13(f1) and (f7). The experimental results agree with the theoretical simulations very well.

6.4 Three-Photon Correlation via Third-Order Nonlinear Optical Processes

Entangled photon sources are essential for various quantum information processing protocols [33]. Compared to the generation of correlated photon pairs from the spontaneous parametric down-conversion (SPDC) process employing nonlinear optical crystals [34], the methods of generating photon pairs using atomic vapors as parametric amplifying media with third-order nonlinearities have attracted more attention in recent years [35–40]. The FWM process tuned near an atomic resonance can be used to produce narrowband correlated photon pairs, which are more appropriate for long-distance quantum communication. Owing to EIT [6, 7] in such multi-level atomic systems, strong enhancement in MWM processes can be expected [1]. In our previous work [1], by carefully aligning the incident beams, several MWM processes with the same frequency and direction are shown to coexist in the multi-level atomic system. However, the correlation properties among the generated FWM signals were not considered previously.

In this section, by employing an improved experimental setup as described in the subsequent text, two coexisting FWM processes in a double Λ-type atomic system are obtained in different directions and frequencies according to their respective phase-matching conditions and energy conservation (Figure 6.14). Here, we show that the two coexisting FWM signals and the probe beam are strongly correlated or anticorrelated with each other in temporal relative intensity, which can be considered as a light source with three-photon correlation. Also, we investigate the interplay between these two coexisting FWM processes in such an atomic system by varying the power of each incident beam.

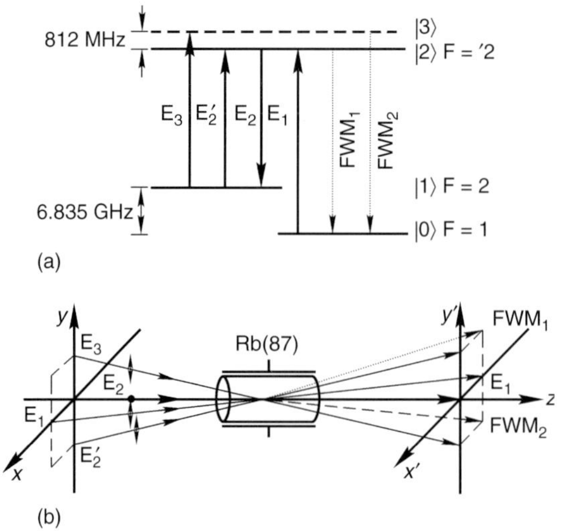

Figure 6.14 (a) Diagram of the relevant energy levels of the ^{87}Rb atom. (b) Schematic of the experimental setup. Double-headed arrows and filled dots denote the horizontal polarization and the vertical polarization of the incident beams, respectively.

6.4.1 Theory and Experimental Scheme

Our experimental system is schematically illustrated in Figure 6.14. Three real energy levels ($|0\rangle$ ($5S_{1/2}$, F = 1), $|1\rangle$ ($5S_{1/2}$, F = 2), and $|2\rangle$ ($5P_{1/2}$, F = 2) in ^{87}Rb atom) and one virtual energy level ($|3\rangle$, which is 816 MHz from the energy level $|2\rangle$) are involved in the experiment as shown in Figure 6.14(a), which form a double-Λ quasi-four-level atomic system.

The laser beam $\boldsymbol{E}_1$ (with frequency ω_1, Rabi frequency Ω_1, and wave vector $\boldsymbol{k}_1$) emitted from an ECDL connects the transition between the energy levels $|0\rangle-|2\rangle$ and serves as the probe beam. A single-frequency ring cavity laser (Coherent 899-21) was tuned to the $|1\rangle-|2\rangle$ transition of the ^{87}Rb atom at 794.98 nm, which splits into three beams $\boldsymbol{E}_2$, $\boldsymbol{E}_2'$, and $\boldsymbol{E}_3$ (ω_i, Ω_i, and $\boldsymbol{k}_i$) serving as the coupling beams, as shown in Figure 6.14(a). The beams $\boldsymbol{E}_2$ and $\boldsymbol{E}_2'$ with the same frequency ω_2 connect the transition between $|1\rangle$ and $|2\rangle$. The $\boldsymbol{E}_3$ beam connecting the transition between the energy levels $|1\rangle$ and $|3\rangle$ was downshifted $\omega_{32} = 816$ MHz by an AOM, thereby ω_3 and the frequency ω_2 of the coupling beam $\boldsymbol{E}_2$ (or $\boldsymbol{E}_2'$) always satisfy $\omega_3 - \omega_2 = \omega_{32}$. These four laser beams noncollinearly propagate in the same direction with small angles (~0.3°) between them and are overlapped at the center point inside a glass cell of length 7.5 cm (Figure 6.14(b)), containing Rb atoms without any buffer gas with a typical atomic density of 2×10^{11} cm^{-3}. When ω_1 was scanned around the transition $|0\rangle-|2\rangle$ (D_1 absorption line), an EIT window formed by the Λ-type three-level subsystem [6, 7] ($|0\rangle-|2\rangle-|1\rangle$) was observed, as well as the generations of two coexisting FWM signals, that is, FWM$_1$ from the Λ-type three-level subsystem ($|0\rangle-|2\rangle-|1\rangle$) satisfying the phase-matching condition

$\boldsymbol{k}_{F1} = \boldsymbol{k}_1 + \boldsymbol{k}_2' - \boldsymbol{k}_2$ and FWM$_2$ from the double-Λ quasi-four-level atomic system ($|0\rangle - |1\rangle - |2\rangle - |3\rangle$) satisfying the phase-matching condition $\boldsymbol{k}_{F2} = \boldsymbol{k}_1 + \boldsymbol{k}_3 - \boldsymbol{k}_2$. They were detected by two avalanche photo diodes (APDs), whereas the probe beam was detected by a photon diode detector.

6.4.2
Theory and Experimental Results

Figure 6.15 shows the probe and two FWM signal spectra as a function of ω_1. In such double-Λ configuration, coherence between the two atomic ground states will be built, which enhances the efficiency of the FWM processes [1]. With the especially arranged spatial pattern of the four incident beams, these three signals (two generated FWM and the probe beams) can be detected separately in space as shown in Figure 6.14(b), which is different from the previous cases [3, 41].

With the coexisting FWM processes, we can investigate the interplay between them by varying the powers of the probe beam $\boldsymbol{E}_1$ and the coupling beams $\boldsymbol{E}_2'$ and $\boldsymbol{E}_3$, respectively. The power of the $\boldsymbol{E}_2$ beam will always be fixed at 0.8 mW with a diameter of 0.8 mm. Figure 6.16(a) shows the power dependence of the FWM signals on the probe beam $\boldsymbol{E}_1$, while the powers of the coupling beams $\boldsymbol{E}_2'$ and $\boldsymbol{E}_3$ are both fixed at 7 mW with a diameter of 1.0 mm. As both FWM processes are related to the probe beam $\boldsymbol{E}_1$, the evolutions of the two FWM signals have the same trend, that is, they both increase as the power of the $\boldsymbol{E}_1$ beam increases. When taking a close look at the evolution of the FWM$_1$ signal, one can find the spectrum profile which has different structures at different power sets of beam $\boldsymbol{E}_1$, as shown in Figure 6.17. The AT splitting observed at the resonance in the FWM$_1$ spectrum (Figure 6.17(a)) is due to the weak probe field of 2 mW, which disappears

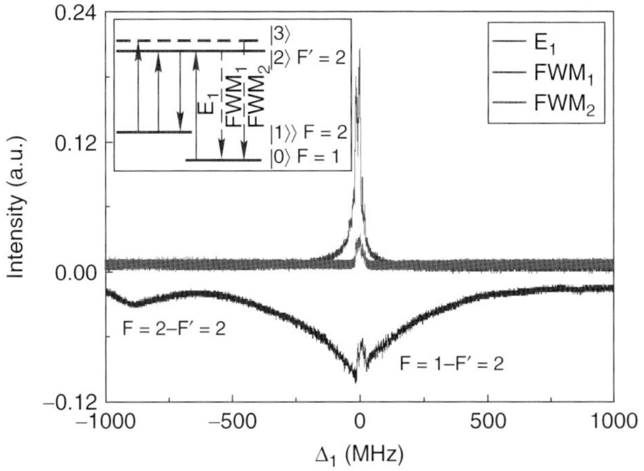

Figure 6.15 Typical coexisting FWM signal spectra, that is, FWM$_1$ (top curve) and FWM$_2$ (middle curves), and the probe spectrum with EIT window (bottom curve), as a function of the probe frequency detuning in the case of Figure 6.14(a). Source: Adapted from Ref. [42].

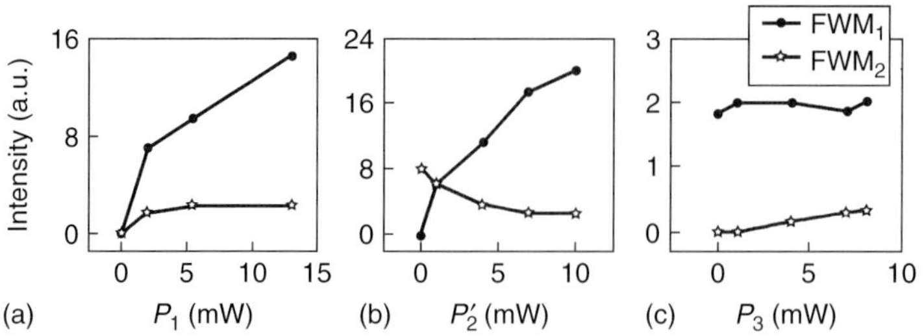

Figure 6.16 Power dependences of the coexisting FWM signals. Filled dots and stars stand for the FWM_1 signal and the FWM_2 signal, respectively. (a) Dependence on the $\boldsymbol{E}_1$ beam power while the powers of the coupling beams $\boldsymbol{E}_2'$ and $\boldsymbol{E}_3$ are both fixed at 7 mW. (b) Dependence on the $\boldsymbol{E}_2'$ beam power while the powers of the probe beam $\boldsymbol{E}_1$ and the coupling beam $\boldsymbol{E}_3$ are fixed at 4 and 7 mW, respectively. (c) Dependence on the $\boldsymbol{E}_3$ beam power while the powers of the probe beam $\boldsymbol{E}_1$ and the coupling beam $\boldsymbol{E}_2'$ are fixed at 4 and 7 mW, respectively.

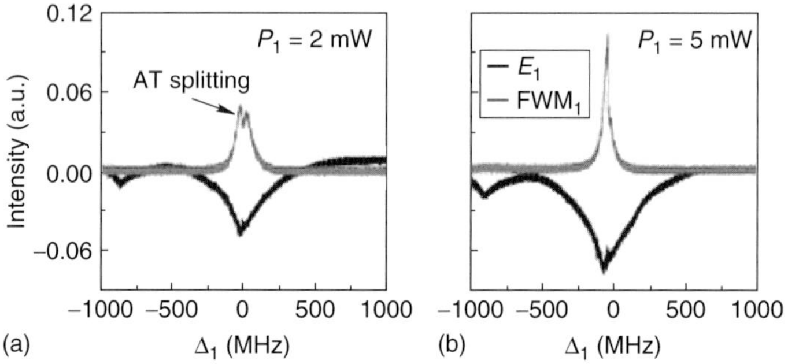

Figure 6.17 Typical AT splitting observed in the FWM_1 spectrum (a) compared with the non-AT splitting FWM_1 spectrum (b) at different power of the probe beam $\boldsymbol{E}_1$.

(Figure 6.17(b)) at the strong probe case ($P_1 \geq 4$ mW), corresponding to the case that the dressed effect will be cancelled gradually as the power of the probe field increases [4]. Next, we study the dependence of the FWM signal powers on the $\boldsymbol{E}_2'$ (or $\boldsymbol{E}_3$) beam, while the powers of the probe beam $\boldsymbol{E}_1$ and the coupling beam $\boldsymbol{E}_3$ (or $\boldsymbol{E}_2'$) are fixed at 4 and 7 mW, respectively. Differing from this case, the coupling beam $\boldsymbol{E}_2'$ (or $\boldsymbol{E}_3$) just contributes to the FWM_1 (or FWM_2) process. One can find that as the power of the coupling beam $\boldsymbol{E}_2'$ (or $\boldsymbol{E}_3$) increases, the FWM_1 (or FWM_2) process also increases gradually, whereas the FWM_2 (or FWM_1) process decreases, as shown in Figure 6.16(b) (or Figure 6.16(c)). We can conclude that competition exists between these two coexisting FWM processes. In addition, Figure 6.16 also shows that the efficiency of the resonant FWM_1 process is characteristically higher than the nonresonant FWM_2 process.

Finally, we would like to show that the three individually detected signals, that is, the two coexisting FWM signals and the probe signal are strongly intensity correlated or anticorrelated with each other. Figure 6.18(a) shows single-shot

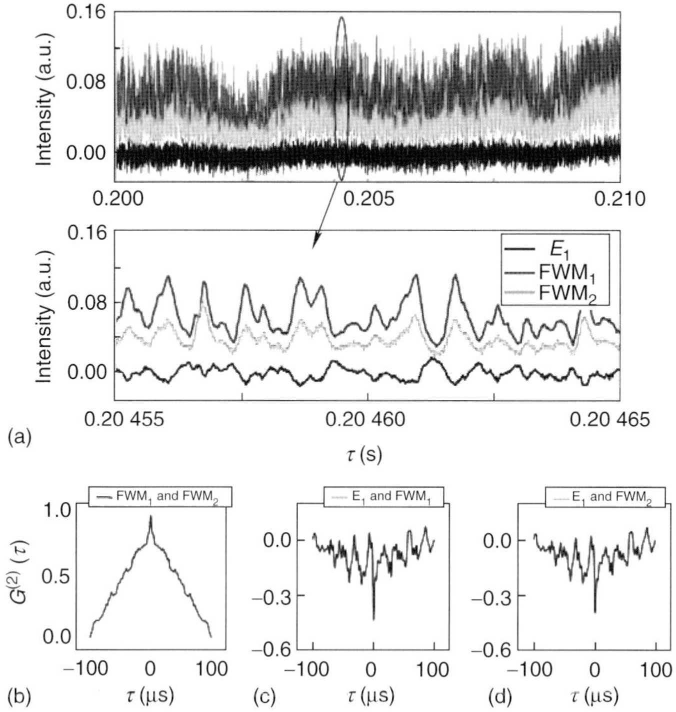

Figure 6.18 (a) Typical temporal waveforms of the probe field (bottom curves) and the coexisting FWM signal fields, that is, FWM_1 field (top curves) and FWM_2 field (middle curves). Cross-correlation $G^{(2)}(\tau)$ for (b) FWM_1 and FWM_2; (c) the probe field $\boldsymbol{E}_1$ and the FWM; and (d) the probe field $\boldsymbol{E}_1$ and the FWM_2, respectively. Source: Adapted from Ref. [42].

temporal waveforms of these three signals that are taken by three independent detectors, showing clear correlation and anticorrelation in the fluctuations of the three beams, respectively. The corresponding cross-correlation function $G^{(2)}(\tau)$ between intensities of two optical beams can be calculated by Ref. [43].

$$G^{(2)}(\tau) = \frac{\langle \delta I_m(t)\delta I_n(t+\tau)\rangle}{\sqrt{\langle [\delta I_m(t)]^2\rangle\langle [\delta I_n(t+\tau)]^2\rangle}} \tag{6.19}$$

where averaging over time is defined as $\langle \delta I_i(t)\rangle = \int_t^{t+T} \delta I_i(t)dt/T$; $\langle I_i\rangle$ is the average intensity of each laser beam and $\delta I_i(t)$ gives the time-dependent intensity fluctuations shown in Figure 6.18(a). τ is the selected time delay between the recorded signals; T is the time of integration (in this case, $T = 100$ μs). Figure 6.18(b) shows the cross-correlation of the two coexisting FWM signals and the $G^{(2)}(0)$ is calculated to be 0.994 using the experimental data, implying that the two generated FWM signals are strongly correlated. Meanwhile, the $G^{(2)}(\tau)$ function between the probe beam $\boldsymbol{E}_1$ and each of the FWM signals are shown in Figure 6.18(c) and (d), respectively. As different types of detectors were used in the experiment, the results only give about -0.5 anticorrelation between the generated FWM signal and the probe beam. We believe that they have strong anticorrelation with each other by carefully selecting and controlling the detectors. From this analysis, one can conclude that these three detected signals have three-photon correlation mediated by the third-order nonlinearity, which can be used for three-photon entanglement in quantum information processing.

6.4.3 Conclusion

We discuss three-photon correlation in a double-Λ-type quasi-four-level atomic system via FWM processes. By carefully aligning the incident beams, two FWM processes can coexist in such a system, which can be detected individually according to their respective phase-matching conditions. Owing to the improved experimental setup, these two FWM signals and the probe beam are found to be strongly (temporally) correlated or anticorrelated with each other.

6.5 Vacuum Rabi Splitting and Optical Bistability of MWM Signal Inside a Ring Cavity

6.5.1 Introduction

VRS [44] has been reported in strongly coupled single two-level atom-cavity system [45], where the frequency distance of the VRS is given by $2g$ with single-atom coupling strength g. When N two-level atoms collectively interact with the cavity mode [46], the coupling strength can be increased as $G = g\sqrt{N}$ and the distance of VRS for the collectively coupled atom-cavity system will be $2G$ [47]. Recently, studies on atom-cavity interactions are extended to a more composite system with an optical cavity and coherently prepared multi-level atoms [48], in which a narrow central peak was observed besides two broad sidebands (representing VRS) and can be well explained by the intracavity dispersion properties [49]. When the atom-cavity interaction reaches the "superstrong-coupling" condition with atom-cavity coupling strength G to be near or larger than the cavity free spectral range (FSR), multi-normal-mode splitting can be observed and well explained by the linear-dispersion enhancement due to the largely increased atomic density in cavity [50, 51].

Inspired by the intracavity phenomenon that quantum bright correlated light beams can be produced by driving an OPO above its threshold [52, 53], the bright correlated beams with an off-resonance FWM process have also been experimentally demonstrated [54]. In addition, experimental studies of coherently prepared atoms confined in a cavity lead to the observations of the EIT line width narrowing [55], coherent control of OB and multistability [56].

We discuss the relationship of VRS and OB of cavity MWM signals and achieve the goal to control VRS and OB simultaneously through the coherent control of dark and bright states and get the inclined VRS. Studies on the multi-dressed VRS and OB behavior of the zero- and high-order transmitted cavity MWM signal in the coupled system consisting of a specific ring cavity and reverse Y-type four-level atoms assemble are presented, with VRS coming from the atom-cavity collective effect induced by high atom density, while OB behavior comes from the self-Kerr nonlinearity effect (feedback effect). And the influences of dark and bright states are involved in detail. In addition, we show the linear gains and thresholds for the generated bright correlated light beams in lambda (Λ) and cascade (Ξ) subsystems in an OPO process and further discuss the OB thresholds in our OPA MWM process, which can bring about interesting applications in quantum information processing.

6.5.2 Basic Theory

We theoretically study on a cavity-atom coupling system consisting of rubidium atoms confined in the four-mirror-formed mode volume with a length of $L_c = 17$ cm (Figure 6.19(a)). The mirrors M3 and M1 are input and output mirrors with a radius of 50 mm, whose reflectance $r_3(r_1)$ and transmittance $t_3(t_1)$ fulfill the condition $r_i^2 + t_i^2 = 1$ $(i = 1, 3)$, while the mirrors M2 and M4 are high-quality reflectors. Cavity frequency scanning and locking can be implemented by a piezoelectric transducer (PZT) behind M4. The length of the rubidium vapor cell including the Brewster windows is $L_a = 7$ cm. The cell is wrapped in μ-metal sheets to shield from external magnetic fields and a heat tape is placed outside the sheets for controlling the temperature to influence the atomic density. As we do not consider Doppler effects in this section, our analysis is also suitable for the ring or standing-wave cavity.

A reverse-Y energy-level system, as shown in Figure 6.19(b), is constructed by four energy levels ($|0\rangle 5S_{1/2}(F=3)$, $|1\rangle 5P_{3/2}$, $|2\rangle 5D_{5/2}$, $|3\rangle 5S_{1/2}(F=2)$). In this atomic system, a horizontally polarized weak probe field E_1 (frequency ω_1, wave vector $\mathbf{k}_1$, and Rabi frequency G_1) probes the lower transition $|0\rangle \rightarrow |1\rangle$. Two

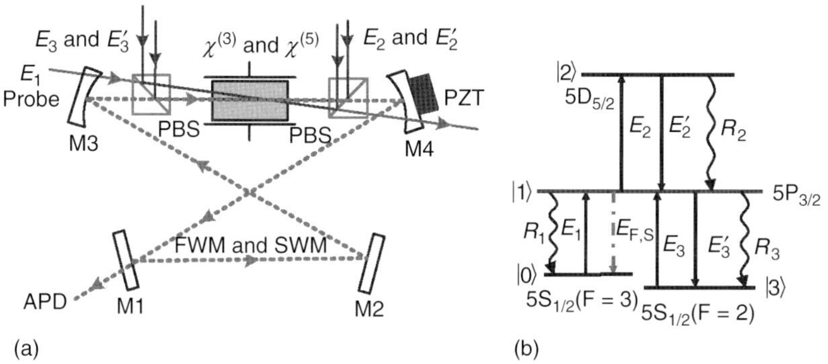

Figure 6.19 (a) A schematic diagram of a ring cavity containing the four-level atoms coupled with coherent external probe and control fields. The optical cavity length, which is fixed on atomic transition frequency, can be adjusted by PZT mounted on the mirror M4. The frequency of the input laser (as probe light) is scanned to measure the transmission spectra. (b) Scheme of the four-level atomic system.

vertically polarized coupling fields E_2 (ω_2, $\mathbf{k}_2$, and G_2) and E_2' (ω_2, $\mathbf{k}_2'$, and G_2') propagate in the direction opposite to E_1, and drive the upper transition $|1\rangle \to |2\rangle$. Two additional vertically polarized coupling fields E_3 (ω_3, $\mathbf{k}_3$, and G_3) and E_3' (ω_3, $\mathbf{k}_3'$, and G_3') propagate in the same direction as E_1, and drive the transition $|3\rangle \to |1\rangle$. There will be two EIT windows in ladder-type subsystems $|0\rangle \to |1\rangle \to |2\rangle$ and $|0\rangle \to |1\rangle \to |3\rangle$, both satisfying the two-photon Doppler-free condition. Moreover, there will be two FWM processes E_{F2} (satisfying $\mathbf{k}_{F2} = \mathbf{k}_1 + \mathbf{k}_2 - \mathbf{k}_2'$), E_{F3} ($\mathbf{k}_{F3} = \mathbf{k}_1 + \mathbf{k}_3' - \mathbf{k}_3$), four SWM processes E_{S2} ($\mathbf{k}_{S2} = \mathbf{k}_1 + \mathbf{k}' - \mathbf{k}_{33} + \mathbf{k}_2 - \mathbf{k}_2$), E_{S2}' ($\mathbf{k}_{S2}' = \mathbf{k}_1 + \mathbf{k}_3' - \mathbf{k}_3 + \mathbf{k}_2' - \mathbf{k}_2'$), E_{S3} ($\mathbf{k}_{S3} = \mathbf{k}_1 + \mathbf{k}_2 - \mathbf{k}_2' + \mathbf{k}_3 - \mathbf{k}_3$), and E_{S3}' ($\mathbf{k}_{S3}' = \mathbf{k}_1 + \mathbf{k}_2 - \mathbf{k}_2' + \mathbf{k}_3' - \mathbf{k}_3'$) generated in this system, which are all horizontally polarized. Such MWM signals propagate in the same direction as E_3', and can circulate in the ring cavity according to the cavity configuration. In our model, the total electromagnetic field can be written as $E = E_p e^{i\omega_1 t} + E_2 e^{i\omega_2 t} + E_2^* e^{i\omega_2 t} + E_3 e^{i\omega_3 t} + E_F e^{i\omega_1 t} + E_S e^{i\omega_1 t} + \text{c.c.}$. Generally, the density matrix elements related to FWM (SWM) signals can be obtained by solving the density matrix equations. Especially, when $E_3(E_3')$ are blocked, via the simple perturbation chain $\rho_{00}^{(0)} \xrightarrow{\omega_1} \rho_{10}^{(1)} \xrightarrow{\omega_2} \rho_{20}^{(2)} \xrightarrow{-\omega_2} \rho_{10}^{(3)}$, we can obtain the simple third-order density element $\rho_{F2}^{(3)} = G_{F2}/[d_2 d_1^2]$ for E_{F2} (its intensity $I_{F2} \propto |\rho_{F2}^{(3)}|^2$), where $G_{F2} = -iG_1 G_2 (G_2')^* \exp(i\mathbf{k}_{F2} \cdot \mathbf{r})$, $d_1 = \Gamma_{10} + i\Delta_1$, $d_2 = \Gamma_{20} + i(\Delta_1 + \Delta_2)$; $\Delta_i = \Omega_i - \omega_i$ is frequency detuning with resonance frequency ω_i and Γ_{ij} is the transverse relaxation rate between states $|i\rangle$ and $|j\rangle$. If $E_2(E_2')$ is strong enough, via the singly-dressed perturbation chain $\rho_{00}^{(0)} \xrightarrow{\omega_1} \rho_{G_2+0}^{(1)} \xrightarrow{\omega_2} \rho_{20}^{(2)} \xrightarrow{-\omega_2} \rho_{G_2\pm 0}^{(3)}$, we can obtain the singly-dressed third-order density element $\rho_{SDF2}^{(3)} = G_{F2}/[d_2(d_1 + |G_2|^2/d_2)^2]$ for E_{F2} (the intensity $I_{F2} \propto |\rho_{SDF2}^{(3)}|^2$). In the dressed FWM processes E_{F2}, $E_2(E_2')$ not only generates but also dresses the signal E_{F2}, so we refer to this dressing effect as *internal dressing*. Then, with E_3 turned on, the simple SWM process can be described by the perturbation chain $\rho_{00}^{(0)} \xrightarrow{\omega_1} \rho_{10}^{(1)} \xrightarrow{-\omega_3} \rho_{30}^{(2)} \xrightarrow{\omega_3} \rho_{10}^{(3)} \xrightarrow{\omega_2} \rho_{20}^{(4)} \xrightarrow{-\omega_2} \rho_{10}^{(5)}$, and the corresponding density element is $\rho_{S3}^{(5)} = G_{S3}/[d_2 d_3 d_1^3]$ for E_{S3} (its intensity $I_{S3} \propto |\rho_{S3}^{(5)}|^2$), where $G_{S3} = iG_1 G_2 (G_2')^* G_3 G_3^* \exp(i\mathbf{k}_{S3} \cdot \mathbf{r})$ and $d_3 = \Gamma_{30} + i(\Delta_1 - \Delta_3)$. If $E_2(E_2')$ and E_3 are all strong enough, the doubly-dressed E_{F2} (internal- and external-dressing effects of $E_2(E_2')$ and E_3, respectively) and E_{S3} (internal-dressing effect of both $E_2(E_2')$ and E_3) can be generated simultaneously. Via the perturbation chains $\rho_{00}^{(0)} \xrightarrow{\omega_1} \rho_{(G_2\pm G_3\pm)0}^{(1)} \xrightarrow{\omega_2} \rho_{20}^{(2)} \xrightarrow{-\omega_2} \rho_{(G_2\pm G_3\pm)0}^{(3)}$ and $\rho_{00}^{(0)} \xrightarrow{\omega_1} \rho_{(G_2\pm G_3\pm)0}^{(1)} \xrightarrow{-\omega_3} \rho_{30}^{(2)} \xrightarrow{\omega_3} \rho_{(G_2\pm G_3\pm)0}^{(3)} \xrightarrow{\omega_2} \rho_{20}^{(4)} \xrightarrow{-\omega_2} \rho_{(G_2\pm G_3\pm)0}^{(5)}$, we can obtain the doubly-dressed density elements $\rho_{DDF2}^{(3)} = G_{F2}/[d_2(d_1 + |G_2|^2/d_2 + |G_3|^2/d_3)^2]$ for E_{F2} (its intensity $I_{F2} \propto |\rho_{DDF2}^{(3)}|^2$), and $\rho_{DDS3}^{(5)} = G_{S3}/[d_2 d_3 (d_1 + |G_2|^2/d_2 + |G_3|^2/d_3)^3]$ for E_{S3} (the intensity $I_{S3} \propto |\rho_{DDS3}^{(5)}|^2$). The two doubly-dressed MWM signals can be distinguished in different EIT windows. Next, we set the rubidium atomic gas cell in the ring cavity (Figure 6.19(a)), in which the cavity polariton is formed in the interaction between the cavity mode and the N identical atoms with four energy levels. And the identical atoms assembled are most easily described as a homogeneously broadened medium in the small-gain limit. Set a as the cavity field, coupled with the transition $|0\rangle \to |1\rangle$, to form cavity mode with the generated FWM (SWM) signal according to the cavity configuration. Under the weak-cavity field limitation and with all the atoms initially in the ground state $|0\rangle$ and if the system is in equilibrium state, the transmitted cavity mode induced by FWM (SWM) and doubly dressed by $E_2(E_2')$ and E_3 can be obtained by Eq. (6.20) as

$$a_{\text{FWM}} = \frac{-g\sqrt{N}G_{\text{F(S)}}}{d_4(d_1 + g^2N/d_4 + |G_2|^2/d_2 + |G_3|^2/d_3)} \tag{6.20}$$

where $d_4 = i(\Delta_1 - \Delta_{ac}) + \gamma$.

6.5.3
VRS of Zero-Order Mode

In this section, we discuss multi-dressed VRS of zero-order cavity mode (single mode) of the MWM process in the coupled atom-cavity system based on the master equation formalism theory. This phenomenon is derived from the atom-cavity coupling effect, reflected mainly by g^2N in the denominator in Eq. (6.20), which shows a self-dressing effect for both coming from the atom-cavity coupling effect and dressing the cavity field in return. Meanwhile, we also analyze the related suppression (dark state) and enhancement (bright state) phenomena by scanning the frequency detuning of one dressing field.

6.5.3.1 Multi-Dressed VRS

In this part, we show the normal VRS and the AT splitting based on VRS of the transmitted cavity MWM signal by scanning the probe detuning. Figure 6.20 is the transmitted cavity spectra versus the probe frequency detuning, while the cavity field is tuned to resonate with the atomic transition $|0\rangle \rightarrow |1\rangle$ ($\Delta_{ac} = 0$). First, when E_2 and E_2' are turned on but their powers are relatively weak (i.e., not considering the dressing effect of $E_2(E_2')$) as shown in Figure 6.20(a1), the transmitted cavity FWM spectrum exhibits two peaks (representing two cavity-atom polaritons), forming the normal VRS of the atom-cavity system with separation $2g\sqrt{N}$ in frequency due to the self-dressing effect of g^2N (atom-cavity coupling effect). The two eigenvalues corresponding to dressed states $|\pm\rangle$ induced by g^2N are $\lambda_\pm = \pm g\sqrt{N}$ (due to $\Delta_{ac} = 0$, measured from $|1\rangle$) after diagonalizing the interaction Hamiltonian. And the two peaks of the transmitted cavity FWM signal (Figure 6.20(a1)) correspond, from left to right, to the dressed states $|+\rangle$ and $|-\rangle$, respectively (Figure 6.20(c1)). Next, when the power of E_2 becomes so strong that its dressing effect must be taken into account and its frequency matches the polariton resonance ($\Delta_2 = -g\sqrt{N} = -100\Gamma_{20}$ and $\Delta_2 = g\sqrt{N} = 100\Gamma_{20}$ in Figure 6.20(a2) and (a3), respectively), the AT splitting effect based on the VRS is induced. Owing to the dressing effects of g^2N and E_2, the two-peak spectrum in Figure 6.20(a1) turns into the three-peak spectrum in Figure 6.20(a2) and (a3). When E_2 resonates with the dressed state $|-\rangle$ with $\Delta_2 = -g\sqrt{N} = -100\Gamma_{20}$, a pair of new dressed states $|-\pm\rangle$ appear as shown in Figure 6.20(c2), and the corresponding eigenvalues are $\lambda_{-\pm} = \pm G_2$ (measured from $|-\rangle$) with separation

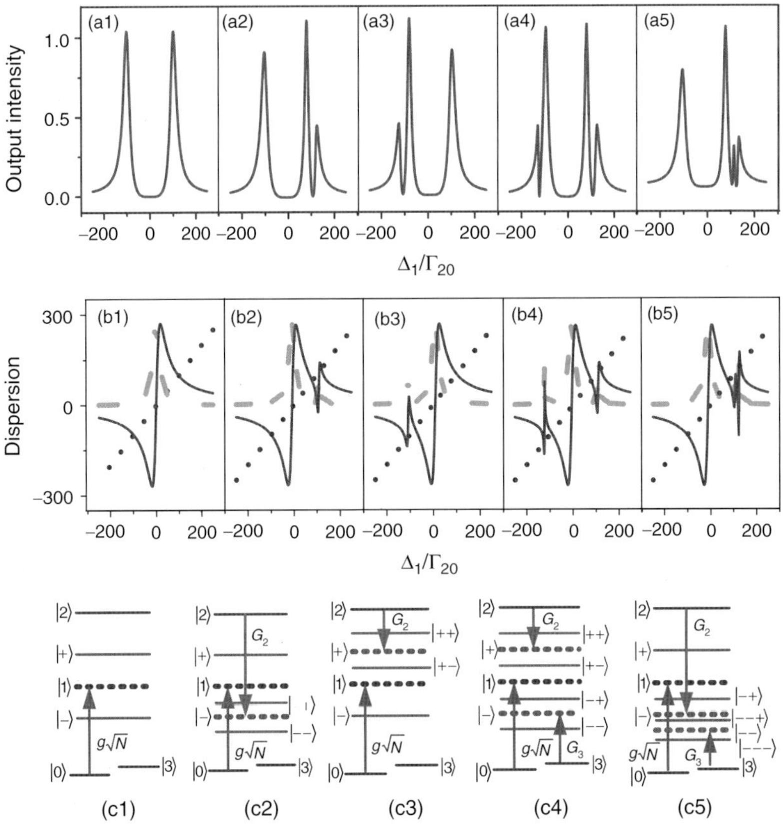

Figure 6.20 (a) Theoretically calculated transmitted cavity FWM (SWM) spectra versus Δ_1 for $\Delta_{ac}=0$. (a1) Normal VRS when the power of coupling fields E_2 and E'_2 are relatively weak and E_3 and E'_3 are blocked, (a2) and (a3) The AT splitting based on normal VRS when the power of E_2 (or E'_2) is strong enough and E_3 and E'_3 are blocked for (a2) $\Delta_2/\Gamma_{20}=-100$ and (a3) $\Delta_2/\Gamma_{20}=100$, respectively. (a4) and (a5) The AT splitting effect of the transmitted cavity FWM (SWM) when the powers of E_2 (or E'_2) and E_3 are strong sufficiently and E'_3 is blocked for (a4) $\Delta_2/\Gamma_{20}=-100$, $\Delta_3/\Gamma_{20}=-100$, and (a5) $\Delta_2/\Gamma_{20}=-100$, $\Delta_3/\Gamma_{20}=120$, respectively. (b) Theoretical plots of intracavity dispersion curves (solid curves); absorption (dashed curves); and detuning lines (dotted lines) versus Δ_1 corresponding to (a). (c) Dressed-state energy level diagrams corresponding to (a). Source: Adapted from Ref. [57].

$2G_2$. When E_2 couples with the dressed state $|+\rangle$ with $\Delta_2 = g\sqrt{N} = 100\Gamma_{20}$, the dressed states $|+\pm\rangle$ appear as shown in Figure 6.20(c3), and the corresponding eigenvalues are $\lambda_{+\pm} = \pm G_2$ (measured from $|+\rangle$) with separation $2G_2$. The three peaks of the transmitted cavity FWM signal in Figure 6.20(a2) and (a3) correspond, from left to right, to the dressed states $|+\rangle$, $|-+\rangle$, $|--\rangle$ (Figure 6.20(c2)) and $|++\rangle$, $|+-\rangle$, $|-\rangle$ (Figure 6.20(c3)), respectively. At last, with the field E_3 turned on and set as strong field, the dressing effect of E_3 must be considered. As a result, the triply-dressed transmitted cavity FWM (SWM) signal with four peaks is obtained, as shown in Figure 6.20(a4) and (a5). The four peaks in Figure 6.20(a4), from left to right, correspond to the dressed states $|++\rangle$, $|+-\rangle$, $|-+\rangle$ and $|--\rangle$, where the dressed states $|+\rangle$ induced by g^2N is split into $|++\rangle$, $|+-\rangle$ by E_3 ($\Delta_3 = -g\sqrt{N} = -100\Gamma_{20}$) with eigenvalues $\lambda_{+\pm} = \pm G_3$ (measured from $|+\rangle$), while $|-\rangle$ into $|-+\rangle$, $|--\rangle$ by E_2 ($\Delta_2 = -g\sqrt{N} = -100\Gamma_{20}$) with eigenvalues $\lambda_{-\pm} = \pm G_2$ (measured from $|-\rangle$), respectively, shown as Figure 6.20(c4). However, the four peaks in Figure 6.20(a5), from left to right, correspond to the dressed states $|+\rangle$, $|-+\rangle$, $|--+\rangle$, and $|---\rangle$, where $|-\rangle$ is split into $|-+\rangle$, $|--\rangle$ by E_2 ($\Delta_2 = -g\sqrt{N} = -100\Gamma_{20}$) with $\lambda_{-\pm} = \pm G_2$ (measured from $|-\rangle$), and further $|--\rangle$ is split into $|--+\rangle$ and $|---\rangle$ by E_3 ($\Delta_3 = g\sqrt{N} + G_2 = 120\Gamma_{20}$) with $\lambda_{--\pm} = \pm G_3$ (measured from $|--\rangle$), as shown in Figure 6.20(c5). Here, we only give the resonant case, in which the dressing field resonates with either naked energy level or the dressed energy level dressed by another dressing field, and other cases without resonant splitting can also be obtained by adjusting frequency detuning of the dressing fields.

Moreover, as the transmitted cavity FWM (SWM) signal can be determined by the corresponding absorption and dispersion characteristics, now we analyze the cavity transmission signals via the intracavity dispersion curves (solid), absorption curves (dashed), and detuning lines (dotted) shown in Figure 6.20(b1) to (b5), corresponding to Figure 6.20(a1) to (a5). Generally, the intersecting points of the detuning line and dispersion curve represent corresponding peaks in transmission spectra. In Figure 6.20(b1), there are three intersecting points, while in the transmitted cavity spectrum in Figure 6.20(a1), there are only two peaks corresponding to the bilateral intersections in Figure 6.20(b1). The disappearance of the peak corresponding to the middle intersecting point results from the larger absorption (dashed curve in Figure 6.20(b1)) at $\Delta_1/\Gamma_{20} = 0$. For other cases shown in Figure 6.20(b2) to (b5) corresponding to Figure 6.20(a2) to (a5), the explanations are the same as that for Figure 6.20(b1). The above-mentioned analysis indicates that the explanation from the dressing energy diagram in Figure 6.20(c) is in accordance with that from the intracavity dispersion and absorption properties in Figure 6.20(b).

6.5.3.2 **Avoided Crossing Plots**

In order to investigate the transmitted cavity signals influenced by the probe field and the other coupling field simultaneously, we present the avoided crossing plots in detail.

Figure 6.21 shows the transmitted cavity FWM (SWM) signal as a simultaneous function of Δ_1 and different frequency detunings of dressing fields. The typical avoided crossing shape in each plot clearly expresses the dark and bright states in the transmitted process in the atom-cavity system. For instance, when scanning Δ_1 and Δ_2 with other detunings fixed, as illustrated in Figure 6.21(a), the cavity field induces the primary splitting (VRS); E_3 induces the secondary splitting (AT splitting, corresponding to $\Delta_1/\Gamma_{20} = 200$); and E_2 induces the triple splitting with minimal distance at $\Delta_2 = -\Delta_1 = g\sqrt{N}$, $-g\sqrt{N} + G_3$, and $-g\sqrt{N} - G_3$ from left to right by scanning Δ_1 (horizontal axis). Simultaneously, by scanning Δ_2 (vertical axis), the transmitted cavity FWM (SWM) signal is suppressed (dark state) along the

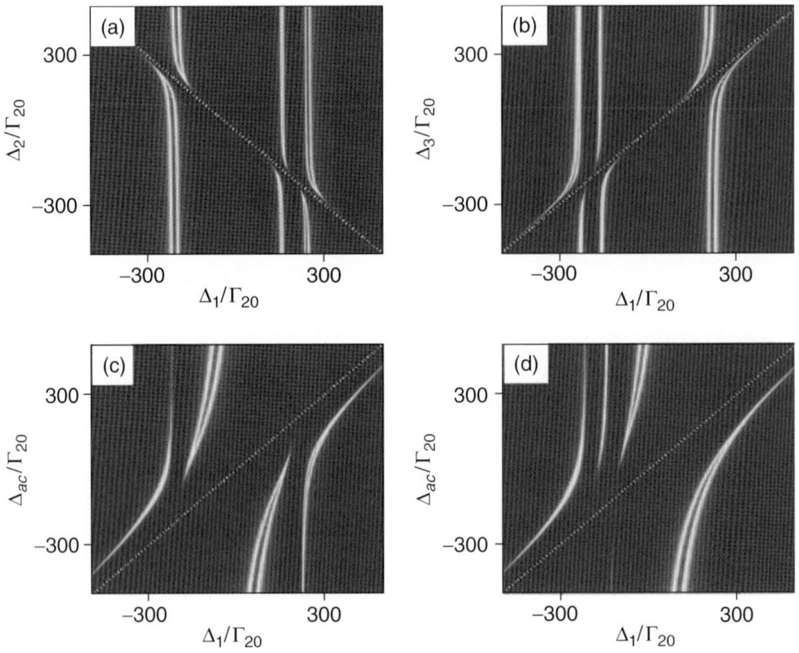

Figure 6.21 The transmitted cavity FWM (SWM) signal as a function of Δ_1 and different frequency detuning of dressing field for Δ_2 (a), Δ_3 (b), and Δ_{ac} (c) and (d), respectively when E_3' is blocked. The parameters used in the theoretical plots are (a) $\Delta_3/\Gamma_{20}=200$, (b) $\Delta_2/\Gamma_{20}=200$, (c) $\Delta_2/\Gamma_{20}=\Delta_3/\Gamma_{20}=200$, and (d) $\Delta_2/\Gamma_{20}=200$, $\Delta_3/\Gamma_{20}=-120$.

dotted dividing line $\Delta_1+\Delta_2=0$, which exactly satisfies the suppression condition resulting from the dressing effect of E_2 in this process. Similarly, by scanning Δ_1 and Δ_3 (Figure 6.21(b)), and Δ_1 and Δ_{ac} (Figure 6.21(c) and (d) with different Δ_2 and Δ_3), the transmitted cavity FWM (SWM) signals are suppressed significantly at $\Delta_1-\Delta_3=0$ (suppression condition from the dressing effect of E_3) and $\Delta_1-\Delta_{ac}=0$ (from the self-dressing effect of g^2N) along the dividing lines in Figure 6.21(b) to (d), respectively.

6.5.3.3 Suppression and Enhancement of MWM

Then we study the suppression (dark state) and enhancement (bright state) effect of the transmitted cavity signal of doubly-dressed FWM and triply-dressed FWM (SWM) in the atom-cavity system by scanning Δ_2 or Δ_3 at different Δ_1/Γ_{20}, illustrated in Figure 6.22(a1) to (a3). The baselines represent the transmitted cavity signal undressed by the scanning field. We call the dips lower than the baselines as suppression and peaks higher as enhancement.

Figure 6.22(a1) shows the suppression and enhancement of the transmitted cavity signal dressed by g^2N and $E_2(E_2')$ when Δ_2 is scanned at different Δ_1 within the atom-cavity subsystem $|0\rangle-|1\rangle-|2\rangle$ (FWM signal E_{F2}) with E_3 and E_3' blocked.

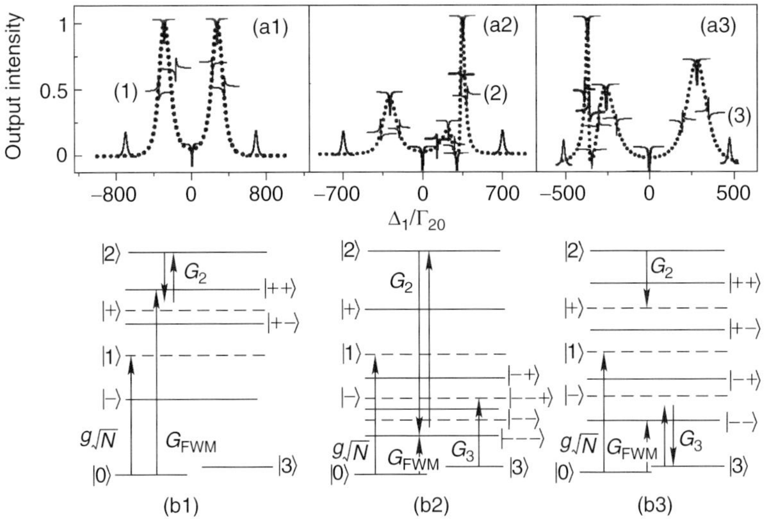

Figure 6.22 (a) The transmitted cavity spectra of the generated FWM (or SWM) dressed by fields a and $E_2(E_2')$ (a1), and fields a, $E_2(E_2')$, and E_3(a2) and (a3) versus Δ_2(a1) and (a2) and Δ_3(a3) with different Δ_1. The dashed profiles are cavity transmission signals versus Δ_1. (b1)–(b3) The dressed energy-level diagrams of the cavity transmission signal corresponding to (a1)–(a3).

The dashed double-peak global profile in Figure 6.22(a1), induced by g^2N, is the cavity-transmitted FWM signal without the dressing effect of $E_2(E_2')$. If Δ_1 is set from negative to positive in order, the cavity-transmitted FWM signal shows the evolution from all enhancement ($\Delta_1/\Gamma_{20} = -700$), to left enhancement and right suppression ($\Delta_1/\Gamma_{20} = -331$), to all suppression ($\Delta_1/\Gamma_{20} = -257$), to left suppression and right enhancement ($\Delta_1/\Gamma_{20} = -200$), to all suppression ($\Delta_1/\Gamma_{20} = 0$), to left enhancement and right suppression ($\Delta_1/\Gamma_{20} = 198$), to all suppression ($\Delta_1/\Gamma_{20} = 257$), to left suppression and right enhancement ($\Delta_1/\Gamma_{20} = 325$), to all enhancement ($\Delta_1/\Gamma_{20} = 700$), as shown in Figure 6.22(a1). Such an evolution is caused by the interaction between the dressing fields g^2N and $E_2(E_2')$. Here, we only take curve (1) in Figure 6.22(a1) for a detailed analysis. Under the self-dressing effect of g^2N, $|1\rangle$ will be split into two dressed states $|\pm\rangle$, and $|+\rangle$ will be split into two secondary-dressed states $|+\pm\rangle$ in the region $\Delta_1 < 0$ by $E_2(E_2')$. For the curve (1) ($\Delta_1/\Gamma_{20} = -331$), the transmitted cavity signal can resonate with $|++\rangle$ when Δ_2 is scanned if enhancement condition $\Delta_1 + \lambda_+ + \lambda_{++} = 0$ is satisfied, where $\lambda_+ = g\sqrt{N}$ and $\lambda_{++} = (\Delta_2 - \lambda_+)/2 + \sqrt{(\Delta_2 - \lambda_+)^2/4 + |G_2|^2}$, and then two-photon resonance of the transmitted cavity signal occurs if the suppression condition $\Delta_1 + \Delta_2 = 0$ is satisfied. As a result, the transmitted cavity signal shows left enhancement and right suppression, as illustrated by curve (1) and Figure 6.22(b1). Other curves in Figure 6.22(a1) showing different suppression and enhancement can be explained similarly to curve (1), so we only give their suppression (due to two-photon resonance) and enhancement (due to the resonance with dressed states) conditions as $\Delta_1 + \Delta_2 = 0$ and $\Delta_1 + \lambda_+ + \lambda_{+\pm} = 0$ for

$\Delta_1 < 0$, and $\Delta_1 + \Delta_2 = 0$ or $\Delta_1 + \lambda_- + \lambda_{-\pm} = 0$ for $\Delta_1 > 0$, where $\lambda_\pm = \pm g\sqrt{N}$ and $\lambda_{\pm\pm} = (\Delta_2 - \lambda_\pm)/2 \pm \sqrt{(\Delta_2 - \lambda_\pm)^2/4 + |G_2|^2}$. The results of Figure 6.22(a1) in $\Delta_1 < 0$ and $\Delta_1 > 0$ regions have symmetric centers $\Delta_1/\Gamma_{20} = \mp 257$, respectively, where the transmitted cavity signals are all purely suppressed.

Next, we study the transmitted cavity signal triply dressed by g^2N, $E_2(E_2')$, and E_3 when Δ_2 is scanned at different Δ_1 within the atom-cavity system (Figure 6.19(b)) with E_3' blocked as shown in Figure 6.22(a2). The dashed triple-peak global profile is the transmitted cavity FWM (SWM) signal induced by g^2N and E_3 when scanning Δ_1 without the dressing effect of $E_2(E_2')$. The three peaks correspond to the dressed states $|+\rangle$ and $|-\pm\rangle$, where the dressing effect of E_3 splits the state $|-\rangle$, dressed by g^2N, into $|-\pm\rangle$. The results by scanning Δ_2 at different Δ_1 show evolution similar to Figure 6.22(a1), except that two symmetric centers ($\Delta_1/\Gamma_{20} = 226$ and $\Delta_1/\Gamma_{20} = 360$) exist in the region $\Delta_1 > 0$, caused by the interaction of the three dressing fields g^2N, E_3, and $E_2(E_2')$. As the analysis method is similar to that used in Figure 6.22(a1), here we give a dressed energy diagram (Figure 6.22(b2)) corresponding to curve (2) to prove the left suppression and right enhancement. The enhancement conditions are $\Delta_1 + \lambda_+ + \lambda_{+\pm} = 0$ for $\Delta_1 < 0$ with $\lambda_+ = g\sqrt{N}$ and $\lambda_{+\pm} = (\Delta_2 - \lambda_-)/2 \pm \sqrt{(\Delta_2 - \lambda_+)^2/4 + |G_2|^2}$, and $\Delta_1 + \lambda_- + \lambda_{-+} + \lambda_{-+\pm} = 0$ or $\Delta_1 + \lambda_- + \lambda_{--} + \lambda_{--\pm} = 0$ for $\Delta_1 > 0$ with $\lambda_- = -g\sqrt{N}$, $\Lambda_{-\pm} = -(\Delta_3 + \lambda_-)/2 \pm \sqrt{(\Delta_3 + \lambda_-)^2/4 + |G_3|^2}$, and $\lambda_{-\pm\pm} = (\Delta_2 - \lambda_{-\pm})/2 \pm \sqrt{(\Delta_2 - \lambda_{-\pm})^2/4 + |G_2|^2}$ due to the cavity-transmitted signal resonating with the dressed state. And the suppression condition is $\Delta_1 + \Delta_2 = 0$ for both $\Delta_1 < 0$ and $\Delta_1 > 0$ due to two-photon resonance.

At last, we study the triply-dressed transmitted cavity FWM (SWM) process when Δ_3 is scanned to discuss the suppression and enhancement effect, as shown in Figure 6.22(a3). When Δ_1 is scanned without the dressing effect of E_3, the transmitted cavity FWM (SWM) spectrum with three peaks induced by g^2N and $E_2(E_2')$ is obtained, as the dashed curve in Figure 6.22(a3). These three peaks correspond to the dressed states $|+\pm\rangle$ and $|-\rangle$, where the dressing effect of $E_2(E_2')$ splits the state $|+\rangle$, induced by g^2N, into $|+\pm\rangle$. The results by scanning Δ_3 at different Δ_1 in Figure 6.22(a3) show an evolution opposite to that in Figure 6.22(a2), because the transmitted cavity FWM (SWM) field forms a Λ-type subsystem (Figure 6.22(a3)) with E_3 and a ladder-type subsystem (Figure 6.22(a2)) with $E_2(E_2')$. They have two symmetric centers ($\Delta_1/\Gamma_{20} = -357$ and $\Delta_1/\Gamma_{20} = -270$) in the region $\Delta_1 < 0$ caused by the interaction between the three dressing fields g^2N, $E_2(E_2')$, and E_3, and one symmetric center ($\Delta_1/\Gamma_{20} = 286$) in the region $\Delta_1 > 0$ caused by the interaction between g^2N and E_3. In this case, the enhancement condition is $\Delta_1 + \lambda_+ + \lambda_{++} + \lambda_{++\pm} = 0$ or $\Delta_1 + \lambda_+ + \lambda_{+-} + \lambda_{+-\pm} = 0$ for $\Delta_1 < 0$ with $\lambda_+ = g\sqrt{N}$ (induced by g^2N), $\lambda_{+\pm} = (\Delta_2 - \lambda_+)/2 \pm \sqrt{(\Delta_2 - \lambda_+)^2/4 + |G_2|^2}$ (induced by $E_2(E_2')$), and $\lambda_{+\pm\pm} = -(\Delta_3 + \lambda_{+\pm})/2 \pm \sqrt{(\Delta_3 + \lambda_{+\pm})^2/4 + |G_3|^2}$ (induced by E_3), and $\Delta_1 + \lambda_- + \lambda_{-\pm} = 0$ for $\Delta_1 > 0$ with $\lambda_- = -g\sqrt{N}$ (induced by g^2N) and $\lambda_{-\pm} = -(\Delta_3 + \lambda_-)/2 \pm \sqrt{(\Delta_3 + \lambda_-)^2/4 + |G_3|^2}$ (induced by E_3). The suppression condition is $\Delta_1 - \Delta_3 = 0$ for both $\Delta_1 < 0$ and $\Delta_1 > 0$. Here, we also give the corresponding energy-level diagram to explain the curve (3) in Figure 6.22(a3).

6.5.4 VRS of High-Order Modes

In order to study the splitting of high-order cavity modes (multi-normal mode), we adopt the intensity transmission coefficient of generated FWM (SWM) field for the coupled atoms-cavity system to discuss the splitting positions by the dressing state theory.

The cavity transmission coefficient of FWM (SWM) field is given by

$$T = \frac{(t_3 t_1)^2 e^{-\alpha L_a}}{(1 - r_3 r_1 e^{-\alpha L_a/2})^2 + 4 r_3 r_1 e^{-\alpha L_a/2} \sin^2(\phi/2)} \tag{6.21}$$

where $\phi(\omega_{F(S)}) = 2\pi(\Delta_{ac} - \Delta_1)/\Delta_{FSR} + (n-1)L_a\omega_{F(S)}/c$ is the round-trip phase shift experienced by the intracavity field around the cavity with FSR of the empty optical cavity $\Delta_{FSR} = 2\pi/(L_c/c)$ and the speed of light in vacuum c. The terms $\alpha = 2(\omega_{F(S)}/c)\mathrm{Im}[(1+\chi)^{1/2}]$ and $n = \mathrm{Re}[(1+\chi)^{1/2}]$ are the absorption coefficient and reflective index of the medium, respectively, with the susceptibility χ of the medium. In this section, we only consider the linear susceptibility, which can be derived by the master equation as

$$\chi = \frac{2g^2 N L_c}{L_a \omega_{F(S)}} \frac{i}{d_1 + |G_2|^2/d_2 + |G_3|^2/d_3} \tag{6.22}$$

The theory of these equations can be used to discuss the splitting positions of high-order cavity modes together with the dressing state theory in the coupled atoms-cavity system.

Figure 6.23 gives the transmission spectra, containing the splitting positions and the height of the multi-mode, of the generated FWM (SWM) when $\Delta_{ac} = 0$ and the power of the coupling fields ($E_2(E_2')$ and E_3) are relatively weak (i.e., not considering their dressing effects) with the increment of the atomic density of the medium. For an empty cavity, the cavity transmission peaks are Lorentzian in shape and have equal mode spacings (Δ_{FSR}), as the dashed curves in Figure 6.23(a). When the coupling strength $g\sqrt{N}$ increases to near or larger than Δ_{FSR} induced by the increased atomic density of the medium in cavity, the transmission spectra can be modified significantly: not only is the zero-order longitudinal mode ($m = 0$) split ($0_\pm$) with symmetrical center $\Delta_1 = 0$ but the high-order modes ($m = \pm 1, \pm 2, \ldots$) are also split ($1_\pm$, $-1_\pm$, $2_\pm$, $-2_\pm$, ...) with symmetrical center $\Delta_1 = -m\Delta_{FSR}/2$ by the cavity field as the solid curves shown in Figure 6.23(a). We introduce the dressing state theory in order to understand the splitting positions of the cavity modes, still considering g^2N from the atom-cavity coupling effect as a dressing field. When a is coupled with a random cavity mode ($m = 0, \pm 1, \pm 2, \ldots$), and after diagonalizing the interaction Hamiltonian for this atom-cavity system, the two eigenvalues corresponding to the splitting positions of peaks $m_\pm$ are derived as $\lambda_{m\pm}^{(m,\Delta_{ac})} = -(\Delta_{ac} + m\Delta_{FSR})/2 \pm \sqrt{(\Delta_{ac} + m\Delta_{FSR})^2/4 + g^2N}$ measured from cavity mode m, which means the two peaks locating at $\Delta_1^{(m,\Delta_{ac})} = -m\Delta_{FSR} - \lambda_{m\pm}^{(m,\Delta_{ac})}$ are relative to the position of m. Especially, when a resonates with zero-order mode ($m = 0$, $\Delta_{ac} = 0$), the splitting positions become $\Delta_1^{(0,0)} = -\lambda_{0\pm}^{(0,0)} = \mp g\sqrt{N}$

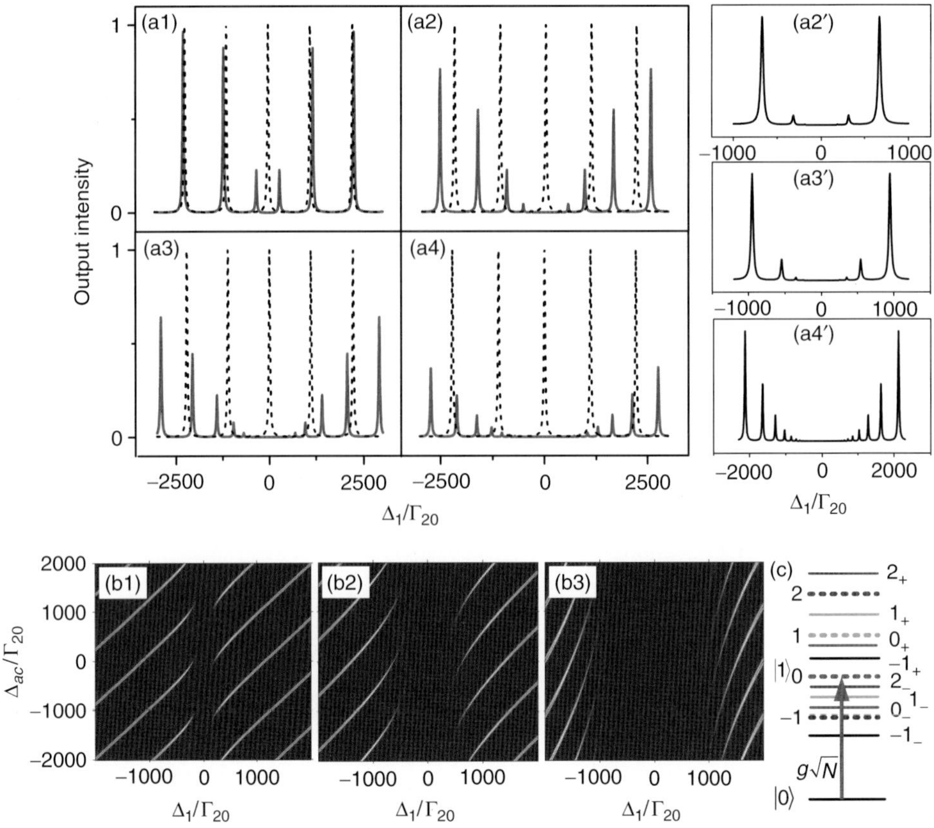

Figure 6.23 (a1)–(a4) Solid curves are transmission spectra of the generated FWM (SWM) when $\Delta_{ac} = 0$ and the power of coupling fields (E_2 (E_2') and E_3) are relatively weak with the atomic density of the medium increasing. Dashed curves are the transmission spectra of empty cavities. The illustrations (a2′)–(a4′) show MWM transmissions in smaller regions corresponding to (a2)–(a4). (b1)–(b3) Avoided crossing plots with increment of the atomic density of the medium. (c) The corresponding dressed energy-level diagram of the system.

and $\Delta_1^{(1,0)} = -\Delta_{FSR} - \lambda_{1\pm}^{(1,0)} = -\Delta_{FSR}/2 \mp \sqrt{\Delta_{FSR}^2/4 + g^2 N}$ corresponding to the peaks of the zero-order cavity mode and first-order mode, which has obvious symmetrical centers $\Delta_1 = 0$ and $\Delta_1 = -\Delta_{FSR}/2$, respectively, and is in accordance with the results in Figure 6.23(a). In addition, when the cavity field a resonates with other cavity modes such as $m = 1$ ($\Delta_{ac} = -\Delta_{FSR}$), the splitting positions are $\Delta_1^{(0,-\Delta_{FSR})} = -\lambda_{0\pm}^{(0,-\Delta_{FSR})} = -\Delta_{FSR}/2 \mp \sqrt{\Delta_{FSR}^2/4 + g^2 N}$ and $\Delta_1^{(1,-\Delta_{FSR})} = -\Delta_{FSR} - \lambda_{1\pm}^{(1,-\Delta_{FSR})} = -\Delta_{FSR} \mp g\sqrt{N}$ corresponding to the two peaks of zero-order and first-order cavity modes. Such results demonstrate that $\Delta_1^{(1,0)} = \Delta_1^{(0,-\Delta_{FSR})}$, which means the positions of the two splitting peaks of first-order mode when a resonates with zero-order mode are the same as that of zero-order mode when a resonates with first-order mode, as shown in Figure 6.23(b). Figure 6.23(b) presents the cavity-transmitted FWM (SWM) signal as a function of Δ_1 and Δ_{ac}, with the increment of the atomic density of the medium from (b1) to (b3). They are typical avoided crossing plots, which show the splitting of the multi-mode due to the cavity field scanned along the horizontal axis. The smallest splitting distance is at $\Delta_{ac}/\Gamma_{20}=0$ for zero-order mode, $\Delta_{ac}/\Gamma_{20}\approx 1000$ for first-order mode, and so on, and all of them increase with the accretion of the atomic density of the medium.

Next, we discuss the same case as Figure 6.23 except that the power of $E_2(E_2')$ is sufficiently strong to reveal its dressing effect on cavity transmission signal with $\Delta_2=0$, as shown in Figure 6.24. Compared with the cavity transmission spectrum in Figure 6.23, the result in Figure 6.24 shows an additional peak at $\Delta_1\approx 0$ due to the frequency pulling and absorption suppression corresponding to the intracavity dark state at $\Delta_1\approx -\Delta_2=0$. With the coupling strength $g\sqrt{N}$ increasing to near or larger than Δ_{FSR}, not only is the zero-order mode ($m=0$) split into three peaks ($0_\pm$ and 0_0), but the high-order modes ($m=\pm 1, \pm 2, \ldots$)

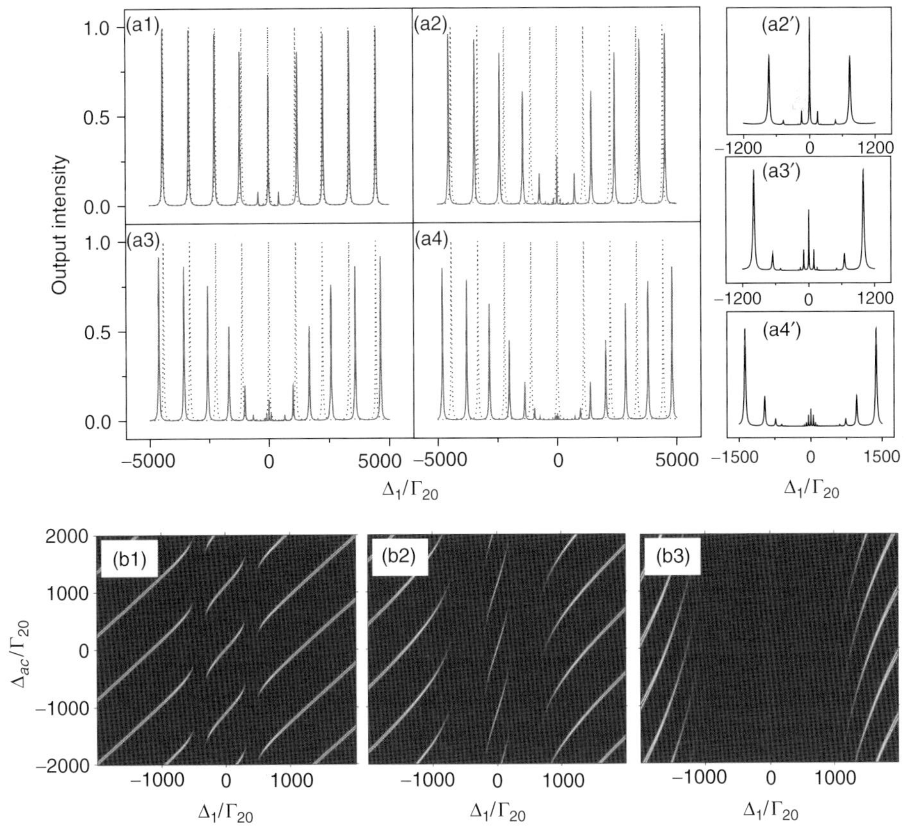

Figure 6.24 The figure setup is as Figure 6.23(a) and (b), except that the power of $E_2(E_2')$ is strong sufficiently. Source: Adapted from Ref. [57].

are also split into three peaks ($1_\pm$ and 1_0, $-1_\pm$ and -1_0, $2_\pm$ and 2_0, $-2_\pm$ and $-2_\pm$, etc.) by g^2N and $E_2(E_2')$, and every pair of $m_\pm$ has a symmetrical center $\Delta_1 = -m\Delta_{\mathrm{FSR}}/2$, shown as the solid lines in Figure 6.24(a). We can apply the dressing state theory to explain the positions of the splitting peaks as well as for the random cavity mode m. After diagonalizing the interaction Hamiltonian for this atom-cavity system, in which g^2N and $E_2(E_2')$ are considered as dressing fields and coupled with the random cavity mode ($m = 0, \pm 1, \pm 2, \ldots$), we can obtain the three eigenvalues as $\lambda_{m\pm}^{(m,\Delta_{ac})} = -(\Delta_{ac} + m\Delta_{\mathrm{FSR}})/2 \pm \sqrt{(\Delta_{ac} + m\Delta_{\mathrm{FSR}})^2/4 + g^2N + |G_2|^2}$ and $\lambda_0^{(m,\Delta_{ac})} = -(\Delta_{ac} + m\Delta_{\mathrm{FSR}})$ measured from cavity mode m. Thus, the m-order cavity mode can be split into three peaks including the two side peaks $m_\pm$ locating at $\Delta_1^{(m,\Delta_{ac})} = -m\Delta_{\mathrm{FSR}} - \lambda_{m\pm}^{(m,\Delta_{ac})}$ and a narrow peak m_0 locating at $\Delta_1^{(m,\Delta_{ac})} = -m\Delta_{\mathrm{FSR}} - \lambda_0^{(m,\Delta_{ac})}$ under superstrong-coupling condition. Similarly, we can also deduce $\Delta_1^{(1,0)} = \Delta_1^{(0,-\Delta_{\mathrm{FSR}})}$, which means the positions of the two side peaks of the first-order mode when a resonates with zero-order mode are the same as that of the zero-order mode when a resonates with first-order mode, as can be demonstrated in Figure 6.24(b). Figure 6.24(b) gives the cavity-transmitted FWM (SWM) signal as a function of Δ_1 and Δ_{ac} with the atomic density increasing from (b1) to (b3). Similarly, they are avoided crossing plots of multi-mode except an additional peak at $\Delta_1 \approx 0$ due to the dressing effect of $E_2(E_2')$, compared with that in Figure 6.23(b).

6.5.5 Steady-State Linear Gain and OPO Threshold

The four-level inverted-Y system in Figure 6.19(b) also may lead to three possibilities of fluorescence generation corresponding to the lambda (Λ) subsystem ($|0\rangle \to |1\rangle \to |3\rangle$ with E_1 and E_3 on) and two cascade (Ξ) subsystems ($|0\rangle \to |1\rangle \to |2\rangle$ with E_1 and E_2 on; $|3\rangle \to |1\rangle \to |2\rangle$ with E_3 and E_2 on). When the fluorescence interacts with cavity mode, the bright correlated anti-Stokes and Stokes light beams are generated. We set up the theoretical model to obtain the steady-state linear gain for these generated paired beams in the above-mentioned atom-cavity subsystems and get corresponding thresholds.

First, we discuss the Λ-type subsystem in which E_1 resonates with transition $|0\rangle \to |1\rangle$, E_3 resonates with $|3\rangle \to |1\rangle$, and the cavity field a is coupled with the two transitions so that the bright correlated light beams R_3 from $|1\rangle$ to $|3\rangle$ and R_1 from $|1\rangle$ to $|0\rangle$ can circulate in cavity. Similar to the generation of a single bright beam, the two beams can couple and influence each other because of the crossed energy level for stimulation and generation. With a further consideration of the degeneracy and the collisions effect between $|3\rangle$ and $|0\rangle$ for preventing the optically pumping to $|3\rangle$, and under the steady-state approximation, we can obtain the linear gain from the coefficient of a in the expression $ig\sqrt{N}\rho_{10}$ and $ig\sqrt{N}\rho_{13}$ as

$$G_{10}^{\Lambda} = \frac{-2g^2N\{2G_1^2\gamma_{30}t + G_3(G_1+G_3)[2\Gamma_{30}(\gamma_{10}+\gamma_{30}-\gamma_{03}) - \gamma_{03}t] + \Gamma_{13}\Gamma_{30}\gamma_{30}t\}}{\begin{array}{c} n\{[2sG_1 + \Gamma_{10}h](G_1^2 + \Gamma_{13}\Gamma_{30}) + [2qG_1^2 + \Gamma_{13}h](G_3^2 + \Gamma_{10}\Gamma_{30}) \\ -x + G_1^2G_3^2(-6\gamma_{30} + 12\Gamma_{30} + 2t - 6\gamma_{03})n\}G_1 \end{array}} \tag{6.23}$$

$$G_{13}^{\wedge} = \frac{-g^2N\{3G_3^2\gamma_{03}t + G_1(G_1+2G_3)[2\Gamma_{30}(\gamma_{03}+\gamma_{13}-\gamma_{30})-\gamma_{30}t] + \Gamma_{10}\Gamma_{30}\gamma_{03}t\}}{\begin{array}{c}[2sG_1+\Gamma_{10}h](G_1^2+\Gamma_{13}\Gamma_{30}) + [2qG_1^2+\Gamma_{13}h](G_3^2+\Gamma_{10}\Gamma_{30}) \\ -x + G_1^2G_3^2(-6\gamma_{30}+12\Gamma_{30}+2t-6\gamma_{03})\end{array}} \tag{6.24}$$

where $h=\gamma_{03}\gamma_{10}+\gamma_{03}\gamma_{13}+\gamma_{10}\gamma_{30}+\gamma_{13}\gamma_{30}$, $q=\gamma_{10}+2\gamma_{03}+\gamma_{30}$, $s=\gamma_{13}+\gamma_{03}+2\gamma_{30}$, $x=\gamma_{10}\gamma_{13}\gamma_{30}h$, and $t=\gamma_{10}+\gamma_{13}$. There will be linear gain for both the bright correlated light beams when $G_{10}^{\wedge}>0$ and $G_{13}^{\wedge}>0$ are simultaneously satisfied, which means that the two beams can both oscillate above these two thresholds with narrowed line widths when we adjust G_1 and G_3. We conclude a possible method to control the experimental conditions to satisfy the thresholds condition from Eq. (6.23) and Eq. (6.24), in which Γ_{30} should be very small so that it can be neglected and $\gamma_{10}+\gamma_{13}+6\Gamma_{30}>3(\gamma_{03}+\gamma_{30})$ has to guarantee positive denominators. Moreover, a crucial requirement is that $[2p^2/(1+p)]<[\gamma_3/\gamma_{30}]<[(p^2+2p)/3]$, where $p=G_1/G_3$ ($0<p<1$ or $p>2$), which provides two possible ways by which it can be experimentally achieved. When in the boundary of thresholds, we can hold G_1 as a constant while increasing G_3 with $G_3>G_1$, or we can hold G_3 as a constant while increasing G_1 with $G_1>2G_3$.

A Ξ-type subsystem $|0\rangle\to|1\rangle\to|2\rangle$, in which E_1 resonates with $|0\rangle\to|1\rangle$, E_2 resonates with $|2\rangle\to|1\rangle$, and the cavity field a couples with the two transitions, leads to the similar result that bright correlated light beams R_2 from $|2\rangle$ to $|1\rangle$ and R_1 from $|1\rangle$ to $|0\rangle$ can circulate and affect each other in the cavity. Meanwhile, we also assume the conditions $\rho_{00}+\rho_{11}+\rho_{22}=1$, $\rho_{10}=-\rho_{01}$, and $\rho_{21}=-\rho_{12}$.

And considering an approximation of the neglected crossed influence of the cavity field, we get the steady-state linear gain as

$$G_{10}^{\Xi} = \frac{g^2N(3\gamma_{21}G_1^2n + \Gamma_{21}\Gamma_{20}\gamma_{21}n + G_2^2(\gamma_{03}(\gamma_{10}-\gamma_{21}) + 2\Gamma_{20}n))}{\begin{array}{c}[2(\gamma_{03}+2\gamma_{21})G_1^2+\Gamma_{10}l](G_1^2+\Gamma_{21}\Gamma_{20}) + [2uG_2^2+\Gamma_{21}l] \\ (G_2^2+v) - \Gamma_{21}vl + G_2^2G_1^2(-6\gamma_{03}+12\Gamma_{20}+2j)\end{array}} \tag{6.25}$$

$$G_{21}^{\Xi} = \frac{g^2N(3\gamma_{03}G_2^2j + \Gamma_{20}\Gamma_{10}\gamma_{03}j + G_1^2(\gamma_{21}(2\gamma_{03}-\gamma_{10}) + 2\Gamma_{20}(2\gamma_{03}-\gamma_{21})))}{\begin{array}{c}[2(2\gamma_{03}+3\gamma_{21})G_1^2+\Gamma_{10}l](G_1^2+v) + [2uG_2^2+\Gamma_{21}l] \\ (G_2^2+\Gamma_{20}\Gamma_{10}) - \Gamma_{10}vl + G_1^2G_2^2(16\Gamma_{20}-8\gamma_{03}+2j)\end{array}} \tag{6.26}$$

where $l=\gamma_{10}\gamma_{03}+\gamma_{10}\gamma_{21}+\gamma_{03}\gamma_{21}$, $u=2\gamma_{03}+\gamma_{10}$, $v=\Gamma_{21}\Gamma_{20}$, $j=\gamma_{10}-\gamma_{21}$, and $n=\gamma_{03}-\gamma_{10}$ for simplification. Similarly, to observe the two bright correlated light beams, it requires $G_{10}^{\Xi}>0$ and $G_{21}^{\Xi}>0$ and then it can be easily concluded that $\gamma_{03}>\gamma_{10}>\gamma_{21}$ is necessary with $6\gamma_{20}+\gamma_{10}>3\gamma_{03}+\gamma_{21}$ for $G_{10}^{\Xi}>0$ while $8\Gamma_{20}+\gamma_{10}>\gamma_{21}+4\gamma_{03}$ for $G_{21}^{\Xi}>0$. It indicates that when it meets the demands of these thresholds, two bright correlated light beams can oscillate simultaneously as increasing G_1 and G_2 or just increasing one of them. Analogically, we can also get the gain and threshold of the other Ξ subsystem $|3\rangle\to|1\rangle\to|2\rangle$, so we do not show more details here.

It is known that the three pairs of fluorescence signals can amplify the injected vacuum field in free space. Similarly, if an MWM signal is injected and interacts with the fluorescence signals, it can be amplified as well. In our atom-cavity system (Figure 6.19), if the bright correlated light beams are considered, an OPA process will occur when we inject the FWM and SWM signals. The FWM and SWM signals can be amplified and squeezed better in the interaction with the bright correlated light beams, which can lead to a two-mode squeezing process between FWM and SWM signals, and this kind of squeeze can be used in quantum entanglement processing.

6.5.6
OB Behavior of MWM

In this section, in order to get a more comprehensive understanding of the intracavity influence of dark and bright states in the MWM process, we further study the OB behavior resulting from the self-Kerr nonlinear effect in the atom-cavity system as shown in Figure 6.19(a). Considering the relatively large self-Kerr nonlinearity susceptibility, which mainly functions as an unneglected feedback effect (also a self-dressing effect) of the cavity in the MWM process, we investigate the OB phenomenon based on master equation formalism and cavity transmission coefficient and conclude the coexistence and cascade competition of the VRS and OB behavior.

6.5.6.1 OB of Zero-Order Mode

First, we analyze OB behavior of the zero-order mode based on the master equation formalism theory. As the generated FWM (SWM) field can circulate in the ring cavity while the probe field and other coupling fields cannot, we only consider the generated FWM (SWM) field to form the cavity mode. Clearly, the generated FWM (SWM) cavity mode has a self-dressing effect of $|G_F|^2$, which is derived from the relatively strong feedback effect. This self-dressing effect has an influence similar to that of the self-dressing effect of g^2N, internal-dressing effects of $|G_2|^2$, and external-dressing effect of $|G_3|^2$, which can together result in a close interaction between VRS and OB. Here, we take the transmitted cavity FWM signal, for example, which can be obtained from Eq. (6.20) as

$$a_{\text{FWM}} \propto \frac{ig\sqrt{N}G_1G_2(G_2')^*}{d_4d_2(d_1)^2(d_1+g^2N/d_4+|G_2|^2/d_2+|G_3|^2/d_3+|G_F|^2/\Gamma_{00})} \tag{6.27}$$

where the internal-dressing effect of E_2 and external-dressing effect of E_3 are both included when the powers of E_2 and E_3 are sufficiently strong. For a perfectly tuned ring cavity, the output intensity I_o (a function in proportion, not the experimental output intensity) of the transmitted cavity FWM field is proportional to $|a_{\text{FWM}}|^2$, while the input intensity I_i of the incident field is proportional to $|G_1|^2$. So, the input–output relationship can be expressed as:

$$\frac{I_o}{I_i} \propto \left| ig\sqrt{N}G_2(G_2')^* \left[d_4d_2{d_1}^2 \left(d_1 + \frac{g^2N}{d_4} + \frac{|G_2|^2}{d_2} + \frac{|G_3|^2}{d_3} + \frac{I_o}{\Gamma_{00}} \right) \right]^{-1} \right|^2 \tag{6.28}$$

where I_o on the right-hand side describes the feedback effect of cavity for the transmitted cavity signal. There is a cascade relationship between I_o and g^2N on the right-hand side of Eq. (6.28), with I_o, coming from $|G_F|^2$, serving as the essential feedback effect in the origin of OB and g^2N serving as the essential atom-cavity coupling effect in the origin of VRS. Such a relationship indicates the cascade coexistence and competition between OB and VRS.

We numerically study the input–output intensity relationship under steady-state condition with a relatively weak E_2 field and without theE_3 field as shown in Figure 6.25, which displays OB behavior of the transmitted cavity FWM signal influenced by I_i and Δ_1. Here, the mode splitting is different from the normal VRS (only from g^2N in Section 6.3) because the modes, after splitting, are all inclined for the feedback effect (self-dressing effect); however, the major splitting law just stays the same with VRS. The inclined VRS results from the relatively large feedback effect ($|G_F|^2$) and g^2N, leading to the conclusion that there exists a strong cascade interaction between VRS and OB. Figure 6.25(a) illustrates the intensity of the output-transmitted cavity FWM signal as the functions of I_i and Δ_1, which reveals the OB threshold of the OPA FWM process and hysteresis cycle obviously

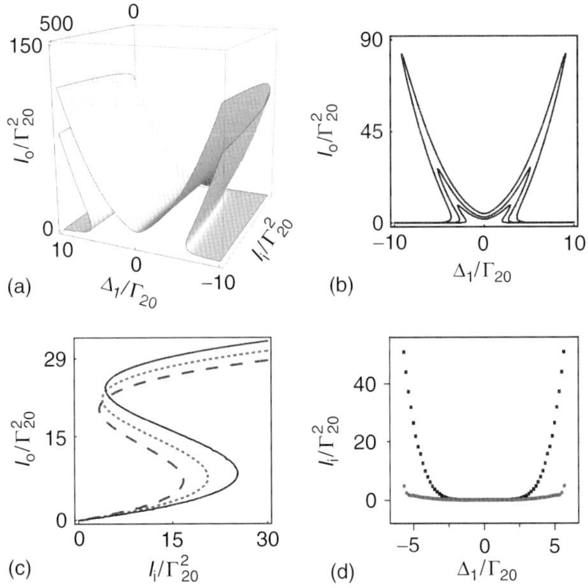

Figure 6.25 Observed input–output intensity relationship of the system with probe detuning Δ_1/Γ_{20} scanned when only zero-order cavity mode is considered and $E_2(E_2')$ is relatively weak while E_3 and E_3' are blocked. (a) FWM transmission output changes with both probe detuning and probe input intensity. (b) FWM transmission output with probe detuning at different probe input intensity $I_i/\Gamma_{20}{}^2$= 0.1,1,10 from inside to outside. (c) OB at different probe detunings Δ_1/Γ_{20}= 4.5 (dashed), Δ_1/Γ_{20}= 4.7 (dotted), and Δ_1/Γ_{20}= 4.9 (solid). (d) Probe input intensity versus Δ_1/Γ_{20} with the dots of the up branch standing for the right OB threshold of the OPA FWM process and dots of the down branch standing for the left one.

with the variation of Δ_1. These modulations of the VRS in frequency domain and input–output relationship (OB behavior) simultaneously result from the significant change in the absorption and dispersion characteristics of the medium. In detail, Figure 6.25(b) shows the transmitted cavity FWM signal expanding rapidly with I_i increasing when Δ_1 is scanned. Increased $|\Delta_1|$ results in a significant change of the OB behavior and the increasing of the right OB threshold as in Figure 6.25(c). Figure 6.25(d) displays the OB threshold values versus Δ_1, in which the left OB threshold value shifts slowly, while the right one shifts sharply. Moreover, Figure 6.25(d) also demonstrates that the OB always exists in certain sideband regions corresponding to the generated inclined peaks after splitting and there is no OB at or close to the point $\Delta_1/\Gamma_{20}=0$ (dark state) whatever the probe input intensity is, which results from the linear dispersion at the line center $(\Delta_1/\gamma_{20}=0)$ that disappeared by the interference between two possible absorption channels $|0\rangle \rightarrow |\pm\rangle$ induced by g^2N. This interesting result led us to make the following study to verify the primary conclusion that OB does not exist at or near all the positions of dark states induced by the dressing fields.

Analogically, we analyze the similar dependence of OB behavior when the power of E_2 is sufficiently strong to reveal its dressing effect, as shown in Figure 6.26. The basic OB characteristic is in accordance with the previous analysis. As a consequence of the dressing effect of E_2, the absorption and dispersion coefficient of the medium can be modified dramatically, which results in the AT splitting based on the VRS and the three different frequency (Δ_1) regions of the OB, shown as Figure 6.26(a) and (b). Figure 6.26(c) illustrates the fierce decreasing of the two OB threshold values of the OPA FWM process when increasing the Rabi frequency of E_2 as $G_2/\Gamma_{20}=1,2,3$, especially the right one. Figure 6.26(d) displays the decreasing OB threshold values with decreasing $|\Delta_1|$, the same as Figure 6.25(c). Also, there is no OB at the position of dark state $(\Delta_1/\Gamma_{20}=0$ and $\Delta_1+\Delta_2=0)$ because of the lack of linear dispersion induced by g^2N and the suppression effect induced by E_2, as shown in Figure 6.26(e). The phenomena in this part demonstrate the changeable properties of OB behavior influenced by the dark state induced by dressing effect and we believe this will provide a new kind of experimental control of VRS and OB simultaneously.

6.5.6.2 OB of High-Order Modes

On the other hand, we consider the OB of high-order modes theoretically based on the traditional cavity transmission coefficient as contrast and also take the transmitted FWM signal, for example. The total susceptibility can be expressed as $\chi=\chi^{(1)}+\chi^{(3)}|E_i|^2=N\mu_{10}^2(\tilde{\rho}_{10}^{(1)}+\tilde{\rho}_{10}^{(3)})/[\varepsilon_0\hbar G_F]$, in which $\chi^{(1)}$ and $\chi^{(3)}$ are the linear susceptibility and the third-order nonlinear susceptibility, respectively. Here, we can obtain $\tilde{\rho}_{10}^{(1)}$ and $\tilde{\rho}_{10}^{(3)}$ by solving the coupled density matrix equations as

$$\tilde{\rho}_{10}^{(1)}=\frac{iG_F}{d_1+|G_2|^2/d_2+|G_3|^2/d_3} \tag{6.29}$$

$$\begin{aligned}\tilde{\rho}_{10}^{(3)} = {} & \frac{-iG_F|G_F|^2}{[d_1+(|G_2|^2/d_2)+(|G_3|^2/d_3)]^3}+\frac{-iG_F|G_1|^2}{[d_1+(|G_2|^2/d_2)+(|G_3|^2/d_3)]^3}\\ & +\frac{-iG_F|G_2|^2}{[d_1+(|G_2|^2/d_2)+(|G_3|^2/d_3)]^2d_2}+\frac{-iG_F|G_3|^2}{[d_1+(|G_2|^2/d_2)+(|G_3|^2/d_3)]^2d_3}\end{aligned} \tag{6.30}$$

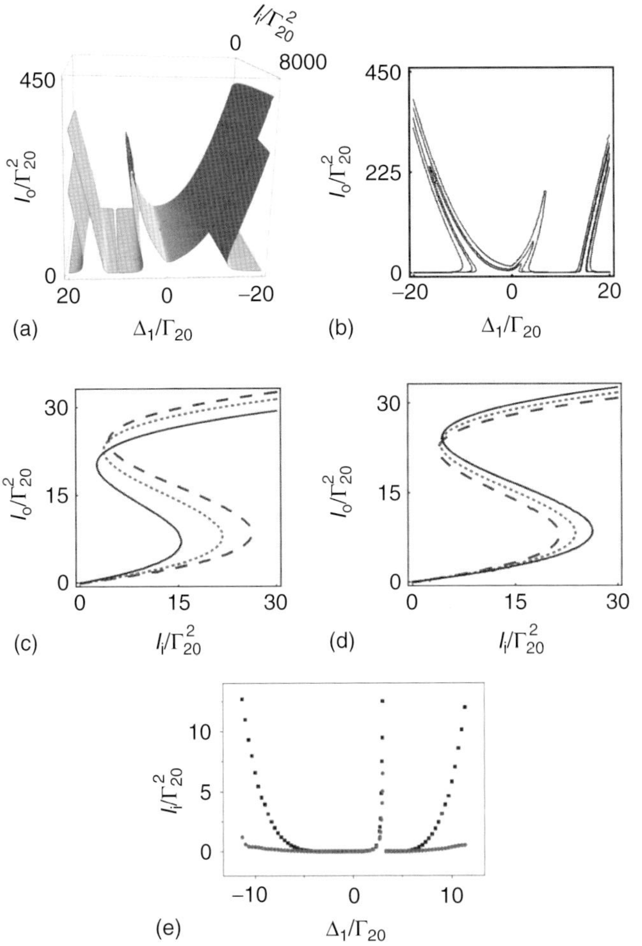

Figure 6.26 Observed input–output intensity relationship characteristics of the system with probe detuning Δ_1/Γ_{20} scanned when only zero-order cavity mode is considered; $E_2(E_2')$ is sufficiently strong, while E_3 and E_3' are blocked. (a) FWM transmission output changes with both probe detuning Δ_1/Γ_{20} and probe input intensity $I_i/\Gamma^2{}_{20}$ when G_2 is strong sufficiently. (b) FWM transmission output with probe detuning at $I_i/\Gamma^2{}_{20} = 10,50,680$ from inside to outside. (c) FWM transmission output intensity with probe input intensity when $G_2/\Gamma_{20} = 1$ (dashed), $G_2/\Gamma_{20} = 2$ (dotted), and $G_2/\Gamma_{20} = 3$ (solid). (d) FWM transmission output intensity with probe input intensity at $\Delta_1/\Gamma_{20} = 4.8$ (dashed), $\Delta_1/\Gamma_{20} = 4.9$ (dotted), and $\Delta_1/\Gamma_{20} = 5$ (solid). (e) the OB threshold of the OPA FWM process versus Δ_1/Γ_{20} for $G_2/\Gamma_{20} = 2$ at $\Delta_1/\Gamma_{20} = -10$ with the dots of the up branch standing for the right OB threshold and dots of the down branch standing for the left one. Source: Adapted from Ref. [57].

In Eq. (6.30), the first term in $\tilde{\rho}_{10}^{(3)}$ indicates the self-Kerr effect, while the last three terms indicate the cross-Kerr effect in medium. Consider that $|G_F|^2$ is proportional to the intensity of transmitted cavity FWM field when the feedback effect is considered. So, the susceptibility can be revised as

$$\chi = \frac{2g^2 N L_c}{L_a \omega_1} \left\{ \begin{array}{c} \frac{i}{d_1+(|G_2|^2/d_2)+(|G_3|^2/d_3)} + \frac{-iI_o}{[d_1+(|G_2|^2/d_2)+(|G_3|^2/d_3)]^3} + \frac{-i|G_1|^2}{[d_1+(|G_2|^2/d_2)+(|G_3|^2/d_3)]^3} \\ + \frac{-i|G_2|^2}{[d_1+(|G_2|^2/d_2)+(|G_3|^2/d_3)]^2 d_2} + \frac{-i|G_3|^2}{[d_1+(|G_2|^2/d_2)+(|G_3|^2/d_3)]^2 d_3} \end{array} \right\} \tag{6.31}$$

Then, on the basis of Eq. (6.31), OB can be investigated.

By setting the relevant parameters at appropriate values, we can increase $g\sqrt{N}$ to be near or larger than Δ_{FSR} as to observe the OB of high-order modes, and OB still only appears at the position where absorption and dispersion are appropriate under the influence of splitting. Now we succeed in controlling their OB thresholds of the OPA FWM process and curve shape by adjusting several related parameters. When there is one dressing field and only the self-Kerr nonlinearity effect is considered, the corresponding results are shown in Figure 6.27. When N decreases as in Figure 6.27(a), the left OB threshold increases slowly while the right one increases sharply, leading to a dramatically expansion of the bistable region. Similarly, if we increase the intensity of the dressing field as in Figure 6.27(b), both the OB thresholds decrease and the size of the bistable region nearly stays the same. At last, we discuss the OB at $\Delta_1 + \Delta_2 = 0, 0.5, -0.5$, as shown in Figure 6.27(c). As we have assumed according to the analysis previously, there is no OB effect at dark-state position and OB will gradually appear when the frequency detuning deviates from $\Delta_1 + \Delta_2 = 0$ step by step. All the results reveal that the origin of OB is related to the splitting of cavity modes to some extent, and we can apply this kind of control of the OB threshold of the OPA FWM process into the development of the OB switch.

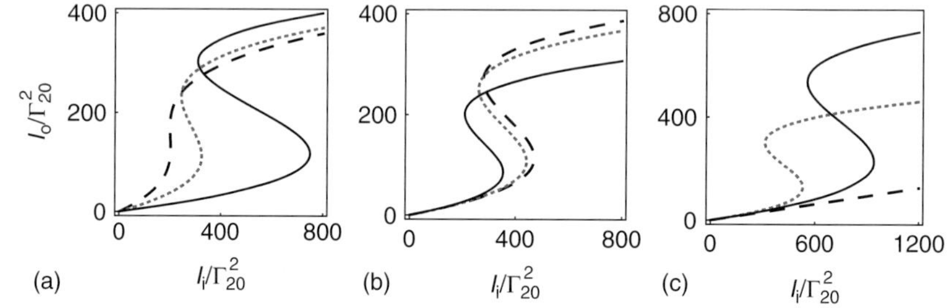

Figure 6.27 FWM transmission output intensity versus probe input intensity when $E_2(E_2')$ is sufficiently strong with E_3 and E_3' blocked. The dashed, dotted and solid curves correspond to (a) N decreases as $N/\Gamma_{20} = 100,10,1$. (b) The Rabi frequency of dressing field increases as $G_2/\Gamma_{20} = 0,2,4$. (c) Different detuning combinations $\Delta_1 + \Delta_2 = 0,0.5,-0.5$ are adopted when $G_2/\Gamma_{20} = 2$.

6.5.7 Conclusion

We have discussed the control of dark state on the multi-dressed VRS and OB behavior of zero- and high-order transmitted cavity MWM signal in the coupled system consisting of a specific ring cavity and reverse Y-type four-level atoms assemble. The numerical calculations are based on the master equation formalism and the traditional cavity transmission coefficient. We have demonstrated the coexistence and cascade competition between VRS and OB behavior, where the VRS results from the atom-cavity collective effect induced by high atom density, while the OB behavior results from the sufficiently strong feedback effect. Owing to the complicated generation and operation process of the MWM signal, the inclined VRS is obtained under the interaction with OB, and such OB behavior cannot appear at or near the position of the dark state induced by the dressing effect. We also demonstrate the suppression and enhancement of the multi-dressed MWM signal by scanning the frequency detuning of the dressing field and the linear gains and thresholds for the paired bright correlated light beams in the OPO process with OB threshold of the OPA MWM process. Such control of dark and bright states on the VRS, OB, and so on can be achieved by adjusting the cavity field and other coupling fields, and provides potential applications in optical devices and quantum information projects.

Problems

6.1 Explain why the two-photon fluorescence with high coherence and monochromaticity is ultranarrow in frequency domain, when compared with the traditional fluorescence.

6.2 Discuss the influence of changing the dressing effects of pump field on probe transmission, FWM, and fluorescence signals.

6.3 Explain why MWM is a more controllable way to entangled photon sources, when compared with SPDC.

6.4 Discuss the competition between OB and VRS with the MWM process in a coupled atom-cavity system and how to control them.

References

1. Zhang, Y., Brown, A.W., and Xiao, M. (2007) Opening four-wave mixing and six-wave mixing channels via dual electromagnetically induced transparency windows. *Phys. Rev. Lett.*, **99**, 123603.
2. Zhu, C.J., Senin, A., Lu, Z.H., Gao, J., Xiao, Y., and Eden, J. (2005) Polarization of signal wave radiation generated by parametric four-wave mixing in rubidium vapor: ultrafast (~150-fs) and nanosecond time scale excitation. *Phys. Rev. A*, **72**, 023811.
3. Li, Y. and Xiao, M. (1996) Enhancement of nondegenerate four-wave mixing based on electromagnetically induced transparency in rubidium atoms. *Opt. Lett.*, **21**, 1064–1066.
4. Zhang, Y., Nie, Z., Wang, Z., Li, C., Wen, F., and Xiao, M. (2010) Evidence of Autler-Townes splitting in high-order

nonlinear processes. *Opt. Lett.*, **35**, 3420–3422.
5. Wielandy, S. and Gaeta, A.L. (1998) Investigation of electromagnetically induced transparency in the strong probe regime. *Phys. Rev. A*, **58**, 2500–2505.
6. Harris, S.E. (1997) Electromagnetically induced transparency. *Phys. Today*, **50**, 36.
7. Xiao, M., Li, Y.-Q., Jin, S.-Z., and Gea-Banacloche, J. (1995) Measurement of dispersive properties of electromagnetically induced transparency in rubidium atoms. *Phys. Rev. Lett.*, **74**, 666–669.
8. Moseley, R.R., Shepherd, S., Fulton, D.J., Sinclair, B.D., and Dunn, M.H. (1995) Spatial consequences of electromagnetically induced transparency: observation of electromagnetically induced focusing. *Phys. Rev. Lett.*, **74**, 670–673.
9. Li, Y., Jin, S., and Xiao, M. (1995) Observation of an electromagnetically induced change of absorption in multilevel rubidium atoms. *Phys. Rev. A*, **51**, 1754–1757.
10. Hemmer, P., Katz, D., Donoghue, J., Cronin-Golomb, M., Shahriar, M., and Kumar, P. (1995) Efficient low-intensity optical phase conjugation based on coherent population trapping in sodium. *Opt. Lett.*, **20**, 982–984.
11. Zhang, Y., Li, P., Zheng, H., Wang, Z., Chen, H., Li, C., Zhang, R., and Xiao, M. (2011) Observation of Autler-Townes splitting in six-wave mixing. *Opt. Express*, **19**, 7769–7777.
12. Qi, J., Spano, F.C., Kirova, T., Lazoudis, A., Magnes, J., Li, L., Narducci, L.M., Field, R.W., and Lyyra, A.M. (2002) Measurement of transition dipole moments in lithium dimers using electromagnetically induced transparency. *Phys. Rev. Lett.*, **88**, 173003.
13. Qi, J., Lazarov, G., Wang, X., Li, L., Narducci, L.M., Lyyra, A.M., and Spano, F.C. (1999) Autler-Townes splitting in molecular lithium: prospects for all-optical alignment of nonpolar molecules. *Phys. Rev. Lett.*, **83**, 288–291.
14. Qi, J. and Lyyra, A.M. (2006) Electromagnetically induced transparency and dark fluorescence in a cascade three-level diatomic lithium system. *Phys. Rev. A*, **73**, 043810.
15. Zhao, Z., Wang, Z., Li, P., Huang, G., Li, N., Zhang, Y., Yan, Y., and Zhang, Y. (2012) Opening fluorescence and four-wave mixing via dual electromagnetically induced transparency windows. *Laser Phys. Lett.*, **9**, 802.
16. Khadka, U., Zhang, Y., and Xiao, M. (2010) Control of multitransparency windows via dark-state phase manipulation. *Phys. Rev. A*, **81**, 023830.
17. Li, C., Zheng, H., Zhang, Y., Nie, Z., Song, J., and Xiao, M. (2009) Observation of enhancement and suppression in four-wave mixing processes. *Appl. Phys. Lett.*, **95**, 041103.
18. Li, N., Zhao, Z., Chen, H., Li, P., Li, Y., Zhao, Y., Zhou, G., Jia, S., and Zhang, Y. (2012) Observation of dressed odd-order multi-wave mixing in five-level atomic medium. *Opt. Express*, **20**, 1912–1929.
19. Wang, Z., Zhang, Y., Zheng, H., Li, C., Wen, F., and Chen, H. (2011) Switching enhancement and suppression of four-wave mixing via a dressing field. *J. Mod. Opt.*, **58**, 802–809.
20. Hahn, K.H., King, D.A., and Harris, S.E. (1990) Nonlinear generation of 104.8-nm radiation within an absorption window in zinc. *Phys. Rev. Lett.*, **65**, 2777–2779.
21. Li, P., Zhao, Z., Wang, Z., Zhang, Y., Lan, H., Chen, H., and Zheng, H. (2012) Phase control of bright and dark states in four-wave mixing and fluorescence channels. *Appl. Phys. Lett.*, **101**, 081107–081107-4.
22. Wang, H., Goorskey, D., and Xiao, M. (2001) Enhanced kerr nonlinearity via atomic coherence in a three-level atomic system. *Phys. Rev. Lett.*, **87**, 073601.
23. Agrawal, G.P. (1990) Induced focusing of optical beams in self-defocusing nonlinear media. *Phys. Rev. Lett.*, **64**, 2487–2490.
24. Bennink, R.S., Wong, V., Marino, A.M., Aronstein, D.L., Boyd, R.W., Stroud, C. Jr.,, Lukishova, S., and Gauthier, D.J. (2002) Honeycomb pattern formation by laser-beam filamentation in atomic sodium vapor. *Phys. Rev. Lett.*, **88**, 113901.

25. Stentz, A.J., Kauranen, M., Maki, J.J., Agrawal, G.P., and Boyd, R.W. (1992) Induced focusing and spatial wave breaking from cross-phase modulation in a self-defocusing medium. *Opt. Lett.*, **17**, 19–21.
26. Lukin, M., Yelin, S., Fleischhauer, M., and Scully, M. (1999) Quantum interference effects induced by interacting dark resonances. *Phys. Rev. A*, **60**, 3225–3228.
27. Yan, M., Rickey, E.G., and Zhu, Y. (2001) Observation of doubly dressed states in cold atoms. *Phys. Rev. A*, **64**, 013412.
28. Zhang, Y., Nie, Z., Zheng, H., Li, C., Song, J., and Xiao, M. (2009) Electromagnetically induced spatial nonlinear dispersion of four-wave mixing. *Phys. Rev. A*, **80**, 013835.
29. Zhang, Y., Zuo, C., Zheng, H., Li, C., Nie, Z., Song, J., Chang, H., and Xiao, M. (2009) Controlled spatial beam splitter using four-wave-mixing images. *Phys. Rev. A*, **80**, 055804.
30. Zhang, Y., Wang, Z., Zheng, H., Yuan, C., Li, C., Lu, K., and Xiao, M. (2010) Four-wave-mixing gap solitons. *Phys. Rev. A*, **82**, 053837.
31. Zhang, Y., Wang, Z., Nie, Z., Li, C., Chen, H., Lu, K., and Xiao, M. (2011) Four-wave mixing dipole soliton in laser-induced atomic gratings. *Phys. Rev. Lett.*, **106**, 93904.
32. Sang, S., Wu, Z., Sun, J., Lan, H., Zhang, Y., and Zhang, X. (2011) Observation of angle switching of dressed four wave mixing images. *IEEE Photonics J.*, **4**, 1973–1986.
33. Bouwmeester, D., Ekert, A.K., and Zeilinger, A. (2001) *The Physics of Quantum Information*, Springer, Berlin.
34. Kwiat, P.G., Mattle, K., Weinfurter, H., Zeilinger, A., Sergienko, A.V., and Shih, Y. (1995) New high-intensity source of polarization-entangled photon pairs. *Phys. Rev. Lett.*, **75**, 4337–4341.
35. Balić, V., Braje, D.A., Kolchin, P., Yin, G., and Harris, S. (2005) Generation of paired photons with controllable waveforms. *Phys. Rev. Lett.*, **94**, 183601.
36. Boyer, V., McCormick, C., Arimondo, E., and Lett, P. (2007) Ultraslow propagation of matched pulses by four-wave mixing in an atomic vapor. *Phys. Rev. Lett.*, **99**, 143601.
37. Harada, K., Ogata, M., and Mitsunaga, M. (2007) Four-wave parametric oscillation in sodium vapor by electromagnetically induced diffraction. *Opt. Lett.*, **32**, 1111–1113.
38. Okuma, J., Hayashi, N., Fujisawa, A., Mitsunaga, M., and Harada, K. (2009) Parametric oscillation in sodium vapor by using an external cavity. *Opt. Lett.*, **34**, 698–700.
39. Boyer, V., Marino, A.M., Pooser, R.C., and Lett, P.D. (2008) Entangled images from four-wave mixing. *Science*, **321**, 544–547.
40. Pooser, R., Marino, A., Boyer, V., Jones, K., and Lett, P. (2009) Quantum correlated light beams from non-degenerate four-wave mixing in an atomic vapor: the D1 and D2 lines of 85Rb and 87Rb. *Opt. Express*, **17**, 16722–16730.
41. Lü, B., Burkett, W., and Xiao, M. (1998) Nondegenerate four-wave mixing in a double-Lambda system under the influence of coherent population trapping. *Opt. Lett.*, **23**, 804–806.
42. Zheng, H., Khadka, U., Song, J., Zhang, Y., and Xiao, M. (2011) Three-field noise correlation via third-order nonlinear optical processes. *Opt. Lett.*, **36**, 2584–2586.
43. Sautenkov, V.A., Rostovtsev, Y.V., and Scully, M.O. (2005) Switching between photon-photon correlations and Raman anticorrelations in a coherently prepared Rb vapor. *Phys. Rev. A*, **72**, 065801.
44. Zhu, Y., Gauthier, D.J., Morin, S.E., Wu, Q., Carmichael, H.J., and Mossberg, T.W. (1990) Vacuum Rabi splitting as a feature of linear-dispersion theory: analysis and experimental observations. *Phys. Rev. Lett.*, **64**, 2499–2502.
45. Thompson, R., Rempe, G., and Kimble, H. (1992) Observation of normal-mode splitting for an atom in an optical cavity. *Phys. Rev. Lett.*, **68**, 1132–1135.
46. Agarwal, G. (1984) Vacuum-field Rabi splittings in microwave absorption by Rydberg atoms in a cavity. *Phys. Rev. Lett.*, **53**, 1732–1734.

47. Tavis, M. and Cummings, F.W. (1968) Exact solution for an N-molecule-radiation-field Hamiltonian. *Phys. Rev.*, **170**, 379–384.

48. Wu, H., Gea-Banacloche, J., and Xiao, M. (2008) Observation of intracavity electromagnetically induced transparency and polariton resonances in a Doppler-broadened medium. *Phys. Rev. Lett.*, **100**, 173602.

49. Wu, H., Xiao, M., and Gea-Banacloche, J. (2008) Evidence of lasing without inversion in a hot rubidium vapor under electromagnetically-induced-transparency conditions. *Phys. Rev. A*, **78**, 041802.

50. Yu, X., Xiong, D., Chen, H., Wang, P., Xiao, M., and Zhang, J. (2009) Multi-normal-mode splitting of a cavity in the presence of atoms: a step towards the superstrong-coupling regime. *Phys. Rev. A*, **79**, 061803.

51. Yu, X. and Zhang, J. (2010) Multi-normal mode-splitting for an optical cavity with electromagnetically induced transparency medium. *Opt. Express*, **18**, 4057–4065.

52. Reynaud, S., Fabre, C., and Giacobino, E. (1987) Quantum fluctuations in a two-mode parametric oscillator. *J. Opt. Soc. Am. B*, **4**, 1520–1524.

53. Lane, A., Reid, M., and Walls, D. (1988) Absorption spectroscopy beyond the shot-noise limit. *Phys. Rev. Lett.*, **60**, 1940–1942.

54. Boyer, V., Marino, A., and Lett, P. (2008) Generation of spatially broadband twin beams for quantum imaging. *Phys. Rev. Lett.*, **100**, 143601.

55. Wang, H., Goorskey, D., Burkett, W., and Xiao, M. (2000) Cavity-linewidth narrowing by means of electromagnetically induced transparency. *Opt. Lett.*, **25**, 1732–1734.

56. Joshi, A. and Xiao, M. (2003) Optical multistability in three-level atoms inside an optical ring cavity. *Phys. Rev. Lett.*, **91**, 143904.

57. Yuan, J., Feng, W., Li, P., Zhang, X., Zhang, Y., Zheng, H., and Zhang, Y. (2012) Controllable vacuum Rabi splitting and optical bistability of multi-wave-mixing signal inside a ring cavity. *Phys. Rev. A*, **86**, 063820.

7
Coherent Modulation of Photonic Band Gap in FWM Process

Highlights

Enhanced nonlinear Kerr effect can lead to electromagnetically induced grating, which is the reason for the spatial "shift" and "splitting" of MWM, as well as the formation of vortex solitons and surface solitons.

When multiple laser beams interact with multi-level atomic systems, interesting spatial effects for the MWM can be induced. In this chapter, the formation of EIG and the EIL in FWM and their various effects are introduced. First, we discuss the shift and splitting of the two FWM beams, which can be effectively modulated by the frequency detuning, the angles between beams, as well as the powers of the pump fields. Then, the existence and interesting evolution of vortices in nonlinear atomic vapor are introduced, with the linear, cubic, and quintic susceptibilities considered simultaneously. After that, we focus on the formation and stability of the multi-component dipole and vortex vector solitons when the FWM signal is generated under the modulating effects of the one-dimensional (1D) EIG. In the following, by creating a 2D lattice state with spatial periodic atomic coherence, 2D surface solitons of the FWM signal are presented, and the surface solitons can be well controlled by different experimental parameters, such as probe frequency, pump powers, and grating period. Last, the Talbot effects in the propagations of the FWM and SWM signals are introduced. Studies of such controlled beam shift and spatial splitting can be useful in understanding image storage, spatial soliton formation and dynamics, and in device development for spatial signal processing, such as switches and routers.

7.1
Spatial Interplay of Two FWM Images

7.1.1
Introduction

Spatially shifting and splitting of the weak laser beam caused by another stronger beam in Kerr nonlinear optical media have been theoretically predicted and

Quantum Control of Multi-Wave Mixing, First Edition. Yanpeng Zhang, Feng Wen, and Min Xiao.

experimentally demonstrated in recent years [1, 2]. The spatial variations in the pump laser strength can lead to the electromagnetically induced focusing of the probe beam [3] due to laser-induced atomic coherence [4, 5]. It was shown that self- and cross-Kerr nonlinearities [6] can be significantly enhanced and modified, which are essential in generating the efficient FWM processes and large refractive index modulation. It plays an important role in lasing without inversion [7], slow light generation [8], photon controlling and information storage [9–11], quantum communications [12], nonlinear optics, wave mixing process [13, 14], and fiber lasers [15, 16], to name a few. In addition, EIG resulted from a strong standing wave (SW) interacting with the atomic medium, which can diffract a weak incident field into high-order direction and is profoundly developed in multi-level atoms [17, 18]. Recently, we have observed spatial shift, spatial splitting, gap solitons, and dipole solitons of the FWM beams generated in multi-level atomic systems [19, 20], which can be well controlled by additional dressing laser beams via the XPM.

In this section, we discuss the interplay between two components of FWM beams in a ladder-type three-level atomic system. First, we show that by arranging laser beams in a certain spatial configuration, the spatial splitting of one of the FWM beams will appear in the x-direction due to XPM or in the y-direction induced by EIG. Simultaneously, the other exhibits the opposite result. Second, the mode of the splitting affected by the detuning of the probe field and the angle of the pumping fields is discussed thoroughly. Third, we investigate the interplay including the intensity modulation, shift, and splitting of two FWM beams, which depends on the detuning, angles, and powers of pumping fields and the temperature in the experiment. Finally, we theoretically demonstrate that the splitting of the FWM signals can be well controlled by strong pumping laser beams. Such studies can be useful in better understanding the formation, propagation, and interplay of spatial solitons, which will contribute to the spatial manipulations of the FWM beams in optical imaging storage and making optical devices, such as beam splitter, router, and switching.

7.1.2
Theoretical Model and Experimental Scheme

A ladder-type three-level system [$|0\rangle$ ($3S_{1/2}$), $|1\rangle$ ($3P_{3/2}$), $|2\rangle$ ($4D_{3/2}$)] from Na atom (with Na vapor in an 18 cm long heat-pipe oven), as shown by Figure 7.1(a), is involved in the experiment. Five laser beams are applied to the atomic system with the spatial configuration given in Figure 7.1(b) and (c). The field $\boldsymbol{E}_1$ (frequency ω_1, wave vector $\mathbf{k}_1$, the Rabi frequency G_1, and frequency detuning $\Delta_1 = \Omega_1 - \omega_1$, where Ω_1 is the resonant frequency of transition $|0\rangle$ to $|1\rangle$) and $\boldsymbol{E}_1'$ (ω_1, $\mathbf{k}_1'$, G_1', and Δ_1) are the pump fields propagating in one direction with a small angle of θ_1 between them. The projections of θ_1 onto the x–z- and y–z-planes are θ_{1x0}($\approx$0.3°) and θ_{1y}($\approx$0.05°), respectively. The probe field $\boldsymbol{E}_3$ (ω_1, $\mathbf{k}_3$, G_3, Δ_1) propagates in the direction opposite to $\boldsymbol{E}_1$. Fields $\boldsymbol{E}_1$, $\boldsymbol{E}_1'$, and $\boldsymbol{E}_3$, with a diameter of about 0.8 mm, are from the one near-transform-limited dye laser (10 Hz repetition rate, 5 ns pulse width, and 0.04 cm^{-1} line width) connecting the transition $|0\rangle$–$|1\rangle$ and generate

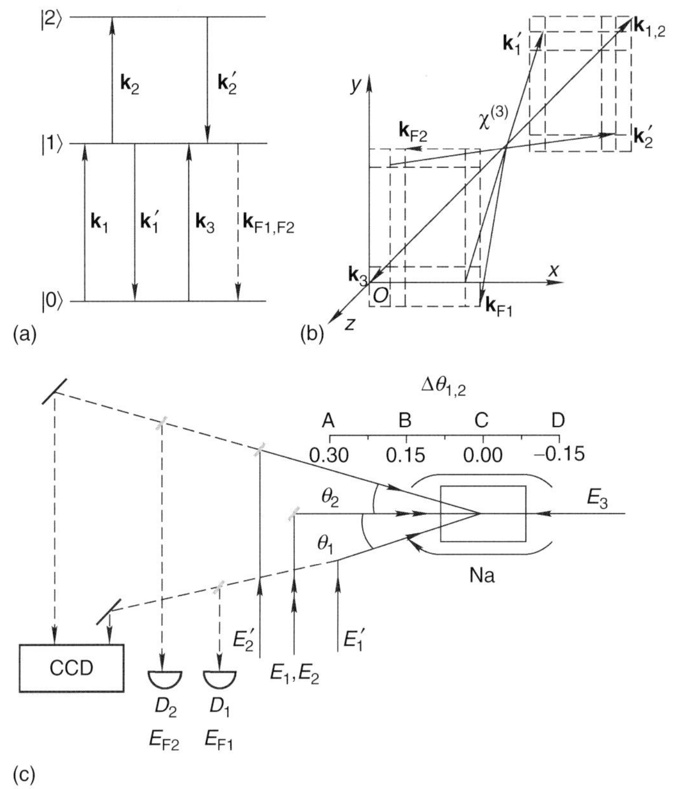

Figure 7.1 (a) Two FWM processes to generate $\boldsymbol{E}_{F1}$ and $\boldsymbol{E}_{F2}$ in a ladder-type three-level atomic system. (b) and (c) Spatial geometry for the laser beams in the experiment. The scale gives the angle value of the corresponding position in heat-pipe oven.

an efficient DFWM signal $\boldsymbol{E}_{F1}$ (G_{F1}) in the direction shown in Figure 7.1(b) and (c). Another pair of pump fields, $\boldsymbol{E}_2$ (ω_2, $\mathbf{k}_2$, G_2, Δ_2) and $\boldsymbol{E}_2'$ (ω_2, $\mathbf{k}_2'$, G_2', Δ_2), are also applied, with $\boldsymbol{E}_2$ propagating in the same direction of $\boldsymbol{E}_1$ and $\boldsymbol{E}_2'$ having a small angle θ_2. The projections of θ_2 onto the y–z- and x–z-planes are $\theta_{2y0}(\approx 0.3^\circ)$ and $\theta_{2x}(\approx 0.05^\circ)$, respectively. Driving the transition $|1\rangle-|2\rangle$, $\boldsymbol{E}_2$, and $\boldsymbol{E}_2'$ with a diameter of 1.1 mm are generated from another dye laser with the same pulse parameters as the former one. An NDFWM signal $\boldsymbol{E}_{F2}$ (G_{F2}) and a DFWM $\boldsymbol{E}_{F1}$ are produced simultaneously, satisfying the phase-matching conditions $\mathbf{k}_{F1} = \mathbf{k}_1 - \mathbf{k}_1' + \mathbf{k}_3$ and $\mathbf{k}_{F2} = \mathbf{k}_2 - \mathbf{k}_2' + \mathbf{k}_3$, respectively, as shown in Figure 7.1. In this experiment, we define $\Delta\theta_1 = \theta_{1x} - \theta_{1x0}$, where θ_{1x} is the angle of $\boldsymbol{E}_1$ and $\boldsymbol{E}_1'$ in the x–z-plane and θ_{1x0} is the angle of $\boldsymbol{E}_1$ and $\boldsymbol{E}_1'$ in the x–z-plane when they are set at the point C (Figure 7.1(c)). Using the same method, we get $\Delta\theta_2 = \theta_{2y} - \theta_{2y0}$ for $\boldsymbol{E}_2$ and $\boldsymbol{E}_2'$ in the y–z-plane. As shown in Figure 7.1(c), when the pumping fields set from D (the back of the heat-pipe oven) to B (the front), the value of $\Delta\theta_{1,2}$ changes from −0.15 to 0.15. And if $\Delta\theta_{1,2}$ increases to 0.3, the pumping fields already set at A (outside of the heat-pipe oven). In order to optimize the beam shift and splitting effects, the fields $\boldsymbol{E}_1'$ and $\boldsymbol{E}_2'$ (energy 20 μJ) are set to be the strongest, approximately 5 times larger than $\boldsymbol{E}_1$ and $\boldsymbol{E}_2$ (energy 4 μJ), and 50 times larger than the weak probe beam $\boldsymbol{E}_3$ (energy 0.4 μJ), which is as weak as the generated FWM beams. These weak beams are split into two equal components by a 50% beam splitter before being detected. One is captured by a CCD camera, and the other is detected by photomultiplier tubes ($\mathrm{D}_{1,2}$) and a fast gated integrator (gate width is ~50 ns).

To understand the observed spatial splitting and shift of the probe and FWM beams, we consider the SPM and XPM processes. The propagation equations that give the mathematical description of the SPM- and XPM-induced spatial interplay for the probe and FWM beams are

$$\begin{aligned}\frac{\partial u_3(x,y,z)}{\partial z} &- \frac{i}{2k_3}\left[\frac{\partial^2 u_3(x,y,z)}{\partial x^2} + \frac{\partial^2 u_3(x,y,z)}{\partial y^2}\right] \\ &= \frac{ik_0}{2}[(2n_0 - 1) + i\chi_I{}^{(1)} + n_2^{S1}|u_3(x,y,z)|^2 \\ &+ n_2^{X1}|u_1'(x,y,z)|^2 + n_2^{X2}|u_2'(x,y,z)|^2]u_3(x,y,z)\end{aligned} \tag{7.1}$$

$$\begin{aligned}\frac{\partial u_{\mathrm{F1}}(x,y,z)}{\partial z} &- \frac{i}{2k_3}\left[\frac{\partial^2 u_{\mathrm{F1}}(x,y,z)}{\partial x^2} + \frac{\partial^2 u_{\mathrm{F1}}(x,y,z)}{\partial y^2}\right] \\ &= \frac{ik_0}{2}[(2n_0 - 1) + i\chi_I{}^{(1)} + n_2^{S2}|u_{\mathrm{F1}}(x,y,z)|^2 \\ &+ n_2^{X3}|u_1'(x,y,z)|^2\cos^2(\pi\xi/\Lambda) + n_2^{X4}|u_2'(x,y,z)|^2]u_{\mathrm{F1}}(x,y,z)\end{aligned} \tag{7.2}$$

$$\begin{aligned}\frac{\partial u_{\mathrm{F2}}(x,y,z)}{\partial z} &- \frac{i}{2k_3}\left[\frac{\partial^2 u_{\mathrm{F2}}(x,y,z)}{\partial x^2} + \frac{\partial^2 u_{\mathrm{F2}}(x,y,z)}{\partial y^2}\right] \\ &= \frac{ik_0}{2}[(2n_0 - 1) + i\chi_I{}^{(1)} + n_2^{S3}|u_{\mathrm{F2}}(x,y,z)|^2 + n_2^{X5}|u_1'(x,y,z)|^2 \\ &+ n_2^{X6}|u_2'(x,y,z)|^2\cos^2(\pi\xi/\Lambda)]u_{\mathrm{F2}}(x,y,z)\end{aligned} \tag{7.3}$$

where z is the longitudinal coordinate in the weak beam propagation direction and $\mathbf{k}_3 = \mathbf{k}_{\mathrm{F1}} = \mathbf{k}_{\mathrm{F2}} = \omega_1 n_0/c$. n_0 is the linear refractive index at ω_1; n_2^{S1-S3} are the self-Kerr coefficients of $\boldsymbol{E}_{3,\mathrm{F1,2}}$; n_2^{X1-X6} are the cross-Kerr coefficients of $\boldsymbol{E}_{3,\mathrm{F1,F2}}$ induced by $\boldsymbol{E}_1'$ and $\boldsymbol{E}_2'$. $\chi^{(1)}$ is the linear susceptibilities. $u_{3,\mathrm{F1,F2}}$ and $u_{1,2}'$ are the amplitudes of the beams $\boldsymbol{E}_{3,\mathrm{F1,F2}}$ and $\boldsymbol{E}_{1,2}'$. Here, on the left-hand side of these equations, the second terms give the diffraction of the beams during propagation. On the right-hand side, the first term is linear refractive index, the second the linear absorption, the third the nonlinear self-Kerr effect, and the last two terms the nonlinear cross-Kerr effects. Equation (7.1), Equation (7.2), and Equation (7.3) can be numerically solved by the commonly employed split-step Fourier method.

The Kerr nonlinear coefficient is negative for a self-defocusing medium and positive for a self-focusing one, and the general expressions are $n_0 = \sqrt{1 + \chi^{(1)}}$ and $n_2 \approx \mathrm{Re}\chi^{(3)}/(\varepsilon_0 c n_0)$. The linear and nonlinear susceptibilities are expressed as $\chi^{(1)} = N\mu_{10}^2\rho_{10}^{(1)}/(\hbar\varepsilon_0 G_{3,\mathrm{F1,F2}})$ and $\chi^{(3)} = D\rho_{10}^{(3)}$, where $D = N\mu_{10}^4/(\hbar^3\varepsilon_0 G_{3,\mathrm{F1,F2}}|G_{3,\mathrm{F1,F2}}|^2)$ for n_2^{S1-S3} and $D = N\mu_{10}^4/(\hbar^3\varepsilon_0 G_{3,\mathrm{F1,F2}}|G_i|^2)$ for n_2^{X1-X6} [19], where N

is the atomic density of the medium (determined by the cell temperature) and μ_{i0} is the dipole matrix element between energy levels $|0\rangle$ and $|i\rangle$. The first and third coherence terms $\rho_{10}^{(1)}$ and $\rho_{10}^{(3)}$ can be obtained by solving the coupled density matrix equations.

The periodic term $\cos^2(\pi\xi/\Lambda)$ in Eq. (7.1), Eq. (7.2), and Eq. (7.3) describes the spatial periodical modulation effect of EIG1 in the x-direction and EIG2 in the y-direction (or EIG3 in the y-direction and EIG4 in the x-direction) created by the interference of beams $\boldsymbol{E}_1$ and $\boldsymbol{E}_1'$ ($\boldsymbol{E}_2$ and $\boldsymbol{E}_2'$). In such gratings, the refractive index of work beams is modified to vary periodically along the x-axis (or y-axis) due to the cross-Kerr effect of the interference pattern. Because such a modification is derived from atomic-enhanced nonlinearity, the index, that is, $n(\Delta, I, x) = n_1(\Delta) + \delta n(x)$, strongly depends on both the frequency detuning and intensity of the inducing beams, where $\Delta n = n_2 I \cos^2(\pi x/\Lambda)$ accounts for the periodic property of the index. Here, Λ is the period of the interference pattern and can be given by $\Lambda = \lambda/\theta$, where θ is the angle between $\boldsymbol{E}_1$ and $\boldsymbol{E}_1'$ (or $\boldsymbol{E}_2$ and $\boldsymbol{E}_2'$).

If we neglect the diffraction term, the SPM contribution, the small XPM contribution, and the EIG modulation term, Eq. (7.1), Eq. (7.2), and Eq. (7.3) can be solved for an approximately analytical solution $u_{3,F1,F2}(x,y,z) = u_{3,F1,F2}(x,y,0)\exp(-l_{loss} + i\phi_{NL})$, where the nonlinear phase shift $\phi_{NL} = k_0[n_2^{X1,3,5}|u_1'(x,y)|^2 + n_2^{X2,4,6}|u_2'(x,y)|^2]z/2$ and the linear absorption $l_{loss} = k_0\chi^{(1)}z/2$. So, the intensity of $\boldsymbol{E}_{3,F1,F2}$ can be expressed as

$$I_{3,F1,F2}(x,y,z) = \cos^2\phi_{NL}\left\{\exp\left[\frac{-(x^2+y^2)}{2}\right] + \exp\left[\frac{-(x-x_0)^2-(y-y_0)^2}{2}\right]\right\}^2 \times u_{3,F1,F2}^2(x,y,0) \tag{7.4}$$

The additional transverse propagation wave vector can be obtained as $\delta\mathbf{k}_x = (\partial\phi_{NL}/\partial x)\boldsymbol{x}$, where $\boldsymbol{x}$ is the unit vector along the horizontal axis. The direction of $\delta\mathbf{k}_x$ determines the horizontal propagation of the beam, while the phase-front curvature $\partial^2\phi_{NL}/\partial x^2 < 0$ ($\partial^2\phi_{NL}/\partial x^2 > 0$) leads to local focusing (defocusing) of the beams imposed on the weak field by the strong field. When $n_2 > 0$, ϕ_{NL} has a positive Gaussian profile and $\delta\mathbf{k}_x$ is always toward the beam center of the strong field. Therefore, the weak field is shifted to the center of the strong field with a global splitting, while the local defocusing and focusing due to positive and negative phase-front curvatures split the weak field into more parts locally. And the weak field will shift away from the strong field when $n_2 < 0$, which also brings about the global and local splitting. Similarly, the spatial variation in the high-order derivative sign of ϕ_{NL} ($\partial^n\phi_{NL}/\partial x^n$) leads to more split parts. The expression of the nonlinear phase shift shows that stronger spatial splitting can occur if the intensity of strong fields I, n_2 increases and the intensity of weak fields I_0 decreases. For $\boldsymbol{E}_3$, $I_0 \propto |\rho_{10}^{(1)}|^2$ and $\rho_{10}^{(1)} = iG_3/(d_1 + |G_2|^2/d_2)$. For $\boldsymbol{E}_{F1}$ and $\boldsymbol{E}_{F2}$, $I_0 \propto |\rho_{10}^{(3)}|^2$, where $\rho_{10}^{(3)} = g/[\Gamma_{00}(d_1 + |G_2|^2/d_2)^2]$ for $\boldsymbol{E}_{F1}$ and $\rho_{10}^{(3)} = g/[(d_1 + |G_1^2|^2/\Gamma_{00})d_2]$ for $\boldsymbol{E}_{F2}$, with $g = -iG_3G_1G_1'^*$, $d_1 = i\Delta_1 + \Gamma_{10}$, $d_2 = \Gamma_{20} + i(\Delta_1 + \Delta_2)$, and Γ_{ij} being the transverse relaxation rate between the states $|i\rangle$ and $|j\rangle$. In the experiment, I_0 can be controlled via the suppression effect that is induced by the dressing effect.

Specially, as dressing fields, $\boldsymbol{E}_1$ and $\boldsymbol{E}_1'$ (or $\boldsymbol{E}_2$ and $\boldsymbol{E}_2'$) can effectively suppress the intensity of $\boldsymbol{E}_{F2}$ ($\boldsymbol{E}_{F1}$) when the condition $\Delta_1 + \Delta_2 = 0$ is satisfied.

7.1.3
The Interplay of Two FWM Beams

First, we change the power of $\boldsymbol{E}_1'$ and $\boldsymbol{E}_2'$ to investigate the interplay of two FWM beams at certain frequency detuning and angles of the corresponding fields at an optical depth of 6.35×10^6. As shown in Figure 7.2(a2), the spatial splitting of $\boldsymbol{E}_{F1}$ in the x-direction appears with $\Delta_1 = -30$ GHz and $\Delta\theta_1 = 0.15^\circ$. As the power of $\boldsymbol{E}_1'$ increases, the beam splits from one part to two parts. In order to understand these results, we consider the relative positions of $\boldsymbol{E}_{F1}$ and $\boldsymbol{E}_1'$. Because $\boldsymbol{E}_1'$ has a small angle with $\boldsymbol{E}_1$ either in the x–z-plane or the y–z-plane as shown in Figure 7.1(b), $\boldsymbol{E}_{F1}$ overlaps in both the x- and y-directions with $\boldsymbol{E}_1'$ due to the phase-matching condition $\mathbf{k}_{F1} = \mathbf{k}_1 - \mathbf{k}_1' + \mathbf{k}_3$. When $\boldsymbol{E}_1$ and $\boldsymbol{E}_1'$ are set at $\Delta\theta_1 = 0.15^\circ$ or $\Delta\theta_1 = -0.15^\circ$, the x-directional overlap between them is dominant as the angle in the x–z-plane is far larger than that in the y–z-plane. At $\Delta\theta_1 = 0$, $\boldsymbol{E}_{F1}$ exactly overlaps with $\boldsymbol{E}_1'$ in the y-direction. Furthermore, owing to the larger n_2 around $\Delta_1 = 0$, the horizontal aligned EIG1 is induced by $\boldsymbol{E}_1$ and $\boldsymbol{E}_1'$ in the x-direction with interference fringes spacing $\Lambda(\theta_{1x} \approx 0.3^\circ)$. Meanwhile, a vertical aligned EIG2 is induced in the y-direction with interference fringes spacing $\Lambda(\theta_{1y} \approx 0.05^\circ)$. As θ_{1x} is far larger than θ_{1y}, the horizontal fringe spacing is too small to be observed. Therefore, the x-directional splitting in Figure 7.2(a2) is obviously observed because the overlap in the x-direction and the nonlinear cross-Kerr effect of $\boldsymbol{E}_1'$ are dominating. In contrast, the y-directional splitting in Figure 7.2(a3) is due to the EIG2. In order to study the dependence of the beam splitting on the power of $\boldsymbol{E}_1'$, we consider the Eq. (7.4). According to $I_{3,F1,F2}(z,\xi) \propto \cos^2\phi_{NL}$, the splitting number is proportional to ϕ_{NL}, which is directly proportional to the intensity of $\boldsymbol{E}_1'(I)$. As a result, stronger splitting will be observed when I increases as shown in Figure 7.2(a2). For $\boldsymbol{E}_{F1}$, the splitting in the y-direction due to EIG2 is strong when the power of $\boldsymbol{E}_1'$ increases because of the large profile of $\boldsymbol{E}_{F1}$, as shown in Figure 7.2(a3). For $\boldsymbol{E}_{F2}$, similar results are obtained because of the same reason when the power of $\boldsymbol{E}_2'$ increases, as shown in Figure 7.2(b). However, because $\theta_{2y0} \approx 0.3^\circ$ and $\theta_{2x} \approx 0.05^\circ$ for $\boldsymbol{E}_{F2}$, the splitting results of $\boldsymbol{E}_{F1}$ and $\boldsymbol{E}_{F2}$ are just the opposite.

Next, the interplay of two FWM beams is studied when we change the power of $\boldsymbol{E}_1'$ or $\boldsymbol{E}_2'$ as shown in Figure 7.2(c). Figure 7.2(c1) and (c2) displays the interplay between $\boldsymbol{E}_2'$ and $\boldsymbol{E}_{F1}$. When the power of $\boldsymbol{E}_2'$ increases, $\boldsymbol{E}_{F1}$ will be more suppressed (the suppression condition $\Delta_1 + \Delta_2 = 0$ is satisfied) and the intensity of $\boldsymbol{E}_{F1}$ will be weaker. According to ϕ_{NL}, the x-directional splitting of $\boldsymbol{E}_{F1}$ becomes stronger as shown in Figure 7.2(c1). The same results are obtained for $\boldsymbol{E}_{F2}$ as shown in Figure 7.2(c3). The y-directional splitting of $\boldsymbol{E}_{F1}$ due to EIG2 or the x-directional splitting of $\boldsymbol{E}_{F2}$ due to EIG4 is determined by the intensity and size of the spot when the frequency and angle of the pump fields are constant. If the intensity of FWM increases and the size of FWM becomes wider, the spatial splitting is stronger, as shown in Figure 7.2(a3), (b2), and (c2).

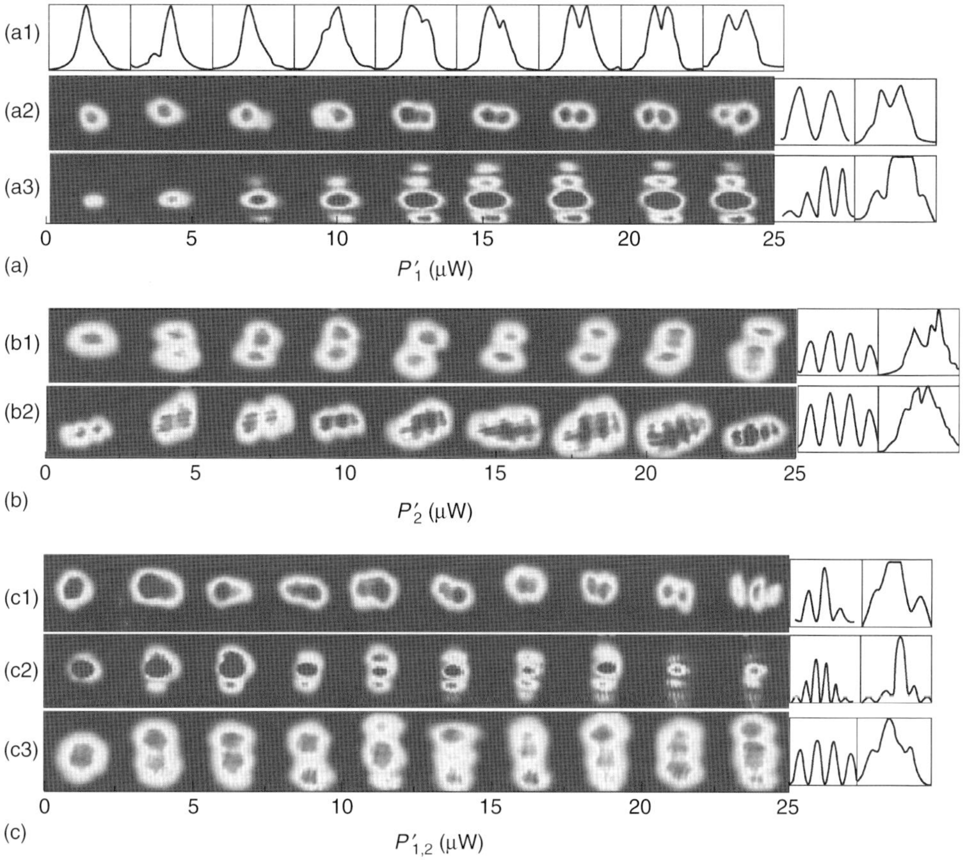

Figure 7.2 Images of $\boldsymbol{E}_{F1}$ or $\boldsymbol{E}_{F2}$ versus the power of the pump fields at an optical depth of 6.35×10^6. (a) $\boldsymbol{E}_{F1}$ versus P'_1 when $\boldsymbol{E}_2$ and $\boldsymbol{E}'_2$ are blocked: (a1) the transverse section profile, (a2) the images with $\Delta_1 = -30$ GHz and $\Delta\theta_1 = 0.15^\circ$. (a3) The images with $\Delta_1 = -20$ GHz and $\Delta\theta_1 = 0$. (b) $\boldsymbol{E}_{F2}$ versus P'_2 when $\boldsymbol{E}_1$ and $\boldsymbol{E}'_1$ are blocked. $\boldsymbol{E}_2$ and $\boldsymbol{E}'_2$ are set (b1) at $\Delta\theta_2 = 0.15^\circ$ with $\Delta_2 = -15$ GHz and (b2) at $\Delta\theta_2 = 0$ with $\Delta_2 = -10$ GHz. (c) $\boldsymbol{E}_{F1}$ versus P'_2 when $\boldsymbol{E}_1$ and $\boldsymbol{E}'_2$ are set (c1) at $\Delta\theta_1 = 0.15^\circ$ with $\Delta_1 = -25$ GHz and $\Delta_2 = 25$ GHz (c2) at $\Delta\theta_1 = 0$ with $\Delta_1 = -20$ GHz and $\Delta_2 = 20$ GHz. (c3) $\boldsymbol{E}_{F2}$ versus P'_1 when $\boldsymbol{E}_2$ and $\boldsymbol{E}'_2$ are set at $\Delta\theta_1 = 0$ with $\Delta_1 = 15$ GHz and $\Delta_2 = -15$ GHz. The curves of the left column are the experimental results and those of the right column are the calculated beam profiles. The horizontal coordinate is $x(mm)$ for $\boldsymbol{E}_{F1}$ and $y(mm)$ for $\boldsymbol{E}_{F2}$. The vertical one is the FWM beam profile. Source: Adapted from Ref. [21].

Now, let us investigate the spatial interplay between the two FWM beams dependent on the angle of the pump fields. Figure 7.3(a) demonstrates that the beam profile of $\boldsymbol{E}_{F1}$ changes when $\Delta\theta_1$ is changed, with $\boldsymbol{E}_2$ and $\boldsymbol{E}'_2$ being blocked at an optical depth of 6.35×10^6. As shown in Figure 7.3(a1), the transformation between the x- and y-directional spatial splitting will be easily observed at $\Delta_1 = -25$ GHz.

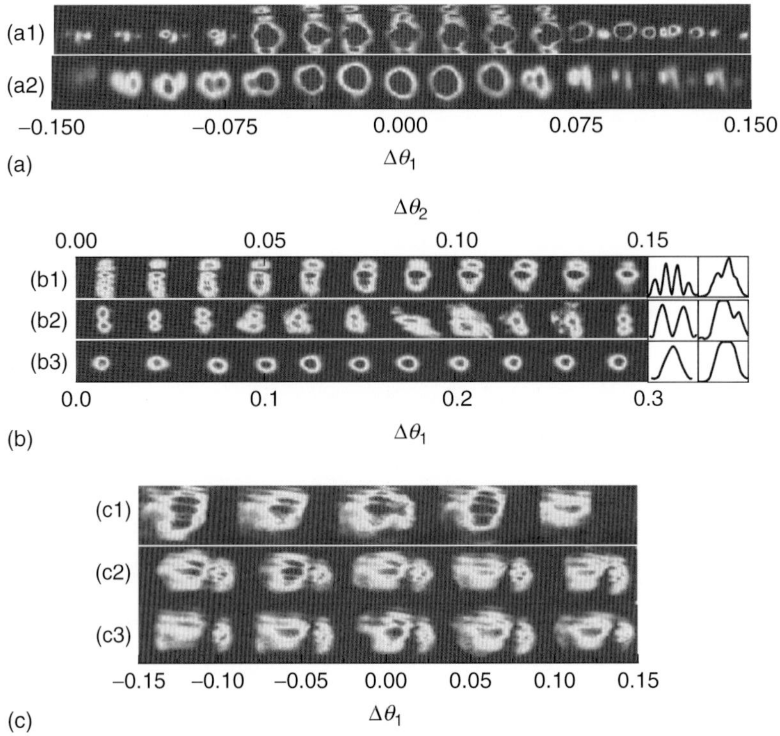

Figure 7.3 Images of $\boldsymbol{E}_{F1}$ or $\boldsymbol{E}_{F2}$ versus the angle between the corresponding two pump fields at an optical depth of 6.35×10^6. (a) $\boldsymbol{E}_{F1}$ versus $\Delta\theta_1$ when both $\boldsymbol{E}_2$ and $\boldsymbol{E}_2'$ are blocked with (a1) $\Delta_1 = -25$ GHz and (a2) $\Delta_1 = -50$ GHz. (b1) $\boldsymbol{E}_{F1}$ versus $\Delta\theta_2$ with $\Delta_1 = -20$ GHz and $\Delta_2 = 20$ GHz when $\boldsymbol{E}_1$ and $\boldsymbol{E}_1'$ are set at $\Delta\theta_1 = 0$. (b2) $\boldsymbol{E}_{F2}$ versus $\Delta\theta_1$ with $\Delta_1 = -25$ GHz and $\Delta_2 = -25$ GHz when $\boldsymbol{E}_2$ and $\boldsymbol{E}_2'$ are set at $\Delta\theta_2 = 0.15$. (b3) $\boldsymbol{E}_3$ versus $\Delta\theta_1$. The condition is similar to that of (b2). (c) $\boldsymbol{E}_{F1}$ versus $\Delta\theta_1$ with $\Delta_1 \approx -20$ GHz and $\Delta_2 = 20$ GHz, (c1) both $\boldsymbol{E}_1$ and $\boldsymbol{E}_2'$ off, (c2) only $\boldsymbol{E}_2'$ on, (c3) both $\boldsymbol{E}_2$ and $\boldsymbol{E}_2'$ on. The setup of the right columns is similar to those in Figure 7.2.

The x-directional splitting is obtained from $\Delta\theta_1 = -0.15^\circ$ to $\Delta\theta_1 = -0.075^\circ$ and from $\Delta\theta_1 = 0.075^\circ$ to $\Delta\theta_1 = 0.15^\circ$, while the y-directional splitting is obtained from $\Delta\theta_1 = -0.075^\circ$ to $\Delta\theta_1 = 0.075^\circ$. At $\Delta_1 = -50$ GHz, although the experimental condition in Figure 7.3(a2) is the same as that in Figure 7.3(a1), the y-directional splitting cannot be observed.

The observation of the transformation between the x- and y-directional splitting can be attributed to the changing of the spatial overlapping mode between the weak signal $\boldsymbol{E}_{F1}$ and the strong beam $\boldsymbol{E}_1'$. As mentioned in the preceding text, $\boldsymbol{E}_{F1}$ only shows the x-directional splitting because the overlap of $\boldsymbol{E}_{F1}$ and $\boldsymbol{E}_1'$ in the x-direction is more dominant than that in the y-direction when $\Delta\theta_1 = -0.15^\circ$, and the fringe spacing of EIG1 is too small to be observed. When $\Delta\theta_1$ changes from -0.15° to 0.15°, the y-directional overlap of $\boldsymbol{E}_{F1}$ and $\boldsymbol{E}_1'$ increases but the x-directional overlap decreases because of the phase-matching condition $\mathbf{k}_{F1} = \mathbf{k}_1 - \mathbf{k}_1' + \mathbf{k}_3$. So, the x-direction al splitting disappears gradually and EIG2 modulation becomes significant. Finally, when $\Delta\theta_1 = 0$, $\boldsymbol{E}_1'$ overlaps with $\boldsymbol{E}_{F1}$ exactly in the y-direction. Thus, the spatial modulation of $\boldsymbol{E}_{F1}$ is only in the y-direction and its multi-spots caused by the periodic modulation of EIG2 are shown in Figure 7.3(a1). Compared to the splitting due to pure XPM, the EIG modulation needs larger Kerr nonlinearity. At $\Delta_1 = -50$ GHz, the pump fields are further away from the resonant state; thus, n_2 becomes very small compared to that at $\Delta_1 = -25$ GHz. So, the EIG modulation cannot be observed, which makes the y-directional splitting induced by EIG2 unobtainable when $\Delta\theta_1 = 0$, as shown in Figure 7.3(a1).

Figure 7.3(b1) clearly shows that the y-directional splitting, shift, and intensity of the $\boldsymbol{E}_{F1}$ beam all change when we move $\Delta\theta_2$ from 0° to 0.15°. At first, when $\theta_{2x} = 0$, the $\boldsymbol{E}_{F1}$ has the weakest intensity and its y-directional splitting is the strongest because the dressing suppressed effect of $\boldsymbol{E}_2'$ (the suppression condition $\Delta_1 + \Delta_2 = 0$ is satisfied) is the strongest. When $\Delta\theta_2 = -0.15^\circ$, the intensity of $\boldsymbol{E}_{F1}$ increases because the weakened suppression effect and therefore the y-directional splitting will be weaker. During the movement of $\boldsymbol{E}_2'$, we also observe a slight shift of $\boldsymbol{E}_{F1}$ because $\boldsymbol{E}_2'$ can attract $\boldsymbol{E}_{F1}$ via the XPM at $\Delta_1 = -20$ GHz.

As $\boldsymbol{E}_1'$ moves from $\Delta\theta_1 = 0$ to $\Delta\theta_1 = 0.3^\circ$, $\boldsymbol{E}_{F2}$ and $\boldsymbol{E}_3$ shift as shown in Figure 7.3(b2) and (b3). First, both $\boldsymbol{E}_3$ and $\boldsymbol{E}_{F2}$ are shifted toward to $\boldsymbol{E}_1'$ because the strong attraction due to the XPM of $\boldsymbol{E}_1'$ at $\Delta_1 = -25$ GHz. Then, when $\boldsymbol{E}_1'$ moves to $\Delta\theta_1 = 0.3^\circ$, it gradually transfers to the right side of $\boldsymbol{E}_3$ and $\boldsymbol{E}_{F2}$, so they get a right-direction shift. It is noted that there is additional reflector on the propagating path of $\boldsymbol{E}_3$, so it shows an opposite-direction shift. However, for $\boldsymbol{E}_{F2}$, its overlap with $\boldsymbol{E}_1'$ is not only in the x-direction but also in the y-direction, and the y-directional distance from $\boldsymbol{E}_{F2}$ to $\boldsymbol{E}_1'$ varies from long to short, and then to long when $\boldsymbol{E}_1'$ is moved from $\Delta\theta_1 = 0$ to $\Delta\theta_1 = 0.3^\circ$. As a result, $\boldsymbol{E}_{F2}$ will first be shifted below and then pulled back to its original position because of the attraction of $\boldsymbol{E}_1'$ at $\Delta_1 = -25$ GHz. It is worth mentioning that the splitting of $\boldsymbol{E}_{F2}$ is affected by $\boldsymbol{E}_1'$. In Figure 7.3(b2), when $\boldsymbol{E}_1'$ moves from $\Delta\theta_1 = 0$ to $\Delta\theta_1 = 0.3^\circ$, the inclined splitting of $\boldsymbol{E}_{F2}$ is obtained at the same time. The intensities of the$\boldsymbol{E}_{F2}$ beams also change as a different spatial state of $\boldsymbol{E}_1'$ is adopted. Because we adjust the detuning to make the dressing effect of $\boldsymbol{E}_1'$ on $\boldsymbol{E}_{F2}$ be suppression, the $\boldsymbol{E}_{F2}$ intensity will be the weakest when $\boldsymbol{E}_1'$ arrives at $\Delta\theta_1 = 0.15^\circ$.

By changing $\Delta\theta_1$ with different dressing fields, we study the spatial dynamics of $\boldsymbol{E}_{F1}$. From $\Delta\theta_1 = -0.15^\circ$ to $\Delta\theta_1 = 0.15^\circ$ in Figure 7.3(c), the rows of experimental spots successively correspond to $\boldsymbol{E}_{F1}$ without dressing fields, with $\boldsymbol{E}_2'$ dressing and with both $\boldsymbol{E}_2$ and $\boldsymbol{E}_2'$ dressing, respectively. It is obvious that $\boldsymbol{E}_{F1}$ beams are split into more parts along the x-direction when both $\boldsymbol{E}_2$ and $\boldsymbol{E}_2'$ beams are on under the suppression condition ($\Delta_1 + \Delta_2 = 0$). Its intensity shows more significant suppression, compared to the case without dressing or only with $\boldsymbol{E}_2'$ dressing. This difference is due to the fact that the splitting will become stronger with a weaker $\boldsymbol{E}_{F1}$.

Furthermore, we investigate the interplay between two FWMs dependent on frequency detuning. Figure 7.4(a2) presents the $\boldsymbol{E}_{F1}$ images versus Δ_1 with $\Delta\theta_1 = 0.075^\circ$ as well as $\boldsymbol{E}_2$ and $\boldsymbol{E}_2'$ being blocked. The transformation between the splitting in the x- and y-directions is observed clearly. The signal intensity of

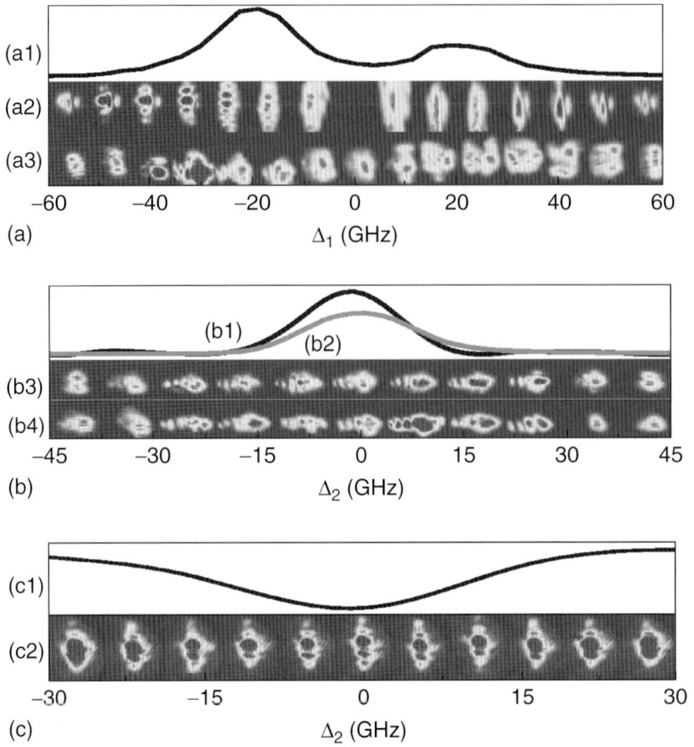

Figure 7.4 (a1) The curve of $\mathbf{E}_{F1}$ signal intensity versus Δ_1. (a2) Images of $\mathbf{E}_{F1}$ beam versus Δ_1. (a3) Images of $\mathbf{E}_{F2}$ beam versus Δ_1 with $\Delta\theta_1 = 0.075$ and both $\mathbf{E}_2$ and $\mathbf{E}_2'$ are blocked. (b) The intensity curves (b1, b2) or images (b3, b4) of $\mathbf{E}_{F2}$ signal intensity versus Δ_2 with $\Delta\theta_2 = 0.075$, (b1, b3) both $\mathbf{E}_1$ and $\mathbf{E}_1'$ are on, (b2, b4) both $\mathbf{E}_1$ and $\mathbf{E}_1'$ are off. (c1) The intensity curve and (c2) images of $\mathbf{E}_{F1}$ signal versus Δ_2 with $\Delta\theta_1 = 0$. All the cases are at an optical depth of 6.35×10^6. Source: Adapted from Ref. [21].

$\boldsymbol{E}_{F1}$ is shown in Figure 7.4(a1). Comparing Figure 7.4(a1) with (a2), we can first conclude that the splitting is approximately symmetrical with respect to $\Delta_1 = 0$. Second, the splitting in the x-direction is observed far away from the resonance, while that in the y-direction is near the resonant state. And third, the number of splitting is determined by Δ_1. A similar splitting transformation process is also observed for $\boldsymbol{E}_{F2}$, as shown in Figure 7.4(a3).

Various SPM and XPM processes are involved to understand the spatial splitting of FWM beams observed. For $\boldsymbol{E}_{F1}$, first, because the absolute Kerr nonlinear coefficient $|n_2|$ is symmetrical with respect to $\Delta_1 = 0$, the derivative spatial splitting has the same symmetry. Second, the transformation between the x- and y-directional splitting can be explained by the changing of relative positions between the weak and strong beams. When $\Delta\theta_1 = 0.075^\circ$, $\boldsymbol{E}_1'$ and $\boldsymbol{E}_{F1}$ overlap mainly in the x-direction, which leads to the splitting of the $\boldsymbol{E}_{F1}$ beam in the x-direction due to the nonlinear cross-Kerr effect. In the two sides far away from resonance where $|n_2|$ is low and the beam size is small, the x-directional splitting is observed. While $|n_2|$ becomes large, the beam size becomes wide and the FWM intensity will be strengthened. So $\boldsymbol{E}_1'$ and $\boldsymbol{E}_{\mathrm{F1}}$ mainly overlap in the y-direction and lead to the splitting in the y-direction due to the EIG2 induced by $\boldsymbol{E}_1$ and $\boldsymbol{E}_1'$. As a result, the splitting in the x-direction is obtained in the two sides and that in the y-direction is observed around the two intensity peaks. Moreover, around the two peaks, $|n_2|$ gradually arrives at its maximum, so we can obtain the periodic FWM beam profile by more significant spatial splitting. While around $\Delta_1 = 0$, n_2 is always close to zero, so the splitting of FWM beam cannot appear.

Figure 7.4(b1) to (b4) shows the intensities of $\boldsymbol{E}_{\mathrm{F2}}$ with or without the dressing fields $\boldsymbol{E}_1$ and $\boldsymbol{E}_1'$ versus Δ_2. The transformation between x- and y-directional splitting is obtained by scanning Δ_2 as shown in Figure 7.4(b3) and (b4). Comparing Figure 7.4(b1) with (b2), we observe that $\boldsymbol{E}_{\mathrm{F2}}$ is suppressed by $\boldsymbol{E}_1$ and $\boldsymbol{E}_1'$. A similar result is also observed on $\boldsymbol{E}_{\mathrm{F1}}$ which is suppressed by $\boldsymbol{E}_2$ and $\boldsymbol{E}_2'$, as shown in Figure 7.4(c).

Finally, we investigate the interplay between two FWM signals that depend on the sample temperature, that is, the optical depth D. It is well known that the temperature can effectively influence the atomic density N in the vapor medium. We theoretically obtain $n_2 = \mathrm{Re}\,\chi^{(3)}/(\varepsilon_0 c n_0)$, $\chi^{(3)} = D\rho_{10}^{(3)}$, and $D = N\mu_{10}^4/(\hbar^3\varepsilon_0 G_{3,\mathrm{F1,F2}}|G_i|^2)$, so increasing of D will result in increasing of N, nonlinear coefficient n_2 and further bring larger nonlinear phase shift ϕ_{NL}, which indicates more spatial splitting and shift in the FWM beam profile shown in Figure 7.5(a) to (c). So the beam profile of the FWM signal becomes more vivid at higher temperatures. However, for larger optical depth, the linear absorption is larger too, which dissipates the FWM signal significantly. Therefore, the observed results are caused by the two opposite factors. Figure 7.5(a) clearly shows the change in the splitting, shift, and suppression of the $\boldsymbol{E}_{\mathrm{F2}}$ signal when the spatial position of $\boldsymbol{E}_1'$ is controlled to move from $\Delta\theta_1 = 0$ to $\Delta\theta_1 = 0.15^\circ$ at different temperatures. First, when $\boldsymbol{E}_{\mathrm{F1}}$ and $\boldsymbol{E}_{\mathrm{F2}}$ are set $E_{\mathrm{F1}} = 0.15^\circ$ and $\boldsymbol{E}_{\mathrm{F2}}$ overlaps with 6.35×10^6 at $\boldsymbol{E}_{\mathrm{F1}}$ with largest atomic density, $\boldsymbol{E}_{\mathrm{F2}}$ suffers from strongest suppression of $\boldsymbol{E}_2$ and therefore it is the weakest, leading to the strongest y-directional splitting and shifting of $\boldsymbol{E}_2'$. Then, if $\boldsymbol{E}_1'$ is moved from $\Delta\theta_1 = 0$ to $\Delta\theta_1 = 0.15^\circ$, its overlap with $\Delta_1 = -30$ GHz will decrease and the suppression of $\boldsymbol{E}_{\mathrm{F2}}$ signal by $\Delta_1 = -20$ GHz will become weak gradually, which leads to the splitting due to the nonlinear cross-Kerr effect being weak. The decrease in this overlap, however, leads to the attraction of $\Delta\theta_1 = 0$ from $\boldsymbol{E}_{\mathrm{F2}}$. It can be observed in Figure 7.5(a) that $\boldsymbol{E}_{\mathrm{F2}}$ shifts upward with the increasing of $\boldsymbol{E}_1$.

The change in splitting, shift, and suppression versus the powers of the dressing fields is shown in Figure 7.5(b) and (c). At higher temperatures, the transformation between the x- and y-directional splitting is obtained for $\boldsymbol{E}_1'$, which is shifted to the left side by $\boldsymbol{E}_2$ and overlaps with $\boldsymbol{E}_2'$ to induce the x-directional splitting as shown in Figure 7.5(c).

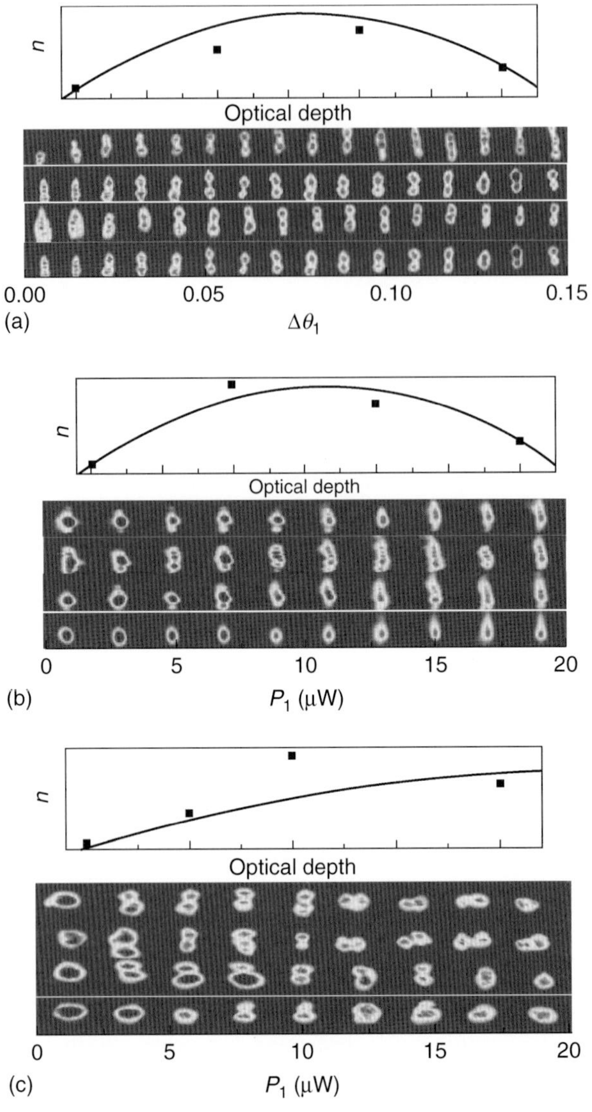

Figure 7.5 For the images in (a, b, c), from bottom to top, the optical depth is 6.35×10^6, 9.97×10^6, 1.54×10^7, and 2.33×10^7 successively. The vertical axis label (n) is the splitting number of FWM beam. (a) Images of $\mathbf{E}_{F2}$ versus $\Delta\theta_1$ with $\Delta_1 = -20$ GHz, $\Delta_2 = 20$ GHz, and $\Delta\theta_2 = 0.15$. (b) The power dependence of the dressing effect of $\mathbf{E}_2'$ on $\mathbf{E}_{F1}$ beams with $\Delta\theta_1 = 0$ at $\Delta_1 = -25$ GHz and $\Delta_2 = 25$GHz. (c) The power dependence of the dressing effect of $\mathbf{E}_1'$ on $\mathbf{E}_{F2}$ beams when $\Delta\theta_2 = 0.15$ at $\Delta_1 = -20$ GHz and $\Delta_2 = 20$ GHz. The above-mentioned panels are the splitting number of FWM (square dots) and the fitted curve versus different optical depth.

7.2
Optical Vortices Induced in Nonlinear Multi-Level Atomic Vapors

7.2.1
Introduction

An optical vortex is an interesting structure that possesses a phase defect at a point (called *the vortex core*) and a rotational energy flow around it. In recent decades, optical vortices and vortex solitons attracted a lot of attention from research groups all over the world, for their potential applications in optical data storage [22], distribution [23], and processing [24] as well as in the study of optical tweezers [25], trapping and guiding of cold atoms [26], and entanglement states of photons [27]. To date, research on optical vortices in many kinds of media, such as bulk nonlinear media [28], discrete systems [29], atomic vapors [30], dissipative optical systems [31], and Bose–Einstein condensates [32], has been reported. In Ref. [33], the authors discussed the properties of multi-dimensional beams via atomic coherence. Still, there are many interesting topics to be researched.

In this section, we demonstrate the existence and curious evolution of vortices in a ladder-type three-level nonlinear atomic vapor with linear, cubic, and quintic susceptibilities considered simultaneously with the dressing effect. We find that the number of beads and topological charge of the incident beam, as well as its size, greatly affect the formation and evolution of vortices. To determine the number of induced vortices and the corresponding rotation direction, we give common rules associated with the initial conditions coming from various incident beams.

7.2.2
Theoretical Model and Numerical Simulation

We consider paraxial propagation of probe vortex beams in a ladder-type three-level atomic system formed by $3S_{1/2}$, $3P_{1/2}$, and $5S_{1/2}$ levels of sodium. The model is described by the nonlinear Schrodinger equation (NLSE) of the form [31, 33],

$$i\partial_z\psi + \frac{1-i\beta}{2k}\nabla_\perp^2\psi + \frac{k}{2}\chi\psi = 0 \qquad (7.5)$$

where $\nabla_\perp^2 = \partial xx + \partial yy$ is the transverse Laplacian, k is the wave number, ψ is the amplitude of the beam, β is the diffusion coefficient, and χ is the total susceptibility of the atomic vapor system. We consider a cubic–quintic NLSE as the underlying propagation model that adequately describes the physics of atomic vapor systems. The total susceptibility can be obtained through the Liouville pathway using perturbation theory [34]

$$\chi = \frac{\eta}{K} - \left(\frac{\eta}{K^2}\right)\left(\frac{|G_1|^2}{d_1} + \frac{|G_2|}{d_2}\right) + \left(\frac{\eta}{K^3}\right)\left(\frac{|G_1|^4}{{d_1}^2} + \frac{|G_2|^4}{d_2^2}\right)$$

with $\eta = iN\mu_{10}^2/(\hbar\varepsilon_0)$, $d_2 = \Gamma_{20} + i(\Delta_1 + \Delta_2)$, and $K = (\Gamma_{10} + i\Delta_1) + |G_2|^2/d_2$, where $G_1 = \mu_{10}\Psi/\hbar$ is the Rabi frequency of the probe field, G_2 is the Rabi frequency of the coupled field, N is the atomic density, and $\Delta_{1,2}$ is the detuning of the probe (coupled) field. Γ_{ij} denotes the population decay rate between the corresponding energy levels $|i\rangle$ and $|j\rangle$, and μ_{10} is the electric dipole moment. The second and the third terms in χ represent the cubic and the quintic contributions to the total susceptibility. Equation (7.5) is similar to the complex Ginzburg–Landau equation, with the diffusion coefficient β coming from the models of laser cavities. We launch an incident beam in Eq. (7.5) of the form (in polar coordinates),

$$\psi(z=0,r,\theta) = A\,\mathrm{sech}\left(\frac{r-R_0}{r_0}\right) \times (\cos(n\theta) + iB\sin(n\theta))\exp(il\theta) \tag{7.6}$$

with $B=0$, $n\neq 0$ for a necklace, $B\neq 0$, $n\neq 0$ for an azimuthon [35], and $B=1$ or $n=0$ for a vortex. Here, R_0 is the mean radius, r_0 is the width, A is the amplitude, B is the modulation coefficient ($1-B$ is the modulation depth), l is the input topological charge, and $2n$ is the number of necklace beads. The beam in Eq. (7.6) can be viewed as a superposition of two vortices with net topological charges (NTCs) [36] $l+n$ and $l-n$, respectively. The vortex with a larger magnitude of NTC will dissipate faster during propagation, due to the diffusion term in Eq. (7.5) [37]. Thus, the vortex with a smaller NTC will remain stable for longer during propagation, and the overall NTC will correspond to that charge.

First of all, we set the parameters to be $N=10^{13}\,\mathrm{cm}^{-3}$, $\mu_{10}=3\times 10^{-29}\,\mathrm{cm}$, $\Delta_1=1\,\mathrm{MHz}$, $\Delta_2=-1\,\mathrm{MHz}$, $G_2=40\,\mathrm{MHz}$, $\Gamma_{10}=2\pi\times 4.86\,\mathrm{MHz}$, $\Gamma_{20}=2\pi\times 0.485\,\mathrm{MHz}$, $\lambda=600\,\mathrm{nm}$, $r_0=100\,\mu\mathrm{m}$, $R_0=200\,\mu\mathrm{m}$, and $\beta=0.5$ in our numeric for convenience without special statement [38]. The evolution of a vortex with $l=0$ is shown in the insets in Figure 7.6(a). The input notch of the vortex soon disappears and the vortex changes into a super-Gaussian-like pulse during propagation, with the width and amplitude growing dramatically. For comparison, if we set $l=1$ and redo the evolution, the width and the amplitude still grow, but the notch at the origin remains, as shown in the insets in Figure 7.6(b). Even though the amplitude increases during propagation, there exists a saturable maximum at $\sim 10\,\mathrm{V\,m}^{-1}$, as seen from the radial intensity profiles in Figure 7.6(a) and (b); these results are quite similar to the results from Refs. [28] and [33]. The saturation phenomenon can be explained by the competition between the cubic and the quintic nonlinearity [39]. The nonlinearity is defocusing, and as such can support stable vortices [28]. The beam spreads during propagation, because the nonlinearity is too weak to balance diffraction and form a soliton. If l is set to 3 and 6, there will be 3 and 6 notches around the origin, as shown in Figure 7.6(c) and (d). The right panels there present the corresponding phases that demonstrate every notch is a vortex.

If we set $n=0.5$, we can investigate the evolution of a unique "necklace" with crescent shape. In Figure 7.7, we exhibit a series of evolution snapshots corresponding to a different value of l. By comparing Figure 7.7(b) with (e), and Figure 7.7(c) with (d), we can conclude that the number of notches that appear in the beam is determined by $\min\{|l\pm n|\}$, which is also the absolute value of the NTC of the survived vortex component. Even though Figure 7.7(a) and (f) have the same number of vortices, they rotate in opposite directions because the corresponding NTCs are -5 and 5, respectively. The common rule on how to calculate the number

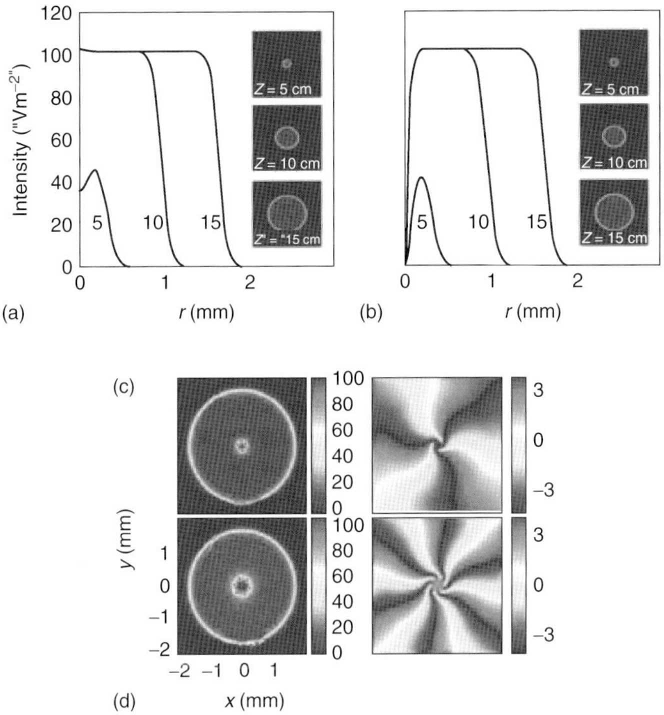

Figure 7.6 (a) and (b) Evolution of vortex at the same time the beads fuse on account of the diffusion term β; incidences, for several propagation distances and for $l = 0$ and $l = 1$, respectively. (c) and (d) Output intensities (left panels) and phases (right panels) of vortices with $l = 3$ and $l = 6$, respectively. The color bars and physical scales of the transverse plane shown here are the same in all other figures. Source: Adapted from Ref. [40].

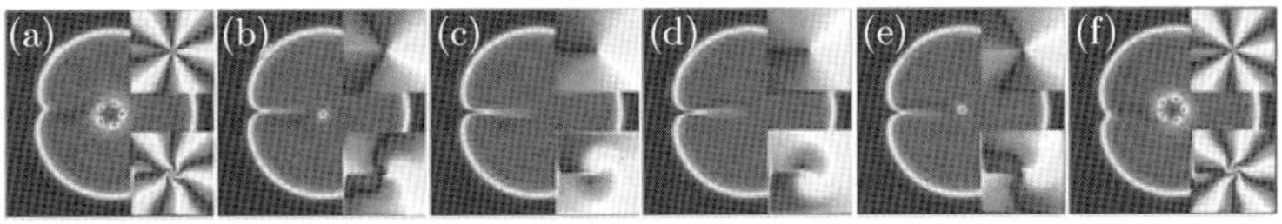

Figure 7.7 (a)–(f) Evolution outputs at $z = 15$ cm, from a crescent input ($n = 0.5$) corresponding to $l = -5.5$, -1.5, -0.5, 0.5, 1.5, and 5.5, respectively. The top and bottom insets present the input and output phases, respectively.

of vortices and how to determine their rotation directions is presented in Table 7.1, where the circular arrows represent rotation senses of the vortices.

Numerical experiments indicate that there are two critical values $r_{cr1} \approx 42.5$ μm and $r_{cr1} \approx 36.1$ μm for the beam width r_0. The evolution of quadrupole azimuthons with $n = 2$ and different l corresponding to $r_0 > r_{cr1}$ are shown in Figure 7.8. Following the above-mentioned rule, the number of notches at the origin is determined by NTC; so, from (a) to (f) in Figure 7.8, the number of notches is 4, 0, 1, 1, 0, and 4. However, in Figure 7.8(d) to (f), there are four additional notches

Table 7.1 Properties of the induced vortices from necklace and azimuthon incidences with different l and n.

Necklace inputs							
	$l<-n$	$l=-n$	$-n<l<0$	$l=0$	$0<l<n$	$l=n$	$l>n$
No.	↻	—	↺	—	↻	—	↺
	$-n-l$	0	$n+l$	0	$n-l$	0	$-n+l$
Azimuthon inputs when $r_0 > r_{cr1}$							
	$l<-n$	$l=-n$	$-n<1<0$	$l=0$	$0<l<n$	$l=n$	$l>n$
Outer	—	—	—	—	↺	↺	↺
Inner	↻	—	↺	↺	↻	—	↺
No.	$-n-l$	0	$n+l$	n	$3n-l$	$2n$	$n+l$
Azimuthon inputs when $r_0 < r_{cr2}$ with $l>n$							
	$n=1$		$n=2(l\leq 5\|l>5)$		$n\geq 3$		
Outer	↺		↻|↺		—		
Inner	↺		↻|↺		↺		
No.	$l+n$		$l+n\|l-n$		$l-n$		

Adapted from Ref. [40].

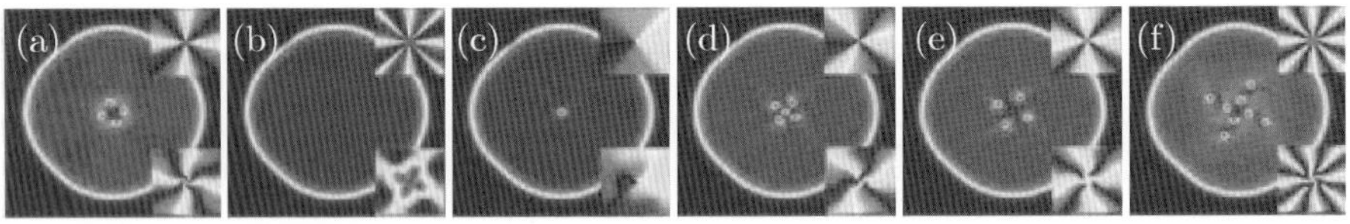

Figure 7.8 Evolutions of azimuthon incidences with $n=2$, and $B=0.5$ for (a) $l=-6$, (b) -2, (c) -1, (d) 1, (e) 2, and (f) 6, respectively. The figure setup is as in Figure 7.7.

around the origin, which equal the number of beads. The explanation is that there is energy flow around the phase singularities, and at the same time the beads fuse because of the diffusion term β. Therefore, when the speed of energy flow is greater than the fusion speed of beads, new vortices form at the twist of the beads. It is worth mentioning that the vortices around the origin appear as vortex pairs (with charges of +1 and −1), so the NTC of the beam is still conserved in all other figures. The common rule for this case is also shown in Table 7.1, under the "azimuthon inputs when $r_0 > r_{cr1}$" case.

A hexapole azimuthon with $l>n$ is presented in Figure 7.9(a) to (c); $r_0 = 40\ \mu m$ is chosen to fall in between r_{cr2} and r_{cr1}. According to the rule from Table 7.1, the total number of induced vortices should equal $n+l=12$. However, from Figure 7.9 we see that there are more and more vortices appearing during propagation. The reason is that the speed of energy flow is the largest, as compared to those of fusion and spreading. From the phases at different propagation distances, we see

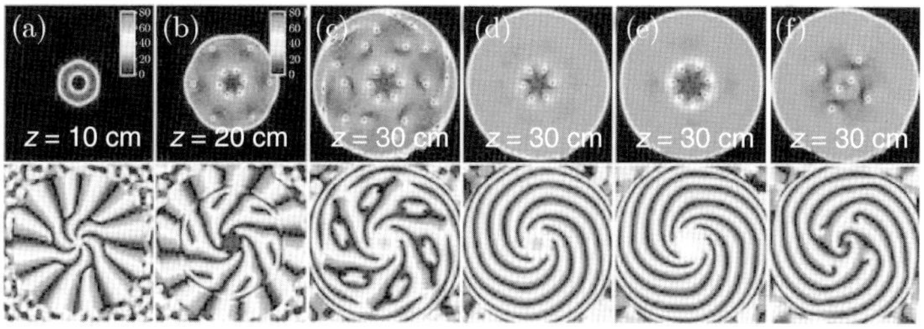

Figure 7.9 (a)–(c) Evolution of an azimuthon with $l = 9$, $n = 3$, $B = 0.5$, $r_0 = 40$ μm at different propagation distances. (d)–(f) show azimuthons with $r_0 = 30$ μm at 30 cm with ($l = 9$, $n = 3$), ($l = 9$, $n = 2$), and ($l = 4$, $n = 2$), respectively. The top row are the intensities, the bottom row are the phases. Source: Adapted from Ref. [40].

that the energy flow brought by the vortices around the origin of the fused beam is always faster than that at the edge of the beam; therefore, the asynchrony preferably forms new phase singularities at the edge of the beam, and new vortices are induced correspondingly. But we cannot give a certain rule for this case because the number of induced vortices is greatly affected by the beam width. What one sees in Figure 7.9(c) is 6 vortices at the core plus 12 pairs of ± 1 vortices induced at the rim of the beam.

If we set $r_0 = 30$ μm, which fulfills the condition $r_0 < r_{cr2}$, and redo the propagation of the beam used in Figure 7.9(a) to (c), we find that the number of induced vortices is 6, which can be calculated from $l - n = 6$, as shown by the output intensities and phases in Figure 7.9(d). The common rule for this case is exhibited in Table 7.1 under "azimuthon inputs when $r_0 < r_{cr2}$ with $l > n$" (other cases are the same as those under "azimuthon inputs when $r_0 > r_{cr1}$"). The lower the number of beads, the more energy in each of the beads, which will strengthen the energy flow. That is why the number of induced vortices is $l + n$ if $n = 1$. However, the exact rule for $n = 2$ is not certain because the energy flow is weakened, and whether it can produce more vortices or not, it is up to the value of l set for the initial beam. If l is bigger, the fusion will be accelerated, so the production of new vortices will be limited. Here, $l = 5$ is a boundary for this case: the rule is $l - n$ for $l > 5$ and $l + n$ for $l \leq 5$. Figure 7.9(e) and (f) display two numerical experiments corresponding to $l = 9$ and $l = 4$, in which the number of vortices are $l - n = 7$ and $l + n = 6$, respectively.

7.2.3
Conclusion

We have demonstrated that optical vortices can form from vortex, necklace, and azimuthon incidences with different topological charges, in multi-level atomic vapors when linear, cubic, and quintic susceptibilities are considered simultaneously. The appearance of vortices results from a combined action of the number of topological charges, the beam width of incidences, the diffusion effect, cubic–quintic nonlinearities, and the loss/gain in the medium, simultaneously. We have formulated common rules of finding the number as well as the rotation direction of the induced vortices. The effects can be observed in sodium as well as other atomic vapors with incident beams produced by using multi-wave interference [41] or phase mask [42] methods in a similar experimental setup as that in Ref. [21]. Our findings open a new venue and introduce a new method for studying vortices, and may broaden the field of their applications.

7.3
Multi-Component Spatial Vector Solitons of FWM

In recent years, spatial optical solitons and their interactions have attracted considerable attention because of their potential applications in many areas including all-optical switching, optical data storage, distribution, and processing [43]. The interaction between solitons can result in soliton fusion, fission, spiraling [44, 45], and the formation of more complicated localized states. It has been shown that several beams can interact with each other to produce multi-component vector solitons. The three-component dipole vector solitons have been studied theoretically [46] and observed in photorefractive crystals [47]. Dipole solitons in nonlocal nonlinear media have also been investigated [48–51]. It was suggested that the nonlocality can significantly modify the interaction between solitons. In the past decade, vector solitons in periodic photonic lattices have become an active field of research [52, 53]. These lattices can be formed by either interfering pairs of optical beams or by using amplitude masks. It will exhibit more novel propagation phenomena and lead to many novel spatial solitons due to the periodic refractive index in the lattice. Very recently, we have observed FWM three-component dipole vector solitons in laser-induced atomic gratings [20], in which the linear and nonlinear index modulations enhanced by the atomic coherence play an important role. The easy controls of experimental parameters in the interaction between the multi-level atom system and multi-beams ensure the current system can be used to observe the formation of multi-component spatial solitons.

Composite vortex solitons have been widely studied in self-focusing and self-defocusing media [28], such as radially symmetric vortex, rotating soliton clusters [54], and azimuthally modulated vortex solitons [55, 56]. In self-focusing nonlinear media, vortex solitons may undergo azimuthal instability and they decay into fundamental solitons during propagation [57]. Several mechanisms are exploited to overcome such instabilities. The saturation nonlinearity can arrest the instability [58]. Mutual coupling between the components of vector solitons can suppress the azimuthal instability [59]. Many theoretical studies of vortex vector solitons have been reported [60–64]. Some studies show that if the components of a vector soliton are made sufficiently incoherent in the transverse dimension, the instability can be eliminated [62, 65]. Counter-rotating vortex vector solitons can be stable in self-focusing saturable media [63]. Furthermore, the existence of N-component ($N > 3$) vector solitons that carry different topological charges has been predicted theoretically [64]. The study on the interactions of vector solitons [66, 67] allows us to understand the formation of more complicated solitons. However, the experimental observation reports of vortex vector solitons are very rare till now. The first experimental observation of the two-component vortex vector solitons with hidden vortices was reported in nematic liquid crystals recently [68].

In this section, we discuss dipole and vortex multi-component vector solitons in generated FWM signals in a two-level atomic system. It includes two degenerate and six nondegenerate FWM components. We analyze the interactions between such copropagating FWM soliton components and the formation of multi-component dipole- and vortex-mode vector solitons. The controllability of these multi-component solitons by the frequency detuning of input laser beams is also investigated.

7.3.1 Basic Theory and Experimental Scheme

Our experiments are carried out in a Na atom vapor oven. Energy levels $|0\rangle(3S_{1/2})$ and $|1\rangle(3P_{3/2})$ form a two-level atomic system interacting with six laser beams. These laser beams are all adjusted to connect the transition between $|0\rangle$ and $|1\rangle$, of which the resonant frequency is Ω (Figure 7.10(a)). Laser beams are spatially aligned in the configuration shown in Figure 7.10(b). Two laser beams E_1 (with frequency ω_1, wave vector $\mathbf{k}_1$, and Rabi frequency G_1) and E_1' (ω_1, $\mathbf{k}_1'$, G_1'), with a small angle $\theta_1 \approx 0.3^\circ$ between them, propagate in the opposite direction of the weak probe beam E_3 (ω_3, $\mathbf{k}_3$, G_3). These three laser beams come from the same dye laser DL1 (10 Hz repetition rate, 5 ns pulse width, and 0.04 cm^{-1} line width) with the frequency detuning $\Delta_1 = \omega_1 - \Omega$, and their wave vectors are all nearly in the x–o–z-plane. The other three laser beams E_2 (ω_2, $\mathbf{k}_2$, G_2), E_2'

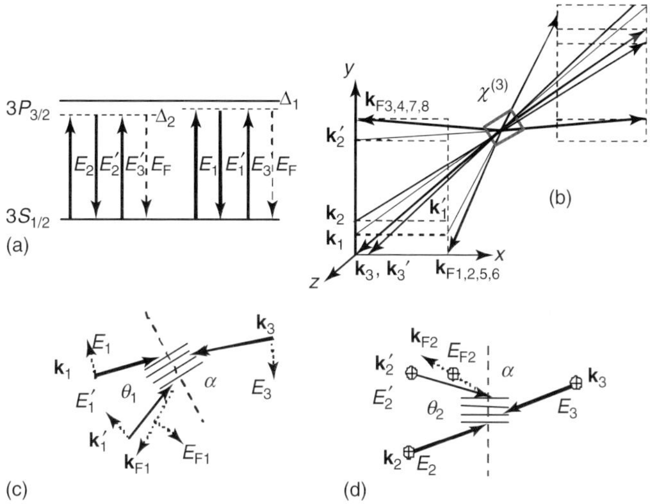

Figure 7.10 (a) Two-level system with six laser beams tuned to the same transition. (b) Spatial beam geometry used in the experiment. (c) The horizontally and (d) vertically aligned EIG1 and EIG2, respectively.

$(\omega_2, \mathbf{k}'_2, G'_2)$, and E'_3 $(\omega'_3, \mathbf{k}'_3, G'_3)$ are from another dye laser DL2 (which has the same characteristics as the DL1) with a frequency detuning $\Delta_2 = \omega_2 - \Omega$. Among them, E_2 copropagates with E_1; E'_3 copropagates with E_3, while E'_2 propagates with a small angle $\theta_2 \approx 0.3^\circ$ between E_2 and its wave vector in the y–o–z-plane. In this case, there will be eight FWM signals coexisting in the same atomic systems, E_{F1} $(\omega_1, \mathbf{k}_{F1} = \mathbf{k}_1 - \mathbf{k}'_1 + \mathbf{k}_3)$, E_{F2} $(\omega_2, \mathbf{k}_{F2} = \mathbf{k}_1 - \mathbf{k}'_1 + \mathbf{k}'_3)$, E_{F5} $(\omega_2, \mathbf{k}_{F5} = \mathbf{k}_2 - \mathbf{k}'_1 + \mathbf{k}_3)$, E_{F6} $(2\omega_2 - \omega_1, \mathbf{k}_{F6} = \mathbf{k}_2 - \mathbf{k}'_1 + \mathbf{k}'_3)$, E_{F3} $(\omega_1, \mathbf{k}_{F3} = \mathbf{k}_2 - \mathbf{k}'_2 + k_3)$, E_{F4} $(\omega_2, \mathbf{k}_{F4} = \mathbf{k}_2 - \mathbf{k}'_2 + \mathbf{k}'_3)$, E_{F7} $(2\omega_1 - \omega_2, \mathbf{k}_{F7} = \mathbf{k}_1 - \mathbf{k}'_2 + \mathbf{k}_3)$, E_{F8} $(\omega_1, \mathbf{k}_{F8} = \mathbf{k}_1 - \mathbf{k}'_2 + \mathbf{k}'_3)$. According to the phase-matching conditions, $E_{F1,2,5,6}$ $(E_{F3,4,7,8})$ propagate in the direction opposite to $\mathbf{k}'_2$ $(\mathbf{k}'_2)$ with very small angles among them. The stronger dressing beams $E'_{1,2}$ are approximately 10 times stronger than the pump beams $E_{1,2}$, and 1000 times stronger than the weak probe beam E_3, E'_3 and the generated FWM beams. The coherence length of the FWM signals is $L^c_F = \pi/|\Delta\mathbf{k}|$, where $\Delta \boldsymbol{k}$ is the mismatching in the wave mixing process. According to the phase-matching conditions, we can obtain the coherence lengths of these FWM signals, $L^c_{F1} = 2\pi c\omega_1/[n_1\omega_1|\omega_1 - \omega_2|\theta^2] \to \infty$, $L^c_{F2} = 2\pi c\omega_1/[n_1\omega_1 |\omega_1 - \omega_2|\theta^2] = 4.58$ km, $L^c_{F3} = 2\pi c\omega_1/[n_1\omega_1|\omega_1 - \omega_2|\theta^2] = 3.1$ cm, $L^c_{F4} = 2\pi c\omega_2/ [n_1\omega_1|\omega_1 - \omega_2|\theta^2] \to \infty$, $L^c_{F5} = 2\pi c(2\omega_1 - \omega_2)/n_1|(\omega_1 - \omega_2)(8\omega_1 - 4\omega_2 + \omega_1\theta^2)| = 3.1$ cm, $L^c_{F6} = L^c_{F7} = \pi c/[2n_1|\omega_1 - \omega_2|] = 3.14$ cm, $L^c_{F8} = 2\pi c(2\omega_2 - \omega_1)/n_1 |(\omega_2 - \omega_1)(8\omega_2 - 4\omega_1 + \omega_2\theta^2)| = 4.58$ km. The diffraction length of the generated FWM beams is defined as $L_D = k_F w_0^2$, where w_0 is the spot size of the FWM beams. In the two-level system, the diffraction length of the dipole soliton is about 1.5 mm.

In this system, several laser beams and FWM signals pass through an atomic medium, and, therefore, the XPM and SPM can effectively affect the propagation and spatial patterns of the propagating FWM signals. When the spatial diffraction is balanced by the XPM or SPM Kerr nonlinearity, many types of spatial solitons, such as dipole-mode solitons and vortex solitons, will be observed.

In the experiment, the key to observe dipole-mode solitons is to create a laser-induced index gratings with sufficiently high index contrast (via Kerr non-linearity $n_2 I$) in the atomic medium [20]. The sodium atomic density needs to reach $2.9 \times 10^{13}\text{cm}^{-3}$ $(T = 250\,^\circ\text{C})$, which can produce the needed variation in the nonlinear index of $\Delta n = 1.94 \times 10^{-4}$. The strong beams E_1 and E'_1 (E_2 and E'_2) can induce atomic coherence, which can modify the linear susceptibility and nonlinear Kerr effect significantly. Therefore, the spatial periodic intensity derived from the interference of E_1 and E'_1 (E_2 and E'_2) leads to periodic variation in the linear and nonlinear refractive index, and finally induces an electromagnetic-induced grating EIG1 (EIG2). The fringe spacing of EIG1 and EIG2 are determined by $\Lambda_i = \lambda_i/\theta_i$ ($i = 1$ for EIG1, and 2 for EIG2). The periodically modulated total linear and nonlinear refractive index is given by the expression $n(\zeta) = n_0 + \delta n_1 \cos(2\pi\xi/\Lambda_i) + \delta n_2 \cos(4\pi\xi/\Lambda_i)$, where $n_0 = n_{01} + n_{02} = (1 + G_a)G_{F2}N\mu_{10}/(\varepsilon_0 E_3 \Delta_1)$ is the spatial uniform refractive index; $\delta n_1 = G_a(n_{01}^2 - 1)/4n_{01} + (G_a^2/2 + G_a)(n_{02}^2 - 1)/2n_{02}$ and $\delta n_2 = (n_{02}^2 - 1)G_a^2/16n_{02}$ are the coefficients of the spatially varying terms in the modulated index, and $G_a = G_2^2/[(\Delta_1 + \Delta_2)\Delta_1]$. Similar to the photonic crystal, the spatial periodic index in the grating can lead to photonic band gap. The width of such a gap is given by

$\Delta\Omega = 2\Omega_0 n_2 I/\pi n_1$, where Ω_0 is its center frequency. If Δ_1 (or Δ_2) falls within a certain range, the Bragg reflection signals of incident beams can be significantly enhanced and the corresponding transmission signals can be suppressed greatly. Therefore, interacting with these gratings, the probe beam E_3 or E_3' will experience intensive Bragg reflection. The FWM signals $E_{F1,2}$ and $E_{F3,4}$ can be considered as the results of the electromagnetically induced diffraction (EID) of the probe beam E_3 or E_3' by the horizontally and vertically aligned EIG1 and EIG2 Figure 7.10(c) and (d), respectively. When the diffraction of the FWM signal is balanced by the cross-Kerr nonlinearity in propagation, the dipole-mode soliton generated from $E_{F1,2}$ and $E_{F3,4}$ will be produced. Moreover, the beams E_2 and E_1' (or beams E_1 and E_2') can also induce their moving grating EIG3 (or EIG4). The wave vector of the grating is $\mathbf{k}_g = \mathbf{k}_2 - \mathbf{k}_1'$ (or $\mathbf{k}_g = \mathbf{k}_1 - \mathbf{k}_2'$) with the phase velocity $v = \Delta\omega/|\mathbf{k}_g|$, where $\Delta\omega = \omega_2 - \omega_1$ is the frequency difference between the two induced beams [69]. Similar to the signals $E_{F1,2}$ and $E_{F3,4}$, the FWM signals $E_{F5,6}$ and $E_{F7,8}$ are the EID results of the probe beam E_3 or E_3' by the moving gratings EIG3 and EIG4, respectively.

For simplicity, we take the plane wave assumption and express the two classes of signals induced by the horizontally and vertically aligned EIG as $E_{Fp} = A_{Fp}(\zeta)\exp(ik_{Fp}z)$, $p = 1,2,5,6$ and $E_{Fq} = A_{Fq}(\zeta)\exp(ik_{Fq}z)$, $q = 3,4,7,8$, respectively. These FWM fields can couple to each other and their propagation satisfies the following evolution equations in the medium with Kerr nonlinearity

$$\sum_{p=1,2,5,6} \frac{\partial E_{Fp}}{\partial z} - \frac{i\nabla_\perp^2 E_{Fp}}{2k_{Fp}} = \sum_{p=1,2,5,6} \frac{ik_{Fp}}{n_1}\left(n_2^{Sp}\left|E_{Fp}\right|^2 + 2n_2^{Fp}\right)E_{Fp} + \eta_1 E_1 (E_1')^* \sum_{q=3,4,7,8} E_{Fq} \tag{7.7}$$

$$\sum_{q=3,4,7,8} \frac{\partial E_{Fq}}{\partial z} - \frac{i\nabla_\perp^2 E_{Fq}}{2k_{Fq}} = \sum_{q=3,4,7,8} \frac{ik_{Fq}}{n_1}\left(n_2^{Sq}\left|E_{Fq}\right|^2 + 2n_2^{Fq}\right)E_{Fq} + \eta_2 E_2 (E_2')^* \sum_{p=1,2,5,6} E_{Fp} \tag{7.8}$$

where $n_2^{Sp,Sq}$ are self-Kerr nonlinear coefficients of E_{Fq} and E_{Fq}, $n_2^{Fp} = \sum_{k=1}^{6} \Delta n_2^{Xpk} = n_2^{Xp1}|E_3|^2 + n_2^{Xp2}|E_1|^2 + n_2^{Xp3}|E_2|^2 + n_2^{Xp4}|E_1'|^2 + n_2^{Xp5}|E_2'|^2 + n_2^{Xp6}|E_3'|^2$ and $n_2^{Fq} = \sum_{k=1}^{6} \Delta n_2^{Xqk} = n_2^{Xq1}|E_3|^2 + n_2^{Xq2}|E_1|^2 + n_2^{Xq3}|E_2|^2 + n_2^{Xq4}|E_1'|^2 + n_2^{Xq5}|E_2'|^2 + n_2^{Xq6}|E_3'|^2$, $n_2^{Xp1-Xp6}$ and $n_2^{Xq1-Xq6}$ are cross-Kerr nonlinear coefficients of $E_{1,2,3}$ and $E_{1,2,3}'$ to E_{Fq} and E_{Fq}, respectively. All the Kerr nonlinear coefficients are generally defined as $n_2 = \mathrm{Re}\chi^{(3)}/(\varepsilon_0 c n_1)$. The dressed third-order nonlinear susceptibility is $\chi^{(3)} = D\rho_{10}^{(3)}$, where $D = N\mu_{10}^4/(\hbar^3\varepsilon_0 G_{Fp,Fq} G_k^2)$ ($G_{Fp,Fq}$ are Rabi frequencies of $E_{Fp,Fq}$, respectively). N is the atomic density and μ_{10} is the dipole matrix element between $|0\rangle$ and $|1\rangle$. The density matrix element $\rho_{10}^{(3)}$ can be obtained by solving the density matrix equations of the two-level atomic system. $\eta_1 = \xi\mu_{10}^5 F_a$ and $\eta_2 = \xi\mu_{21}^5 F_b$ with $\xi = i4\pi\omega_1 N/ch^4$, where F_a and F_b related to the Rabi frequencies of the involved fields, the frequency detuning Δ_1 (Δ_2) and the relaxation rates of the system [70]. The differences in the last terms on the right side of Eq. (7.6) and Eq. (7.7) show that the signals $E_{F1,2,5,6}$ mainly suffer the cross-Kerr effect from the dressing field E_1' and EIG1, while $E_{F3,4,7,8}$ suffer that from E_2' and EIG2. The in-phase dipole mode of an FWM signal can be created by its origin EIG, and the two poles in the dipole are trapped jointly in neighbor photonic fringes in the EIG.

The N-component vector soliton can be constructed from simple soliton components. For interactions between soliton components in medium with saturation nonlinearities, the critical angle $\theta_c = (n_{max} - n_{min})/n_{max}$ [71] plays a key role, where n_{max} and n_{min} are the maximum and minimum values of the nonlinear refractive index induced by the soliton, respectively. When the collision angle θ is less than θ_c, a beam can be coupled into a waveguide induced by other beams, so two soliton components can undergo fusion or fission. In our experiment, in each class of FWM signal, four soliton beams copropagate with very small angles among them, and therefore they can fuse with each other to form a new soliton under certain conditions. The superposition of the four horizontally aligned dipole components $E_{F1,2,5,6}$ (with topological charge $m_{F1,2,5,6} = -1$) generates a new horizontally oriented dipole-mode soliton (four-component dipole-mode soliton) with total amplitude $\sum_p E_{Fp}$ ($p = 1, 2, 5, 6$). Similarly, the superposition of four vertically aligned dipole components $E_{F3,4,7,8}$ ($m_{F3,4,7,8} = 1$) generates a vertically oriented four-component soliton with amplitude $\sum_q E_{Fq}$ ($q = 3, 4, 7, 8$). Two nodeless probe beams E_3 and E_3' ($m_{k3,k3'} = 0$) act as the fundamental components, providing an attractive force tightly binding both the dipoles and the multi-component ($N = 10$) structure. For the vector soliton composed of all these components, the zero total angular momentum $m_{k3,k3'} + m_{F1,2,5,6} + m_{F3,4,7,8} = 0$ makes the structure stable. Therefore, the total intensity ($I = |E_{k3,k3'}|^2 + |\sum_p E_{Fp}|^2 + |\sum_q E_{Fq}|^2$) reaches a steady state in propagation after a long distance. On the other hand, the multi-component structure can also be stabilized by using optical lattices or induced gratings. The lattice creates an optical waveguide, which could prevent the constituents of the multi-component soliton from expansion or contraction.

The vortex soliton can also be created in this two-level system in which six nearly degenerate frequency waves ($E_{1,2,3}$ and $E_{1,2,3}'$) exist. Specifically, when three or more plane waves overlap in the medium, the interference patterns can induce vortex-like index modulation with phase singularities. Furthermore, the diffraction of a light beam can be compensated for by the nonlinearity and the FWM modulated solitons can be created. The propagation equations of vortex solitons in cylindrical coordinate are written as follows [72],

$$\sum_{p=1,2,5,6} \frac{\partial E_{Fp}}{\partial z} - \frac{i}{2k_{Fp}} \left(\frac{1}{r}\frac{\partial E_{Fp}}{\partial r} + \frac{\partial^2 E_{Fp}}{\partial r^2} + \frac{1}{r^2}\frac{\partial^2 E_{Fp}}{\partial \varphi^2} \right) = \sum_{p=1,2,5,6} \frac{ik_{Fp}}{n_1} \left(n_2^{Sp} \left| E_{Fp} \right|^2 + 2n_2^{Fp} \right) E_{Fp} \tag{7.9}$$

$$\sum_{q=3,4,7,8} \frac{\partial E_{Fq}}{\partial z} - \frac{i}{2k_{Fq}} \left(\frac{1}{r}\frac{\partial E_{Fq}}{\partial r} + \frac{\partial^2 E_{Fq}}{\partial r^2} + \frac{1}{r^2}\frac{\partial^2 E_{Fq}}{\partial \varphi^2} \right) = \sum_{q=3,4,7,8} \frac{ik_{Fq}}{n_1} \left(n_2^{Sq} \left| E_{Fq} \right|^2 + 2n_2^{Fq} \right) E_{Fq} \tag{7.10}$$

The modulated vortex and dipole solitons are characterized by two independent integer numbers, topological charge m, and the number of intensity peaks M. In our experiment, they are created jointly by the interference of multiple beams and the XPM of the dressing and pump fields. The soliton solutions can be written as $E_{\mathrm{F}p} = u_p \sec h[u_p(k_{\mathrm{F}p} n_2^{Sp}/n_0)^{1/2}(r - r_p)] \cos(M\varphi/2) \exp(im_{\mathrm{F}p}\varphi + i\phi_p)$ $\exp(ik_{\mathrm{F}p}z)$ and $E_{\mathrm{F}q} = u_q \sec h[u_q(k_{\mathrm{F}q} n_2^{Sq}/n_0)^{1/2}(r - r_q)] \cos(M\varphi/2) \exp(im_{\mathrm{F}q}\varphi + i\phi_q)$ $\exp(ik_{\mathrm{F}q}z)$, where $u_{p,q}$ are the soliton amplitudes; $r_{p,q}$ are the initial peak positions; and $\phi_{p,q} = 2k_{\mathrm{F}p,q} n_2 z I_{2,1} \mathrm{e}^{-r^2/2}/(n_0 I_{\mathrm{F}p,q})$ are the nonlinear phase shifts introduced by the Kerr effect. For rotating dipole solitons, the number of intensity peaks $M = 2$. For modulated vortex solitons, $M \geq 3$. Furthermore, if $m_{\mathrm{F}}\varphi + \phi = 0$, the angular velocity of the modulated soliton becomes $\omega = 0$ and the rotation of the soliton cancels. The topological charge m of the vortex soliton is determined by $\delta_{r,i}$. $\delta_r = -\arctan\left(\sum_{i=1}^{n} E_i n_i k_{iy} \sin T_{\psi i} / \sum_{i=1}^{n} E_i n_i k_{ix} \sin T_{\psi i}\right)$, $\delta_i = -\arctan\left(\sum_{i=1}^{n} E_i n_i k_{iy} \cos T_{\psi i} / \sum_{i=1}^{n} E_i n_i k_{ix} \cos T_{\psi i}\right)$, with n being the number of laser beams that create the spiral phase plate and n_i being the nonlinear refractive index; $T_{\psi i} = \psi_{i0} + n_i(\mathbf{k}_{ix} T_x + \mathbf{k}_{iy} T_y)$, with (T_x, T_y) being the coordinates of the singularity point. When $\delta_r > \delta_i$ ($\delta_r < \delta_i$), $m = 1$ ($m = -1$), which means the phase changes clockwise (anticlockwise). To observe vortex soliton, the ideal temperature is around 265 °C (atomic density needs to reach $5.6 \times 10^{13}\mathrm{cm}^{-3}$). Under suitable conditions, several copropagating vortex solitons also can fuse to form a new multi-component vector soliton.

7.3.2
Experimental Observation of Multi-Component Solitons

When a probe beam and two pump beams are turned on, a single FWM signal can be obtained. Figure 7.11 presents the dipole-like patterns of eight FWM signals in the self-focusing region. Because beams E_1 and E_1' (or E_2 and E_1') are horizontally aligned in the x–z-plane, the grating EIG1 and EIG3 have horizontal orientation (Figure 7.10(b)). Therefore, dipole components $E_{\mathrm{F}1,2,5,6}$ induced by EIG1 and EIG3 have their two humps horizontally along the x-axis. For the same reason, the dipole components $E_{\mathrm{F}3,4,7,8}$ induced by vertically oriented grating EIG2 and EIG4 have their two humps along the vertical y-axis. The probe beam E_3 (or E_3') is deviated from the FWM signals with a small angle ($\theta_1 \approx 0.3°$), so it is employed as the fundamental soliton. The interaction between the nodeless probe beam and the arbitrary dipole-like FWM signal forms a basic dipole vector soliton, denoted as (0,−1) for $E_{\mathrm{F}1,2,5,6}$ and (0,1) for $E_{\mathrm{F}3,4,7,8}$. When the frequency detuning Δ_2 is scanned from the large negative values to the resonant frequency, except $E_{\mathrm{F}1}$, other dipole-mode components decay into a nodeless fundamental one at resonance or large frequency detuning. This detuning dependence can be explained by the nonlinear refractive index n_2 (Figure 7.12(d)) and the nonlinear phase shift $\phi_{p,q} = 2k_{\mathrm{F}p,q} n_2 z I_{2,1} e^{-r^2/2}/(n_0 I_{\mathrm{F}p,q})$, which determines the spatial splitting of the FWM signal. When Δ_2 is set at resonance or the large negative detuning, n_2 and $\phi_{p,q}$ become minimum, so the index contrast of EIGs is not sufficiently high

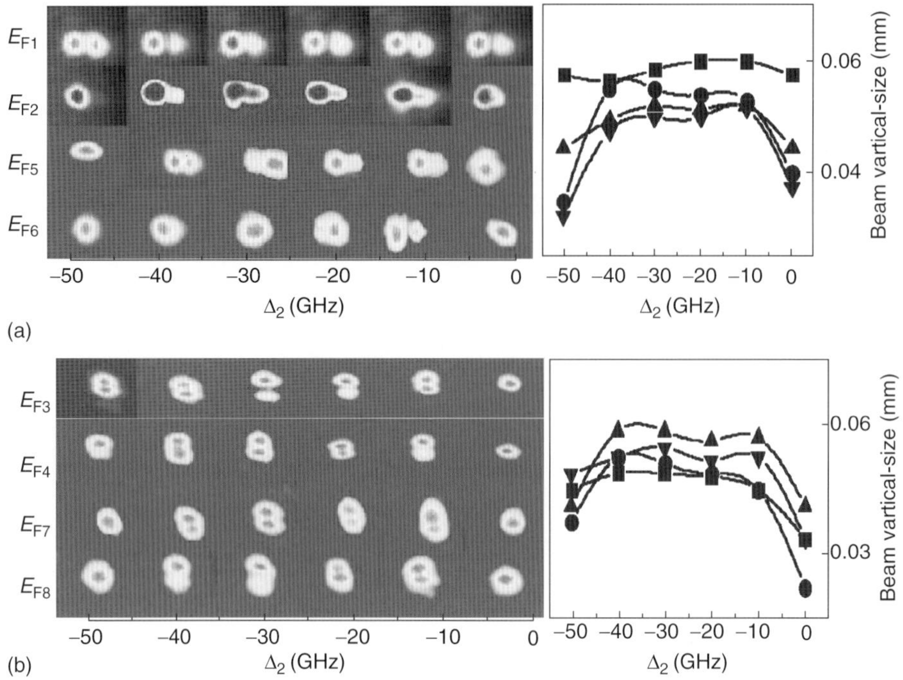

Figure 7.11 (a) Images (left) and vertical-sizes (right) of FWM signals $E_{F1,2,5,6}$ at different detuning Δ_2, E_{F1} (squares), E_{F2} (circles), E_{F5} (triangles), and E_{F6} (reverse triangles). (b) Images (left) and vertical-sizes (right) of FWM signals $E_{F3,4,7,8}$ at different detuning Δ_2, E_{F3} (squares), E_{F4} (circles), E_{F7} (triangles), and E_{F8} (reverse triangles). Source: Adapted from Ref. [21].

to maintain the dipole-mode pattern of the FWM signal and they decay into a nodeless one. The field E_{F1} always retains a dipole-like pattern because its cross-Kerr coefficient is not affected by E_2, E_2', and E_3'. The right panels of Figure 7.11 describe the horizontal (vertical) size of the beams $E_{F1,2,5,6}$ ($E_{F3,4,7,8}$). It is shown that the beam sizes of these eight dipole components remain constant when the frequency detuning Δ_2 is changed from −40 to −10 GHz. This indicates that the dipole soliton is formed in this region.

In order to observe the influence of other experimental conditions on the pattern formation of the FWM signals, the powers, positions, and frequency detuning of the involved laser beams are adjusted. Figure 7.12(a) gives the beam profiles of the FWM signal E_{F7} versus the input powers of E_1 and E_2'. The spatial splitting of signal E_{F7} is determined by the nonlinear phase shift introduced by the EIG4, which is created by the beams E_1 and E_2'. At low power $I_{1,2}$, the beam E_{F7} presents a nodeless spot. With increasing $I_{1,2}$, E_{F7} splits into two coherent spots. Now the dipole-mode soliton can be formed if the spatial diffraction is balanced by the XPM or SPM Kerr nonlinearity. The vertical size of beam E_{F7} remains unchanged in a specific region. Further with increasing $I_{1,2}$, the nonlinear phase shift ϕ_7 becomes large, so the splitting of E_{F7} become significant, and the vertical size of E_{F7} increases.

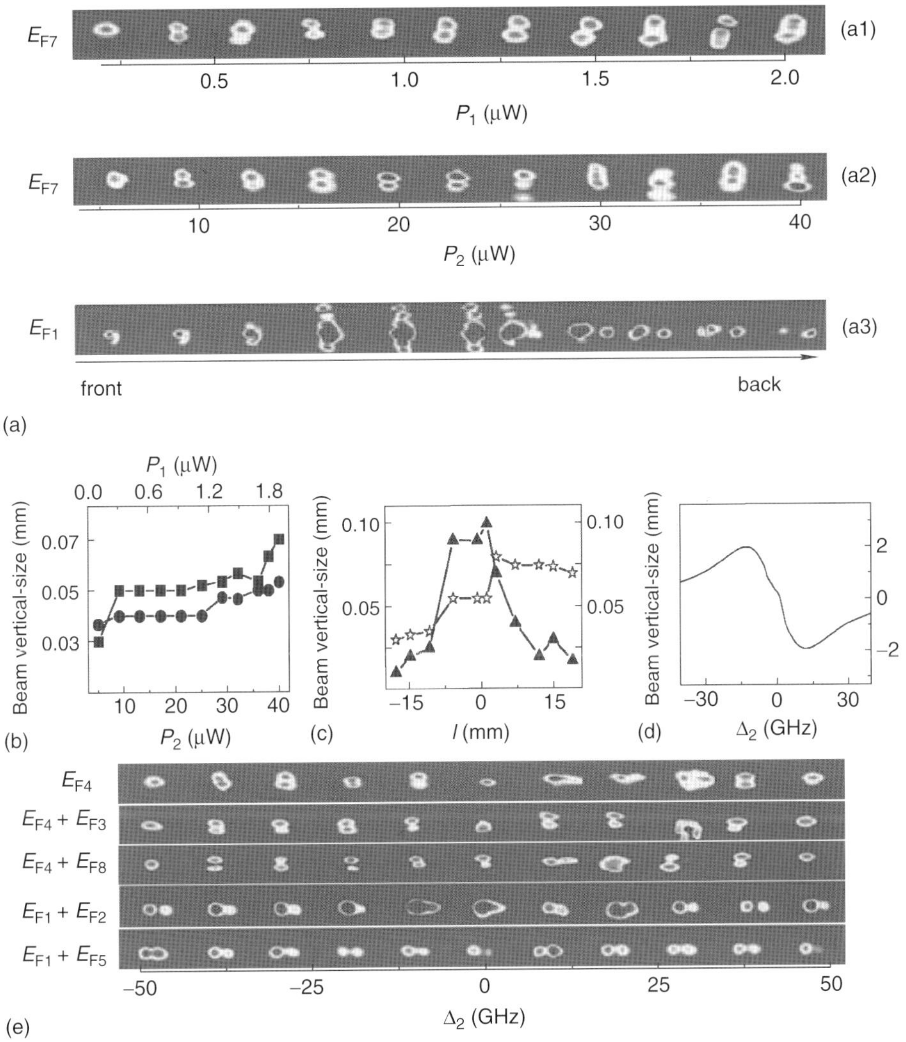

Figure 7.12 (a1, a2) Images of E_{F7} at different powers of (a1) E_1, (a2) E_2'. (a3) Images of E_{F1} for different spatial configuration of E_1 and E_1'. From left to right, the intersection position of E_1 and E_1' is adjusted to change from the front to the back of the heat-pipe oven. (b) Beam vertical-size of E_{F7} versus P_1 (squares) and P_2 (circles). (c) Beam vertical-size (triangles) and horizontal-size (asterisks) of E_{F1} versus the intersection position of E_1 and E_1'. Here, the distance between the intersection position and the oven center is denoted as l. (d) Nonlinear refractive index n_2 of E_{F4}. (e) Images of the pure FWM signal E_{F4} and coexisting signals $E_{F4}+E_{F3}$, $E_{F4}+E_{F8}$, $E_{F1}+E_{F2}$, and $E_{F1}+E_{F5}$, respectively. Source: Adapted from Ref. [21].

Besides the pump field's power, spatial configuration of laser beams also can affect the spatial splitting. In Figure 7.12(a), the intersection of laser beams E_1 and E_1' is adjusted from the front to the back of the heat-pipe oven. We can see that E_{F1} is split in x-direction when E_1 and E_1' set at the front and back of the oven, but in y-direction when they are at the middle of the oven. Figure 7.12(b) shows the variation of the vertical-size of EF7, which is increased in the process of increasing the power P2 and P1. Figure 7.12(c) shows the variation of the horizontal and vertical size of E_{F1} in this transformation. This transformation can be explained by the different spatial position of E_1' overlapped with E_{F1}.

Figure 7.12(e) shows the splitting of the pure FWM signal E_{F4} and coexisting signals $E_{F4} + E_{F3}$ and $E_{F4} + E_{F8}$ when Δ_2 is changed. It is theoretically obtained in Figure 7.12(d) that the Kerr coefficient is negative (positive) with $\Delta_2 > 0$ ($\Delta_2 < 0$), which will lead to the defocusing (focusing) of the FWM signal. It can be seen that the beam E_{F4} splits along vertical direction in the self-focusing region, but it becomes horizontal splitting in the self-defocusing region with small Δ_2. Further increasing Δ_2, the beam spot becomes large and presents three spots. Subsequently, it becomes vertical dipole mode and decays into a nodeless spot when Δ_2 far away from resonance. Such evolution can be explained by the follows. As mentioned, in the self-focusing region, the dipole component E_{F4} induced by vertically aligned EIG2 splits along the y-axis. But in $\Delta_2 > 0$ region, the FWM E_{F4} is shifted along the down-right direction due to the repulsion effect ($n_2 < 0$) of the cross-Kerr nonlinearity of the strong beam E_2', and thus, the beam E_{F4} splits in x-direction [18]. The spatial shift of the FWM signal is proportional to $|n_2|$. At a proper value, horizontal and vertical splittings appear simultaneously. With further increasing of Δ_2 ($|n_2|$ deceases), the spatial shift of E_{F4} becomes small. Therefore, the beam E_{F4} gets back to the unshifted position and suffers from vertical splitting. It decays into single spot when $|n_2|$ is close to zero.

As the probe beam E_3 added, there exist two dipole-mode components E_{F3} and E_{F4} ($m_{F3,4} = 1$). They copropagate in the opposite direction of $\mathbf{k}_2'$ with a very small angle between them. Because they have different frequencies, there exists an incoherent attraction force between them [44]. The soliton interaction has an affinity for real particles [73], and the nonlinear interaction of the solitons is analogous to the Coulomb interaction. Thus the attraction force between two dipole solitons is electric dipole-like [74, 75], and can be written as $F = \left(C_{20} \int \sigma_{p1} v_{20}^1(r, \varphi)\, r dr d\varphi \int \sigma_{p2} v_{20}^2(r, \varphi) r dr d\varphi\right) / R^7$, where $v_{20}^i(r, \varphi) = C_{20}(\sqrt{2}r/\omega_{0s})^2 e^{-r^2/\omega_{0s}^2} \cos 2\varphi$ ($i = 1, 2$) is the intensity distribution profile of one dipole soliton, ω_{0s} the waist radius of Gaussian beams, σ_{pi} the average power in unit area of soliton beam, C_{20} the interaction coefficient of dipole solution, and R distance between the centers of the two dipole components. This attraction force can prevent soliton beams from diverging and makes the beams approach each other. When the distance between the solitons is too small, the two interacting solitons can exchange energy by coupling light into the waveguide induced by each other and eventually fuse. Therefore, the dipole components E_{F3} and E_{F4} fuse to form a new dipole-mode vector soliton which have their two humps along the vertical y-axis. We denote this multi-component soliton structure as (0, 1, and 1). The horizontal splitting of the FWM signal in self-defocusing region disappears when E_3 is on. This indicates that the field E_3 could suppress the spatial shift and thus the horizontal splitting of E_{F4}. When the pump beam E_1 is turned on, there exist two incoherent dipole-mode components E_{F4} and E_{F8} ($m_{F4,8} = 1$). They also can fuse to form a dipole-mode vector soliton. Figure 7.12(e) also presents the coexisting signals $E_{F1} + E_{F2}$ and $E_{F1} + E_{F5}$, which have topological charge $m_F = -1$. The generated multi-component solitons have their two humps horizontally along the x-axis. We denote this multi-component soliton structure as (0, −1, and −1).

Figure 7.13(a) depicts the superposition of four dipole-mode components. When five laser beams $E_{1,2,3}$ and $E'_{1,3}$ are turned on (dressing field E'_2 is blocked) at the same time, there are four horizontal-splitting dipole-mode components $E_{F1,2,5,6}$ copropagating in the opposite direction of $\mathbf{k}'_1$ with very small angles among them. We can see four dipole-mode components also fuse to generate a four-component dipole-mode soliton with total amplitude $\sum_p E_{Fp}$. The beam profile of $\sum_p E_{Fp}$ retains horizontal splitting. Similarly, when $E_{1,2,3}$ and $E'_{2,3}$ are turned on (dressing field E'_1 is blocked), the vertically aligned four-component dipole-mode soliton $\sum_q E_{Fq}$ is obtained with the interaction among four vertical-splitting dipole components $E_{F3,4,7,8}$. When six laser beams are all turned on, $\sum_p E_{Fp}$ and $\sum_q E_{Fq}$ can coexist in the same atomic systems and be influenced by the dressing field E'_2 and E'_1, respectively. Compared with the case of absence of the dressing fields, the intensities of $\sum_p E_{Fp}$ and $\sum_q E_{Fq}$ become weak. It is attributed to the suppression effect in the dressing of E'_1 or E'_2 (with $\Delta_1 + \Delta_2 \approx 0$ satisfied). Moreover, the dipole-like splitting of beams $\sum_p E_{Fp}$ and $\sum_q E_{Fq}$ becomes more distinct because of the increasing of

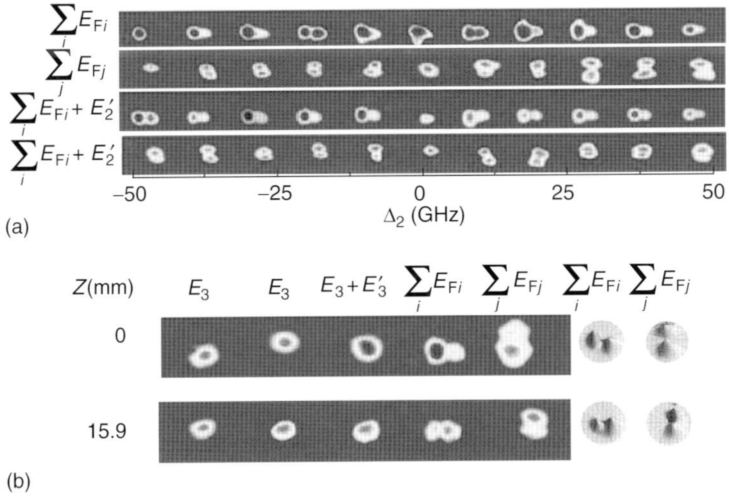

Figure 7.13 (a) Images of the four-component dipole-mode soliton $\sum_i E_{Fi}$ and $\sum_j E_{Fj}$ at different Δ_2 with and without dressing fields E'_2 and E'_1, respectively. (b) Experimental (left) and numerical (right) results of 10-component dipole-mode soliton with the propagation length $Z = 0, 15.9$ mm at $\Delta_2 = -15$ GHz.

the nonlinear phase shift induced by the dressing fields. At the same time, there are two fundamental components E_3 and E'_3 copropagating in the same direction. They fuse to form a new fundamental soliton ($\sum E_0 = E_3 + E'_3$) with circular cross section (as seen in Figure 7.13(b)). The fundamental component $\sum E_0$ and two mutually perpendicular dipole components $\sum_p E_{\mathrm{F}p}$ and $\sum_q E_{\mathrm{F}q}$ copropagate with a small angle ($\theta_1 \approx 0.3^\circ$) between $\sum E_0$ and $\sum_p E_{\mathrm{F}p}$ (or $\sum E_0$ and $\sum_q E_{\mathrm{F}q}$). The interaction force between fundamental soliton and dipole soliton can be written as $F = \left[C_{12} \int \sigma_{p1} v_{00}^1 (r, \varphi)\, r\mathrm{d}r\mathrm{d}\varphi \int \sigma_{p2} v_{20}^2 (r, \varphi) r\mathrm{d}r\mathrm{d}\varphi\right] / R^4$, where C_{12} is the interaction coefficient between fundamental soliton and dipole soliton. From the expressions of interaction forces we can see, as the separation of solitons is increased, that the force ($F \propto 1/R^7$) of two dipole solitons decays more quickly than the force ($F \propto 1/R^4$) between fundamental soliton and dipole soliton. It indicates that the fundamental soliton has more significant long-range constraint on dipole soliton, and plays an important role in the formation of the multi-component vector soliton. Therefore, the fundamental component $\sum E_0$ and two dipole components $\sum_p E_{\mathrm{F}p}$ and $\sum_q E_{\mathrm{F}q}$ constitute a new 10-component dipole-mode vector soliton.

The atomic density N_0 is determined by the temperature of the atomic vapor, the change in which leads to the change in the propagation distance z for the involved beams. In the experiment, with the temperature increasing, the corresponding propagation distance increases about $Z = aL_\mathrm{h} = 15.9$ mm, where $L_\mathrm{h} = 19.4$ mm is the half-length of the heat-pipe oven and a is the temperature increasing multiple. This propagation distance is 10.6 times longer than the diffraction length ($L_\mathrm{D} = 1.5$ mm) of the FWM signal or the probe beam. We can see (in Figure 7.13(b)) the spatial profiles of the beams $\sum_p E_{\mathrm{F}p}$ and $\sum_q E_{\mathrm{F}q}$ change very little in the propagation distance. This indicates that steady propagation of multi-component dipole-mode vector soliton is achieved.

In the two-level system, the input laser beams ($E_{1,2,3}$ and $E'_{1,2,3}$) have nearly degenerate frequency, so the spatial interference of these beams can create a stationary beam pattern with a phase singularity, resulting in a vortex soliton at proper temperature (around 265 °C), suitable detuning and configurations of input laser beams. Figure 7.14 presents the images of the FWM signals $E_{\mathrm{F}1,2,5,6}$ versus the detuning Δ_1. In the $\Delta_1 > 0$ region, all the four FWM signals have a self-defocusing character and show vortex patterns, then they decay into fundamental spots when $\Delta_1 = 0$. In the defocusing media, a diffracting core of an optical vortex may get self-trapped and generate a vortex soliton. Specifically, for the DFWM signal E_{F1}, the interference between E_1, E'_1, and E_3 forms a spiral phases plate. In this case, the parameters are

$$\delta_r = -\arctan \frac{E_1 n_1 k_{1y} \sin T_{\psi 1} + E'_1 n_{1'} k'_{1y} \sin T_{\psi 1'} + E_3 n_3 k_{3y} \sin T_{\psi 3}}{E_1 n_1 k_{1x} \sin T_{\psi 1} + E'_1 n_{1'} k'_{1x} \sin T_{\psi 1'} + E_3 n_3 k_{3x} \sin T_{\psi 3}}$$

$$\delta_i = -\arctan \frac{E_1 n_1 k_{1y} \cos T_{\psi 1} + E'_1 n_{1'} k'_{1y} \cos T_{\psi 1'} + E_3 n_3 k_{3y} \cos T_{\psi 3}}{E_1 n_1 k_{1x} \cos T_{\psi 1} + E'_1 n_{1'} k'_{1x} \cos T_{\psi 1'} + E_3 n_3 k_{3x} \cos T_{\psi 3}}$$

For NDFWM signals $E_{\mathrm{F}2,5,6}$, we can see that three nearly degenerate frequency waves also can create spiral phase plate. For $E_{\mathrm{F}2}$, the spiral phase plate is formed

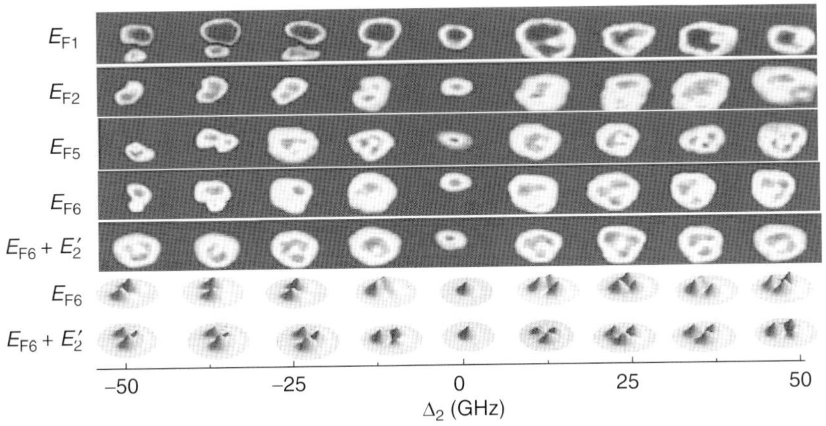

Figure 7.14 Images (the five top rows) of the pure FWM signals $E_{F1,2,5,6}$ and E_2' dressed signal E_{F6} versus Δ_1. The two bottom rows give the numerical results of E_{F6} with and without dressed field E_2', respectively.

by E_1, E_1', and E_3', so

$$\delta_r = -\arctan\frac{E_1 n_1 k_{1y}\sin T_{\psi 1} + E_1' n_{1'} k_{1y}'\sin T_{\psi 1'} + E_3' n_{3'} k_{3y}'\sin T_{\psi 3'}}{E_1 n_1 k_{1x}\sin T_{\psi 1} + E_1' n_{1'} k_{1x}'\sin T_{\psi 1'} + E_3' n_{3'} k_{3x}'\sin T_{\psi 3'}}$$

$$\delta_i = -\arctan\frac{E_1 n_1 k_{1y}\cos T_{\psi 1} + E_1' n_{1'} k_{1y}'\cos T_{\psi 1'} + E_3' n_{3'} k_{3y}'\cos T_{\psi 3}}{E_1 n_1 k_{1x}\cos T_{\psi 1} + E_1' n_{1'} k_{1x}'\cos T_{\psi 1'} + E_3' n_{3'} k_{3x}'\cos T_{\psi 3}}$$

At the same time, the XPM of the strong field E_1' [18] separates the FWM beam into three spots along a ring, forming a modulated vortex soliton. The modulated vortex beam has angular momentum $M \propto mP$, where $P = \int |E|^2 dr$ is the total power of vortex.

In the $\Delta_1 < 0$ region, vortex solitons become instabile due to the self-focusing nature of nonlinearity. We can see that the modulated vortex pattern of $E_{F1,2,5,6}$ transforms into a rotating dipole-type pattern. The tilted dipole pattern indicates the spatial twist of the beam, which carries nonzero angular momentum. The rotation of the dipole soliton is because the initial angular momentum of the vortex beam transfers to the orbital angular momentum of dipole splinters. The conservation of angular momentum is satisfied in this interaction process. The stable rotating dipole soliton is defined as dipole azimuthon [76], and the astigmatic transformations of vortex beams into spiraling dipole azimuthons also has been observed recently in nematic liquid crystals [77]. When a stronger dressing field E_2' is turned on, the incoherent coupling between the fundamental beam E_2' and the vortex beam E_{F6} can lead the vortex pattern of E_{F6} to appear in the self-focusing region. It is attributed to the enhanced cross-Kerr nonlinear modulation by E_2'. E_{F6} is deflected closer to E_2' because of the attraction of E_2'. The attraction force between the fundamental beam E_2' and the vortex beam E_{F6} is $F = \left[C_{30}\int \sigma_{p1} v_{00}^1(r,\varphi)\, r\mathrm{d}r\mathrm{d}\varphi \int \sigma_{p2} v_{30}^2(r,\varphi) r\mathrm{d}r\mathrm{d}\varphi\right]/R^7$, where $v_{30}^i(r,\varphi) = C_{30}(\sqrt{2}r/\omega_{0s})^3 \mathrm{e}^{-r^2/\omega_{0s}^2}\cos 3\varphi$ is the intensity distribution profile of one vortex soliton, and C_{30} is the interaction coefficient of the vortex and fundamental solution. More importantly, the interaction between E_2' and E_{F6} could counterbalance the self-focusing effect of E_{F6}, which results in the instability of vortex beam. This phenomenon is coincident with the result reported in vortex vector solitons [60], and it provides an effective method to stabilize the vortex beam in a self-focusing medium.

Figure 7.15 gives the superposition of two or four vortex beams. When $E_{1,2,3}$ and E_1' are turned on, a stronger DFWM signal E_{F1} and a weak nondegenerate signal E_{F5} are generated and copropagate with a very small angle. The experimental results detected in the $\mathbf{k}_1'$ direction can be regard as the superposition of two mutually incoherent vortex beams. If the intensity of the signal E_{F1} is far larger than that of E_{F5}, the vortex soliton E_{F1} induces a linear waveguide with an effective potential [78]. Moreover, for mutually incoherent vortex solitons in self-defocusing medium, the potential is attractive independent of their relative phase. This waveguide can support both vortex mode ($m=1$) and fundamental mode ($m=0$) [79, 80]. If the intensity of the signal E_{F5} tends to that of E_{F1}, where the SPM and XPM effects operate together, the waveguide induced by solitons is nonlinear and gives rise to vortex vector soliton. In Figure 7.15(a) we can see the beam profiles of the

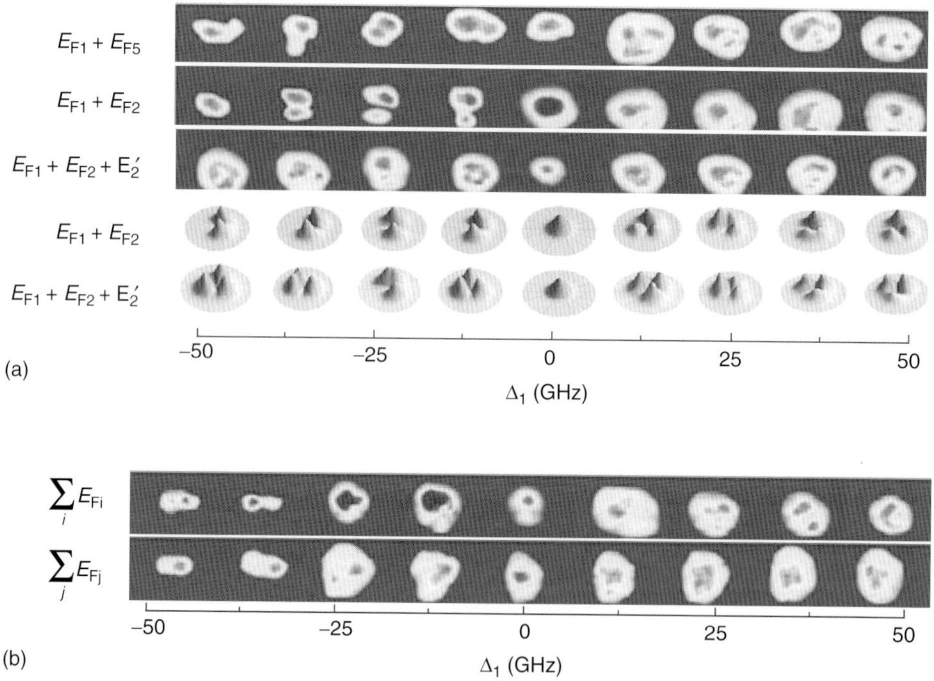

Figure 7.15 Images of the superposition vortex signal at different Δ_1. (a) Two-component vortex solitons $E_{F1}+E_{F5}$ and $E_{F1}+E_{F2}$, respectively. The two bottom rows give the numerical results of $E_{F1}+E_{F2}$ and E_2' dressed $E_{F1}+E_{F2}$. (b) Four-component vortex solitons $\sum_i E_{Fi}$ and $\sum_j E_{Fj}$. Source: Adapted from Ref. [21].

superposition signal show vortex-type in the self-defocusing region; it indicates that two vortex components merge together to form a new vortex vector soliton.

In the self-focusing region, the radial symmetric character of this vortex vector soliton is broken. Similar to the scalar vortex soliton, it transforms into a rotating dipole pattern. It is indicated that two vortex components can couple with each other to form a rotating dipole vector soliton in the self-focusing region. Moreover, the orientation of the dipole vector soliton changes with the frequency detuning. Similar results are obtained in the case of $E_{F1} + E_{F2}$. However, when a very strong dressing field E_2' is turned on, the vortex vector soliton can be stabilized in the self-focusing region. As discussed in the preceding text, the enhanced cross-Kerr nonlinearity by E_2' could counterbalance the self-focusing effect and suppress the breakup of the vortex soliton. Therefore, the stable vortex vector soliton can be formed in the self-focusing region by the enhanced XPM. It has been reported that two vortex components with opposite topological charges also can form a stable vortex vector soliton in the self-focusing region [68]. Figure 7.15(b) is the superposition results of four copropagating vortex beams. In the self-defocusing region, one stronger vortex beam E_{F1} (or E_{F4}) and three weak vortex beams $E_{F2,5,6}$ (or $E_{F3,7,8}$) also can fuse to form a new four-component vortex beam $\sum_p E_{Fp}$ (or $\sum_q E_{Fq}$). In the self-focusing region, they decay into rotating dipole-type solitons.

7.3.3 Conclusion

We have experimentally demonstrated multi-component dipole and vortex solitons generated in FWM in two-level atomic system. The composite dipole solitons contain four horizontal-splitting and four vertical-splitting dipole-mode components. They can propagate stably in the medium. The composite vortex solitons include four copropagating vortex components. In the self-defocusing region, these vortex components fuse to form a new vortex vector soliton. In the self-focusing region, both scalar and vector vortex solitons transform into rotating dipole-mode solitons. However, if a very strong dressing field is turned on, vortex solitons can be stabilized in the self-focusing region. The interaction forces among components of dipole and vortex vector solitons are also investigated. This study will help us understand the fundamental mechanisms in vector soliton formations and the interactions between different components, and open the door for the development of the device in all-optical communication and signal processing.

7.4 Surface Solitons of FWM in EIL

When multiple laser beams propagate in a Kerr-type nonlinear medium, optical solitons or modulation instabilities can occur because of the XPM in self-defocusing regime [1, 81–83]. As the localized modes propagate at the interface between two media with different properties, surface waves could display certain intriguing features in various physical systems from the honeycomb lattice of carbon atoms [84] and optical periodically layered media [85] to the plasmon-polariton waves [86]. In recent years, different kinds of spatial solitons at the interface of different media have been studied both theoretically and experimentally [87–89]. One interesting and important type is the surface soliton forming at the edge of a periodic structure and a bulk medium [88–90]. Different from the case with two uniform media, the trapping mechanism for optical surface waves formed at the interface of uniform media and periodical refractive index media is that the propagation eigenvalues fall within the photonic band gap of the system [90, 91]. As an example, one-dimensional in-phase surface solitons that have been demonstrated in the AlGaAs array [90]. Two-dimensional discrete surface solitons, forming at the corner of a finite optically induced two-dimensional waveguide lattice and a continuous medium, have also been experimentally observed [92, 93]. In general, there are many new interesting features in optical surface waves with different kinds of nonlinearities, such as nonlocal surface solitons [87], polychromatic surface solitons [94], and spatiotemporal surface light bullets [95]. To systematically investigate these interesting phenomena, it is important to create media with flexible periodic indices and nonlinearity. Recently, gap solitons in a multi-level atomic medium, where the strong periodic refractive index can be generated and controlled, were experimentally demonstrated.

In this section, we show the formations of one- and two-dimensional surface solitons in the EIGs and EIL. Compared with the surface solitons created in the photorefractive crystals and other systems, the ones formed in the atomic medium can be easily controlled by many experimental parameters, such as the probe detuning, the pump powers, and the incident angles of the relevant beams, as the indices in the generated gratings and lattice can be changed by altering the periodically modulated atomic coherence and nonlinearities [82]. We also investigate the soliton dynamics under the competition between the generation of surface solitons due to the nonlinear localization and the Bragg reflection due to photonic band gap. With the flexible controls in the experimental parameters, we can easily explore various parameter spaces and observe interesting phenomena in forming the two-dimensional surface solitons. These studies will be useful for understanding the fundamental mechanisms in soliton formations and their dynamics, and can open new ways to control the diffraction of optical beams and develop new schemes for spatial optical switching [96] and pattern formation [97] in optical communication and all-optical image processing.

7.4.1
Basic Theory and Experimental Scheme

The relevant energy levels for Na atoms (in heat pipe) are shown in Figure 7.16(a). The pulse laser beams are aligned spatially as shown in Figure 7.16(b). The energy levels $|0\rangle$ ($3S_{1/2}$), $|1\rangle$ ($3P_{3/2}$), and $|2\rangle$ ($4D_{3/2,5/2}$) form a ladder-type atomic system. Two beam geometric configurations have been used in the investigations, and the first one is shown in Figure 7.16(b). The pump beams denoted by the wave vectors

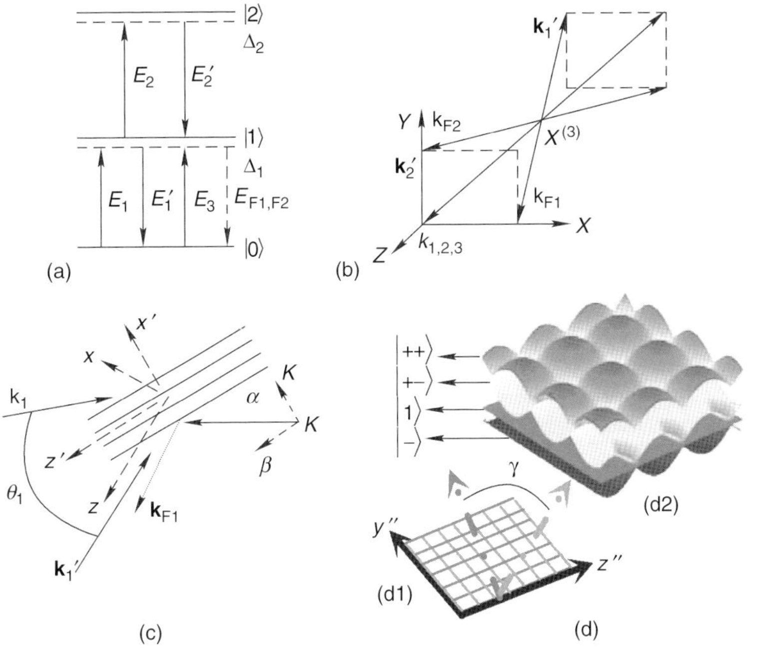

Figure 7.16 (a) The ladder-type three-level atomic system. (b) The first spatial beam geometric configuration used in the experiment for the investigation of 1D surface soliton. (c) Generation of FWM1 via Bragg scattering of the probe beam on the EIG1 created by E_1 and E_1'. (d1) The illustration of the EIL composed of EIG1 and EIG2 and (d2) the lattice state.

$\mathbf{k}_1$ and $\mathbf{k}_1'$ (with same Rabi frequency G_1 and frequency ω_1, connecting the transition $|0\rangle$ to $|1\rangle$ with resonant frequency Ω_1) propagate in the opposite direction of the weak probe beam $\mathbf{k}_3$ (G_3, ω_1) with small angles θ_1 ($= 0.3^\circ$) between them. The three beams $\mathbf{k}_1$, $\mathbf{k}_1'$, and $\mathbf{k}_3$ are from the same near-transform-limited dye laser (10 Hz rate, 5 ns pulse width, and 0.04 cm^{-1} line width). Two other pump beams $\mathbf{k}_2$ and $\mathbf{k}_2'$ (G_2, ω_2) (connecting the transition $|1\rangle$ to $|2\rangle$ with resonant frequency Ω_2) from another dye laser also propagate along the opposite direction of $\mathbf{k}_3$ with angle θ_2 ($= 0.3^\circ$) between them. We define the frequency detuning as $\Delta_i = \Omega_i - \omega_i$ ($i = 1, 2$). Under such a condition, a one-photon resonant DFWM process (Figure 7.16(c)) can occur between the two energy levels $|0\rangle$ and $|1\rangle$, satisfying the phase-matching condition of $\mathbf{k}_{F1} = \mathbf{k}_3 + \mathbf{k}_1 - \mathbf{k}_1'$ and propagating nearly opposite to the beam $\mathbf{k}_1'$. Also, a two-photon NDFWM signal (with $\mathbf{k}_{F2} = \mathbf{k}_3 + \mathbf{k}_2 - \mathbf{k}_2'$) can be generated, propagating nearly opposite to the beam $\mathbf{k}'_2$. Both these FWM signals are sampled by CCD cameras.

In Figure 7.16(b), the two strong pumping beams $\mathbf{k}_1$ and $\mathbf{k}_1'$ ($\mathbf{k}_2$ and $\mathbf{k}_2'$) interfere with each other to induce a horizontally aligned grating in X–Z (Y–Z)-plane with a period $\Lambda_1 = \lambda_1/2\sin(\theta_1/2)$ ($\Lambda_2 = \lambda_2/2\sin(\theta_2/2)$) [82]. With these gratings, the incident beam can have intense Bragg reflection. The generated FWM signals $\boldsymbol{E}_{F1}$

(transverse-magnetic (TM) polarization with respect to EIG1) and $\boldsymbol{E}_{F2}$ (transverse-electric (TE) polarization with respect to EIG2) are the Bragg reflected signals of the probe beam $\boldsymbol{E}_3$ (launched obliquely into the gratings EIG1 and EIG2), respectively [98]. For instance, as shown in Figure 7.16(c), the incident probe beam has an angle α with respect to the tangential direction of EIG1, which can be decomposed into two components [73]. One component propagates tangentially to the interface between the EIGs and the uniform medium and its propagation characteristics can be represented by the surface-wave propagation constant β. The other component propagates in the direction normal to EIGs, which can be characterized by the Bloch wave vector κ. Similar to the Snell's law, the $\boldsymbol{E}_{F1}$ beam (as the Bragg reflected probe beam), is symmetric to the incident probe beam with respect to the normal direction of EIG1, and, therefore, it is also the syntheses of the tangential surface-wave component and the normal Bragg reflected component. The surface-wave component of the probe beam propagates along the z' direction in Figure 7.16(c), parallel to the interface between the EIGs and the uniform medium, while FWM1 propagates along the z-axis in Figure 7.16(c). The coordinate's rotation transformation components between these two frames are $\xi' = \xi\cos(\theta_i/2) - z\sin(\theta_i/2)$ and $z' = \xi\sin(\theta_i/2) + z\cos(\theta_i/2)$, where $i = 1$ and $\xi = x$ are for EIG1 and $i = 2$ and $\xi = y$ for EIG2, respectively. Since θ_i is very small, the surface-wave component is the dominant one in the generated FWM signals $\boldsymbol{E}_{F1}$ and $\boldsymbol{E}_{F2}$.

To force the incident probe beam to excite the surface mode and form the surface soliton, we should adjust the Bloch wave vector κ to fall into the forbidden band, so the normal Bragg reflected component is attenuated. When the laser beam propagates at the edge of the EIG and homogeneous medium, the FWM signals can be expressed as $E^S_{F1,2}(\xi') = E_\kappa(\xi')\exp[i(\kappa\xi' + \beta z)]$ for $\xi' > 1$ and $E^S_{F1,2}(\xi') = \alpha\exp(i\beta z + q_\alpha\xi')$ for $\xi' \leq 1$, where ξ' runs over x' and y', and $\xi' = 1$ defines the boundary of the EIG and homogeneous media. Here, in the forbidden band, there must exist $\kappa = m\pi/\Lambda_i + i\kappa'$ with m being the positive integer, in which the nonzero imaginary part κ' is the attenuation coefficient along the ξ'-axis in the EIGs for the normal Bragg reflected component. Another parameter, $q_a = \sqrt{\beta^2 - [(\omega_i/c)n_\alpha]^2}$ is the attenuation coefficient along the transverse axis in the homogeneous medium, in which n_α is the refractive index ($n_\alpha = n_1 + n_2 I$ with $n_1 \approx 1$). Propagation constant β of the surface-wave component can be obtained by solving the equation $m\pi/\Lambda_i + i\kappa' = \cos^{-1}\{\mathrm{Tr}[T_\Lambda(\beta,\Delta_i)]\}/\Lambda_i$ in the forbidden band, which is derived by the transfer-matrix method and Bloch theorem in medium with periodically modulated refractive index [91], and T_Λ is the transfer matrix of one period in the EIGs. In the Kerr medium, the periodic susceptibility can be expressed by $\chi(\xi) \approx \chi^{(1)}(\xi) + \chi^{(3)}(\xi)|E_2|^2\sin^2(\theta_i/2)$. For one period, the transfer matrix can be expressed by $T_\Lambda(\Delta_1,\beta) = \prod_{j=1}^{N} T_j(\Delta_1,\beta) = \left[T_{\Lambda1}(\Delta_1,\beta) \;\; T_{\Lambda2}(\Delta_1,\beta)\right]^{\mathrm{T}}$, $T_{\Lambda1}(\Delta_1,\beta) = \left[T_{\Lambda11}(\Delta_1,\beta) \;\; T_{\Lambda12}(\Delta_1,\beta)\right]$, and $T_{\Lambda2}(\Delta_1,\beta) = \left[T_{\Lambda21}(\Delta_1,\beta) \;\; T_{\Lambda22}(\Delta_1,\beta)\right]$, where T_j is the transfer matrix of the jth layer when we divide the whole period into N layers, and the dependencies on the frequency (Δ_1) and propagation constant β are obvious.

7.4.2
Fluorescence and FWM via EIT Windows

The two EIGs are in different planes, so the interaction region between them is small and they can be considered to be isolated. The experimental results presented in Figure 7.17 and Figure 7.18 are obtained under such conditions. The experimental configuration can also be changed to have a second configuration in which EIG1 and EIG2 are both in the Y–Z-plane and their orientations are deviated from the Z-axis by 22.5°, anticlockwise and clockwise, respectively, defined as y'' and z''. In such a case, the two EIGs have a considerable overlapping area to construct a 2D EIL with a 2D periodic refractive index due to the periodically modulated atomic coherence. Specifically, the level splitting due to the dressing by the strong pump beams can significantly affect the susceptibility and further modify the refractive index [70]. At the position where the antinodes of the SW created by fields $\boldsymbol{E}_1$ and $\boldsymbol{E}'_1$ come across that of the SW created by $\boldsymbol{E}_2$ and $\boldsymbol{E}'_2$, the naked level $|1\rangle$ will be split into three dressed levels $|++\rangle$, $|+-\rangle$, and $|-\rangle$ after being doubly dressed by $\boldsymbol{E}_1$ and $\boldsymbol{E}'_1$ and $\boldsymbol{E}_2$ and $\boldsymbol{E}'_2$, as shown in Figure 7.16(d1); therefore, the refractive index is strongly modified by these two field pairs [82]. In contrast, at the position where the nodes of the two SWs overlap, level $|1\rangle$ is not dressed and the index will not be modified. When the antinodes of one SW encounter the nodes of another SW,

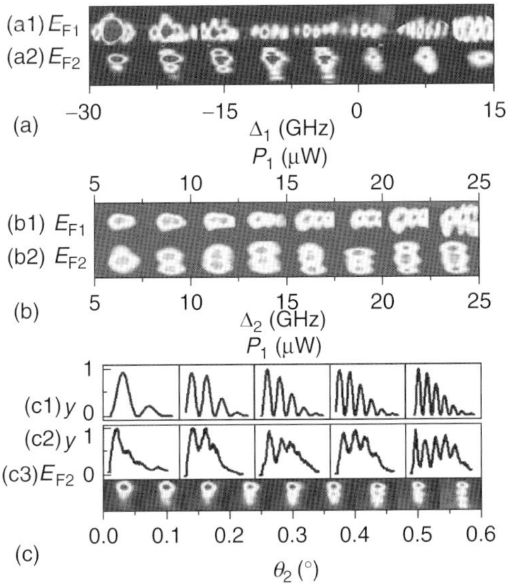

Figure 7.17 The images of FWM1 (a1, b1) and FWM2 (a2, b2), when (a) Δ_1 is scanned from −30 to 15 GHz, and (b) when the pump powers P_1 (in b1) of the pumping fields E_1 and E'_1 and P_2 (in b2) of E_2 and E'_2 are changed from 5 to 25 μW, respectively. (c1) Theoretical calculations of the y-axis transverse section profile of FWM2 when θ_2 is changed from small to large discretely. (c2) Experimental results corresponding to (c1). (c3) The experimental captured images of FWM2, in which the first, third, fifth, seventh, and ninth correspond to (c2). Δ_2 is set to be 0 GHz.

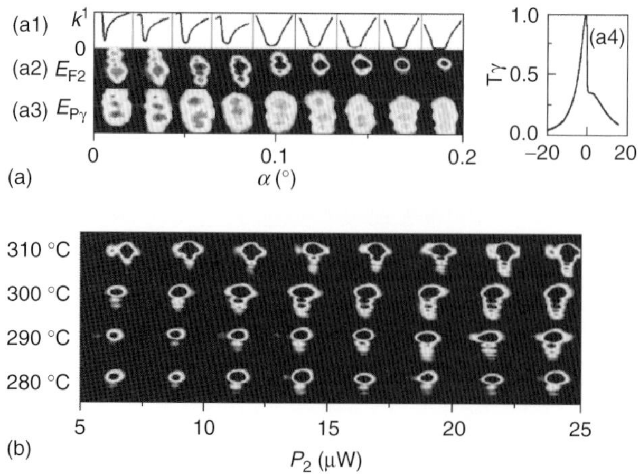

Figure 7.18 (a1) Theoretically calculated dispersion curves of EIG2. In each curve, Δ_1 is scanned from −20 to 20 GHz continuously; for figures from left to right, α is increased discretely from 0 to 0.2°. (a2) Experimentally observed formation and evolution of the FWM2 surface solitons versus α. (a3) The probe transmission versus α. (a4) The probe transmission spectrum with $\alpha = 0.2°$. (b) The evolutions of the surface solitons of FWM2 at four discrete temperatures of 280, 290, 300 and 310 °C when the powers of $\mathbf{E}_2$ and $\mathbf{E}_2'$ change from 5 to 25 μW with a step size of 2.5 μW.

level $|1\rangle$ will be split into two dressed levels $|+\rangle$ and $|-\rangle$; therefore, the index will be modified only by a single field pair. The spatial periodical distribution of these three cases can lead to a lattice state as shown in Figure 7.16(d2) and a further distinguished 2D periodic index modulation as illustrated in Figure 7.16(d1). The sensitivity of such an index modulation to the detuning of the probe field and powers of the dressing fields leads to the tunability for the soliton formation by multi-parameters.

The propagations of FWM1 and FWM2 under either 1D or 2D modulations with diffraction can be expressed by

$$\frac{\partial E_{F1}}{\partial z} - \frac{i\nabla_T^2 E_{F1}}{2k_{F1}} = \frac{ik_{F1}}{n_1}\left\{n_2^{S1}|E_{F1}|^2 + 2[n_2^{X1}|E_1'|^2[\varsigma\mu(z'') + (1-\varsigma)\mu(x')] + \varsigma n_2^{X2}|E_2'|^2\mu(y'')]\right\} E_{F1}, \quad (7.11)$$

$$\frac{\partial E_{F2}}{\partial z} - \frac{i\nabla_T^2 E_{F2}}{2k_{F2}} = \frac{ik_{F2}}{n_1}\left\{n_2^{S2}|E_{F2}|^2 + 2[n_2^{X3}|E_1'|^2\varsigma\mu(z'') + n_2^{X4}|E_2'|^2[\varsigma\mu(y'') + (1-\varsigma)\mu(y')]]\right\} E_{F2} \quad (7.12)$$

where ∇_T^2 is the Laplace operator with respect to the x- and y-axes and describes the diffraction. The y''- and z''-axes denote the orientations of EIG1 and EIG2, respectively. In these equations, $\mu(\xi) = \cos^2(\pi\xi/\Lambda_i)$ for $\xi \geq 1$ and $\mu(\xi) = 1$ for $\xi < 1$ ($\xi = x', y', y'', z''$) represent the periodically modulated refractive index pattern inside the EIGs ($\xi \geq 1$) and the uniform index outside ($\xi < 1$) the EIGs

($\xi = x', y''$ for EIG1 and $\xi = y', z''$ for EIG2), respectively. Surface solitons can be generated at the interface of these two regions with a high refractive index contrast. The factor $\varsigma = 0$ or 1 is introduced to describe the assembling of EIG1 and EIG2 into EIL, that is, $\varsigma = 1$ for sufficiently large overlapping area as in Figure 7.16(d), and $t\varsigma = 0$ when the overlapping area is not sufficiently large as in Figure 7.16(b). With $\varsigma = 1$, both FWM signals suffer from the modulations of the lattice, with the indices varying periodically in two directions according to $\mu(y'')$ and $\mu(z'')$. In Eq. (7.11), n_1 is the linear refractive index at ω_1; $n_2^{S1,S2}$ are the self-Kerr coefficients of $\boldsymbol{E}_{\mathrm{F1}}$ and $\boldsymbol{E}_{\mathrm{F2}}$, and n_2^{X1-X4} are the cross-Kerr coefficients from $\boldsymbol{E}_{1,2}$ and $\boldsymbol{E}'_{1,2}$, respectively. These Kerr nonlinear coefficients can all be calculated by the general expression $n_2 = \mathrm{Re}\chi^{(3)}/(\varepsilon_0 c n_1)$. The third-order nonlinear susceptibility is given by $\chi^{(3)} = D\rho_{10}^{(3)}$, where $D = N\mu_{10}^4/(\hbar^3 \varepsilon_0 G_{\mathrm{F1,F2}} G_i^2)$, with N and μ_{10} being the atomic density and the dipole matrix element of the transition $|0\rangle$ to $|1\rangle$, respectively. Here, $G_{\mathrm{F1,F2}}$ are the Rabi frequencies of $\boldsymbol{E}_{\mathrm{F1,F2}}$. The third-order density matrix element $\rho_{10}^{(3)}$ can be obtained by solving the density matrix equation [70]. Under our experimental condition, the sodium vapor (heated to have an atomic density of $2.9 \times 10^{13} \mathrm{cm}^{-3}$) has quite a large Kerr nonlinearity [82]. The pump beams $\boldsymbol{E}_{1,2}$ and $\boldsymbol{E}'_{1,2}$ are approximately 100 times stronger than the weak probe beam E_3 and the two generated FWM beams $\boldsymbol{E}_{\mathrm{F1,F2}}$.

A qualitative analysis of Eq. (7.11) is available when the diffraction and SPM terms are neglected, in which the phase profile distortion of FWM by the EIL or EIG is considered [1]. This will allow the linearization of Eq. (7.11) with respect to x and y, and obtain a set of solutions with phases for $\boldsymbol{E}_{\mathrm{F1}}$ and $\boldsymbol{E}_{\mathrm{F2}}$ as $\phi_{\mathrm{F1}} = 2\mathrm{k}_{\mathrm{F1}}\{n_2^{X1}|E'_1|^2[\varsigma\mu(y'') + (1-\varsigma)\mu(x')] + n_2^{X4}|E'_2|^2\varsigma\mu(z'')\}z/n_1$ and $\phi_{\mathrm{F2}} = 2\mathrm{k}_{\mathrm{F2}}\{n_2^{X3}|E'_2|^2\varsigma\mu(z'') + n_2^{X4}|E'_1|^2[\varsigma\mu(y'') + (1-\varsigma)\mu(y)]\}z/n_1$, both of which is 2D (1D) periodic at $\varsigma = 1$ ($\varsigma = 0$). The curvature of such a distorted phase front is periodically alternative to be positive and negative, so the FWM signal is 2D or 1D periodically modulated. We also can obtain the modulation periodic of FWM signal, FWM1 (FWM2) has period $\Lambda_{\mathrm{F1,F2}}^{x',y'} = n_1\Lambda_1^2/\mathrm{k}_{\mathrm{F1,F2}}\pi|E'_{1,2}|^2 n_2^{X1,X4} z$ along x' (y') axis at $\varsigma = 0$, and $\Lambda_{\mathrm{F1,F2}}^{y''} = n_1\Lambda_1^2/\mathbf{k}_{\mathrm{F1,F2}}\pi|E'_1|^2 n_2^{X1,X3} z$ $\Lambda_{\mathrm{F1,F2}}^{z''} = n_1\Lambda_1^2/k_{\mathrm{F1,F2}}\pi|E'_2|^2 n_2^{X2,X3} z$ along y''- and z''-axes at $\varsigma = 1$.

The geometric configuration in Figure 7.16(b) is first employed. By changing the probe detuning, pump powers, and angles between pump beams, the spatial shapes of EIGs can be controlled to change the spatial pattern formation of the FWM signal. First, by scanning the frequency detuning Δ_1 (Figure 7.17(a)), we can effectively control the spatial contrast of EIG via the dark or bright dressed-state resonances, which can significantly affect the sign and value of the Kerr nonlinear index n_2 [70]. In Figure 7.17(b), EIG1 and EIG2 are responsible for the periodic horizontal stripes of the FWM1 and vertical ones of FWM2, respectively. In the general expression of n_2, with Δ_1 scanned toward $\Delta_1 = 0$ from the negative region, n_2 will increase and reach its maximum at $\Delta_1 \approx -7\mathrm{GHz}$ associated with the resonance of one dressed state (Figure 7.17(a)). Obviously, the FWM1 and FWM2 beams get the most clear periodic stripes near that resonant point, at which n_2 gets larger leading to higher contrast for the EIGs (further bringing larger phase distortion and modulation depth). At other Δ_1 values, n_2 is not sufficiently large and modulation depth becomes low, so the modulated periodic pattern suffers from diffusion and tunneling between adjacent sites, due to the diffraction in the propagation [89]. The investigation of $\boldsymbol{E}_{F1}$ by scanning Δ_1 in its positive region also reveals a similar result, in which another dressed-state resonant frequency is at $\Delta_1 = 7$ GHz. In this region, the pattern of $\boldsymbol{E}_{F2}$ becomes nearly single spot, which is mainly attributed to the considerable focusing effect from the XPM caused by large negative n_2 [1]. Similarly, by increasing the powers of the pump fields $\boldsymbol{E}_1$ and $\boldsymbol{E}'_1$ ($\boldsymbol{E}_2$ and $\boldsymbol{E}'_2$) for $\boldsymbol{E}_{F1}$ ($\boldsymbol{E}_{F2}$) to improve the contrast of EIG1 (EIG2), the two signals evolve from blurry single spots to clear periodically modulated structures, as shown in Figure 7.17(b). Third, in Figure 7.17(c), when the angle θ_2 between the two pump fields $\boldsymbol{E}_2$ and $\boldsymbol{E}'_2$ is increased, the pattern of FWM2 gradually evolves from a single spot to an asymmetrical periodic oscillating structure. This change is due to the decrease in the period of EIG2 with increased θ_2 according to $\Lambda_2 = \lambda_2/2\sin(\theta_1/2)$ and the modulations in the theoretical result agree with the experimental result in Figure 7.17(c).

Then, the competition between the Bragg reflection and surface-wave components is investigated by changing α, the oblique incidence angle of the probe beam into the EIG. By the transfer-matrix method and Bloch theorem described in the preceding text, the dispersion relation of EIG2 between the probe frequency and Bloch wave vector κ under different α is theoretically calculated and shown in Figure 7.18(a1), where the photonic band gap is indicated by a flat region with $\mathrm{Re}[\kappa\pi/\Lambda_i - 1] = 0$, and it also depends on Δ_1 and the propagation constant β. The transmission curve corresponding to the dispersion curve with $\alpha = 0.2^\circ$ is plotted in Figure 7.18(a4), showing a photonic band gap on one side of the EIT window satisfying $\Delta_1 + \Delta_2 = 0$. When α is small, that is, the probe beam launches into the EIG2 nearly parallel to the interface, the forbidden band in EIG2 is very narrow or even destroyed (Figure 7.18(a1)). In this case, the spots in Figure 7.18(a2) show strong probe transmission signal $\boldsymbol{E}_{Pr}$, while the FWM2 beam is weak but has a clear periodic pattern due to the strong surface-wave component. Then, with the increase of α, the transmission of the probe beam decreases; and FWM2 shows an increased intensity and fewer stripes due to the increase of the Bragg reflected component, which can be verified by the dispersion curves in Figure 7.18(a1). When α is further increased, the phase-matching of FWM2 will be lost, and therefore the FWM2 intensity is reduced, as shown in Figure 7.18(a2).

Next, the stability of the surface solitons with respect to the powers of the pump fields $\boldsymbol{E}_2$ and $\boldsymbol{E}'_2$ is investigated. In Figure 7.18(c), the propagation of the FWM beam is monitored by increasing the atomic number density of the sample (by the temperature T), instead of increasing the sample length. The temperature of the cell is initially set at 280 °C and finally increased to 310 °C, with two other points at 290 and 300 °C to capture the images of the FWM2 beam. The equivalent propagation distance in such a temperature range is about 57 cm, which is 60 times larger than the diffraction length given by $L_D = k_{F2}w^2$ and should be sufficient to determine whether the surface solitons are stable. To investigate the stability of the solitons formed by FWM2 with respect to the power P_2, the spots of FWM2 are measured for eight discrete P_2 values from 5 to 25 μW, as shown in Figure 7.18(b). It is obvious that the stable solitons form in the top-right region in this power–temperature phase space (Figure 7.18(b)).

After adjusting the parameters and setting their optimal values, sufficient intersection area becomes essential for the formation of the lattice state and EIL illustrated in Figure 7.16(d2) and (d1), and the generation of 2D surface solitons propagating along the corner of this EIL; so, the beam configuration is reconstructed. In this reconstruction, the orientations EIG1 (EIG2) are anticlockwise (clockwise) rotated in the Y–Z-plane. We define γ as the angle between the orientations of the two EIGs and the maximum is 45°. The images of FWM1 are recorded at 10 discrete positions in the rotation process of EIG1, as shown in Figure 7.19(a1) and (a2). FWM1 in Figure 7.19(a1) only shows modulation in the y''-direction due to the small intersection area when EIG2 is not rotated, although the rotation of EIG1 increases the area to a certain extent. In contrast, in Figure 7.19(a2), there are periodic modulations of FWM1 in two directions, which become increasingly significant with increasing γ because the intersection area has been gradually enlarged and the area with lattice state gradually extends.

Next, the evolution of the FWM1 solitons is investigated when the pump fields $\boldsymbol{E}_1'$ and $\boldsymbol{E}_2'$ are changed, respectively. In Figure 7.19(b1), EIG1 is rotated to have the largest γ, while EIG2 is not rotated. In such a case, one can easily see that the FWM signal remains periodically modulated in the y''-direction. Here, increasing P_1 only has the effect of decreasing the modulation period. In Figure 7.19(b2), EIG2 is rotated, and therefore the 2D modulation of FWM1 is formed. When the power of $\boldsymbol{E}_2'$ increases, the periodic stripes in one direction of such 2D modulations become denser. Such tunability can be well explained by the expression $\Lambda_{F1}^{y_2''} = n_1\Lambda_1^2/k_{F1}\pi|\boldsymbol{E}_2'|^2 n_2^{X2} z$, in which larger P_2 will lead to a smaller

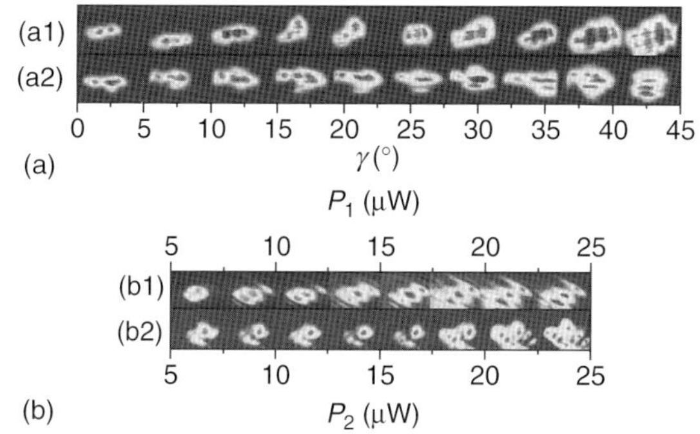

Figure 7.19 (a) The images of the FWM1 signal when EIG1 is rotated by 10 discrete angles with EIG2 not rotated in (a1) and rotated in (a2). (b) The images of the FWM1 signal (b1) when the power P_1 of the pump beam $\boldsymbol{E}_1'$ is successively set at eight discrete values (6, 8, 10, 12, 14, 16, 18 and 20 μW), with EIG1 rotated and EIG2 not rotated, respectively. (b2) The images of the FWM1 signal when the power of $\boldsymbol{E}_2'$ changes as that of $\boldsymbol{E}_1'$ in (b1) and the power of $\boldsymbol{E}_1'$ is fixed at 10 μW, with EIG1 rotated and EIG2 both rotated. Source: Adapted from Ref. [99].

modulation period, and it further verifies that the 2D EIL is composed of two tunable EIGs with different spatial orientations.

7.4.3 Conclusion

We have observed the formation and dynamics of the 2D surface solitons in the FWM signals in the atomic medium. Such surface solitons propagate along the interface between the uniform medium and the EIL, which is composed of two EIGs. Such an investigation will be important in understanding the fundamental mechanisms in soliton formations and their dynamics. It also exploits new ways to flexibly control the diffraction of optical beams and design new devices for optical image storage, processing, and communication.

7.5 Multi-Wave Mixing Talbot Effect

7.5.1 Introduction

The Talbot effect, which was observed by H. F. Talbot in 1836 [100] and first analytically explained by Lord Rayleigh in 1881 [101], is a near-field diffraction phenomenon that the light field spatially imprinted with periodic structure can have self-imaging at certain periodic imaging planes (the so-called Talbot planes). Such self-imaging effect holds a range of applications from image preprocessing and synthesis, photolithography, optical testing, optical metrology, and spectrometry, to optical computing [102].

To date, studies on the Talbot effect have been reported to be associated with atomic waves [103, 104], surface waves [105, 106], nonclassical light [107], or waveguide arrays [108]. Among the research conducted, the nonlinear Talbot effect with second harmonic (SH) were demonstrated both experimentally [109] and theoretically [110] in periodically poled $LiTaO_3$ (PPLT) crystal for the first time. The Talbot effect with beam imprinted with an EIG has been also proposed [111]. However, no progress in the higher nonlinear optical process was reported before, as far as our knowledge goes.

The crucial point and the guiding ideology to observe the Talbot effect is how to produce a spatial periodical incidence in transverse dimension. From this point of view, it is natural to realize that researchers have created spatially periodic FWM and SWM signals due to the periodic atomic coherence induced by SWs [112–115], which can play an important role in lasing without inversion [7], slow light generation [116], photon controlling and information storage [9–11], fiber lasers [117, 118], and so on. The periodic pattern of the MWM signals can be flexibly controllable by adjusting the atomic coherence via changing the beam detuning.

In this section, we investigate the MWM Talbot effect, and the idea can be also executed in some solid crystals such as Pr-doped YSO crystals [119–122] besides atomic vapors.

7.5.2 Theoretical Model and Analysis

We consider the FWM and SWM processes in the reverse Y-type atomic levels system derived by five beams, as shown in Figure 7.20(a) and (b). A candidate for the systems is rubidium atomic vapors. Also, in Pr:YSO crystal, the energy level system with $|0\rangle = |^3H_4(\pm 3/2)\rangle$, $|1\rangle = |^1D_2(\pm 3/2)\rangle$, $|3\rangle = |^3H_4(\pm 1/2)\rangle$ can be used as a subsystem of our proposed system. Here, we note that to guarantee the applicability of our proposal to all the possible candidate systems, we choose parameters in simulation arbitrarily. Figure 7.20(c) and (d) shows the corresponding beam geometric configurations. The transition $|0\rangle \rightarrow |1\rangle$ with resonant frequency Ω_{10} is probed by beam $\boldsymbol{E}_1$ (with frequency ω_1, frequency detuning Δ_1, and wave vector $\mathbf{k}_1$); $|1\rangle \rightarrow |2\rangle$ with resonant frequency Ω_{21} is pumped by beams $\boldsymbol{E}_2$ (with ω_2, Δ_2, and $\mathbf{k}_2$) and $\boldsymbol{E}_2'$ (with ω_2,Δ_2, and $\mathbf{k}_2'$); $|3\rangle \rightarrow |1\rangle$ with resonant frequency Ω_{31} is pumped by beams $\boldsymbol{E}_3$ (with ω_3, Δ_3, and $\mathbf{k}_3$) and $\boldsymbol{E}_3'$ (with ω_3, Δ_3, and $\mathbf{k}_3'$). The frequency detunings are defined as $\Delta_1 = \Omega_{10} - \omega_1$, $\Delta_2 = \Omega_{21} - \omega_2$, and $\Delta_3 = \Omega_{13} - \omega_3$. In the Cartesian coordinate introduced in Figure 7.20(c) and (d), the wave vectors are elaborately designed such that $\boldsymbol{E}_1$ propagates the z-negative with an angle θ_1; $\boldsymbol{E}_2$and $\boldsymbol{E}_3$ propagate along the opposite direction of $\mathbf{k}_1$, deviating from the z-positive direction with θ_1; $\boldsymbol{E}_2'$and $\boldsymbol{E}_3'$ propagate symmetrically to $\boldsymbol{E}_2$ and $\boldsymbol{E}_3$ with respect to the z-negative direction.

As shown in Figure 7.20(a), if the fields $\boldsymbol{E}_2$, $\boldsymbol{E}_2'$, $\boldsymbol{E}_3$, and $\boldsymbol{E}_3'$ are not sufficiently strong, we consider the undressed FWM signal $\boldsymbol{E}_{\mathrm{F}}$ with $\omega_{\mathrm{F}} = \omega_1$ and $\mathbf{k}_{\mathrm{F}} = \mathbf{k}_1 + \mathbf{k}_3 - \mathbf{k}_1'$; if $\boldsymbol{E}_3$ and $\boldsymbol{E}_3'$ are weak but $\boldsymbol{E}_2$ and $\boldsymbol{E}_2'$ are sufficiently strong to induce the dressing effect (to act as the dressing fields), we consider the singly-dressed FWM signal $\boldsymbol{E}_{\mathrm{F1}}$ with $\omega_{\mathrm{F1}} = \omega_1$ and $\mathrm{k}_{\mathrm{F1}} = \mathrm{k}_{\mathrm{F}}$. However, if $\boldsymbol{E}_3$ and $\boldsymbol{E}_3'$ are also sufficiently strong to induce dressing effect, we consider the doubly-dressed FWM signal $\boldsymbol{E}_{\mathrm{F2}}$

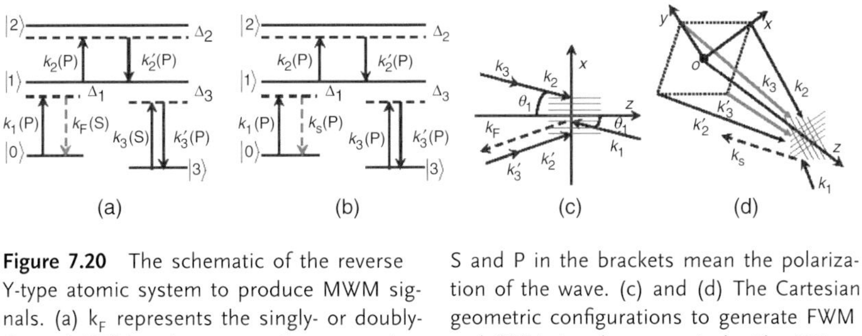

Figure 7.20 The schematic of the reverse Y-type atomic system to produce MWM signals. (a) k_{F} represents the singly- or doubly-dressed FWM and corresponds to a weak or strong $\mathbf{k}_3$, respectively. (b) Emission of the doubly-dressed SWM signal $\mathbf{k}_{\mathrm{S}}$.The letters S and P in the brackets mean the polarization of the wave. (c) and (d) The Cartesian geometric configurations to generate FWM and SWM signals, respectively; The SWs along x and y are constructed by $\mathbf{k}_2$, $\mathbf{k}_2'$ and $\mathbf{k}_3$, $\mathbf{k}_3'$ respectively.

with $\omega_{F2} = \omega_1$ and $k_{F2} = k_F$. Even though $\boldsymbol{E}_2$ and $\boldsymbol{E}'_2$ will induce another FWM signal (P-polarization), we still can get the expected FWM signal via a Polaroid for its polarization selectivity. By using five beams with polarizations as shown in Figure 7.20(b), an SWM signal $\boldsymbol{E}_S$ with $\omega_S = \omega_1$ and $\mathbf{k}_S = \mathbf{k}_1 + \mathbf{k}_2 - \mathbf{k}'_2 + \mathbf{k}_3 - \mathbf{k}'_3$ can be obtained.

To obtain the MWM signal, we solve the density matrix equations and the Liouville pathway for $\rho_{10}^{(3)}$ and $\rho_{10}^{(5)}$, the amplitude of which is proportional to that of the MWM signal. First, for the undressed FWM signal $\boldsymbol{E}_F$, we can obtain $\rho_{10}^{(3)}$ via the Liouville pathway [34] as

$$\rho_{00}^{(0)} \xrightarrow{E_1} \rho_{10}^{(1)} \xrightarrow{(E_3)^*} \rho_{30}^{(2)} \xrightarrow{E'_3} \rho_{10}^{(3)} \tag{7.13}$$

$$\rho_{10}^{(3)} = \frac{-iG_{\text{FWM}}}{d_1^2 d_3} \tag{7.14}$$

where $G_{\text{FWM}} = G_{10}G_{30}(G'_{30})^* \exp(ik_F \cdot r)\exp(-i\omega_1 t)$ $d_1 = \Gamma_{10} + i\Delta_1$, $d_3 = \Gamma_{30} + i(\Delta_1 - \Delta_3)$. The terms $G_{i0} = \mu_{mn}E_{i0}/\hbar$ and $G'_{i0} = \mu_{mn}E'_{i0}/\hbar$ with $i = 1, 2$ are the amplitudes of the Rabi frequencies of E_i and E'_i, respectively, where μ_{mn} is the dipole moment of transition between $|m\rangle$ and $|n\rangle$, which is driven by E_i; E_i; and E'_{i0} are the amplitudes of $\boldsymbol{E}_{i0}$ and $\boldsymbol{E}'_{i0}$, respectively. Γ_{pq} represents the transverse relaxation rate between $|p\rangle$ and $|q\rangle$ for $p \neq q$, and the longitudinal relaxation rate of $|p\rangle$ for $p = q$. It is obvious to see that the amplitude of $\rho_{10}^{(3)}$ in Eq. (7.14) is spatially independent, so $\boldsymbol{E}_F$ is uniform.

Next, we consider the singly-dressed FWM process from $\boldsymbol{E}_2$ and $\boldsymbol{E}'_2$, which are strong sufficiently. In the spatial interaction region, $\boldsymbol{E}_2$ and $\boldsymbol{E}'_2$ will interfere with each other and create an SW, which leads to periodic Rabi frequency amplitude $|G_{2t}(x)|^2 = G_{20}^2 + G'^2_{20} + 2G_{20}^2 G'_{20} \cos\left[2\left(k_2 \sin\theta_1\right)x\right]$. Therefore, the Liouville pathway for $\boldsymbol{E}_{F1}$ will be modified into

$$\rho_{00}^{0} \xrightarrow{E_1} \rho_{|G_{2t}(x)\pm 0|}^{(1)} \xrightarrow{(E_3)*} \rho_{30}^{(2)} \xrightarrow{E'_3} \rho_{|G_{2t}(x)\pm 0\rangle}^{(3)} \tag{7.15}$$

where $|G_{2t}(x)\pm\rangle$ represent two dressed states produced by the spatial periodic dressing effect. According to Eq. (7.15), we can obtain the spatial-dependent dressed density matrix element as,

$$\rho_{F1}^{(3)}(x) = \frac{-iG_{\text{FWM}}}{[d_1 + |G_{2t}(x)|^2/d_2]^2 d_3} \tag{7.16}$$

where $d_2 = \Gamma_{20} + i(\Delta_1 + \Delta_2)$. The amplitude of $\rho_{F1}^{(3)}$ shows obvious periodic variation along direction with a period of $d_x = \lambda_2/(2\sin\theta_1)$.

When $\boldsymbol{E}_3$ and $\boldsymbol{E}'_3$ get strong significantly, we will get the doubly-dressed FWM signal $\boldsymbol{E}_{F2}$. The Liouville pathway and the doubly-dressed density matrix element for this case are

$$\rho_{00}^{(0)} \xrightarrow{E_1} \rho_{|G_{3t}\pm G_{2t}(x)\pm 0\rangle}^{(1)} \xrightarrow{E'_3} \rho_{|G_{3t}\pm G_{2t}(x)\pm 0\rangle}^{(3)} \tag{7.17}$$

$$\rho_{F_2}^{(3)}(x) = \frac{-iG_{\text{FWM}}}{[d_1 + |G_{30}|^2/d_3 + |G_{2t}(x)|^2/d_2]^2 d_3} \tag{7.18}$$

In Figure 7.20(b), the spatially periodic dressing effect from $\boldsymbol{E}_2$ and $\boldsymbol{E}_2'$ remains, and the dressing fields $\boldsymbol{E}_3$ and $\boldsymbol{E}_3'$ with same polarization (compared to the case of Eq. (7.13) with $\boldsymbol{E}_3$ and $\boldsymbol{E}_3'$ having orthogonal polarizations) can also interfere with each other and create periodic Rabi frequency amplitude $|G_{3t}(y)|^2 = G_{30}^2 + G_{30}'^2 + 2G_{30}G_{30}'\cos[2(k_3\sin\theta_2)y]$. Therefore, this SWM will suffer from two dressing effects, which are both spatially periodic with different periods d_x and $d_y = \lambda_3/(2\sin\theta_2)$, respectively. The corresponding Liouville pathway and doubly-dressed density matrix elements are

$$\rho_{00}^{0} \xrightarrow{E_1} \rho_{|G_{3t}(y)\pm G_{2t}(x)\pm 0\rangle}^{(1)} \xrightarrow{(E_3)^*} \rho_{30}^{(2)} \xrightarrow{E_3'} \rho_{|G_{3t}\pm G_{2t}(x)\pm 0\rangle}^{(3)} \xrightarrow{(E_2)^*} \rho_{2\pm 0}^{(4)} \xrightarrow{E_2'} \rho_{|G_{3t}(y)\pm G_{2t}(x)\pm 0\rangle}^{(5)} \tag{7.19}$$

$$\rho_{10}^{(5)} = \frac{iG_{\text{SWM}}}{[d_1 + |G_{2t}(x)|^2/d_2 + |G_{3t}(y)|^2/d_3]^3 d_2 d_3} \tag{7.20}$$

where $G_{\text{SWM}} = G_{10}G_{20}(G_{20}')^* G_{30}(G_{30}')^* \exp(i\mathbf{k}_S \cdot \mathbf{r})\exp(-i\omega_1 t)$

With these expressions, the spatial periodic pattern formation of MWM signals can be investigated. It is clear that the periodical variation of singly-dressed FWM, doubly-dressed FWM and doubly-dressed SWM signals with respect to the x-axis at $z=0$, are all derived from the periodical dressing effects. With different probe detuning, these MWM signals will show profiles with significant differences. So, in order to analyze such dependence, we consider the so-called enhancement and suppression conditions due to the dressing effect before investigating the MWM Talbot effects.

7.5.3 Suppression and Enhancement Conditions

The characteristics of FWM are affected by the dressing effect, which depends on Δ_1, Δ_2, and $|G_{2t}(x)|^2$. As shown in Figure 7.21, because $|G_{2t}(x)|^2$, $|1\rangle$ is split into two dressing states $|G_{2t}(x)\pm\rangle$ with eigenfrequencies $\lambda_{|G_{2t}(x)\pm\rangle} = \Delta_2/2 \pm \sqrt{\Delta_2^2/4 + |G_{2t}(x)|^2}$. Therefore, $|G_{2t}(x)\pm\rangle$ are periodic along x.

Absorption will be enhanced when the probe resonates with the dressing states, that is, $\Delta_1 = -\lambda_{|G_{2t}(x)\pm\rangle}$ which corresponds to the EIA condition. Accordingly, the FWM signal will get enhanced resonantly. Thus, we define $\Delta_1 = -\lambda_{|G_{2t}(x)\pm\rangle}$ as the enhancement condition. However, when the probe reaches two-photon resonance ($\Delta_1 + \Delta_2 = 0$), absorption will be suppressed (the EIT condition), and the FWM signal in such a case will be suppressed correspondingly. Thus, we define $\Delta_1 + \Delta_2 = 0$ as the suppression condition. So, if Δ_1 is set at discrete values orderly from negative to positive and Δ_2 is scanned, we can obtain the periodic enhancement/suppression condition and FWM signals along x. In Figure 7.21(a1) to (e1), we display the periodic FWM signals along x corresponding to different Δ_1, and the insets are the dependence of the FWM signal intensity on Δ_2 at $z=0$ and $x=0$. In Figure 7.21(a2) to (e2), we exhibit the corresponding split energy level diagrams, and in Figure 7.21(a3) to (e3) the corresponding periodic properties of the split energy levels along x. In Figure 7.21(a1), Δ_1 is relatively large and

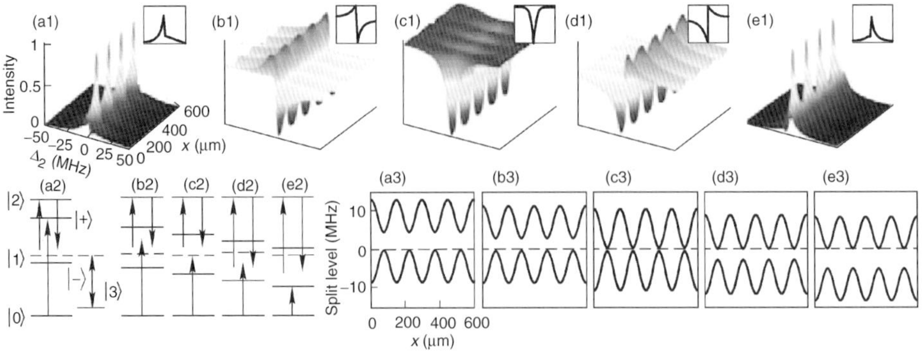

Figure 7.21 The normalized intensities versus Δ_2 and x of singly-dressed FWM with (a1) enhancement effect, (b1) enhancement–suppression effect, (c1) suppression effect, (d1) suppression–enhancement effect, and (e1) enhancement effect, corresponding to $\Delta_1 = -5$, -4, 0, 4, and 5 MHz, respectively. The insets are the incident intensity versus Δ_2. (a2)–(e2) The corresponding split energy level (lines marked $|+\rangle$ and $|-\rangle$ on the right are split energy levels, the dotted line is the initial energy level) diagrams with dressing fields at $x = 0$. (a3)–(e3) The corresponding periodic split energy levels versus x. The other parameters are $G_2 = 15$ MHz, $\Gamma_{10} = 5$MHz, and $\Gamma_{20} = 1$ kHz. Source: Adapted from Ref. [123].

positive, and we only consider the enhancement because of the weakness of the suppression. When Δ_1 decreases but is still positive, Δ_2 will meet the enhancement condition ($\Delta_1 = -\Lambda_{|G_{2t}(x)+\rangle}$) first, and then the suppression condition as shown in Figure 7.21(b1). Because of $\Delta_1 = 0$ in Figure 7.21(c1), Δ_2 can never meet the enhancement condition. The case in Figure 7.21(d1) is opposite to that in Figure 7.21(b1). And in Figure 7.21(e1), the enhancement is dominant again, the same as that in Figure 7.21(a1), for Δ_1 is relatively large and negative.

For the doubly-dressed MWM signals $\boldsymbol{E}_{F2}$ and $\boldsymbol{E}_S$, we take $\boldsymbol{E}_{F2}$ to execute our discussions. First, the interference between $\boldsymbol{E}_3$ and $\boldsymbol{E}'_3$ leads to periodic dressing effect, and therefore splits the naked state into two dressing states $|G_{3t}(y)\pm\rangle$ with eigenfrequencies $\lambda_{|G_{3t}(y)\pm\rangle} = -\Delta_3/2 \pm \sqrt{\Delta_3^2/4 + |G_{3t}(y)|^2}$. $\boldsymbol{E}_2$ and $\boldsymbol{E}'_2$ also act as dressing fields, so the first-order dressing state $|G_{3t}(y)\pm\rangle$ will split into second-order dressing states $|G_{3t}(y) \pm G_{2t}(x)\pm\rangle$ with frequencies

$$\lambda_{|G_{3t}(y)-G_{2t}(x)\pm\rangle} = \frac{-\Delta_3 - \sqrt{\Delta_3^2 + 4|G_{3t}(y)|^2}}{2} + \frac{\Delta'_2 \pm \sqrt{\Delta'^2_2 + 4|G_{2t}(x)|^2}}{2} \tag{7.21}$$

$$\lambda_{|G_{3t}(y)+G_{2t}(x)\pm\rangle} = \frac{-\Delta_3 + \sqrt{\Delta_3^2 + 4|G_{3t}(y)|^2}}{2} + \frac{\Delta''_2 \pm \sqrt{\Delta''^2_2 + 4|G_{2t}(x)|^2}}{2} \tag{7.22}$$

where $\Delta'_2 = \Delta_2 - (-\Delta_3 - \sqrt{\Delta_3^2 + 4|G_{3t}(y)|^2})/2$, $\Delta''_2 = \Delta_2 - (-\Delta_3 + \sqrt{\Delta_3^2 + 4|G_{3t}(y)|^2})/2$, respectively.

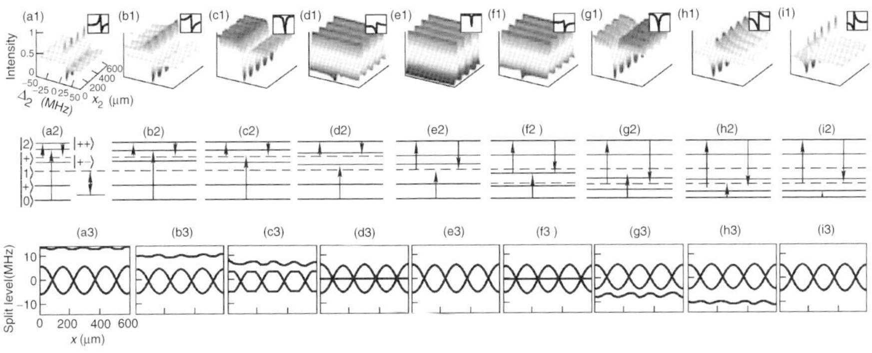

Figure 7.22 The normalized intensities versus Δ_2 and x of (a1) doubly-dressed FWM with enhancement effect, (b1) enhancement-suppression effect, (c1) suppression effect, (d1) suppression-enhancement effect, (e1) suppression effect, (f1) enhancement-suppression effect, (g1) suppression effect, (h1) suppression-enhancement effect, and (i1) enhancement effect, corresponding to $\Delta_1 = -18$ MHz, $\Delta_1 = -18, -8, -5, -2, 0, 2, 5, 8$, and 18 MHz, respectively. The insets are the incident intensity versus Δ_2. (a2)–(i2) The corresponding split energy level diagrams with dressing fields at $x = 0$. (a3)–(i3) The corresponding periodic split energy levels versus x. The other parameters are $\Delta_3 = 0$, $G_2 = G_3 = 15$ MHz, $\Gamma_{10} = 5$ MHz, and $\Gamma_{20} = 1$ kHz.

Similar to the case for $\boldsymbol{E}_{F1}$, we can also investigate the enhancement and suppression for $\boldsymbol{E}_{F2}$ and $\boldsymbol{E}_S$. And their enhancement and suppression correspond to $\Delta_1 = -\lambda_{|G_{3t}(y) \pm G_{2t}(x)\pm\rangle}$ and $\Delta_1 + \Delta_2 = 0$, respectively. The periodic doubly-dressed FWM signals along x corresponding to different conditions are shown by the mesh plots in Figure 7.22(a1) to (i1), with insets being the dependence of the signal intensity on Δ_2 at $z = 0$ and $x = 0$. The corresponding split energy level schematics are shown in Figure 7.22(a2) to (i2), and the corresponding spatial periodic energy levels in Figure 7.22(a3) to (i3). Doubly-dressed SWM enhancement and suppression effects are similar to the results shown in Figure 7.22. In light of the SWM, here is a two-dimensional case, we exhibit two cases of the second-order splitting energy levels as shown in Figure 7.23(a) and (b).

7.5.4
Talbot Effect of MWM Signals

In the perspective of Fourier optics, the transfer function of a Fresnel diffraction system with z as the propagation axis can be expressed as $H_F(\xi) = \exp(ik_z z)\exp(-i\pi\lambda z\xi^2)$ in frequency domain [124], where ξ is the spatial frequency and k_z is the projection of k along z. The field of the MWM signal $g_0(x,y) \propto \rho(x,y)$, so it can be expanded into two-dimensional Fourier series as $g_0(x,y) = \sum_{m,n=-\infty}^{\infty} c_{m,n} \exp[i2\pi(nx/d_x + my/d_y)]$, and in frequency domain the equation can be written as

$$G_0(\xi,\eta) = \sum_{m,n=-\infty}^{\infty} c_{m,n}\delta(\xi - n/d_x)\cdot\delta(\eta - m/d_y) \tag{7.23}$$

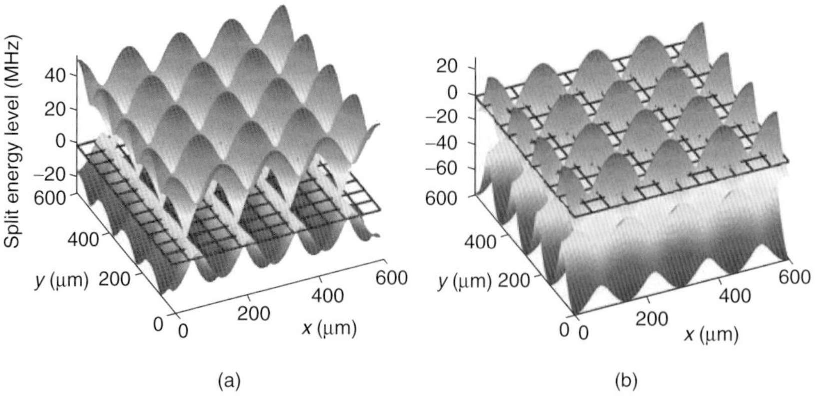

Figure 7.23 (a) The surfaces represent $|G_{3t}(y)+G_{2t}(x)+\rangle$ (top), $|1\rangle$ (grid), and $|G_{3t}(y)+G_{2t}(x)-\rangle$ (bottom), respectively. (b) The surfaces are $|G_{3t}(y)-G_{2t}(x)+\rangle$ (top), $|1\rangle$ (grid), and $|G_{3t}(y)-G_{2t}(x)-\rangle$ (bottom), respectively. $\Delta_2 = 2$ MHz for (a) and -2 MHz for (b). The values of the other parameters are $\Delta_3 = -3$ MHz, $G_{20} = G'_{20} = G_{30} = G'_{30} = 15$ MHz, $\Gamma_{10} = 5$ MHz, $\Gamma_{20} = \Gamma_{30} = 1$ kHz, $\lambda_1 = 776$ nm, $\lambda_2 = \lambda_3 = 780$ nm, and $\theta_1 = \theta_2 = 0.3°$.

where $c_{m,n}$ is the Fourier coefficient. So, considering the Fresnel diffraction, the MWM signal at a z distance is

$$G(\xi,\eta) = G_0(\xi,\eta)\exp(ik_z z)\exp[-i\pi\lambda_1 z(\xi^2+\eta^2)] \tag{7.24}$$

Plugging Eq. (7.23) into Eq. (7.24), we end up with

$$G(\xi,\eta) = \sum_{m,n=-\infty}^{\infty} c_{m,n}\delta\left(\xi-\frac{n}{d_x}\right)\cdot\delta\left(\eta-\frac{m}{d_y}\right)\exp(ik_z z)\exp\left\{-i\pi\lambda_1 z\left[\left(\frac{n}{d_x}\right)^2+\left(\frac{m}{d_y}\right)^2\right]\right\} \tag{7.25}$$

For simplicity, we let $\lambda_2 = \lambda_3$ and $\theta_1 = \theta_2$; thus, the periods along x and y are the same, that is, $d_x = d_y$. Singly- and doubly-dressed FWM signals only concern one SW formed by k_2 and k'_2, and therefore we do not consider the y component. In such a case, $\exp[-i\pi\lambda_1 z(n/d_x)^2] = 1$ if $z = 2qd_x^2/\lambda_1$ ($q = 1, 2, 3, \ldots$), so after inverse Fourier transformation, we get

$$g(x) = g_0(x)\exp(ik_{z_0} z) \tag{7.26}$$

Because of $|g(x)|^2 = |g_0(x)|^2$, we can see the self-imaging of the MWM signals at $z = 2qd_x^2/\lambda_1$, and $z_T = z|_{q=1}$ is the Talbot length. It is worth mentioning that the images on the z_T/N planes are fractional Talbot images [125], where N is an integer bigger than 1.

We first choose $\boldsymbol{E}_{F1}$ as the incidence to execute the simulation. As shown in Figure 7.21, the spatial profile of the incident signal varies along Δ_2 with different Δ_1, and this variation leads to different diffraction process of FWM signal under different conditions. Figure 7.24(a1) and (b1) shows the corresponding Talbot effect carpets under suppression conditions and enhancement conditions, respectively.

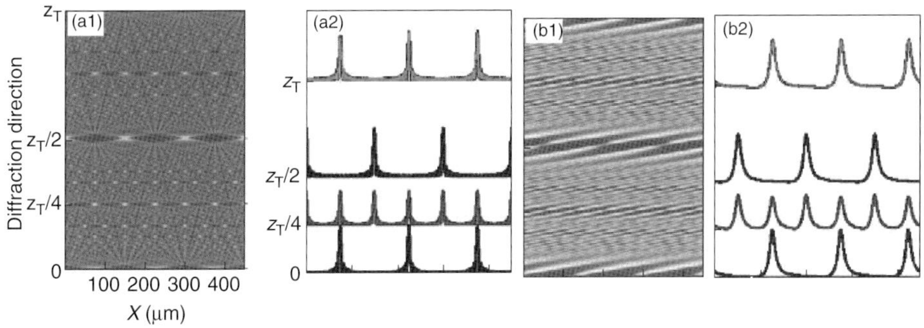

Figure 7.24 (a1) The Talbot effect carpets for $\mathbf{E}_{F1}$ under suppression conditions with $\Delta_1 = 3$ MHz and $\Delta_2 = -3$ MHz. (b1) The Talbot effect carpets for $\mathbf{E}_{F1}$ under enhancement conditions with $\Delta_2 = -3$ MHz and $-\Delta_1 = \Delta_2/2 + \sqrt{\Delta_2^2/4 + |G_{2t}(x)|^2}$. (a2) and (b2) The intensity profiles at $z = 0$, $z = z_T/4$, $z = z_T/2$, and $z = z_T$, respectively. The other parameters are $G_{20} = G'_{20} = 15$ MHz, $\Gamma_{10} = 5$ MHz, and $\Gamma_{20} = \Gamma_{30} = 1$ kHz. Source: Adapted from Ref. [123].

Figure 7.24(a2) and (b2) shows the intensity profiles of the repeated images cut at certain fractional Talbot lengths. From Figure 7.24, we can not only clearly see the periodic singly-dressed FWM can reappear along z but also that the carpet stripes are oblique and such an obliquity is more obvious under enhancement conditions.

For $\boldsymbol{E}_{F2}$, the results are shown in Figure 7.25. Clearly, we obtain the self-imaging of the incident $\boldsymbol{E}_{F2}$ at the Talbot plane again, no matter whether under suppression or enhancement conditions. Contrary to the case in Figure 7.24(a1), which is obtained under suppression conditions, the Talbot effect shown in Figure 7.25(a1) seems oblique. Under enhancement conditions, the carpet stripes are more obvious, as shown in Figure 7.25(b1), but the obliquity is almost the same with that shown in Figure 7.25(a1). By comparing the Talbot carpets shown in Figure 7.24(a1) and (b1), we find the obliquity and the width of the stripes are almost unchanged.

Last but not the least, we discuss the Talbot effect from $\boldsymbol{E}_S$.We consider two orthogonal SWs from two dressing fields $\boldsymbol{E}_2$, $\boldsymbol{E}'_2$ and $\boldsymbol{E}_3$, $\boldsymbol{E}'_3$ simultaneously, to form a 2D lattice, which is periodic both along x and y as shown in Figure 7.20(d); thus, a 2D SWM signal will be excited. In Figure 7.26 and Figure 7.27, we first give the isosurface plots of the Talbot effect of the 2D SWM signal under suppression conditions and enhancement conditions, respectively. And then we choose four intensity plots at certain places during propagation to show the details more clearly. We can see that at the Talbot length shown in Figure 7.26(d) and Figure 7.27(d), the 2D SWM signals are reproduced. At half the Talbot length, as shown in Figure 7.26(c) and Figure 7.27(c), the self-images shifted half period both along x and y. At one quarter of the Talbot length, fractional self-images can be seen as shown in Figure 7.26(b) and Figure 7.27(b), in which the images are twice as many as those in Figure 7.26(a) and Figure 7.27(a). Under the enhancement conditions shown in Figure 7.27, the images are clearer than those under suppression conditions as shown in Figure 7.26. Here, we want to point out the reason for the periods of fractional Talbot effects shown in Figure 7.24(b2), Figure 7.25(a2)

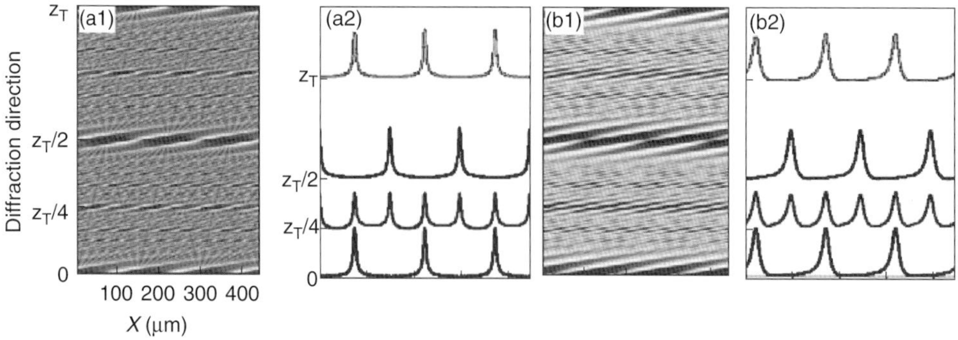

Figure 7.25 (a1) The Talbot effect carpets for $\mathbf{E}_{F2}$ under suppression conditions with $\Delta_1 = 18$MHz, $\Delta_2 = -18$MHz, and $\Delta_3 = 18$MHz. (b1) The Talbot effect carpets for $\mathbf{E}_{F2}$ under enhancement conditions with $\Delta_2 = -18$MHz, $\Delta_3 = 18$MHz, and Δ_1 based on Eq. (7.16). (a2) and (b2) The setup is as Figure 7.24(a2) and (b2). The other parameters are $G_{20} = G'_{20} = 15$ MHz, $G_{30} = 2$ MHz, $\Gamma_{10} = 5$ MHz, and $\Gamma_{20} = \Gamma_{30} = 1$ kHz.

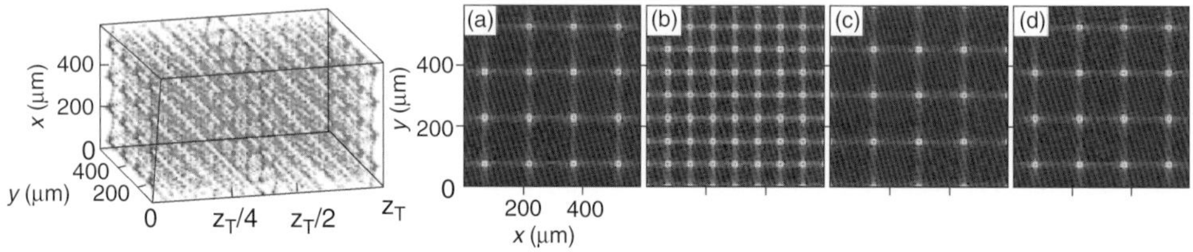

Figure 7.26 The isosurface plot is the Talbot effect for ES under the suppression condition. The four panels are the contour plots of the Talbot effect at (a) $z = 0$, (b) $z = z_T/4$, (c) $z = z_T/2$, and (d) $z = z_T$, respectively. The parameters are $\Delta_1 = 18$ MHz, $\Delta_2 = -18$ MHz, $\Delta_3 = 18$ MHz, $\Gamma_{10} = 5$ MHz, $\Gamma_{20} = \Gamma_{30} = 1$ kHz, and $G_{20} = G'_{20} = G_{30} = G'_{30} = 15$ MHz. Source: Adapted from Ref. [123].

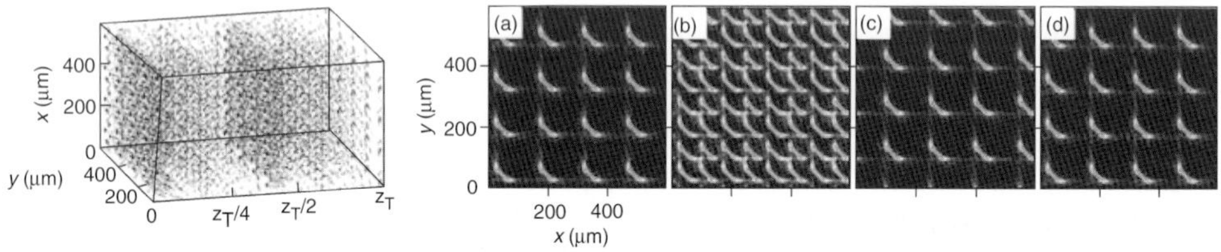

Figure 7.27 The Talbot effect for E_s under the enhancement condition. The setup is as the Figure 7.26, but with Δ_1 according to Eq. (7.22).

and (b2), and Figure 7.27(b) seemingly not being the half period at $z = 0$ is due to the inertial weakness of the numerical simulations. The Talbot effect changes dramatically during propagation, especially around small fractional Talbot lengths. Even a tiny deviation from the exact $z_T/4$ will bring a seemingly imperfect quarter Talbot effect. Calculations with very high resolution will help us approach perfect fractional Talbot effects, but beyond our computer's memory.

7.5.5 Conclusion

We have studied the Talbot effects with completely controllable MWM signals. We have obtained the spatially periodic FWM and SWM signals by the interference between two dressing fields in a reverse Y-type atomic level system. The intensities of the MWM signals can be effectively controlled via enhancement conditions and suppression conditions derived from dressing effect. The Talbot effect from a singly- as well as a doubly-dressed FWM is observed in the numerical experiment. Different from the case of FWM, for SWM, we construct a 2D lattice, from which the SWM is 2D modulated, to investigate the 2D Talbot effect. We find that the numerical simulations agree with the theoretical predictions very well. Our scheme is more advantageous in the controllability, compared to the previous studies, and is mainly attributed to the modulation of the MWM signal by the dressing effect. Our findings not only enrich the understanding of the MWM theory but also offer a different method to investigate the Talbot effect.

Problems

7.1 Explain why spatial shift and splitting of the MWM beam can be controlled by the cross-phase modulation of pump beam.

7.2 Summarize the formation condition of dipole and vortex multi-component vector solitons, as well as two-dimensional surface solitons of MWM in vapor atomic media.

7.3 Explain why the solitons of MWM in atomic medium are more controllable when compared with the solitons in photorefractive crystals.

References

1. Agrawal, G.P. (1990) Induced focusing of optical beams in self-defocusing nonlinear media. *Phys. Rev. Lett.*, **64**, 2487–2490.
2. Hickmann, J.M., Gomes, A., and de Araújo, C.B. (1992) Observation of spatial cross-phase modulation effects in a self-defocusing nonlinear medium. *Phys. Rev. Lett.*, **68**, 3547–3550.
3. Moseley, R.R., Shepherd, S., Fulton, D.J., Sinclair, B.D., and Dunn, M.H. (1996) Electromagnetically-induced focusing. *Phys. Rev. A*, **53**, 408.
4. Harris, S.E. (1997) Electromagnetically induced transparency. *Phys. Today*, **50**, 36.
5. Harris, S. and Yamamoto, Y. (1998) Photon switching by quantum interference. *Phys. Rev. Lett.*, **81**, 3611–3614.
6. Wang, H., Goorskey, D., and Xiao, M. (2001) Enhanced Kerr nonlinearity via atomic coherence in a three-level atomic system. *Phys. Rev. Lett.*, **87**, 073601.
7. Imamoğlu, A. and Harris, S. (1989) Lasers without inversion: interference of dressed lifetime-broadened states. *Opt. Lett.*, **14**, 1344–1346.
8. Hau, L.V., Harris, S.E., Dutton, Z., and Behroozi, C.H. (1999) Light speed reduction to 17 metres per second in an ultracold atomic gas. *Nature*, **397**, 594–598.

9. Imamoglu, A. and Lukin, M.D. (2001) Controlling photons using electromagnetically induced transparency. *Nature*, **413**, 273–276.
10. Liu, C., Dutton, Z., Behroozi, C.H., and Hau, L.V. (2001) Observation of coherent optical information storage in an atomic medium using halted light pulses. *Nature*, **409**, 490–493.
11. Phillips, D., Fleischhauer, A., Mair, A., Walsworth, R., and Lukin, M. (2001) Storage of light in atomic vapor. *Phys. Rev. Lett.*, **86**, 783–786.
12. Duan, L.M., Lukin, M., Cirac, I., and Zoller, P. (2001) Long-distance quantum communication with atomic ensembles and linear optics. *Nature*, **414**, 413–418Arxiv preprint quant-ph/0105105.
13. Kang, H., Hernandez, G., and Zhu, Y. (2004) Resonant four-wave mixing with slow light. *Phys. Rev. A*, **70**, 61804.
14. Lukin, M., Matsko, A., Fleischhauer, M., and Scully, M.O. (1999) Quantum noise and correlations in resonantly enhanced wave mixing based on atomic coherence. *Phys. Rev. Lett.*, **82**, 1847–1850.
15. Ling, H.Y., Li, Y.Q., and Xiao, M. (1998) Electromagnetically induced grating: homogeneously broadened medium. *Phys. Rev. A*, **57**, 1338.
16. Skenderović, H. (2005) Coherent Control by Four-Wave Mixing.
17. Zhang, Y., Nie, Z., Zheng, H., Li, C., Song, J., and Xiao, M. (2009) Electromagnetically induced spatial nonlinear dispersion of four-wave mixing. *Phys. Rev. A*, **80**, 013835.
18. Zhang, Y., Zuo, C., Zheng, H., Li, C., Nie, Z., Song, J., Chang, H., and Xiao, M. (2009) Controlled spatial beam splitter using four-wave-mixing images. *Phys. Rev. A*, **80**, 055804.
19. Zhang, Y., Wang, Z., Zheng, H., Yuan, C., Li, C., Lu, K., and Xiao, M. (2010) Four-wave-mixing gap solitons. *Phys. Rev. A*, **82**, 053837.
20. Zhang, Y., Wang, Z., Nie, Z., Li, C., Chen, H., Lu, K., and Xiao, M. (2011) Four-wave mixing dipole soliton in laser-induced atomic gratings. *Phys. Rev. Lett.*, **106**, 93904.
21. C. Li, S. Sang, Y. Zhang, J. Sun, Z. Zhang, X. Wang, H. Zheng, Y. Li, Spatial interplay of two four-wave mixing images, *J. Opt. Soc. Am. B*, **29**, 3015–3020.
22. Voogd, R.J., Singh, M., and Braat, J.J.M. (2004) The use of orbital angular momentum of light beams for optical data storage. Optical Data Storage Topical Meeting 2004, International Society for Optics and Photonics, pp. 387–392.
23. Tikhonenko, V., Christou, J., and Luther-Daves, B. (1995) Spiraling bright spatial solitons formed by the breakup of an optical vortex in a saturable self-focusing medium. *J. Opt. Soc. Am. B*, **12**, 2046–2052.
24. Crabtree, K., Davis, J.A., and Moreno, I. (2004) Optical processing with vortex-producing lenses. *Appl. Opt.*, **43**, 1360–1367.
25. Grier, D.G. (2003) A revolution in optical manipulation. *Nature*, **424**, 810–816.
26. Kuga, T., Torii, Y., Shiokawa, N., Hirano, T., Shimizu, Y., and Sasada, H. (1997) Novel optical trap of atoms with a doughnut beam. *Phys. Rev. Lett.*, **78**, 4713–4716.
27. Mair, A., Vaziri, A., Weihs, G., and Zeilinger, A. (2001) Entanglement of orbital angular momentum states of photons. *Nature*, **412**, 313–316 arXiv preprint quant-ph/0104070.
28. Desyatnikov, A.S., Kivshar, Y.S., and Torner, L. (2005) Optical vortices and vortex solitons. *Prog. Opt.*, **47**, 291–391.
29. Lederer, F., Stegeman, G.I., Christodoulides, D.N., Assanto, G., Segev, M., and Silberberg, Y. (2008) Discrete solitons in optics. *Phys. Rep.*, **463**, 1–126.
30. Skupin, S., Saffman, M., and Krolikowski, W. (2007) Nonlocal stabilization of nonlinear beams in a self-focusing atomic vapor. *Phys. Rev. Lett.*, **98**, 263902.
31. Mihalache, D., Mazilu, D., Lederer, F., Kartashov, Y., Crasovan, L.C., Torner, L., and Malomed, B. (2006) Stable vortex tori in the three-dimensional cubic-quintic Ginzburg-Landau equation. *Phys. Rev. Lett.*, **97**, 73904.

32. Theocharis, G., Frantzeskakis, D., Kevrekidis, P., Malomed, B.A., and Kivshar, Y.S. (2003) Ring dark solitons and vortex necklaces in Bose-Einstein condensates. *Phys. Rev. Lett.*, **90**, 120403.
33. Michinel, H., Paz-Alonso, M.J., and Pérez-García, V.M. (2006) Turning light into a liquid via atomic coherence. *Phys. Rev. Lett.*, **96**, 23903.
34. Zhang, Y. and Xiao, M. (2009) *Multi-Wave Mixing Processes: From Ultrafast Polarization Beats to Electromagnetically Induced Transparency*, Springer, Berlin.
35. Zhang, Y., Skupin, S., Maucher, F., Pour, A.G., Lu, K., and Królikowski, W. (2010) Azimuthons in weakly nonlinear waveguides of different symmetries. *Opt. Express*, **18**, 27846–27857, arXiv preprint arXiv:1010.3664.
36. Molina-Terriza, G. and Torner, L. (2000) Multicharged vortex evolution in seeded second-harmonic generation. *J. Opt. Soc. Am. B*, **17**, 1197–1204.
37. He, Y., Malomed, B.A., and Wang, H. (2007) Fusion of necklace-ring patterns into vortex and fundamental solitons in dissipative media. *Opt. Express*, **15**, 17502–17508 arXiv preprint arXiv:0712.2899.
38. Steck, D.A. (2000) Sodium D Line Data. Report, Los Alamos National Laboratory, Los Alamos, NM.
39. Liu, S., Ma, C., Zhang, Y., and Lu, K. (2011) Bragg gap solitons in PT symmetric lattices with competing nonlinearity. *Opt. Commun.*, **285**, 1934–1939
40. Y. Zhang, Z. Wu, C. Yuan, X. Yao, K. Lu, M. Belic, Optical vortices induced in nonlinear multi-level atomic vapors, *Opt. Lett.*, **37**, 4507–4509.
41. Masajada, J. and Dubik, B. (2001) Optical vortex generation by three plane wave interference. *Opt. Commun.*, **198**, 21–27.
42. Chen, Z., Segev, M., Wilson, D.W., Muller, R.E., and Maker, P.D. (1997) Self-trapping of an optical vortex by use of the bulk photovoltaic effect. *Phys. Rev. Lett.*, **78**, 2948–2951.
43. Trillo, S. and Torruellas, W. (2001) *Spatial Solitons*, Springer-Verlag, Berlin.
44. Stegeman, G.I. and Segev, M. (1999) Optical spatial solitons and their interactions: universality and diversity. *Science*, **286**, 1518–1523.
45. Tikhonenko, V., Christou, J., and Luther-Davies, B. (1996) Three dimensional bright spatial soliton collision and fusion in a saturable nonlinear medium. *Phys. Rev. Lett.*, **76**, 2698–2701.
46. Desyatnikov, A.S., Kivshar, Y.S., Motzek, K., Kaiser, F., Weilnau, C., and Denz, C. (2002) Multicomponent dipole-mode spatial solitons. *Opt. Lett.*, **27**, 634–636.
47. Motzek, K., Kaiser, F., Weilnau, C., Denz, C., McCarthy, G., Krolikowski, W., Desyatnikov, A., and Kivshar, Y.S. (2002) Multi-component vector solitons in photorefractive crystals. *Opt. Commun.*, **209**, 501–506.
48. L. Ge 2011 Q. Wang, M. Shen, J. Shi, Dipole solitons in nonlocal nonlinear media with anisotropy, *Opt. Commun.*, **284**, 2351–2356.
49. Shen, M., Chen, X., Shi, J., Wang, Q., and Krolikowski, W. (2009) Incoherently coupled vector dipole soliton pairs in nonlocal media. *Opt. Commun.*, **282**, 4805–4809.
50. Rotschild, C., Alfassi, B., Cohen, O., and Segev, M. (2006) Long-range interactions between optical solitons. *Nat. Phys.*, **2**, 769–774.
51. Ye, F., Kartashov, Y.V., and Torner, L. (2008) Stabilization of dipole solitons in nonlocal nonlinear media. *Phys. Rev. A*, **77**, 043821.
52. Yang, J., Makasyuk, I., Bezryadina, A., and Chen, Z. (2004) Dipole solitons in optically induced two-dimensional photonic lattices. *Opt. Lett.*, **29**, 1662–1664.
53. Efremidis, N.K., Hudock, J., Christodoulides, D.N., Fleischer, J.W., Cohen, O., and Segev, M. (2003) Two-dimensional optical lattice solitons. *Phys. Rev. Lett.*, **91**, 213906.
54. Desyatnikov, A.S. and Kivshar, Y.S. (2001) Rotating optical soliton clusters. **88**, 053901. Arxiv preprint nlin/0112039.
55. Desyatnikov, A.S., Sukhorukov, A.A., and Kivshar, Y.S. (2005) Azimuthons:

spatially modulated vortex solitons. *Phys. Rev. Lett.*, **95**, 203904.

56. Firth, W. and Skryabin, D. (1997) Optical solitons carrying orbital angular momentum. *Phys. Rev. Lett.*, **79**, 2450–2453.
57. Minovich, A., Neshev, D.N., Desyatnikov, A.S., Krolikowski, W., and Kivshar, Y.S. (2009) Observation of optical azimuthons. *Opt. Express*, **17**, 23610–23616.
58. Yang, J. and Pelinovsky, D.E. (2003) Stable vortex and dipole vector solitons in a saturable nonlinear medium. *Phys. Rev. E*, **67**, 016608.
59. Desyatnikov, A.S. and Kivshar, Y.S. (2001) Necklace-ring vector solitons. *Phys. Rev. Lett.*, **87**, 33901.
60. Salgueiro, J.R. and Kivshar, Y.S. (2004) Single-and double-vortex vector solitons in self-focusing nonlinear media. *Phys. Rev. E*, **70**, 056613.
61. Desyatnikov, A.S., Neshev, D., Ostrovskaya, E.A., Kivshar, Y.S., Krolikowski, W., Luther-Davies, B., García-Ripoll, J.J., and Pérez-García, V.M. (2001) Multipole spatial vector solitons. *Opt. Lett.*, **26**, 435–437.
62. Jeng, C.C., Shih, M.F., Motzek, K., and Kivshar, Y. (2004) Partially incoherent optical vortices in self-focusing nonlinear media. *Phys. Rev. Lett.*, **92**, 43904.
63. Desyatnikov, A.S., Mihalache, D., Mazilu, D., Malomed, B.A., and Lederer, F. (2007) Stable counter-rotating vortex pairs in saturable media. *Phys. Lett. A*, **364**, 231–234.
64. Musslimani, Z.H., Segev, M., and Christodoulides, D.N. (2000) Multi-component two-dimensional solitons carrying topological charges. *Opt. Lett.*, **25**, 61–63.
65. Anastassiou, C., Soljačić, M., Segev, M., Eugenieva, E.D., Christodoulides, D.N., Kip, D., Musslimani, Z.H., and Torres, J.P. (2000) Eliminating the transverse instabilities of Kerr solitons. *Phys. Rev. Lett.*, **85**, 4888–4891.
66. Musslimani, Z.H., Soljačić, M., Segev, M., and Christodoulides, D.N. (2001) Interactions between two-dimensional composite vector solitons carrying topological charges. *Phys. Rev. E*, **63**, 066608.
67. Pigier, C., Uzdin, R., Carmon, T., Segev, M., Nepomnyaschchy, A., and Musslimani, Z.H. (2001) Collisions between (2+1)D rotating propeller solitons. *Opt. Lett.*, **26**, 1577–1579.
68. Izdebskaya, Y.V., Rebling, J., Desyatnikov, A.S., and Kivshar, Y.S. (2012) Observation of vector solitons with hidden vorticity. *Opt. Lett.*, **37**, 767–769.
69. Liu, Y., Durand, M., Chen, S., Houard, A., Prade, B., Forestier, B., and Mysyrowicz, A. (2010) Energy exchange between femtosecond laser filaments in air. *Phys. Rev. Lett.*, **105**, 55003.
70. Wang, H., Goorskey, D., and Xiao, M. (2001) Enhanced Kerr nonlinearity via atomic coherence in a three-level atomic system. *Phys. Rev. Lett.*, **87**, 73601–73601.
71. Snyder, A.W. and Sheppard, A. (1993) Collisions, steering, and guidance with spatial solitons. *Opt. Lett.*, **18**, 482–484.
72. Zhang, Y., Nie, Z., Zhao, Y., Li, C., Wang, R., Si, J., and Xiao, M. (2010) Modulated vortex solitons of four-wave mixing. *Opt. Express*, **18**, 10963–10972.
73. Oulton, R.F., Sorger, V.J., Zentgraf, T., Ma, R.M., Gladden, C., Dai, L., Bartal, G., and Zhang, X. (2009) Plasmon lasers at deep subwavelength scale. *Nature*, **461**, 629–632.
74. Marinescu, M. and Dalgarno, A. (1995) Dispersion forces and long-range electronic transition dipole moments of alkali-metal dimer excited states. *Phys. Rev. A*, **52**, 311–328.
75. Singer, K., Stanojevic, J., Weidemüller, M., and Côté, R. (2005) Long-range interactions between alkali Rydberg atom pairs correlated to the ns–ns, np–np and nd–nd asymptotes. *J. Phys. B: At. Mol. Opt. Phys.*, **38**, S295.
76. Lopez-Aguayo, S., Desyatnikov, A.S., Kivshar, Y.S., Skupin, S., Krolikowski, W., and Bang, O. (2006) Stable rotating dipole solitons in nonlocal optical media. *Opt. Lett.*, **31**, 1100–1102.
77. Izdebskaya, Y.V., Desyatnikov, A.S., Assanto, G., and Kivshar, Y.S. (2011) Dipole azimuthons and vortex charge

flipping in nematic liquid crystals. *Opt. Express*, **19**, 21457–21466.
78. Carlsson, A.H., Malmberg, J.N., Anderson, D., Lisak, M., Ostrovskaya, E.A., Alexander, T.J., and Kivshar, Y.S. (2000) Linear and nonlinear waveguides induced by optical vortex solitons. *Opt. Lett.*, **25**, 660–662.
79. Sheppard, A. and Haelterman, M. (1994) Polarization-domain solitary waves of circular symmetry in Kerr media. *Opt. Lett.*, **19**, 859–861.
80. Law, C., Zhang, X., and Swartzlander, G. (2000) Waveguiding properties of optical vortex solitons. *Opt. Lett.*, **25**, 55–57.
81. Kivshar, Y.S. and Agrawal, G.P. (2003) *Optical Solitons: From Fibers to Photonic Crystals*, Academic Press, Boston, MA.
82. Yanpeng, Z., Zhiguo, W., Huaibin, Z., Chenzhi, Y., Changbiao, L., Keqing, L., and Min, X. (2010) Four-wave-mixing gap solitons. *Phys. Rev. A*, **82**.
83. Kivshar, Y. (2006) Spatial solitons: bending light at will. *Nat. Phys.*, **2**, 729–730.
84. Neto, A.H.C., Guinea, F., Peres, N., Novoselov, K., and Geim, A. (2009) The electronic properties of graphene. *Rev. Mod. Phys.*, **81**, 109.
85. Yeh, P., Yariv, A., and Cho, A. (1978) Optical surface waves in periodic layered media. *Appl. Phys. Lett.*, **32**, 104.
86. Davoyan, A.R., Shadrivov, I.V., and Kivshar, Y.S. (2009) Self-focusing and spatial plasmon-polariton solitons. *Opt. Express*, **17**, 21732–21737.
87. Alfassi, B., Rotschild, C., Manela, O., Segev, M., and Christodoulides, D.N. (2007) Nonlocal surface-wave solitons. *Phys. Rev. Lett.*, **98**, 213901.
88. Suntsov, S., Makris, K.G., Christodoulides, D.N., Stegeman, G.I., Haché, A., Morandotti, R., Yang, H., Salamo, G., and Sorel, M. (2006) Observation of discrete surface solitons. *Phys. Rev. Lett.*, **96**, 063901.
89. Makris, K.G., Suntsov, S., Christodoulides, D.N., Stegeman, G.I., and Hache, A. (2005) Discrete surface solitons. *Opt. Lett.*, **30**, 2466–2468.
90. Rosberg, C.R., Neshev, D.N., Krolikowski, W., Mitchell, A., Vicencio, R.A., Molina, M.I., and Kivshar, Y.S. (2006) Observation of surface gap solitons in semi-infinite waveguide arrays. *Phys. Rev. Lett.*, **97**, 083901.
91. Yeh, P., Yariv, A., and Hong, C.S. (1977) Electromagnetic propagation in periodic stratified media. I. general theory. *J. Opt. Soc. Am.*, **67**, 423–438.
92. Heinrich, M., Kartashov, Y.V., Ramirez, L.P.R., Szameit, A., Dreisow, F., Keil, R., Nolte, S., Tünnermann, A., Vysloukh, V.A., and Torner, L. (2009) Two-dimensional solitons at interfaces between binary superlattices and homogeneous lattices. *Phys. Rev. A*, **80**, 063832.
93. Wang, X., Bezryadina, A., Chen, Z., Makris, K.G., Christodoulides, D.N., and Stegeman, G.I. (2007) Observation of two-dimensional surface solitons. *Phys. Rev. Lett.*, **98**, 123903.
94. Sukhorukov, A.A., Neshev, D.N., , Dreischuh, A., Fischer, R., Ha, S., Krolikowski, W., Bolger, J., Mitchell, A., Eggleton, B.J., and Kivshar, Y.S. (2006) Polychromatic nonlinear surface modes generated by supercontinuum light. *Opt. Express*, **14**, 11265–11270.
95. Bonaccorso, F., Sun, Z., Hasan, T., and Ferrari, A. (2010) Graphene photonics and optoelectronics. *Nat. Photonics*, **4**, 611–622.
96. Christodoulides, D.N. and Eugenieva, E.D. (2001) Blocking and routing discrete solitons in two-dimensional networks of nonlinear waveguide arrays. *Phys. Rev. Lett.*, **87**, 233901.
97. Bennink, R.S., Wong, V., Marino, A.M., Aronstein, D.L., Boyd, R.W., Stroud, C.R. Jr., Lukishova, S., and Gauthier, D.J. (2002) Honeycomb pattern formation by laser-beam filamentation in atomic sodium vapor. *Phys. Rev. Lett.*, **88**, 113901.
98. Fink, Y., Winn, J.N., Fan, S., Chen, C., Michel, J., Joannopoulos, J.D., and Thomas, E.L. (1998) A dielectric omnidirectional reflector. *Science*, **282**, 1679–1682.
99. Zhang, Y., Yuan, C., Zhang, Y., Zheng, H., Chen, H., Li, C., Wang, Z., and Xiao, M. (2013) Surface solitons of

four-wave mixing in an electromagnetically induced lattice. *Laser Physics Letters*, **10**, 055406.

100. Talbot, H.F. (1836) LXXVI. Facts relating to optical science. No. IV. *London Edinburgh Philos. Mag. J. Sci.*, **9**, 401–407.
101. Lord Rayleigh (1881) XXV. On copying diffraction-gratings, and on some phenomena connected therewith. *London Edinburgh Philos. Mag. J. Sci.*, **11**, 196–205.
102. Patorski, K. (1989) I The self-imaging phenomenon and its applications. *Prog. Opt.*, **27**, 1–108.
103. Chapman, M.S., Ekstrom, C.R., Hammond, T.D., Schmiedmayer, J., Tannian, B.E., Wehinger, S., and Pritchard, D.E. (1995) Near-field imaging of atom diffraction gratings: the atomic Talbot effect. *Phys. Rev. A*, **51**, 14–17.
104. C. Ryu, M. Andersen, A. Vaziri, M. d'Arcy, J. Grossman, K. Helmerson, W. Phillips, High-order quantum resonances observed in a periodically kicked Bose-Einstein condensate, *Phys. Rev. Lett.*, **96** (2006) 160403.
105. Dennis, M.R., Zheludev, N.I., and García de Abajo, F.J. (2007) The plasmon Talbot effect. *Opt. Express*, **15**, 9692–9700.
106. Maradudin, A. and Leskova, T. (2009) The Talbot effect for a surface plasmon polariton. *New J. Phys.*, **11**, 033004.
107. Müller-Ebhardt, H., Rehbein, H., Li, C., Mino, Y., Somiya, K., Schnabel, R., Danzmann, K., and Chen, Y. (2009) Quantum-state preparation and macroscopic entanglement in gravitational-wave detectors. *Phys. Rev. A*, **80**, 043802.
108. Iwanow, R., May-Arrioja, D.A., Christodoulides, D.N., Stegeman, G.I., Min, Y., and Sohler, W. (2005) Discrete Talbot effect in waveguide arrays. *Phys. Rev. Lett.*, **95**, 53902.
109. Zhang, Y., Wen, J., Zhu, S., and Xiao, M. (2010) Nonlinear Talbot effect. *Phys. Rev. Lett.*, **104**, 183901.
110. Wen, J., Zhang, Y., Zhu, S.N., and Xiao, M. (2011) Theory of nonlinear Talbot effect. *J. Opt. Soc. Am. B*, **28**, 275–280.
111. Wen, J., Du, S., Chen, H., and Xiao, M. (2011) Electromagnetically induced Talbot effect. *Appl. Phys. Lett.*, **98**, 081108.
112. Andre, A. and Lukin, M. (2002) Manipulating light pulses via dynamically controlled photonic band gap. *Phys. Rev. Lett.*, **89**, 143602.
113. Artoni, M. and La Rocca, G. (2006) Optically tunable photonic stop bands in homogeneous absorbing media. *Phys. Rev. Lett.*, **96**, 73905.
114. Gao, J.W., Zhang, Y., Ba, N., Cui, C.L., and Wu, J.H. (2010) Dynamically induced double photonic bandgaps in the presence of spontaneously generated coherence. *Opt. Lett.*, **35**, 709–711.
115. Wu, J.H., Artoni, M., and La Rocca, G. (2008) Controlling the photonic band structure of optically driven cold atoms. *J. Opt. Soc. Am. B*, **25**, 1840–1849.
116. Wang, G., Lu, H., and Liu, X. (2012) Dispersionless slow light in MIM waveguide based on a plasmonic analogue of electromagnetically induced transparency. *Opt. Express*, **20**, 20902–20907.
117. Liu, X., Yang, X., Lu, F., Ng, J., Zhou, X., and Lu, C. (2005) Stable and uniform dual-wavelength erbium-doped fiber laser based on fiber Bragg gratings and photonic crystal fiber. *Opt. Express*, **13**, 142–147.
118. Liu, X.M. (2008) Theory and experiments for multiple four-wave-mixing processes with multifrequency pumps in optical fibers. *Phys. Rev. A*, **77**, 043818.
119. Ma, H. and de Araújo, C.B. (1993) Interference between third-and fifth-order polarizations in semiconductor doped glasses. *Phys. Rev. Lett.*, **71**, 3649–3652.
120. Ham, B.S., Hemmer, P., and Shahriar, M. (1997) Efficient electromagnetically induced transparency in a rare-earth doped crystal. *Opt. Commun.*, **144**, 227–230.
121. Klein, J., Beil, F., and Halfmann, T. (2007) Robust population transfer by stimulated Raman adiabatic passage in a Pr^{3+}:Y_2SiO_5 crystal. *Phys. Rev. Lett.*, **99**, 113003.
122. Wang, H.H., Li, A.J., Du, D.M., Fan, Y.F., Wang, L., Kang, Z.H., Jiang, Y.,

Wu, J.H., and Gao, J.Y. (2008) All-optical routing by light storage in a Pr^{3+}:Y_2SiO_5 crystal. *Appl. Phys. Lett.*, **93**, 221112.

123. Zhang, Y., Yao, X., Yuan, C., Li, P., Yuan, J., Feng, W., and Jia, S. (2012) Controllable multi-wave mixing Talbot effect.

124. Goodman, J.W. (1968) *Introduction to Fourier Optics*, McGraw-Hill, New York.

125. Chen, Z., Liu, D., Zhang, Y., Wen, J., Zhu, S., and Xiao, M. (2012) Fractional second-harmonic Talbot effect. *Opt. Lett.*, **37**, 689–691.

8
Optical Routing and Space Demultiplexer of MWM Process

Highlights

All-optical switching/routing and space demultiplexer can be experimentally demonstrated in FWM process, which are essential elements for the next generation of all-optical communication and computing.

As the cross-Kerr and self-Kerr nonlinearities can be greatly enhanced in the multi-level EIT systems, the spatial diffractions of the generated FWM signal beams as well as the incident probe beam can compensate each other to form spatial solitons during their propagations. When multiple laser beams are involved, the spatial patterns can be quite complicated, determined by their overlaps, frequency detuning, atomic density, and relative intensities. In this chapter, the spatial splitting and shift of FWM beam and their various applications are introduced. We first discuss the spatial shift and splitting characteristics of FWM beam spots, which are controlled by the frequency detuning, the intensities, and polarization states of the dressing beams, and can be used for all-optical switching, switching arrays, and routers. Then, we introduce the space demultiplexer and router by the spatial splitting of one FWM spot into multiple spots, which are periodically modulated owing to the spatial periodic atomic coherence effect. This demonstration of all-optically controlled routing has opened the doors to the potential achievement of the hyper-terahertz all-optical router and space demultiplexer.

8.1
Optical Switching and Routing

8.1.1
Introduction

In order to develop the next generation of all-optical communication and computing, certain optical elements are essential, such as all-optical switches and routers. There have been several new schemes reported recently to demonstrate, in

Quantum Control of Multi-Wave Mixing, First Edition. Yanpeng Zhang, Feng Wen, and Min Xiao.

principle, such all-optically controlled switching and routing functions [1–3]. A weak beam was used to selectively turn on/off the spots in the spatial pattern of a stronger laser beam via XPM in a two-level atomic medium [1], showing a spatial switching effect. Also, controlling the linear [2] and nonlinear [3] optical absorptions of one laser beam by another in coherently prepared atomic media was exploited to show all-optically controlled beam switching. Recently, it was shown that an FWM signal beam can be spatially shifted easily by frequency detuning and intensities of the dressing laser beams following a dispersion-like behavior [4]. Such electromagnetically induced spatial dispersion (EISD) is greatly enhanced as for the frequency (linear and nonlinear) dispersions in the EIT systems [5, 6], which can give large and sensitive spatial displacements for the FWM and the probe beams. Also, if one carefully chooses the parametric regime, the probe and FWM beams can have focusing effects in a self-defocusing medium because of the strong XPM [7, 8], which compensates the beam diffraction when propagating through the long atomic medium.

In this section, we show that by making use of the EISD effect in a three-level ladder-type atomic system [4], all-optical switching/routing effects can be experimentally demonstrated. The FWM signals are generated by two coupling beams in the three- or two-level system, with an additional dressing field to shift the spatial location of the generated FWM beams. The intensities of the initial (before shifting) and final (after shifting) spots of the FWM signals correspond to the "off" and "on" states of the switch. Different shift directions and spot locations are studied as functions of experimental parameters. As there are two FWM beams and each beam has more than one final state (spatial locations), it is possible to construct switching arrays in the current system. Such controllable spatial beam spot shifts can provide potential architectures for beam address selection and routing in all-optical communication and networks.

8.1.2
Theoretical Model and Experimental Scheme

The relevant experimental system is shown in Figure 8.1(a) and (b). Three energy levels from sodium atoms (in a heat pipe oven of length 18 cm) are involved in the experimental schemes. The pulse laser beams are aligned spatially as shown in Figure 8.1(c). In Figure 8.1(a), energy levels $|0\rangle$ ($3S_{1/2}$), $|1\rangle$ ($3P_{3/2}$), and $|2\rangle$ ($4D_{3/2}$) form a ladder-type three-level atomic system. Coupling fields $\boldsymbol{E}_2$ (wavelength of 568.8 nm, angular frequency ω_2, detuning $\Delta_2=0$, wave vector $\mathbf{k}_2$, and Rabi frequency $G_2=5.1$ GHz) and $\boldsymbol{E}_2'$ (ω_2, $\Delta_2=0$, $\mathbf{k}_2'$, and $G_2'=15.5$ GHz) connect the transition between level $|1\rangle$ and level $|2\rangle$, which are from the same near-transform-limited dye laser (10 Hz repetition rate, 5 ns pulse width, and 0.04 cm^{-1} line width). The field $\boldsymbol{E}_2$ in beam 1 propagates in the direction opposite to the weak probe field $\boldsymbol{E}_3$ (wavelength of 589.0 nm, ω_1, Δ_1, $\mathbf{k}_3$, and $G_3=4.8$ GHz) in beam 4, as shown in Figure 8.1(c), connecting the transition between $|0\rangle$ and $|1\rangle$. $\boldsymbol{E}_2'$ in beam 3 propagates in the plane (yz) having a small angle (0.3°) with

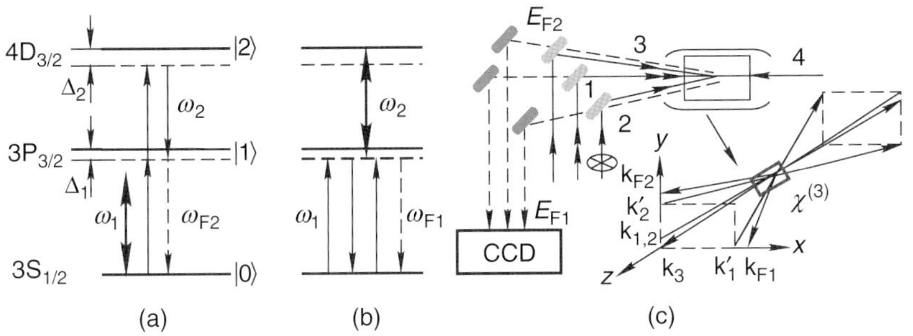

Figure 8.1 (a) and (b) The diagrams of Na energy levels with different coupling schemes. The bold arrows refer to the dressing fields. (c) The experimental scheme and arrangements (inset: the spatial alignments of the incident beams).

$\boldsymbol{E}_2$. With the phase-matching condition, it generates a NDFWM process satisfying $\mathbf{k}_{F2} = \mathbf{k}_3 + \mathbf{k}_2 - \mathbf{k}'_2$ (called E_{F2} for the subsystem $|0\rangle \to |1\rangle \to |2\rangle$). Then, additional fields $\boldsymbol{E}_1$ (ω_1, Δ_1, $\mathbf{k}_1$, and $G_1 = 5.1$ GHz) and $\boldsymbol{E}'_1$ (ω_1, Δ_1, $\mathbf{k}'_1$, and G'_1) are added, which are from the other dye laser with characteristics similar to the first one, also connecting the transition between $|0\rangle$ and $|1\rangle$. $\boldsymbol{E}_1$ adds onto beam 1 and $\boldsymbol{E}'_1$ (beam 2) propagates in another plane (xz), which is perpendicular to the yz-plane with a small angle relative to $\boldsymbol{E}_1$, as shown in the inset of Figure 8.1(c). When $\boldsymbol{E}_1$, $\boldsymbol{E}'_1$, and $\boldsymbol{E}_3$ are turned on simultaneously with blocking $\boldsymbol{E}_2$, $\boldsymbol{E}'_2$ a DFWM process is generated, satisfying the phase-matching condition $\mathbf{k}_{F1} = \mathbf{k}_1 - \mathbf{k}'_1 + \mathbf{k}_3$ (called $\boldsymbol{E}_{F1}$ for the subsystem $|0\rangle \to |1\rangle$) (Figure 8.1(b)). Here, we define detuning $\Delta_i = \Omega_i - \omega_i$ with the atomic resonant frequency Ω_i. The average powers of the laser beams $\boldsymbol{E}_1$, $\boldsymbol{E}'_1$, $\boldsymbol{E}_2$, $\boldsymbol{E}'_2$, and $\boldsymbol{E}_3$ are 3, 100, 5, 95, and 0.14 μW. The laser beams $\boldsymbol{E}_1$ ($\boldsymbol{E}'_1$), $\boldsymbol{E}_2$ ($\boldsymbol{E}'_2$), and $\boldsymbol{E}_3$ (with diameters of about 0.59, 0.82, and 0.59 mm) are horizontally polarized.

When $\boldsymbol{E}_1$, $\boldsymbol{E}'_1$, $\boldsymbol{E}_2$, $\boldsymbol{E}'_2$, and $\boldsymbol{E}_3$ are all turned on simultaneously, the NDFWM process $\boldsymbol{E}_{F2}$ and the DFWM process $\boldsymbol{E}_{F1}$ are generated simultaneously. These two generated FWM signals have the same frequency $\omega_{F1,2}(=\omega_1)$, but propagate in two different directions, which are monitored by a CCD camera (Figure 8.1(c)). In the experiment, the intensity of laser beam $\boldsymbol{E}'_1$ is about five times stronger than the beam $\boldsymbol{E}'_2$, and about 100 times stronger than the beams $\boldsymbol{E}_{1,2,3}$. According to the inset in Figure 8.1(c), with cross-Kerr effect, such horizontal alignment of strong dressing field $\boldsymbol{E}'_1$ and $\boldsymbol{E}'_2$ beams induce the horizontal shift of NDFWM $\boldsymbol{E}_{F2}$ and DFWM $\boldsymbol{E}_{F1}$, respectively [4]. The probe $\boldsymbol{E}_3$ beam is influenced by the combined effect of $\boldsymbol{E}'_1$ and $\boldsymbol{E}'_2$ beams but mainly shifted horizontally by the $\boldsymbol{E}'_1$ beam (Figure 8.3(a)). Thus, a pair of $\boldsymbol{E}_3$ and $\boldsymbol{E}_{F2}$ beams can be switched on and off by the $\boldsymbol{E}'_1$ beam, while one $\boldsymbol{E}_{F1}$ beam can be switched on and off by the $\boldsymbol{E}'_2$ beam at the same time.

The theoretical description of the spatial properties of the beams $E_{3,F1,F2}$ due to self- and cross-Kerr nonlinearities can be given by numerically solving the following propagation equations

$$\frac{\partial E_3}{\partial z} - \frac{i}{2k_3}\frac{\partial^2 E_3}{\partial \xi^2} = \frac{ik_3}{n_0}\left(n_2^{S1}\left|E_3\right|^2 + 2n_2^{X1}|E_1'|^2 + 2n_2^{X2}|E_2'|^2\right)E_3 \tag{8.1}$$

$$\frac{\partial E_{F1}}{\partial z} - \frac{i}{2k_{F1}}\frac{\partial^2 E_{F1}}{\partial \xi^2} = \frac{ik_{F1}}{n_0}\left(n_2^{S2}\left|E_{F1}\right|^2 + 2n_2^{X3}|E_1|^2 + 2n_2^{X4}|E_2'|^2 + 2n_2^{X5}|E_1'|^2\right)E_{F1} \tag{8.2}$$

$$\frac{\partial E_{F2}}{\partial z} - \frac{i}{2k_{F2}}\frac{\partial^2 E_{F2}}{\partial \xi^2} = \frac{ik_{F2}}{n_0}\left(n_2^{S3}\left|E_{F2}\right|^2 + 2n_2^{X6}|E_1'|^2 + 2n_2^{X7}|E_2|^2 + 2n_2^{X8}|E_2'|^2\right)E_{F2} \tag{8.3}$$

where $k_3 = k_{F1} = k_{F2} = \omega_1 n_0/c$. z and ξ are the longitudinal and transverse coordinates, respectively. n_0 is the linear refractive index at ω_1. n_2^{S1-S3} are the self-Kerr coefficients of $\boldsymbol{E}_{3,F1,F2}$ and n_2^{X1-X8} are the cross-Kerr coefficients of $\boldsymbol{E}_{1,2}$ and $\boldsymbol{E}'_{1,2}$, respectively. Generally, the Kerr coefficient can be defined by $n_2 = \mathrm{Re}\chi^{(3)}/(\varepsilon_0 c n_0)$, with the nonlinear susceptibility $\chi^{(3)} = D\rho_{10}^{(3)}$, where $D = N\mu_3^2\mu_{ij}^2/\hbar^3\varepsilon_0 G_3 G_j^2$, $\rho_{10}^{(3)}(E_{F1}) = -iG_{F1}|G_2|^2/\eta$, $\rho_{10}^{(3)}(E_{F2}) = -iG_{F2}|G_1|^2/\eta$, $\rho_{10}^{(3)}(E_3) = -iG_3|G_1|^2/\eta$, and $\eta = D_1^2 D_2$. $D_{1,2}$ are the parameters related to the Rabi frequency of the dressing field, the frequency detuning, and the atomic coherence rate. μ_3 (μ_{ij}) is the dipole matrix element between the states coupled by the probe beam E_3 (between $|i\rangle$ and $|j\rangle$). By assuming Gaussian profiles for the input fields, Eq. (8.1), Eq. (8.2), and Eq. (8.3) are solved by the split-step method.

8.1.3
Optical Switching and Routing via Spatial Shift

When four laser beams ($\boldsymbol{E}'_1$, $\boldsymbol{E}_2$, $\boldsymbol{E}'_2$, and $\boldsymbol{E}_3$) are on, in the presence of the dressing beam $\boldsymbol{E}'_1$, the spatial shift of the $\boldsymbol{E}_{F2}$ beam spot versus probe laser frequency detuning Δ_1 is shown in Figure 8.2(a). The moving trace of the light spot is dispersion-like as in frequency scans [4]. It means that the $\boldsymbol{E}_{F2}$ beam can have a right or left shift. There are two maximal displacements corresponding to the positive maximum nonlinear refraction coefficient and the negative maximum coefficient. Without the $\boldsymbol{E}'_1$ beam, the probe fields $\boldsymbol{E}_3$ and $\boldsymbol{E}_{F2}$ are single strong spots, as shown in Figure 8.3(a). When the dressing field $\boldsymbol{E}'_1$ is on, the intensities of the probe and $\boldsymbol{E}_{F2}$ beams become weaker [10] and are shifted (one to the right and the other to the left of the original position). As we use one more mirror in the probe beam scheme than that of $\boldsymbol{E}_{F2}$, they have a direction opposite to the shift on the CCD screen (Figure 8.3(a)). In fact, in the heated pipe both the beams have a right shift, as shown in Figure 8.3(b) and (c). A larger spatial shift occurs with an increasing $\boldsymbol{E}'_1$ intensity, which can be understood from the expression

$$\varphi_{NL}(z,\xi) = \frac{2k_{3,F2}n_2 I'_1 \exp(-\xi^2)z}{n_0} \tag{8.4}$$

The nonlinear phase shift φ_{NL} is directly proportional to the dressing intensity I'_1. The component of the wave vector of the $\boldsymbol{E}_{F2}$ spot δk_ξ (which we use to measure the shift effect of the optical switch) is the derivative of φ_{NL}, that is, $\delta k_\xi = \partial\varphi_{NL}/\partial\xi$, so the beam spots also move more as the dressing laser intensity increases.

Figure 8.2(b) shows the dressing field dependencies of the spatial shifts based on the numerical calculation and the experimental measurements. Figure 8.2(c)

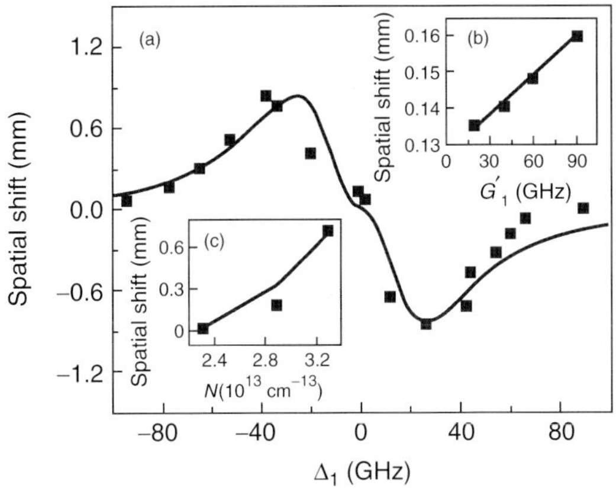

Figure 8.2 (a) Spatial dispersion curves of $\boldsymbol{E}_{F2}$ in the ladder-type three-level system versus Δ_1 with $G'_1 = 52$ GHz at 250 °C. (b) The spatial displacement of $\boldsymbol{E}_{F2}$ versus G'_1 in the ladder-type three-level system at $\Delta_1 = -18$ GHz and 250 °C. (c) The spatial displacement of $\boldsymbol{E}_{F2}$ versus atomic density N with $G'_1 = 52$ GHz at $\Delta_1 = -18$ GHz. The solid lines are theoretically calculated spatial shifts and the scattered points are the experimental results. Source: Adapted from Ref. [9].

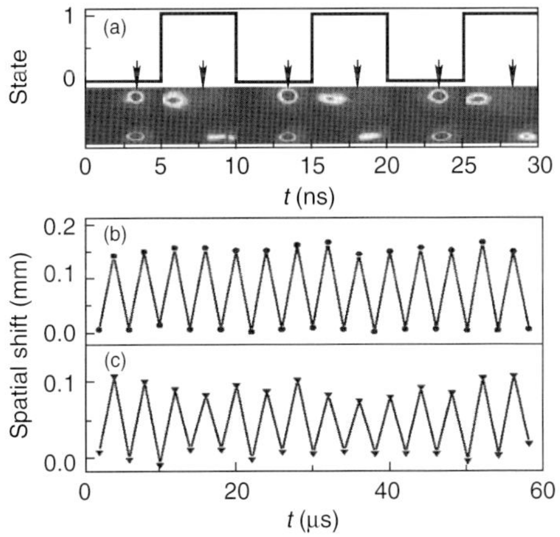

Figure 8.3 (a) Results of the optical switches and the spot shifts of the probe (lower) and $\boldsymbol{E}_{F2}$ (upper) beams obtained from the CCD at $\Delta_1 = -18$ GHz. The arrows are the initial position in x-direction. The spatial shift of (b) the probe and (c) $\boldsymbol{E}_{F2}$ beams in the ladder-type three-level atomic system with $G'_1 = 34$ GHz at $\Delta_1 = -18$ GHz and 250 °C.

presents the temperature dependence (atomic density N) of the shift curves for the theoretical and the experimental results, respectively. We see that increasing the atomic density equals to increasing propagation distance z, and the shift of the spot becomes larger.

So, as shown in the preceding text, the beam spots can have different spatial shifts with different experimental parameters (such as frequency, intensity, and atomic density), which can correspond to different on–off combinations. The switching or routing time is the rising and falling times of the switch-in and switch-out signal. The cross-Kerr refractive index change ($n_2 \propto \mathrm{Re}(\rho_{10}^{(3)})$) limited by the overall spin dephasing time determines the response time of the switch [3, 11, 12]. The estimated switching times of $\boldsymbol{E}_{\mathrm{F1}}$ and $\boldsymbol{E}_{\mathrm{F2}}$ are about 32 and 400 ns, respectively. Here, it should be noted that the overall spin dephasing times of the two-level (Figure 8.1(b)) and ladder-type three-level (Figure 8.1(a)) atomic systems in sodium are determined by the transverse relaxation rates, $1/(2\pi\Gamma_{10})$ and $1/(2\pi\Gamma_{20})$, where $\Gamma_{10} = 4.85$ MHz and $\Gamma_{20} = 398$ kHz for transitions $|0\rangle - |1\rangle$ and $|0\rangle - |2\rangle$, respectively. However, the switching speeds in Figure 8.3 are limited to a microsecond time scale by the speed of the CCD used to take the image.

Figure 8.3(a) shows the two states of the probe and $\boldsymbol{E}_{\mathrm{F2}}$ beams by switching the strong laser beam $\boldsymbol{E}_1'$ off and on as the laser frequency detuning is tuned to get the maximal spatial displacement. When a spot stays at its initial position, it means that the switch is in the "off" state. When the frequencies of the probe and $\boldsymbol{E}_{\mathrm{F2}}$ beams are set at their peak shift positions, the light spots will have their largest shifts, so the switch stands at its "on" state. Such two states form two ports of the optical switch. The upper spot is the $\boldsymbol{E}_{\mathrm{F2}}$ beam and the lower spot is the probe beam. Initially, two spots are set at the same vertical line without the dressing laser beam. As the dressing beam $\boldsymbol{E}_1'$ turns on, the upper spot moves to the left side and the lower spot moves to the right, both of which leave their initial positions completely. The switching contrast can be defined as $C = (I_{\mathrm{off}} - I_{\mathrm{on}})/(I_{\mathrm{on}} + I_{\mathrm{off}})$, where I_{off} is the light intensity at the "off" state and I_{on} is the light intensity at the "on" state. The contrast derived from the experiment is about $C = 92\%$. This experiment provides a physical mechanism to realize an all-optical switching/routing by controlling the dressing laser beam.

A chopper is used to control the dressing field, subtracting the laser pulse repetition time of 0.1 s, which is considered an idle load state. The laser pulse width is 5 ns. The detected switching time is limited by the response time of the CCD, which is about 3 μs, far larger than the laser pulse width. Thus, the switching speed in the current experiment is greatly constrained as shown in Figure 8.3(b) and (c). The on-state just lasts 5 ns, followed by a 3 μs rising time, and then a 5 ns off-state, followed by a 3 μs falling time, and so on. As the spatial displacements of the probe and $\boldsymbol{E}_{\mathrm{F2}}$ beams are mainly determined and controlled by the large cross-Kerr nonlinear coefficients of the strong laser field $\boldsymbol{E}_1'$, the switching speed should be much faster and limited by the atomic coherence time in the nanosecond time scale.

Next, when the five laser beams ($\boldsymbol{E}_1$, $\boldsymbol{E}_1'$, $\boldsymbol{E}_2$, $\boldsymbol{E}_2'$, and $\boldsymbol{E}_3$) are all on, there are interplays between the generated $\boldsymbol{E}_{\mathrm{F1,F2}}$ signals [10] and we can control the shifts of the probe, $\boldsymbol{E}_{\mathrm{F1}}$ and $\boldsymbol{E}_{\mathrm{F2}}$ beams, to achieve a triple binary optical switch. The initial locations of the spots are the "off" states and the switches are considered to come to their "on" states when the spots shift away to new locations. The repetition frequency of the chopper is much longer than the 5 ns pulse width of the dressing laser, so the "on" state lasts several 5 ns intervals and then turns to the "off" state. In Figure 8.4 at $\Delta_1 = -18$ GHz for the self-focusing side and

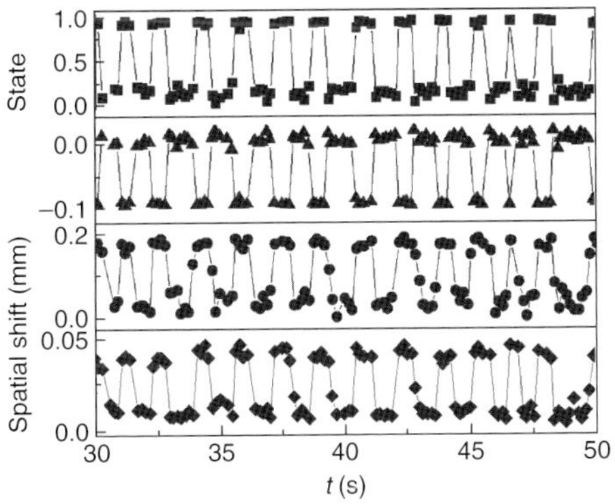

Figure 8.4 The switching processes of the dressing beam E_1' (square), E_{F1} (triangle), E_{F2} (circle), and the probe beam (diamond) in the ladder-type three-level system with $G_1' = 21$ GHz at $\Delta_1 = -18$ GHz and 250 °C. Source: Adapted from Ref. [9].

temperature 250 °C, when E_1' is on, the probe and E_{F2} beams have a right shift due to the E_1' beam via the cross-Kerr nonlinear coefficients. At the same time, the E_{F1} beam is shifted to the left by the dressing field E_2'. When E_1' is off, all the beams come back to their original position ("off" state). As the cross-Kerr nonlinear coefficients n_2^{X4} and n_2^{X6} (n_2^{X1}) of the E_{F1} and E_{F2} (probe) beams induced by the dressing fields E_2' and E_1' are positive, the spots of the E_{F1} and E_{F2} (E_3) beams are shifted to the opposite directions, as shown in Figure 8.4. According to the nonlinear phase shifts $\varphi_{F1} = 2k_{F1}n_2^{X4}|A_2'|^2 z_0/n_0$ and $\varphi_{F2} = 2k_{F2}n_2^{X6}|A_1'|^2 z_0/n_0$ ($\varphi_3 = 2k_3 n_2^{X1}|A_1'|^2 z_0/n_0$), induced by the dressing fields E_2' and E_1', respectively, we can use two controllable parameters, that is, the frequency and intensity of the laser, to control the different shifts of the three spots. Such simultaneous optical switching for three beams can perform the functions of choosing different addresses in data transmissions and can be used as the optical routings, the multiplexer, or all-optical switching arrays for all-optical networks.

In this discussion, we have controlled the probe, E_{F1} and E_{F2} by two dressing fields E_2' and E_1', respectively. In that case, E_{F1} and E_{F2} are shifted toward the opposite directions (Figure 8.4). Actually, such three beams can also be shifted to the same direction when the sign of the cross-Kerr nonlinear coefficient of the E_{F1} signal is opposite to those of the E_{F2} (probe) beams at the proper laser detuning. So, each spot can have left and right locations. Including the initial position, every spot has three possible spatial locations. Totally, there are 3×3 controllable spatial positions. It can achieve a switch array.

The advantages of solids include high density of atoms, compactness, and absence of atomic diffusion, but with relatively broad optical line widths and fast decoherence rates. However, there are still more advantages in studying all-optical switching, especially spatial all-optical switching and routing, using multi-level atomic media via EIT (or atomic coherence)-related effects. The current atomic experiment has several easily tunable experimental parameters (such as laser intensities, frequency detuning, and atomic density), which provide a much better platform (compared to the solid systems) to study the formation and dynamics of the novel spatial optical switch and router protocols. Also, there is a narrow line width (compared to the solid systems) in an atomic media.

8.2 All-Optical Routing and Space Demultiplexer

As the switching speed of traditional electronic switching devices cannot increase along with the communication bandwidth, high-speed all-optical switching devices have been intensively studied. To solve this problem, there have been several new schemes reported recently to demonstrate such all-optically controlled switching and routing functions [1–3, 9, 13]. In coherently prepared atomic media, controlling the linear [2, 7] and nonlinear [3, 5, 8, 9, 11, 14] optical absorptions of one laser beam by another was exploited to show all-optically controlled beam switching [15]. Also, making use of the EISD effect in a three-level ladder-type atomic system, all-optical switching or routing effect is experimentally demonstrated [9, 11]. Currently, most experimental studies on the channel multiplexing technique have been carried out in time [11, 16] or wavelength [17]. In our experiment, the routers can be realized in space through the spatial splitting of FWM and probe beams. Spatially splitting one weak laser beam by another stronger beam in Kerr nonlinear optical media via XPM [4, 8] were predicted and demonstrated by a series of impressive experiments [4]. The number of the splitting beams can be controlled by the intensities, polarization states, and frequency detuning of the involved laser beams, and the atomic density in the EIT systems [5]. Compared with other experiments, the ability we demonstrate to release signals in space is more suitable for routing applications.

In this section, we show the spatial all-optical routing and demultiplexing by making use of the electromagnetically induced spatial splitting (EISS) effect in the two- and three-level atomic systems. Such controllable spatial splitting can provide potential architectures for beam address selecting and space demultiplexing in all-optical communication and networks.

8.2.1 Theoretical Model and Experimental Scheme

Three energy levels ($|0\rangle$ ($3S_{1/2}$), $|1\rangle$ ($3P_{1/2}$), and $|2\rangle$ ($3P_{3/2}$)) from Na atoms (in a heat-pipe oven) are involved in the experiments. Here, we take the three-level V-type system, for example. Coupling fields $\boldsymbol{E}_1$ (frequency ω_1, detuning Δ_1, and wave vector $\mathbf{k}_1$) and $\boldsymbol{E}_1'$ (ω_1, Δ_1, and $\mathbf{k}_1'$) connect the transition between levels $|0\rangle$ and $|1\rangle$, which are from the same near-transform-limited dye laser. Coupling fields $\boldsymbol{E}_2$ (ω_2, Δ_2, and $\mathbf{k}_2$) and $\boldsymbol{E}_2'$ (ω_2, Δ_2, and $\mathbf{k}_2'$) connect the transition between levels $|0\rangle$ and $|2\rangle$ with $\boldsymbol{E}_2$ propagating in the same direction of $\boldsymbol{E}_1$, which are from another near-transform-limited dye laser. From the same laser with $\boldsymbol{E}_1$ and $\boldsymbol{E}_1'$ also connecting the transition between levels $|0\rangle$ and $|1\rangle$, the weak probe field $\boldsymbol{E}_3$ (ω_1,

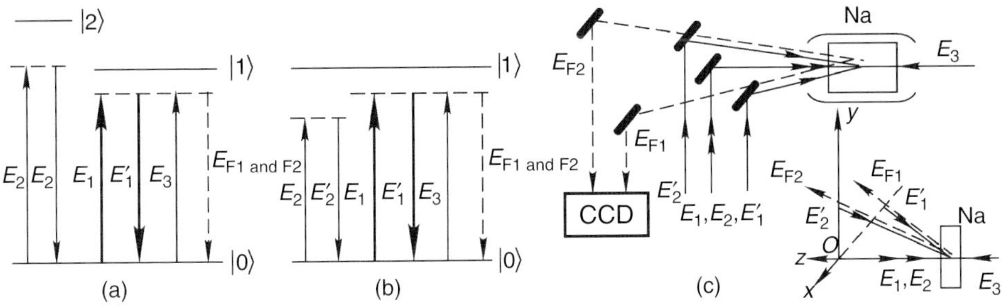

Figure 8.5 The diagrams of Na energy levels with different coupling schemes in (a) the three-level V-type system and (b) the two-level system. (c) Spatial beam geometry used in this experiment, in which the upper part is the setup diagram, and the lower part is the wave vector diagram in a Cartesian coordinate.

Δ_1, and $\mathbf{k}_3$) propagates in the direction opposite to the field E_2. E_1' propagating in the xz-plane deviates from E_1 with a small angle (~0.3°), while E_2' propagating in the yz-plane deviates from E_2 with a small angle (~0.3°). When E_1, E_1', and E_3 are turned on simultaneously with E_2 and E_2' blocked, a DFWM process is generated satisfying the phase-matching condition $\mathbf{k}_{F1} = \mathbf{k}_1 - \mathbf{k}_1' + \mathbf{k}_3$. Similarly, we can get an NDFWM process with $\mathbf{k}_{F2} = \mathbf{k}_2 - \mathbf{k}_2' + \mathbf{k}_3$. Here, we define detuning as $\Delta_i = \Omega_i - \omega_i$ ($i = 1, 2$) with the atomic resonant frequency Ω_i.

When the beams E_2 and E_2' are tuned to $|0\rangle - |1\rangle$, the system turns to the two-level system (Figure 8.5(b)), which generates different DFWM processes E_{F1} and E_{F2} with the phase-matching conditions $\mathbf{k}_{F1} = \mathbf{k}_1 - \mathbf{k}_1' + \mathbf{k}_3$ and $\mathbf{k}_{F2} = \mathbf{k}_2 - \mathbf{k}_2' + \mathbf{k}_3$, respectively. Moreover, the images of E_{F1} and E_{F2} in this system can be captured by CCD.

Under our experimental conditions, the Na vapor is a Kerr medium. The intensity of laser beam E_1 (E_1') is approximately eight times stronger than the beam E_2 (E_2') and 100 times stronger than the weak beam E_3 and two FWM beams. So, beams E_1 (E_1') and E_2 (E_2') are strong enough to control the beams E_{F1}, E_{F2}, and E_3. The theoretical description of the spatial propagation of the beams $E_{3,F1,F2}$ with the self- and cross-Kerr nonlinearities can be given by numerically solving the following propagation Eq. (8.5), Eq. (8.6), and Eq. (8.7) in both the three-level V-type system and two-level systems

$$\frac{\partial A_3}{\partial z} - \frac{i}{2\mathrm{k}_3}\frac{\partial^2 A_3}{\partial \xi^2} = \frac{i\mathrm{k}_3}{n_0}\left(n_2^{S1}|A_3|^2 + 2n_2^{X1}|A_1'|^2 + 2n_2^{X2}|A_1|^2 + 2n_2^{X3}|A_2|^2\right)A_3 \quad (8.5)$$

$$\frac{\partial A_{F1}}{\partial z} - \frac{i}{2\mathrm{k}_{F1}}\frac{\partial^2 A_{F1}}{\partial \xi^2} = \frac{i\mathrm{k}_{F1}}{n_0}\left(n_2^{S2}|A_{F1}|^2 + 2n_2^{X4}|A_1'|^2 + 2n_2^{X5}|A_1|^2\right)A_{F1} \quad (8.6)$$

$$\frac{\partial A_{F2}}{\partial z} - \frac{i}{2\mathrm{k}_{F2}}\frac{\partial^2 A_{F2}}{\partial \xi^2} = \frac{i\mathrm{k}_{F2}}{n_0}\left(n_2^{S3}|A_{F2}|^2 + 2n_2^{X6}|A_1'|^2 + 2n_2^{X7}|A_2'|^2\right)A_{F2} \quad (8.7)$$

where, z and ξ are the longitudinal and transverse coordinates (ξ stands for the x- or y-direction), respectively. n_2^{S1-S3} are the self-Kerr coefficients of $E_{3,F1,F2}$, n_2^{X1-X7}

are the cross-Kerr coefficients of $\boldsymbol{E}_{1,2}$ and $\boldsymbol{E}'_{1,2}$. $A_{3,\mathrm{F1,F2}}$, $A'_{1,2}$, and $A_{1,2}$ are the slowly varying envelope amplitudes of the beams $\boldsymbol{E}_{3,\mathrm{F1,F2}}$, $\boldsymbol{E}_{1,2}$, and $\boldsymbol{E}'_{1,2}$, respectively; $\mathrm{k}_3 = \mathrm{k}_{\mathrm{F1}} = \mathrm{k}_{\mathrm{F2}} = \omega_1 n_0/c$, and n_0 is the linear refractive index at ω_1. Generally, the Kerr coefficient can be obtained as $n_2 = \mathrm{Re}\chi^{(3)}/(\varepsilon_0 c n_0)$, with the nonlinear susceptibility $\chi^{(3)} = D\rho_{10}^{(3)}$, where $D = N\mu_3^2\mu_{ij}^2/\hbar^3\varepsilon_0 G_3 G_j^2$, $\rho_{10}^{(3)}(E_3) = -iG_3|G_1|^2/\eta$, $\rho_{10}^{(3)}(E_{\mathrm{F1}}) = -iG_{\mathrm{F1}}|G_2|^2/\eta$, $\rho_{10}^{(3)}(E_{\mathrm{F2}}) = -iG_{\mathrm{F2}}|G_1|^2/\eta$, and $\eta = D_1^2 D_2$. D_1 and D_2 are the parameters related to the Rabi frequency of the dressing fields, the frequency detuning, and the atomic coherence rates. μ_3 (μ_{ij}) is the dipole matrix element between the states coupled by the probe beam $\boldsymbol{E}_3$ (between $|i\rangle$ and $|j\rangle$). With the assumption that the input lasers are Gaussian, Eq. (8.5), Eq. (8.6), and Eq. (8.7) are solved by the split-step method.

The spatial splitting of the weak beam is caused by the noncollinear propagations of the strong laser beams and the atomic coherence-enhanced cross-Kerr nonlinear indices of refraction. For simplicity, let us only consider the XPM from the two strong controlling beams $\boldsymbol{E}'_{1,2}$. During their propagation through the vapor cell, the wing of the beam $\boldsymbol{E}'_{1,2}$ interacts with the intensity profile of either $\boldsymbol{E}_3$ or $\boldsymbol{E}_{\mathrm{F1,F2}}$ and distorts its phase profile to induce an optical waveguide by XPM. It can be understood from the expression

$$\phi_{\mathrm{NL}}(z,\xi) = \frac{2\mathrm{k}_{\mathrm{F1,2}} n_2 I_{1,2} \mathrm{e}^{-r\xi^2 z}}{n_0 I_0} \tag{8.8}$$

The change of the nonlinear phase shift (ϕ_{NL}) in the propagating expression of a laser determines its spatial displacement and splitting. Specifically, the transverse propagation direction is determined by the additional transverse propagation wave-vector $\delta\mathbf{k}_\xi = (\partial\phi_{\mathrm{NL}}/\partial\xi)\hat{\xi}$ ($\hat{\xi}$ is the unit vector in the transverse axis). Moreover, the differences in $\partial^2\phi_{\mathrm{NL}}/\partial\xi^2$ of different spatial parts of $\boldsymbol{E}_{\mathrm{F1,F2}}$ and $\boldsymbol{E}_3$ beams can lead to the local splitting. When considering the higher order differential $\partial^n\phi_{\mathrm{NL}}/\partial\xi^n$ ($n \geq 3$), we will further get multi-splitting beams. According to the expression of ϕ_{NL}, the amount of spatial splitting is proportional to the cross-Kerr nonlinear coefficient, the field intensity, and the propagation distance.

8.2.2
Optical Switching and Routing

Figure 8.6(a) and (b) shows the creation of two switchable states of both $\boldsymbol{E}_3$ and $\boldsymbol{E}_{\mathrm{F2}}$ beams by controlling the strong laser beam $\boldsymbol{E}_1$ off and on when the beam $\boldsymbol{E}'_1$ is off. In the experiment, we use a chopper with 500 Hz to modulate the switching beams, which is not synchronous with a pulse delay time of 0.1 s. Here, we choose a suitable frequency detuning in self-focusing media to get the obvious spatial displacement. The beam $\boldsymbol{E}_{\mathrm{F2}}$ is spatially located at the upper side of $\boldsymbol{E}_1$ and the beam $\boldsymbol{E}_3$ is located at the lower side of $\boldsymbol{E}_1$.

When $\boldsymbol{E}_1$ is off, the $\boldsymbol{E}_3$ and $\boldsymbol{E}_{\mathrm{F2}}$ beams stay in their initial positions, and here we define these states as the "off" state of the switch. When the dressing field $\boldsymbol{E}_1$ is on, the $\boldsymbol{E}_3$ beam is shifted up while the $\boldsymbol{E}_{\mathrm{F2}}$ beam is shifted down. We define these new states as the "on" state. Each "on" state lasts about 1.2 s, and contains 12

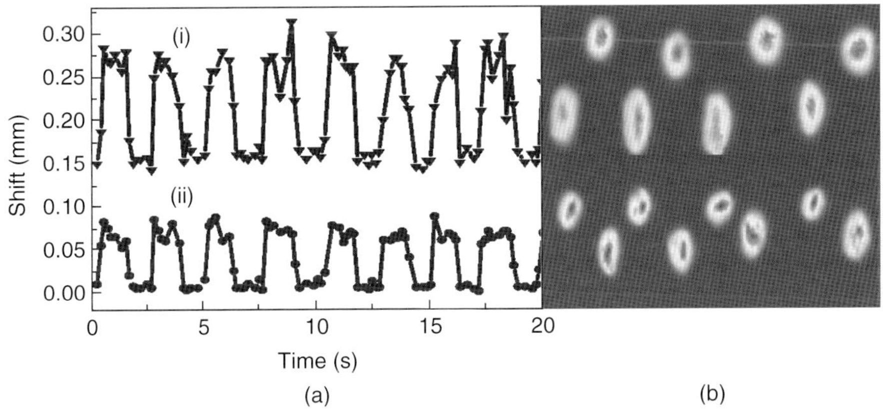

Figure 8.6 (a) Spatial shift of (i) $\boldsymbol{E}_3$ and (ii) $\boldsymbol{E}_{F2}$ in the three-level V-type system with $G_1' = 34$ GHz at 290 °C with $\Delta_1 = -18$ GHz. (b) Results of the optical switches and the spot shifts of the $\boldsymbol{E}_3$ (upper) and $\boldsymbol{E}_{F2}$ (lower) beams obtained from the CCD. Source: Adapted from Ref. [18].

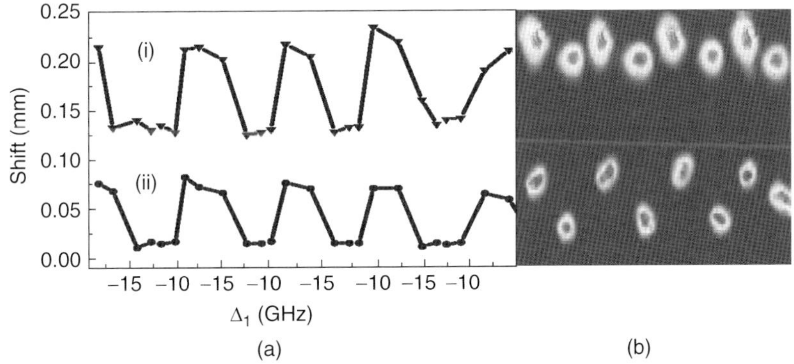

Figure 8.7 (a) Spatial shift of (i) $\boldsymbol{E}_3$ and (ii) $\boldsymbol{E}_{F1}$ in the three-level V-type system with $G_1' = 52$ GHz at 290 °C with $\Delta_1 = -15$ GHz. (b) Results of the optical switches and the spot shifts of the $\boldsymbol{E}_3$ (upper) and $\boldsymbol{E}_{F1}$ (lower) beams obtained from the CCD.

pulses. By such physical mechanism of controlling the dressing beam, all-optical switching is available.

In the presence of the dressing beam $\boldsymbol{E}_1'$, the spatial displacements of $\boldsymbol{E}_3$ and E_{F1} beam spots are shown in Figure 8.7(a) as Δ_1 changing between −15 and − 10 GHz back and forth. $\boldsymbol{E}_3$ and $\boldsymbol{E}_{F1}$ beams are both located at the lower side of $\boldsymbol{E}_1'$ (shown in Figure 8.5(c)). In our experiment, when $\Delta_1 < 0$, the control beam $\boldsymbol{E}_1'$ can attract the weaker beams $\boldsymbol{E}_3$ and $\boldsymbol{E}_{F1}$, as shown in Figure 8.7(b). So, both the beams $\boldsymbol{E}_3$ and $\boldsymbol{E}_{F1}$ are shifted upward to the strong beam $\boldsymbol{E}_1'$. At first, we set $\Delta_1 = -15$ GHz, and define the initial position as the "off" state, which has little displacement. Then we turn the detuning to $\Delta_1 = -10$ GHz, $\boldsymbol{E}_3$ and $\boldsymbol{E}_{F1}$ beams are shifted to the beam center of $\boldsymbol{E}_1'$. Here, the $\boldsymbol{E}_{F1}$ beam has a larger shift, and the switch is set to stand at its "on" state.

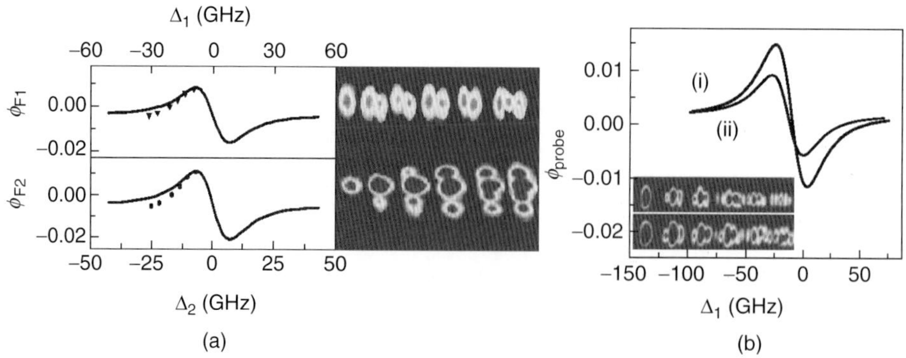

Figure 8.8 (a) The EISS of $\mathbf{E}_{F2}$, $\mathbf{E}_{F1}$ beams with dressing beams $\mathbf{E}'_1$, $\mathbf{E}'_2$ versus different frequency detuning at 300 °C. The scattered points are the experimental results (the triangle points for $\mathbf{E}_{F1}$ and the circle points for $\mathbf{E}_{F2}$) and the solid lines are the theoretically calculated nonlinear phase shift ϕ_{F1} and ϕ_{F2}. The experimental spots in (b) show the splitting of the $\mathbf{E}_{F1}$ and $\mathbf{E}_{F2}$ beams. (c) The EISS of beam $\mathbf{E}_3$ at 290 °C. The solid lines are theoretically calculated nonlinear phase shift ϕ_{probe} in the (i) two- and (ii) three-level systems. The upper and lower insets correspondingly show the experimental spots of probe beam $\mathbf{E}_3$ in the two- and three-level system, respectively. The other parameters are $G'_1 = 47$ GHz and $G'_2 = 5.9$ GHz.

In these two experiments, the spatial displacement feature of the probe and FWM beams can be used to realize spatial optical switching. In the following experiments, we see that the splitting of the probe and FWM beams can be used for all-optical routing and space demultiplexing. Next, we consider the influence of the alternative atomic energy level systems, power, temperature, and polarization configurations on the splitting of probe and FWM beams.

Figure 8.8(a) shows the spatial splitting of the FWM beams in the three-level V-type system. In the self-focusing media, because the $\mathbf{E}_{F1}$ beam overlaps with $\mathbf{E}'_1$ in the x-direction, it is split in this direction. Similarly, the $\mathbf{E}_{F2}$ beam overlaps with $\mathbf{E}'_2$ in the y-direction and then is split in that direction. This phenomenon is derived from the relative positions between the weak beams and the strong dressing (controlling) beams shown in Figure 8.5(c). In our experiment, the frequency detuning Δ_1 and Δ_2 change from −40 to −10 GHz, respectively, which lead to the obvious increase in the nonlinear phase shift ϕ_{NL} due to the increasing nonlinear coefficient n_2. In Figure 8.8(c), we compare the different splitting characteristics of the probe beam $\mathbf{E}_3$ in the two-level and three-level atomic systems. The spots inset show the spatial splitting images of the probe beam. With $\mathbf{E}_1$ ($\mathbf{E}_2$) tuned to the transition between $|0\rangle$ and $|1\rangle$ ($|0\rangle$ and $|2\rangle$) in the three-level V-type system as shown in Figure 8.5, the probe beam splitting shows an EISS curve. When the probe beam $\mathbf{E}_3$ is tuned to the transition between $|0\rangle$ and $|1\rangle$ in the two-level system, the spatial splitting also shows a similar EISS curve. As the dipole moment $\mu_{20} = 1.44 \times 10^{-29}$ C·m is larger than $\mu_{10} = 1.41 \times 10^{-29}$C·m, we can deduce that the cross-Kerr coefficient n_2 in the two-level system is larger than that in the three-level V-type system. From the expression (8.8), we can understand that ϕ_{NL}

is larger in the two-level system. As a result, the probe beam $\boldsymbol{E}_3$ in the two-level system is split into more spots than that in the three-level system. Each splitting state just lasts 5 ns (from a 5 ns single pulse), and the positive edge time and the negative edge time can be limited in 100–300 ps. This provides us a possible way to realize the hyperterahertz all optical switching.

Considering the physical meaning of the expression of the nonlinear phase, we can conclude that the direction of the additional transverse propagation wave vector δk_ξ, determines the transverse propagation direction of the beam. The spatial splitting resulting from the cross-Kerr effect in the propagation could lead to the multi-stripes profile of the probe and FWM beams. As different spatial parts of the weak $\boldsymbol{E}_{\rm F}$ and $\boldsymbol{E}_3$ beams have different δk_ξ, which lead to different transverse propagation speeds (both in value and direction), the global spatial splitting of $\boldsymbol{E}_{\rm F}$ and $\boldsymbol{E}_3$ beams will appear. Then, the differences in $\partial^2\phi_{\rm NL}/\partial\xi^2$ of different parts of $\boldsymbol{E}_{\rm F}$ and $\boldsymbol{E}_3$ beams can further lead to the local splitting and make more peaks in this beam. With some approximations, one can get $u_{\rm F1,F2,E3} \propto A\cos\phi_{\rm NL}$, where A is a function related to the wavelength and intensity of the dressing fields. The extent of splitting spots is influenced by the parameter in $u_{\rm F1,F2,E3}$, which stands for the changing of the beam's field strength. Obviously, the intensities of the beams change periodically according to the item $I_{\rm F1,F2,E3} \propto \cos^2(\phi_{\rm NL})$. When $\phi_{\rm NL}$ increases, the period of the intensities will decrease and the split spots will become denser.

Figure 8.9 shows the enhancement of the splitting of the $\boldsymbol{E}_{\rm F2}$ beam via the increasing intensities and temperatures in the three-level V-type system. We can explain the spatial spitting of the $\boldsymbol{E}_{\rm F2}$ beam, considering the character of the nonlinear phase shift equation in various levels under different experiment temperatures and intensities. First, when $T = 300\,^\circ$C, we increase the intensity I_2 of the $\boldsymbol{E}_2'$ beam gradually from 2 to 18 μW, and the value of $\phi_{\rm NL}$ will significantly increase according to Eq. (8.8). Second, when $I_2 = 15$ μW, we make the temperature rise from 280 to 310 °C to increase the atomic density N, and the nonlinear phase $\phi_{\rm NL}$ will also increase as $n_2 = {\rm Re}\chi^{(3)}/(\varepsilon_0 c n_0)$, $\chi^{(3)} = D\rho_{10}^{(3)}$, and $D = N\mu_3^2\mu_{ij}^2/\hbar^2\varepsilon_0 G_3 G_j^2$. With the expression $I_{\rm F1,F2,E3} \propto \cos^2(\phi_{\rm NL})$, we can predict that the period will be

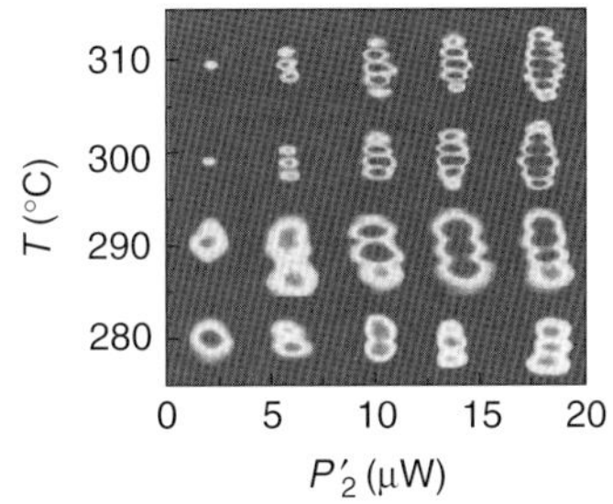

Figure 8.9 The experimental spots showing the different splitting effect by changing the intensity of the dressing beam $\boldsymbol{E}_2'$ and the temperature of the medium in the heat-pipe oven in the three-level V-type system. Source: Adapted from Ref. [18].

decreased and the split spots become denser, which is experimentally verified in Figure 8.9.

By increasing the intensity of the strong control laser beam and the temperature of the medium, enhanced splitting of the FWM beam is observed. For the all-optical routing and space demultiplexer, the conversion efficiency can be easily enhanced by optimizing those parameters that provides us more ways to improve the performance of the all-optical devices.

Next, the dependence of the splitting on the controlling field polarization is investigated. We use one QWP to control the polarizations of the coupling field $\boldsymbol{E}_1$ in the two-level system. The S-polarized component of $\boldsymbol{E}_1$ can then be decomposed into balanced left- and right-hand circularly polarized parts, while the P-polarized component remains linearly polarized. The P- and S-polarized FWM beams generated from different transition pathways among various Zeeman sublevels can be obtained as shown in the polarization configuration in Figure 8.5(c). Then, the generated $\boldsymbol{E}_\mathrm{F}$ beam is decomposed into P- and S-polarized components by a polarization beam splitter (PBS), which is denoted as $\boldsymbol{E}_\mathrm{F}^\mathrm{S}$ and $\boldsymbol{E}_\mathrm{F}^\mathrm{P}$, respectively.

When changing the polarization of $\boldsymbol{E}_1$, we can obtain the expression for $\boldsymbol{E}_\mathrm{F}^\mathrm{P}$ intensity, $I^\mathrm{P} \propto I_\mathrm{F}(\sin^4\theta + \cos^4\theta)$, where $I_\mathrm{F} \propto |\rho_{10}^{(3)}|^2$ with $\rho_{10}^{(3)} = -iG_3G_1(G_1)^*/(F_1^2F_2)$, and θ is the polarization angle of $\boldsymbol{E}_1$. Similarly, there exists $I^\mathrm{S} \propto I_\mathrm{F}\sin^2\theta\cos^2\theta$. Also, one can solve the coupled density matrix equations to obtain $\rho_{10}^{(3)}$ for n_2^P induced by the strong field E', $n_2^\mathrm{P} \propto n_2^a(\sin^4\theta + \cos^4\theta)$, where $n_2^a \propto \mathrm{Re}(-iG_\mathrm{F}^\mathrm{P}G_\mathrm{a}^\mathrm{P})$. Similarly, we can get $n_2^\mathrm{S} \propto n_2^b\sin^2\theta\cos^2\theta$ with $n_2^b \propto \mathrm{Re}(-iG_\mathrm{F}^\mathrm{S}G_a^\mathrm{S})$. The other coefficients in these expressions are $F_a^\mathrm{P} = (G_1^0)^2/F_1^2F$ and $F_a^\mathrm{S} = (G_1^\pm)^2/F_1^2F$ for $\boldsymbol{E}_\mathrm{F}^\mathrm{P}$ and $\boldsymbol{E}_\mathrm{F}^\mathrm{S}$, respectively. Here, F_i ($i = 1, 2$) is the function of the detuning and relaxations for different levels of perturbation chains, and $G_1^\mp$ is the Rabi frequency of the right- and left-hand circularly polarized component of the $\boldsymbol{E}_1$ beam.

In Figure 8.10(b), we can see the different splitting characteristics of the $\boldsymbol{E}_\mathrm{F}$ beam by changing the rotation angle θ and observe that the $\boldsymbol{E}_\mathrm{F}$ beam is split into more parts with certain polarization states of the $\boldsymbol{E}_1$ beam in the period of 90°. Here, we just take $\boldsymbol{E}_\mathrm{F}^\mathrm{S}$ for example. The spot splitting curves have peaks at ±45° (i), and the weakest y-direction splitting appears at the 0° and 90° ($\boldsymbol{E}'$ is located at the lower side of $\boldsymbol{E}_\mathrm{F}$ in the y direction) correspondingly. We can also obtain that the values of $n_2^\mathrm{S}I_1^\mathrm{S}/I_0^\mathrm{S}$ at ±45° are larger than that at 0° and 90°. In the self-focusing medium, $n_2^\mathrm{S}I_1^\mathrm{S}$ is a constant and I_0^S turns to its minimum at ±45°, where $I_1 \propto (E_1)^2$ and $I_0 \propto (E_\mathrm{F}^\mathrm{S})^2$, which makes ϕ reaches its maximum ((ii) in Figure 8.10(b)).

The $\boldsymbol{E}_\mathrm{F}^\mathrm{P}$ beam shows similar periodic splitting as the $\boldsymbol{E}_\mathrm{F}^\mathrm{S}$ beam. However, it can be observed that the intensities and splitting of $\boldsymbol{E}_\mathrm{F}^\mathrm{P}$ and $\boldsymbol{E}_\mathrm{F}^\mathrm{S}$ are different. The nonlinear phases ϕ^S and ϕ^P change with a period of 90° as the polarization states of the $\boldsymbol{E}_1$ beam changing, respectively. Here, the dipole moments of the transitions between $|0\rangle$ and $|1\rangle$ for the S-polarized laser beams are different from those for the P-polarized beams because different transition pathways among various Zeeman sublevels are chosen for these two polarization types, resulting in $I_0^\mathrm{S} < I_0^\mathrm{P}$ for $|\rho_{10}^\mathrm{S}|^2 < |\rho_{10}^\mathrm{P}|^2$. Comparing $\boldsymbol{E}_\mathrm{F}^\mathrm{P}$ with $\boldsymbol{E}_\mathrm{F}^\mathrm{S}$, we can find $n_2^\mathrm{S}I_1^\mathrm{S}I_0^\mathrm{S} > n_2^\mathrm{P}I_1^\mathrm{P}I_0^\mathrm{P}$. Hence, we can easily arrive at the conclusion that the nonlinear phase ϕ^S is larger than ϕ^P, and therefore the splitting of $\boldsymbol{E}_\mathrm{F}^\mathrm{S}$ is stronger than that of $\boldsymbol{E}_\mathrm{F}^\mathrm{P}$. These results match

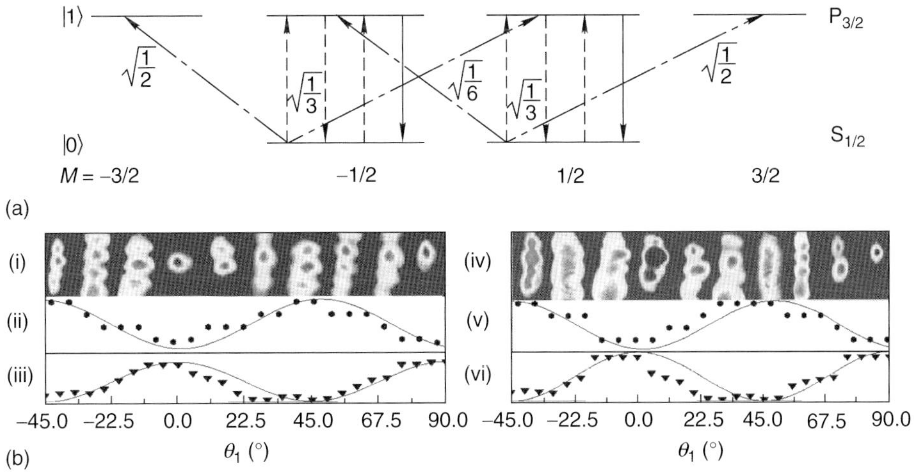

Figure 8.10 (a) The relevant Zeeman levels and various transition pathways. Solid line: polarized beam G_F; dashed lines: the linearly polarized coupling fields G_1 and G_1'; dash-dotted lines: the circularly polarized coupling fields $G_1^{\pm}$; dotted line: the P-polarized probe field G_3. (b) The different splitting images of $\boldsymbol{E}_F^S$ (i) and $\boldsymbol{E}_F^P$ (iv), splitting number ((ii) and (v)), and the intensity of the $\boldsymbol{E}_F^S$ (iii) and $\boldsymbol{E}_F^P$ (vi) components versus different polarization states of $\boldsymbol{E}_1$ beam. The solid dots are the measured results and the solid curves are the fitted theoretical ϕ values. Source: Adapted from Ref. [18].

reasonably well with the theoretical calculations based on the coupled propagation Eq. (8.5), Eq. (8.6), and Eq. (8.7) for the S- and P-polarized FWM components.

By changing the polarization of $\boldsymbol{E}_1$, we get two periodic splitting of $\boldsymbol{E}_F^S$ and $\boldsymbol{E}_F^P$. When the rotation angle θ changes in a period of 90°, the $\boldsymbol{E}_F^S$ ($\boldsymbol{E}_F^P$) beam (optical information) as one light channel can be transferred (or distributed) into three or four different light channels. Such a study of controlling spatial splitting provides a selective way to choose the spatial multiple channels.

Figure 8.11(a) and (b) shows the experimentally measured $\boldsymbol{E}_{F1}$ beam splitting in the y direction in the two-level system. As shown in Figure 8.11(a) and (b), we choose the suitable rotation angle θ of the $\boldsymbol{E}_1$ beam to obtain proper spatial splitting. We use a polarizer with a 10 Hz repetition frequency to change the polarization states of $\boldsymbol{E}_1$ beam in the experiment, but the positive and negative edge times can be 100–300 ps. As θ repeatedly changes between 0° (or 22.5°) and 45°, the $\boldsymbol{E}_{F1}$ beam also is repeatedly transferred between the single-spot state (or two-spot state) and the state of being split into three spots. The splitting number of $\boldsymbol{E}_{F1}$ can be controlled flexibly as shown in Figure 8.11(a) and (b).

When the $\boldsymbol{E}_{F1}$ beam carries information, the original information from one channel (1 × 1) can be distributed into three channels (3 × 1) by changing the polarization states of the strong field in Figure 8.11(b). Such controllable splitting effects give us a way to obtain the space demultiplexer and router with 3 × 1 channels. In addition, we can also change the 2 × 1 channels to 3 × 1 channels easily.

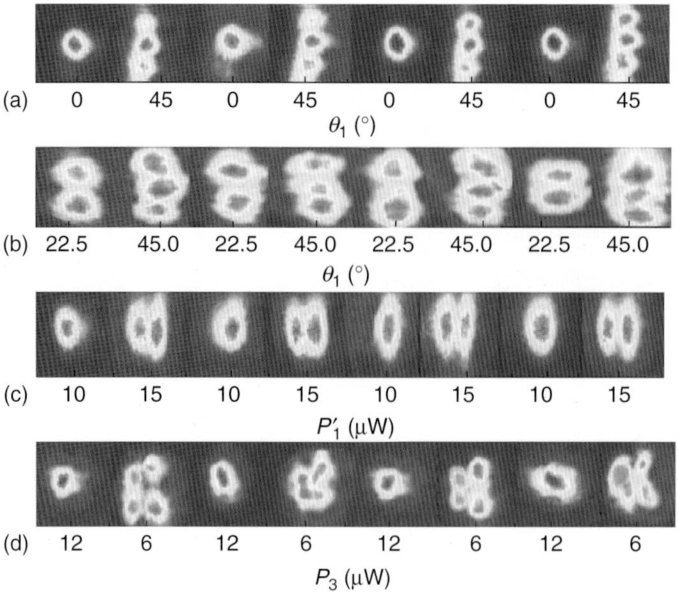

Figure 8.11 (a) and (b) show the spatial splitting of $\boldsymbol{E}_{F1}$ through changing the polarization states of $\boldsymbol{E}_1$ beam in the two-level system. (c) and (d) show the spatial splitting of $\boldsymbol{E}_{F1}$ and $\boldsymbol{E}_3$ beams in the three-level V-type system back and forth, respectively, at 290 °C with $G_1' = 19$ GHz and $G_2' = 12$ GHz.

As shown in Figure 8.11(c), the $\boldsymbol{E}_{F1}$ beam is split from one spot into two spots in the x direction in the three-level V-type system. According to the nonlinear phase shift expression, when the intensity of $\boldsymbol{E}_1'$ increases from 10 to 15 μW, I_1 gets larger, and ϕ_{NL} is correspondingly larger too, the beam is split into two spots. From the formula $I_{F1,F2,E3} \propto \cos^2(\phi_{NL})$, we can see that the period becomes smaller in such cases. For the three-level V-type system as shown in Figure 8.11(d), the probe beam $\boldsymbol{E}_3$ is split from one spot into four spots in two perpendicular directions back and forth. As the intensity of $\boldsymbol{E}_3$ changes from 12 to 6 μW, we can find that both I_1/I_0 and I_2/I_0 get larger. Hence, we can easily observe that ϕ_{NL} gets larger and the $\boldsymbol{E}_3$ beam is split into more spots in both x- and y-directions.

In this experiment, we use an attenuator with a 10 Hz repetition rate to control the intensity of $\boldsymbol{E}_1'$ and $\boldsymbol{E}_3$ beams, so a repeated splitting phenomenon can be observed. The information that the $\boldsymbol{E}_{F1}$ beam carries can be distributed into two channels by changing the intensity of $\boldsymbol{E}_1'$, which can be used as a space demultiplexer with 1×2 channels. What is more, the $\boldsymbol{E}_3$ beam can be split to four spots in two perpendicular directions in Figure 8.11(d), and therefore we can use this characteristic to realize the 2×2 channel routing. Such splitting of a beam in two perpendicular directions has the potential application in all-optical routing and space demultiplexer with $M \times N$ channels, which is very useful in the all-optical communication and networking.

In Figure 8.11, we can see the splits from one spot (or two spots) into more spots. Here, we use the channel equalization ratio $P = 1 - \sqrt{\sum_1^{N-1} (S_i - S)^2/S}$ to measure the demultiplexing effect, where S is the area of one subspot after splitting and S_i is the area of each one in the other subspots, and N is the number of the subspots after splitting. When P is near 100%, we will get more balanced and stable spatial channels. Here, the channel equalization P can approach 80–90%.

Both for the shift and splitting modulations, we can obtain the contrast index η in the experimental spots. It can be expressed as $\eta = (I' - I)/(I' + I)$, where I' is the intensity of the initial spatial state of the spot in the shift modulation or the intensity of the split subspot in the splitting modulation, while I is the intensity of the blank area after the spot is shifted away from the initial state in the shift modulation or the intensity of the blank area in the gap between the subspots in the splitting area. If η is larger, both the shift and splitting will become more clear, and the demultiplexing effect will be better for the all-optical routing and space demultiplexer, which means one channel is less affected by the others and there is little crosstalk between them, so the information can be transmitted with greater accuracy. In our experiment, η can reach about 70–80%.

In these experiments, the periodical splitting phenomenon is observed, and each splitting state just lasts 5 ns, in which the positive and negative edge times can be limited to 100–300 ps. As the spatial splitting of the probe and the two FWM beams is mainly controlled by the large nonlinear cross-Kerr effect of the laser fields $\boldsymbol{E}_1'$ and $\boldsymbol{E}_2'$, the response time is determined only by the atomic coherence time in the Na atomic system and is in the scale less than the nanosecond. In our experiment, the atomic population grating (via perturbation chain $\rho_{00} \rightarrow \rho_{10} \rightarrow \rho_{00}$) rather than the thermal grating is dominant in both the FWM process and the cross-Kerr effect in the Na vapor sample. The atomic population grating has little thermal effect, which can be ignored when ultradense pulse signal is applied. This novel result has opened the doors to the potential achievement of hyperterahertz all-optical router and space demultiplexer.

Compared with the common router made by an electronic device, which has limited speed, our optical routing has a response speed reaching picoseconds time scale. Theoretically, the limiting factor of our switching response speed is the repetition rate of the polarizer and attenuator, but the rise-time of the optical switch can be limited to be several hundred picoseconds. What is more, if the modulation happens in infinite space, the capacity of the all-optical routing will have no upper limit. This demonstration of the all-optical routing scheme has the potential application in the achievement of hyperterahertz all-optical routers and space demultiplexer as it has overcome the fundamental limitations of carrier lifetime experienced in conventional switching methods. Another unique point is that we demonstrate the ability to switch signals in and out in space, which is commonly required for routing applications. The demonstrated all-optical routing and switching could be very useful in all-optical networking and information processing proposed in atomic ensembles.

8.2.3
Conclusion

When we change the existence and frequency detuning of the control laser beams, the spatial displacements of the FWMs and probe laser beams are transferred between two distinguished states. We also experimentally demonstrated the spatial splitting of the FWMs and probe laser beams, which can be used for the space demultiplexer and router. Several ways have been used to realize this spatial splitting, and the property of the spatial splitting is dependent on the temperature of medium, the alternative atomic levels systems, and the intensity and polarization states of the control laser beams. Here, the response time in both the displacement and the splitting can be limited to be in the time scale less than nanosecond, which gives us a new way to realize the hyperterahertz all-optical router and space demultiplexer used in address choosing and information transferring. The current experiment opens the doors for further studies on the spatial manipulations of FWM signal beams in optical imaging storage, all-optical demultiplexer, routers, and future all-optical networking.

Problem

8.1 Explain the mechanism of experimentally demonstrating all-optical switching/routing by EISD effect, and all-optical routing and demultiplexing by the EISS effect.

References

1. Dawes, A.M.C., Illing, L., Clark, S.M., Gauthier, D.J. (2005) All-optical switching in rubidium vapor. *Science*, **308**, 672–674.
2. Brown, A.W., Xiao, M. (2005) All-optical switching and routing based on an electromagnetically induced absorption grating. *Opt. Lett.*, **30**, 699–701.
3. Yan, M., Rickey, E.G., Zhu, Y. (2001) Observation of absorptive photon switching by quantum interference. *Phys. Rev. A*, **64**, 041801.
4. Zhang, Y., Zuo, C., Zheng, H., Li, C., Nie, Z., Song, J., Chang, H., Xiao, M. (2009) Controlled spatial beam splitter using four-wave-mixing images, *Phys. Rev. A*, **80**, 055804.
5. Wang, H., Goorskey, D., Xiao, M. (2001) Enhanced Kerr nonlinearity via atomic coherence in a three-level atomic system, *Phys. Rev. Lett.*, **87**, 73601–73601.
6. Xiao, M., Li, Y.-Q., Jin, S.-Z., Gea-Banacloche, J. (1995) Measurement of dispersive properties of electromagnetically induced transparency in rubidium atoms. *Phys. Rev. Lett.*, **74**, 666–669.
7. Hickmann, J.M., Gomes A., de Araújo, C.B. (1992) Observation of spatial cross-phase modulation effects in a self-defocusing nonlinear medium. *Phys. Rev. Lett.*, **68**, 3547–3550.
8. Agrawal, G.P. (1990) Induced focusing of optical beams in self-defocusing nonlinear media. *Phys. Rev. Lett.*, **64**, 2487–2490.
9. Nie, Z., Zheng, H., Zhang, Y., Zhao, Y., Zuo, C., Li, C., Chang, H., Xiao, M. (2010) Experimental demonstration of optical switching and routing via four-wave mixing spatial shift. *Opt. Express*, **18**, 899–905.

10. Zhang, Y., Khadka, U., Anderson, B., Xiao, M. (2009) Temporal and spatial interference between four-wave mixing and six-wave mixing channels. *Phys. Rev. Lett.*, **102**, 13601.
11. Ham, B.S., Hemmer, P.R. (2000) Coherence switching in a four-level system: quantum switching. *Phys. Rev. Lett.*, **84**, 4080–4083.
12. Zhang, J., Hernandez, G., Zhu, Y. (2008) Optical switching mediated by quantum interference of Raman transitions. *Opt. Express*, **16**, 19112–19117.
13. Yanik, M.F., Fan, S. (2004) Time reversal of light with linear optics and modulators. *Phys. Rev. Lett.*, **93**, 173903.
14. Fleischhauer, M., Imamoglu, A., Marangos, J.P. (2005) Electromagnetically induced transparency: optics in coherent media. *Rev. Mod. Phys.*, **77**, 633.
15. Liu, C., Dutton, Z., Behroozi, C.H., Hau, L.V. (2001) Observation of coherent optical information storage in an atomic medium using halted light pulses. *Nature*, **409**, 490–493.
16. Wang, H.-H., Fan, Y.-F., Wang, R., Du, D.-M., Zhang, X.-J., Kang, Z.-H., Jiang, Y., Wu, J.-H., Gao, J.-Y. (2009) Three-channel all-optical routing in a Pr^{3+}:Y_2SiO_5 crystal. *Opt. Express*, **17**, 12197–12202.
17. Jin, C., Fan, S., Han, S., Zhang, D. (2003) Reflectionless multichannel wavelength demultiplexer in a transmission resonator configuration. *IEEE J. Quantum Electron.*, **39**, 160–165.
18. Li, J., Liu, W., Wang, Z., Wen, F., Li, L., Liu, H., Zheng, H., Zhang, Y. (2012) All-optical routing and space demultiplexer via four-wave mixing spatial splitting. *Appl. Phys. B*, 1–7.

Index

Quantum Control of Multi-Wave Mixing, First Edition. Yanpeng Zhang, Feng Wen, and Min Xiao.